Iron-Containing Enzymes
Versatile Catalysts of Hydroxylation Reactions in Nature

Iron-Containing Enzymes
Versatile Catalysts of Hydroxylation Reactions in Nature

Edited by

Sam P de Visser
The Manchester Interdisciplinary Biocentre, The University of Manchester, Manchester M1 7DN, UK

Devesh Kumar
Molecular Modelling Group, Indian Institute of Chemical Technology, Hyderabad, India

RSCPublishing

ISBN: 978-1-84973-181-2

A catalogue record for this book is available from the British Library

© Royal Society of Chemistry 2011

Published by The Royal Society of Chemistry,
Thomas Graham House, Science Park, Milton Road,
Cambridge CB4 0WF, UK

Registered Charity Number 207890

For further information see our web site at www.rsc.org

Printed and bound by CPI Group (UK) Ltd, Croydon, CR0 4YY

Preface

Heme and non-heme based monoxygenases and dioxygenases are fascinating enzymes that are found in all forms of life. They are involved in vital biochemical processes and are extremely versatile. As such there is huge interest in commercial applications of these systems in biotechnology. Despite the broad interest and applications, though, their structure, mechanism and functions are still not fully understood. However, in recent years a combination of experimental and computational studies has led to significant breakthroughs in the field. In particular, it is now known that many heme and non-heme enzymes use molecular oxygen on an iron center and synthesize an iron(iv)-oxo active species during their catalytic cycle. This key intermediate is involved in oxygen atom transfer reactions to a substrate, hence the heme and non-heme iron based mono- and dioxygenases catalyze aliphatic hydroxylation as well as epoxidation, desaturation and heteroatom oxidations.

This book gives an overview of recent advances in heme and non-heme iron(iv)-oxo chemistry as well as biomimetic analogues. The first five chapters are specifically devoted to the non-heme iron containing dioxygenases, where Chapter 1 (de Visser) gives a general overview on the various mechanisms of the different types of dioxygenases and how computational research assisted in the elucidation of the reaction mechanisms. Chapter 2 (Bugg) then continues with an overview of, mainly, Rieske-type (intradiol and extradiol) dioxygenases, their function and mechanism. The third chapter (Proshlyakov and Hausinger) is completely dedicated to studies performed on taurine/α-ketoglutarate dioxygenases, which are a biological system that essentially has functioned as a template for non-heme iron dioxygenases in general. Chapter 4 (Borowski and Siegbahn) describes detailed computational studies of the Stockholm and Kraków groups on various non-heme iron-containing enzymes and how these studies have further guided experiment. Two chapters in this volume are dedicated to the description of often-used methods in the field: the first is Chapter 5 (Neese and coworkers), where recently developed computational models for the calculation of spectroscopic parameters of iron-based systems

Iron-Containing Enzymes: Versatile Catalysts of Hydroxylation Reactions in Nature
Edited by Sam P de Visser and Devesh Kumar
© Royal Society of Chemistry 2011
Published by the Royal Society of Chemistry, www.rsc.org

are described. The second methods based chapter is Chapter 7 (Dey and Dey) on experimental based techniques, such as magnetic circular dichroism, X-ray absorption spectroscopy and extended X-ray absorption fine structure. Both methods based chapters discuss non-heme iron dioxygenases as well as biomimetic model complexes. A detailed and thorough review on non-heme iron biomimetics is found in Chapter 6 (Costas and coworkers).

The second section of this book describes heme based monoxygenases and dioxygenases. The most common and most thoroughly studied of these are the cytochrome P450 enzymes and Munro and coworkers give a detailed biochemical overview of their structure, mechanism and function (Chapter 8). A computational perspective on the mechanism of substrate activation, as it, for instance, appears in drug metabolizing reactions, is given in Chapter 9 (Kumar), which also details the concept of two-state-reactivity that drives the reaction mechanisms. The cytochromes P450 are highly versatile and hence many bioengineering studies have taken place to make them more biotechnologically relevant. Key studies in this field are reviewed by Mazumdar and coworkers in Chapter 10. Further computational advances have led to the development of quantum mechanics/molecular mechanics techniques for the studies of large (bio)chemical systems. Mulholland and coworkers review the method and describe recent applications in the field of cytochrome P450 chemistry in Chapter 11. The final chapter (Chapter 12, Mowat and coworkers) is focused on the biochemical structure, function and mechanism of heme based dioxygenases, including tryptophan dioxygenases and indoleamine dioxygenases.

Finally, we (the editors) thank all the authors for their contributions and prompt submission of their manuscripts. We are very grateful for the valuable time spent by the authors to create this detailed overview of recent highlights in a fast-moving field of research.

Sam P. de Visser and Devesh Kumar

Contents

Iron-Containing Enzymes: Versatile Catalysts of Hydroxylation Reactions in Nature
Edited by Sam P de Visser and Devesh Kumar
© Royal Society of Chemistry 2011
Published by the Royal Society of Chemistry, www.rsc.org

Chapter 5 Theoretical Spectroscopies of Iron-Containing Enzymes and Biomimetics   **119**

Shengfa Ye, Gemma J. Christian, Caiyun Geng and Frank Neese

Experimental and Computational Studies on the Catalytic Mechanism of Non-heme Iron Dioxygenases

SAM P. DE VISSER

Manchester Interdisciplinary Biocentre and School of Chemical Engineering and Analytical Science, University of Manchester, 131 Princess Street, Manchester M1 7DN, United Kingdom

1.1 Introduction

Dioxygen is essential for human life: it is involved in various bioprocesses such as cellular respiration, the biosyntheses of hormones as well as in defense mechanisms. Since molecular oxygen is needed in most parts of the human body, there are specially designed metalloenzymes with the task of transporting molecular oxygen from the lungs to the source where it is needed. In addition, the body has created metalloenzymes that utilize molecular oxygen for the bio-transformation of compounds both from a biodegradation and biosynthesis point of view. These enzymes are highly versatile, efficient and are known to react stereospecifically and/or regioselectively with substrates. As a consequence, metalloenzymes are of interest to the biotechnology industry, where they are used in bioprocesses, including the synthesis of pharmaceuticals and fine-chemicals.

In nature two types of oxygen-using metalloenzymes have been identified, namely, the monoxygenases and the dioxygenases.[1–9] The former binds

Iron-Containing Enzymes: Versatile Catalysts of Hydroxylation Reactions in Nature
Edited by Sam P de Visser and Devesh Kumar
© Royal Society of Chemistry 2011
Published by the Royal Society of Chemistry, www.rsc.org

molecular oxygen and transfers one atom of O_2 to a substrate, while the other atom is reduced to water. By contrast, in the dioxygenases both atoms of molecular oxygen are relayed to one or two substrate(s). In addition to this the metalloenzymes differ in the way that the metal is bound to the protein backbone. Thus, the enzymes are distinguished as either a heme or a non-heme enzyme. This chapter will focus on non-heme iron-containing dioxygenases, while later chapters (Chapters 8–11) are dedicated to heme-based monoxygenases.

Mononuclear non-heme iron-containing enzymes carry out a large variety of oxygenation reactions in nature with functions that vary from biosynthesis to biodegradation. In many cases the malfunction of these enzymes is correlated to a disease, so that it is important to understand the mechanistic details of dioxygen activation and oxygenation reactions by non-heme iron-containing enzymes. Most mononuclear non-heme iron-containing enzymes utilize molecular oxygen and transfer either one (monoxygenases) or two (dioxygenases) oxygen atoms to one or two substrates. However, there are also systems that oxidize molecular oxygen to two water molecules, thereby reducing the substrate, for instance *via* a ring-closure reaction (as in isopenicillin *N* synthase). Later we will give several examples of each of these processes. Despite the large versatility of non-heme enzymes, they often share a structural similarity where the metal (iron) is bound to the protein backbone *via* two or three protein ligands.[5,10] A common motif observed in mononuclear non-heme iron enzymes is where the metal is bound to two histidine and a carboxylic acid side chain of the protein *via* a 2-His/1-carboxylic acid (Asp/Glu) facial triad.[11,12] An example of this from the prolyl-4-hydroxylase active site is given in Figure 1.1. Generally, dioxygen binding is followed by the generation of a high-valent iron(IV)-oxo species, which has been identified in several enzymes and biomimetic systems as an active oxidant able to abstract hydrogen atoms of strong C–H bonds.[13–15] Thus, in the α-ketoglutarate dependent dioxygenases the formation of an iron(IV)-oxo species occurs concomitant with decarboxylation of α-ketoglutarate to give succinate (Succ). Many excellent reviews have appeared in recent years that describe the structural biology,[5] kinetics,[16] crystal structures and protein

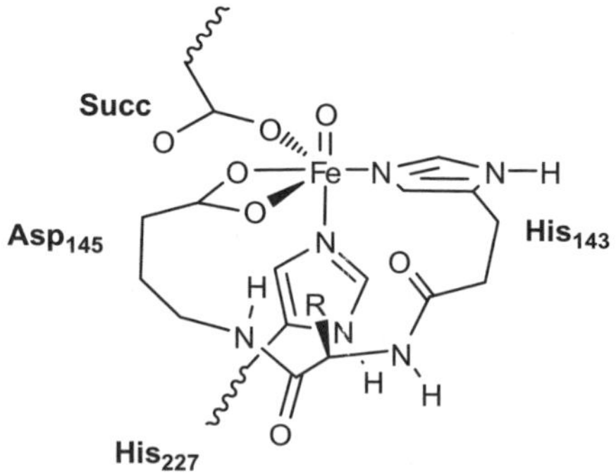

Figure 1.1   Characteristic non-heme iron binding motif with 2His/1Asp ligand system.

folding,[17] and spectroscopic properties[2,18] of enzyme intermediates. In this chapter we give an overview of the mechanism and chemical properties of intermediates in the catalytic cycle of non-heme enzymes, while more detail will follow of several enzymatic systems in subsequent chapters, *e.g.* Chapters 2–4.

1.2 α-Ketoglutarate Dependent Dioxygenases (αKDD) and Halogenases (αKDH)

An extensively studied group of non-heme iron-containing enzymes utilizes α-ketoglutarate as a cofactor and act as a substrate hydroxylase (as in the α-ketoglutarate dependent dioxygenases, αKDD) or halogenase (α-ketoglutarate dependent halogenases, αKDH). In many cases these mononuclear non-heme iron enzymes utilize α-ketoglutarate (sometimes called 2-oxoglutarate) as a cofactor to generate a high-valent iron(IV)-oxo species.[5,19,20] These α-ketoglutarate dependent enzymes are very versatile and catalyze many different reactions in biosystems, ranging from substrate hydroxylation, substrate halogenation, desaturation to ring-closure processes.[5,21–24] The most common reaction, however, is concomitant decarboxylation of α-ketoglutarate combined with substrate monoxygenation to give hydroxylated products as performed by the α-ketoglutarate dependent dioxygenases (αKDD). This is an important function in biosystems, hence αKDD have been found to be involved in the biosynthesis of the vancomycin, fosfomycin and carbapenem group of antibiotics,[25–27] as well as in DNA and RNA base repair mechanisms in mammals.[28–30] These enzymes are also involved in crosslinking of collagen, oxygen sensing and responses to hypoxia.[31–36] Close similarity in function and catalytic mechanism is found between the αKDD with the α-ketoglutarate dependent halogenases (αKDH) that combine decarboxylation of α-ketoglutarate with substrate halogenation.[37]

The α-ketoacid dioxygenases and halogenases undergo a catalytic cycle that shows many similarities; therefore, both are given in Figure 1.2.[38–40] As shown in Figure 1.2, the metal is bound to two histidine and a carboxylic acid (Asp/Glu) side chain of the protein in the dioxygenases, while in the halogenases the carboxylic acid ligand is replaced by a halide anion. The other three ligand sites of the metal are vacant in the resting state but upon cofactor α-ketoglutarate (αKG) binding two of these ligand sites are occupied, namely, with the keto-group of αKG *trans* to the carboxylic acid or halide ligand, while the carboxylic acid group of αKG binds *trans* to a histidine group (structure **A** in Figure 1.2). Upon substrate (SubH) binding the water ligand is displaced from the iron center (**B**) and replaced by molecular oxygen to form structure **C** which is a ferric-superoxo structure. The terminal oxygen atom of the superoxo group attacks the α-keto position of αKG to form a bicyclic ring-structure (**D**). Subsequently, the dioxygen bond breaks and carbon dioxide is released from the complex to give succinate and a high-valent iron(IV)-oxo complex (**E**). This intermediate abstracts a hydrogen atom from the substrate to give the ferric-hydroxo complex (**F**). At this stage the mechanism for the dioxygenases and

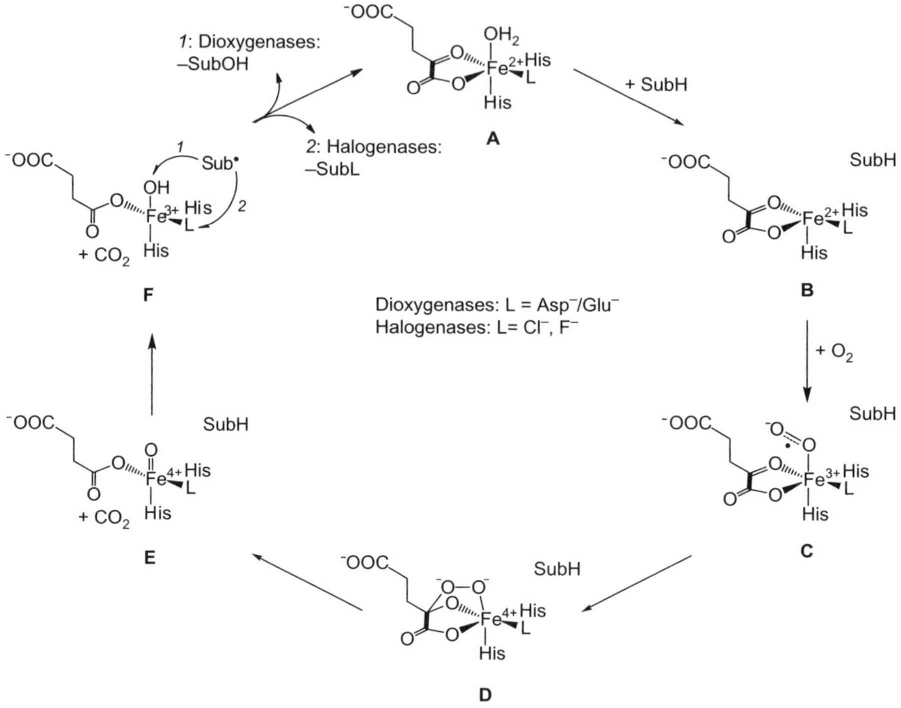

Figure 1.2 Catalytic cycle of substrate (SubH) activation by α-ketoglutarate dependent dioxygenases and halogenases.

halogenases diverges, whereby the hydroxyl group rebounds to the Sub• radical to form hydroxylated products in the α-ketoacid dioxygenases, whereas the halide atom is abstracted in the α-ketoacid halogenases to form halogenated products. In the final step products are released from the metal center and replaced by new reactant molecules to make the catalytic cycle complete. This catalytic cycle has been corroborated with both experimental and theoretical studies and in the next few sections we will highlight key findings of specific αKDD and αKDH enzymes.

1.2.1 Taurine/α-Ketoglutarate Dioxygenase (TauD)

Probably the most extensively studied mononuclear non-heme iron dioxygenase is taurine/α-ketoglutarate dioxygenase (TauD) since it is available in large amounts, highly soluble, and relatively stable.[41] Thanks to this, it was the first enzyme where a high-valent iron(IV)-oxo species was characterized with resonance Raman, Mössbauer and X-ray absorption spectroscopy.[42–44] As a consequence TauD has become the template for mononuclear non-heme iron enzymes and has prompted many additional studies. Here we give a brief overview of recent advances in the field and focus on the consensus mechanism of substrate hydroxylation. A more detailed elaboration of the mechanism and function of TauD enzymes follows in Chapters 2 and 3.

There are several crystal structures available of TauD enzymes, namely, as the protein databank files (pdb) 1GQW,[45] 1GY9,[45] and 1OS7.[46] The latter two structures are substrates (αKG and taurine) bound complexes at 3.0 and 1.9 Å resolution, respectively. Figure 1.3 shows extracts from the 1OS7 pdb file, which is a tetramer. The dimer interface contains two conserved helices of secondary structure that are hydrophobic in nature. The tertiary structure that includes this dimer interface is described by an antiparallel four helical bundle. Each monomer contains a non-heme iron active center that is bound to the protein *via* a 2-His/1-Asp ligand motif through interactions with His_{99}, Asp_{101} and His_{255}. Substrate αKG binds as a bidentate ligand with the carboxylic acid group *trans* to His_{99} and the keto-group *trans* to Asp_{101}. The terminal carboxylic acid tail of αKG is locked in a salt-bridge with Arg_{266} that keeps it in a tight and constraint conformation.[46] The final ligand position of the metal is vacant in the protein databank file (pdb) but will be occupied by molecular oxygen in a later stage in the catalytic cycle. Taurine does not bind directly to the metal but is held in position through many hydrogen bonding interactions in a substrate binding pocket nearby the active site. Key hydrogen bonding interactions of the sulfate group of taurine are with the phenolate group of Tyr_{73}, with the guanidinium group of Arg_{270}, with the imidazole ring of His_{70} and with several crystal water molecules. Site-directed mutants of His70Ala or Arg270Lys failed to bind taurine and implicated the importance of these residues for substrate anchoring through hydrogen bonding interactions.[47] Two phenylalanine rings (of Phe_{159} and Phe_{206}) surround the substrate binding pocket; the former has been identified with substrate binding and orientation.[48,49]

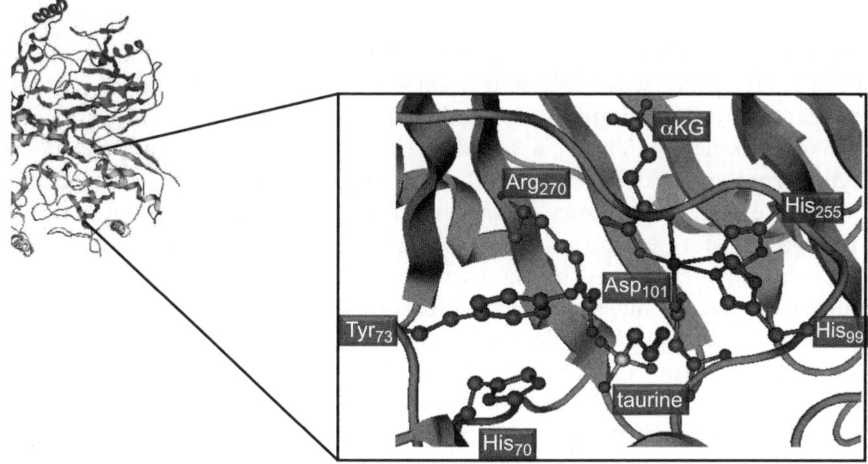

Figure 1.3 Schematic representation of TauD tetramer and an extract of the active site of one of the monomers with labels taken from the pdb file.

Spectroscopic studies on αKG addition to TauD under anaerobic conditions produced an absorption band at around 530 nm, which was attributed to αKG binding to the metal.[50] Addition of taurine to the system increased the spectrum by about 28% and gave an absorption band at 520 nm, while substrate analogues did not give these spectral changes. From stopped-flow UV visible spectroscopy experiments Hausinger *et al.*[50] measured formation rate constants of *ca.* $40\,s^{-1}$ of these αKG–Fe(II)–TauD and taurine–αKG–Fe(II)–TauD chromophores, while a taurine binding rate constant of $>140\ s^{-1}$ was determined, *i.e.* for the conversion of **A** into **B** in Figure 1.2. In addition, these studies provided evidence that **B** is able to react with molecular oxygen with a rate constant of $42\,s^{-1}$, whereas **A** is inactive in a reaction with molecular oxygen. This is important for the enzyme and would prevent uncoupling, whereby αKG is decomposed to succinate and CO_2 without concomitant substrate hydroxylation. Mössbauer spectroscopy studies on αKG and taurine binding provided evidence of a change in metal coordination from five-coordinated to six-coordinated since large Mössbauer parameter changes were observed: the isomer shift δ decreased from 1.27 to 1.16 ± 0.05 mm s^{-1} and the quadrupole splitting ΔE_Q dropped from 3.06 to 2.76 ± 0.05 mm s^{-1}.[42]

The next step in the catalytic cycle of TauD includes binding of molecular oxygen. It was shown that TauD consumes equal amounts of αKG, molecular oxygen and taurine.[51] Using $^{18}O_2$ labeling it was shown that both succinate and hydroxylated taurine carry one atom that originates from molecular oxygen.[52] In the absence of taurine, molecular oxygen reacted with complex **A** to form a yellow chromophore with absorbance band at 408 nm with a formation rate constant of $0.25\ s^{-1}$ and a decay rate constant of 0.5 min^{-1} to give a greenish brown chromophore.[53] Resonance Raman and nanoscale capillary LC/MS/MS studies on the latter provided convincing evidence of an Fe(III)catecholate species formed through self-hydroxylation of Tyr$_{73}$ in the absence of taurine.[53,54] With αKG as co-substrate self-hydroxylation was explained *via* an iron(IV)-oxo oxidant where the oxygen atom exchanges with solvent, while with succinate as a co-substrate probably an iron(III)-hydroperoxo intermediate is formed that rearranges to an iron(V)-oxo-hydroxo complex.[54] These self-hydroxylation reactions may protect the enzyme from more destructive and nonselective oxidations in the absence of substrate. The absorbance in the greenish brown chromophore is due to a charge-transfer transition from catecholate to iron(III). Thus, self-hydroxylation is explained by the formation of an iron(IV)-oxo species (**E** in Figure 1.2) that in the absence of substrate hydroxylates a nearby amino acid residue (Tyr$_{73}$). The detection of a distinct EPR spectrum in the reaction mechanism is explained through the formation of a long-lived Fe(III)OH--Tyr• intermediate in this reaction mechanism.

Stopped-flow absorption spectra provided unambiguous evidence for the accumulation of two intermediates after dioxygen binding.[42] The first of these develops after 20–25 ms, has an absorbance band at 318 nm and decays after about 600 ms. This intermediate also has a strong absorbance feature at

520 nm, which decreases strongly during decay. Mössbauer spectroscopy showed this species to be an Fe(IV) center with an isomer shift $\delta = 0.31$ mm s^{-1} and quadrupole splitting $\Delta E_Q = -0.88$ mm s^{-1}, while the absence of an EPR signal implicates a quintet spin ground state.[42] Subsequent resonance Raman studies of Hausinger *et al.*[43] on $^{16}O_2$ and $^{18}O_2$ binding to TauD unequivocally assigned this first intermediate as an iron(IV)-oxo species. Recently, using density functional theory (DFT) calculations this difference spectrum was reproduced and all peaks were assigned to a characteristic vibration.[40] Thus, the iron-oxo stretch vibration ν_{FeO} is located at 821 cm^{-1} and redshifts by 34 cm^{-1} upon replacement of $^{16}O_2$ by $^{18}O_2$ – hence a negative peak in the difference spectrum at 787 cm^{-1}. The vibration at 859 cm^{-1} is the OOC–C vibration in the succinate group that carries one oxygen atom originating from molecular oxygen, but the down-shift is only 11 cm^{-1}. Two bending vibrations in the carboxylic acid group of succinate are located at 555 and 583 cm^{-1}, the former one is out-of-plane and the latter one in-plane. These two vibrations give a small down-shift of about 6 cm^{-1} upon replacement of $^{16}O_2$ by $^{18}O_2$, but the intensity of the peaks in the spectrum is different in the $^{16}O_2$ as compared to the $^{18}O_2$ spectrum, and as a consequence appears as a positive and negative peak.[40]

Kinetic isotope effect (KIE) studies using taurine and taurine-d_2 showed that the iron(IV)-oxo species reacts with taurine *via* hydrogen abstraction with a large kinetic isotope effect of KIE $= k_H/k_D = 35$.[55] On the other hand, a comparison of the catalytic cycle rate constants in H_2O and D_2O implicated no solvent isotope effect, therefore no protons are relayed during the catalytic cycle of TauD.[47] Rapid freeze-quench X-ray absorption spectroscopy experiments characterized the intermediate with an Fe–O distance of 1.62 ± 0.01 Å.[44] Rate constants for the formation of the iron(IV)-oxo complex and its decay were measured to be 158 000 M^{-1} s^{-1} and 12 s^{-1}, respectively.[47] The kinetics of $^{14}CO_2$ production from quenched-flow experiments using 1-[^{14}C]αKG revealed a second-order rate constant of 1.6×10^5 M^{-1} s^{-1} in excellent agreement with the formation rate constant of the iron(IV)-oxo complex, thus implicating that decarboxylation occurs prior to iron(IV)-oxo formation.[56] A second intermediate after dioxygen binding was detected but its spectroscopic features and lifetime were small to unambiguously assign it to a specific structure. Mössbauer spectroscopy on this structure yielded $\delta = 1.17$ mm s^{-1} and $\Delta E_Q = 2.56$ mm s^{-1}, which are characteristic of an Fe(II) complex.[56] Recent, time-resolved Raman spectra on $^{16}O_2$, $^{16}O^{18}O$ and $^{18}O_2$ supplied to TauD provided evidence of three intermediates in the later stages of the catalytic cycle, namely, the iron(IV)-oxo species (**E** in Figure 1.2), an iron(III)-hydroxo species (**F** in Figure 1.2) and an Fe(II)-product complex.[57] Subsequent, kinetic isotope studies using taurine-d_2 gave a small Fe–O frequency shift (of 2 cm^{-1}) indicative of hydrogen bonding interactions between the iron(IV)-oxo species with hydrogen atoms of the substrate.

Electron spin echo envelope modulation spectroscopy (ESEEM) studies were performed on taurine binding to an NO-bound αKG–Fe(II) TauD using taurine deuterated at the C^1 and/or C^2 positions, *i.e.* $^+H_3N–C^2H_2–C^1H_2–SO_3^-$.[41]

The ESEEM results on taurine deuterated at the C^1 position correspond to an Fe–D distance of 3.4 ± 0.2 Å, while a weaker signal for deuteration at the C^2 positions implicates an average Fe–D distance of 5.1 Å and indicates an orientation that points away from the metal.

DFT studies[58–60] established a detailed mechanistic picture of the catalytic cycle after dioxygen binding (Figure 1.4). Thus, the dioxygen bound complex (**1**) appears in several close-lying states with a septet spin ground state and the quintet spin state somewhat higher. The terminal oxygen atom of **1** attacks the α-keto position and forms a bicyclic ring-structure (**2**). On both spin-state surfaces a barrier of about 10 kcal mol^{-1} is encountered, but this reaction is exothermic on the quintet spin state and slightly endothermic in the septet. As a consequence, the two spin state surfaces approach each other to within 2 kcal mol^{-1}. The subsequent decarboxylation barrier is small and irreversible and gives the Fe(III)-peroxosuccinate complex (**3**). Now the quintet spin state surface drops below the septet one so that a spin-state crossing precedes this complex. The final step is O–O bond breakage in the Fe(III)-peroxosuccinate

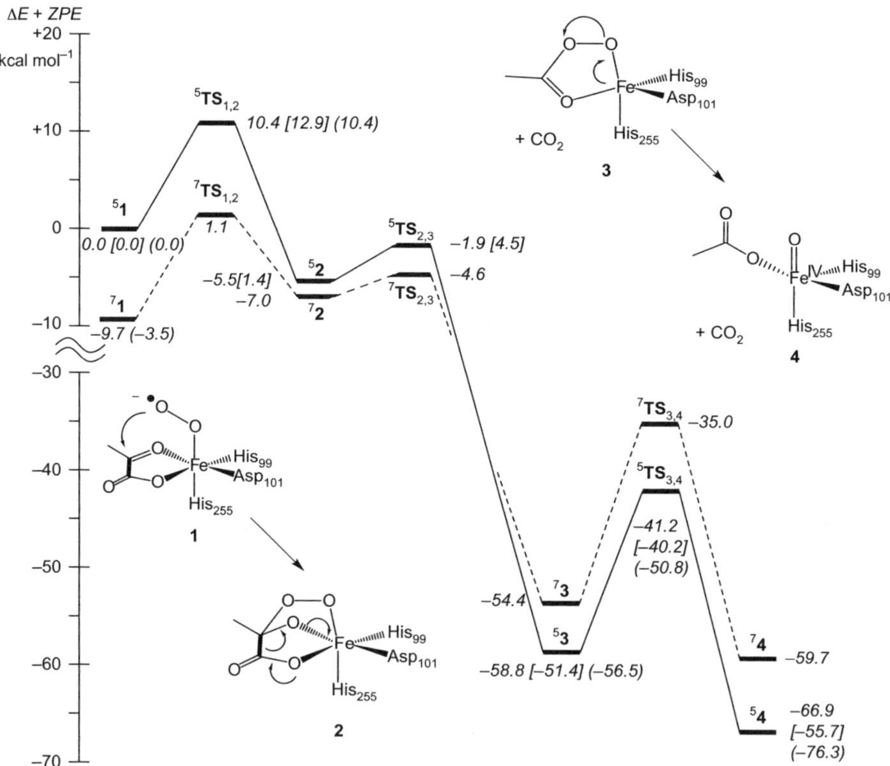

Figure 1.4 DFT calculated potential energy profile of dioxygen activation in the catalytic cycle of TauD. Out of parenthesis are $\Delta E + ZPE$ values from ref. 60, relative energies from ref. 59 are in square brackets and free energies from ref. 58 in parenthesis.

complex to form the iron(IV)-oxo complex (**4**). This process encounters a significant barrier, therefore, it is possible that the Fe(III)-peroxosuccinate complex has a finite lifetime, but it probably will be short. Nevertheless, DFT studies showed that it will not be able to act as an alternative oxidant, since high barriers of substrate epoxidation by ^{5}**3** were found.[60] In a similar way, the lifetime of the bicyclic ring-structure is short and it is not expected to accumulate significantly.

The several DFT studies on the catalytic cycle draw the same set of conclusions: (i) the rate-determining step after dioxygen binding is attack of oxygen on the α-keto position, (ii) decarboxylation is an irreversible process and (iii) the most stable structure is the iron(IV)-oxo species. Recent ^{18}O kinetic isotope effect studies confirmed this mechanism and proved that the rate-determining step in the process is the first step in Figure 1.4.[61] Moreover, ^{5}**4** was shown to abstract a hydrogen atom from taurine with a barrier of only 6.7 kcal mol^{-1} and lead to hydroxylated products with high exothermicity. Quantum mechanics/molecular mechanics (QM/MM) studies on this mechanism gave the same spin-state ordering, relative energies and optimized geometries as small gas-phase DFT models.[62]

As a comparison Figure 1.5 shows the optimized geometry of the iron(IV)-oxo species of TauD calculated with various computational methods and in several research groups.[58,62–65] These optimized geometries are very close and only minor geometric differences are found between the various structures. The calculated Fe–O distance is in good agreement with experimental rapid freeze quench X-ray absorption studies that found an Fe–O distance of 1.62 Å.[44]

DFT studies on a range of different iron(IV)-oxo isomers with either mono- or bidentate carboxylate ligands were carried out and compared structural and Mössbauer data with experiment.[64] The best agreement was found for a bidentate ligated carboxylic acid group with the iron(IV)-oxo in an overall

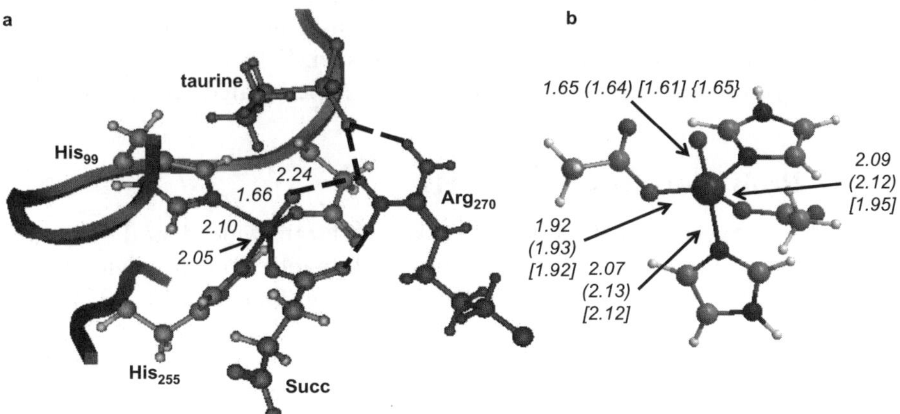

Figure 1.5 Optimized geometry of iron(IV)-oxo species of TauD with bond distances given in ångströms. (a) QM/MM result from ref. 62. (b) DFT optimized structures from refs. 63, (64), [65], and {58}.

quintet spin state, whereas spectroscopic data did not correlate with an overall triplet spin state. Calculated Mössbauer parameters ($\delta = 0.27$ mm s^{-1}, $\Delta E_Q = -0.65$ mm s^{-1} and zero-field splitting parameters $E/D = 0.02$) are in good agreement with experiment. Combined experimental and theoretical studies[64] showed that in the His99Ala mutation the ligand position of the metal is occupied by a water molecule and that the system retains significant activity.

1.2.2 AlkB Repair Enzymes

Another group of non-heme iron dioxygenases that utilize α-ketoglutarate as a cofactor are the AlkB repair enzymes, which recently have been identified to belong to the α-ketoglutarate dependent dioxygenases.[66–68] These enzymes repair methylated DNA (and RNA) bases that have been damaged by intra- or extracellular chemicals, in particular alkylating agents. Thus, one of the most common methylation products is N^3-methyladenine, which has dramatic effects on the biological system since it blocks DNA replication.[28] Compounds such as methyl methanesulfonate and methyl halides have been shown to generate N^1-methyladenine and N^3-methylcytosine, which prevent formation of Watson–Crick base-pairs and as a consequence are toxic. AlkB repair enzymes remove the methyl groups from these positions in N^1-methyladenine and N^3-methylcytosine *via* a mechanism shown in Scheme 1.1.[69] These enzymes contain a non-heme iron center that binds αKG, molecular oxygen and methylated DNA and catalyze the hydroxylation of the methyl group, which in a subsequent step is released as formaldehyde. Isotopic labeling and product distributions show that Fe(II), αKG and molecular oxygen are needed in the process and that formaldehyde and CO_2 are formed.[66,67] Human AlkB repair enzymes were characterized and shown to repair methylated and ethylated DNA strands to give formaldehyde and acetaldehyde, respectively.[70] Crystallographic data provided further evidence that AlkB repair enzymes belong to the αKDD set of enzymes and highlighted a non-heme iron active site, where the metal is bound to a 2-His/1-Asp ligand system.[71] In a similar way, AlkB repair enzymes have been shown to demethylate methylated RNA bases.[72]

1.2.3 Prolyl-4-hydroxylase (P4H)

Prolyl-4-hydroxylase (P4H) is another α-ketoglutarate dependent dioxygenase with relevance for human health, because it is involved in the regioselective hydroxylation of proline groups on the C^4 position (Scheme 1.2).[73] This is an important reaction in biosystems due to crosslinking of collagen helices, oxygen sensing and cellular response to hypoxia through the hypoxia inducible factor (HIF).[31–35] Thus, the crosslinking strand in collagen contains a series of repeated triads, where the first residue usually is proline, the second is often 4-hydroxyproline, while the last residue typically is a glycine.[74] Inhibition of P4H with 3,4-dihydroxybenzoate, on the other hand, protects against neurotoxicity and may be a possible therapeutic treatment of Parkinson's disease.[75]

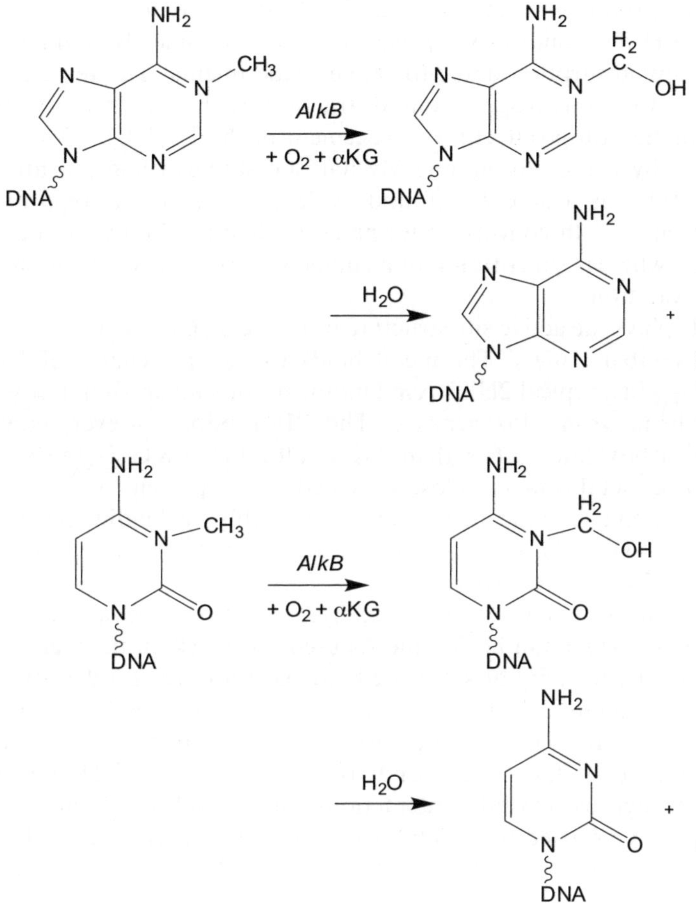

Scheme 1.1   DNA base repair by AlkB enzymes.

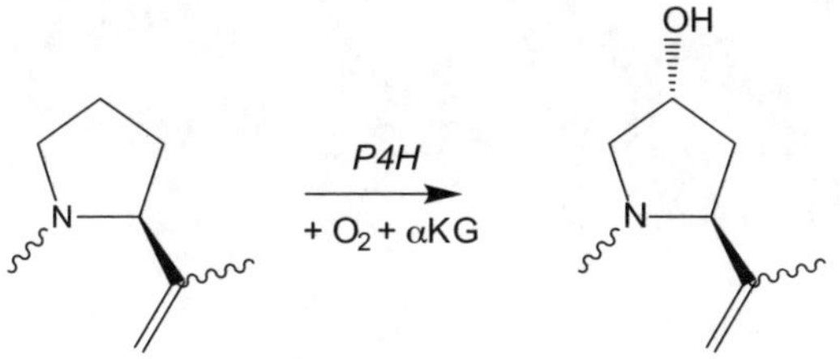

Scheme 1.2   Proline hydroxylation by P4H enzymes.

Studies using a small peptide (Pro-Ala-Pro-Lys)$_3$ as a substrate for P4H enzymes found kinetic and spectroscopic evidence that the proline residues are readily hydroxylated by the enzyme.[76] Isotopic replacement of $^{16}O_2$ by $^{18}O_2$ showed that one oxygen atom of molecular oxygen is incorporated into hydroxo-proline products.[52] Similar to the TauD studies reported above, also in the case of P4H the iron(IV)-oxo species was spectroscopically characterized with stopped-flow absorption and Mössbauer experiments. The reaction was shown to proceed *via* a hydrogen atom abstraction by the iron(IV)-oxo species with a large kinetic isotope effect for replacement of the 2,3,3,4,4,5,5-positions of the substrate by deuterium atoms. Mechanistic studies using 5-oxaproline containing peptides as a substrate found evidence of a Groves-type radical rebound mechanism.[77] In contrast, fluorinated proline residues were found to be unreactive,[76] whereas substitution of a sulfide group at the C^4 position gave substrate sulfoxidation.

Figure 1.6 displays the active site structure of P4H enzymes as taken from the 2JIG protein databank file.[78] The metal binds to the side chains of His_{143}, Asp_{145} and His_{227} in a typical 2His/1Asp binding motif similar to that observed in other non-heme iron dioxygenases. The 2JIG pdb, however, contains pyridine-2,4-dicarboxylate rather than αKG, although αKG is the natural co-substrate. The metal center is closely located to the protein surface, and a water channel is seen above the metal, which is probably the binding position of the substrate.[79–82] Directly above the metal are located a series of aromatic residues, such as Tyr_{140} and Trp_{243}.

Recently, we investigated the substrate hydroxylation mechanism of this important enzyme for human health and focused our work on the factors that determine the regioselectivity of substrate hydroxylation using a density functional theory study on models of the active site of P4H enzymes.[83] In principle, the origin of this regioselectivity can arise from thermodynamic factors, *i.e.* the relative strength of the C^4–H bond to the C^3–H and C^5–H bonds, or alternatively through electrostatic interactions such as substrate approach to the active center. We started our studies with a minimal active site model (1_A) similar to that shown in Figure 1.5(b), where the histidines are abbreviated by

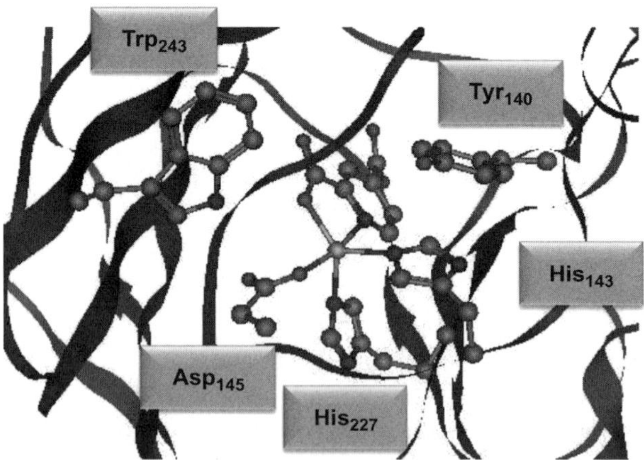

Figure 1.6   Active site of P4H enzyme as taken from the 2JIG protein databank file. Amino acids labeled as in the crystal structure.

imidazole groups and the carboxylic acid groups by acetate. In support of the EPR and Mössbauer studies of Hoffart *et al.*, the iron(IV)-oxo species is in a quintet spin state that is well separated from the lowest lying singlet, triplet and septet spin states. The calculated Mössbauer parameters of $^5\mathbf{1}_A$ give an electric field splitting of $\Delta E_Q = -1.01$ mm s^{-1} and isomer shift $\delta = 0.22$ mm s^{-1}, which are values close to those reported experimentally ($\Delta E_Q = -0.82$ mm s^{-1}, $\delta = 0.3$ mm s^{-1})[76] and well within the error bar of the calculations. The computational studies, therefore, support the experimental assignment of a quintet spin ground state of the iron(IV)-oxo species.

Proline hydroxylation is found to proceed *via* a stepwise mechanism (Figure 1.7) with an initial hydrogen atom abstraction *via* barrier $\mathbf{TS}_{HA,A}$ to form a radical intermediate ($\mathbf{2}_A$) and followed by a rebound barrier *via* $\mathbf{TS}_{reb}$ to give the alcohol product complexes ($\mathbf{3}_A$). The potential energy surface displayed in Figure 1.7 shows that the lowest barrier is obtained for hydrogen abstraction from the C^5 position, whereas that for the C^4 position is actually the highest in energy. These calculations implicate that hydrogen abstraction of proline should occur on the C^5 position preferentially rather than on the C^4 position. Indeed, calculations on the relative strength of the C^3–H, C^4–H and C^5–H bonds of proline as calculated with Equation (1.1) (inset of Figure 1.7) show that the C^5–H bond is the weakest (85.4 kcal mol^{-1}), while the C^4–H bond is the strongest. This BDE_{CH} analysis, therefore, implicates that the energetically most favorable position for substrate hydroxylation should be the C^5 position,

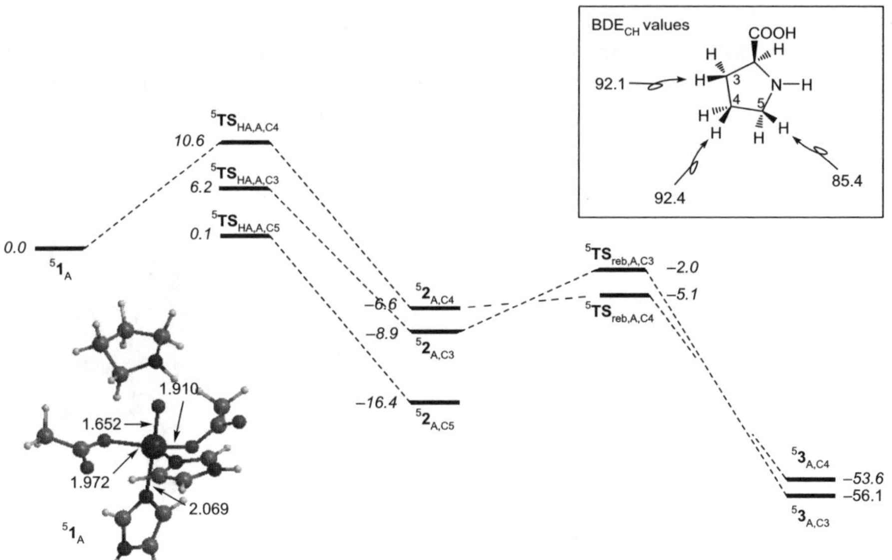

Figure 1.7 Potential energy landscape of proline hydroxylation at the C^3, C^4 and C^5 positions using a minimal active site model **A**. All energies are in kcal mol^{-1} and bond lengths in ångströms. Inset: BDE_{CH} values for hydrogen abstraction at the C^3, C^4 and C^5 positions.

which is indeed supported by the potential energy surface for the small model shown in Figure 1.7:

$$\text{SubH} \rightarrow \text{Sub}^{\bullet} + \text{H}^{\bullet} + \text{BDE}_{\text{CH}} \tag{1.1}$$

Subsequently, we extended our model of the active site of P4H with *p*-cresol for Tyr_{140} and indole for Trp_{243}, which are amino acids that have been implicated with substrate binding orientation. This model is labeled **B** and the quintet spin potential energy surface is given in Figure 1.8. As follows, the lowest hydrogen abstraction barrier using the large active site model is *via* $^{5}\text{TS}_{\text{HA,B,C4}}$ with a barrier of 12.6 kcal mol^{-1}. This barrier is only slightly elevated with respect to the small active site model above (by 2.0 kcal mol^{-1}). By contrast, the hydrogen abstraction barrier at the C^{3} position of proline is raised from 6.2 kcal mol^{-1} for the small active site model without protein perturbations to 13.5 kcal mol^{-1} in an enzyme mimicked environment. Even more dramatic is the change in barrier for hydrogen abstraction at the C^{5} position, which was negligible for the small active site model, but is larger than 30 kcal mol^{-1} for model **B**. The potential energy surfaces displayed in Figures 1.7 and 1.8, therefore, show the dramatic influence of amino acids such as Tyr and Trp that are oriented in such as way that the substrate approach to the active center blocks the weak C–H bonds effectively so that it can only be hydroxylated at the unfavorable C^{4} position.

As can be seen from Figures 1.7 and 1.8, the effect of the protein is only on the reaction barriers, *i.e.* the kinetics of the reaction, and not on the thermodynamics of the reaction. Thus, the reaction energy for hydrogen atom abstraction (ΔH_{r}) by an iron(IV)-oxo species, *i.e.* the reaction step from $^{5}\mathbf{1}$ to $^{5}\mathbf{2}$, is equal to the strength of the C–H bond broken minus the strength of the O–H bond that is formed in the process, Equation (1.2). Indeed, the reaction energies displayed in Figures 1.7 and 1.8 are very similar and the ordering

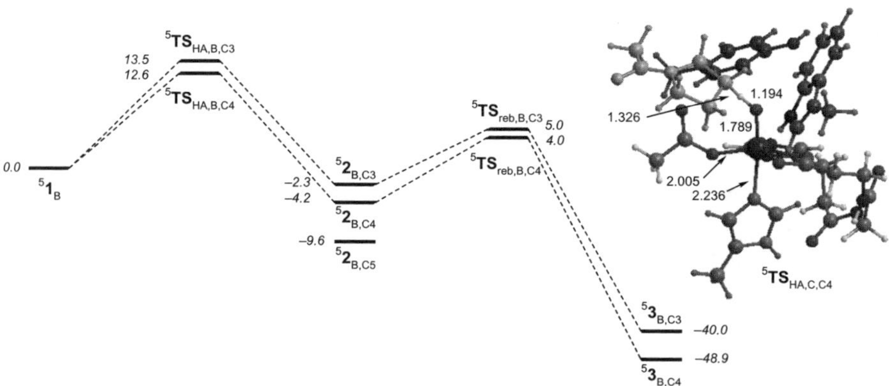

Figure 1.8   Potential energy landscape of proline hydroxylation at the C^{3}, C^{4} and C^{5} positions using a large active site model **B**. All energies are in kcal mol^{-1} and bond lengths in ångströms.

is $^5\mathbf{2}_{C5} < {}^5\mathbf{2}_{C3} < {}^5\mathbf{2}_{C4}$, which is also the ordering in BDE_{CH}. In addition to the raised hydrogen abstraction barriers in the large model, the radical rebound barriers are increased significantly as well. This means that the radical intermediates (**2**) will have a finite lifetime during which rearrangement may occur:

$$\Delta H_r = BDE_{CH} - BDE_{OH} \tag{1.2}$$

Recently, we showed that the C–H bond strength correlates with reaction exothermicity as well as with barrier height.[84–88] Thus, for a range of substrate hydroxylation reactions with either a heme or a non-heme iron(IV)-oxo species a linear correlation was found between hydrogen abstraction barrier and BDE_{CH}. It was further shown that this correlation can be explained with a valence bond curve crossing diagram (details on how to construct valence bond curve crossing diagrams are given in Chapter 9) and that the electron transfer processes in the transition state result in a correlation between barrier height and BDE_{CH}. Figure 1.9 shows the correlation of barrier heights with BDE_{CH} using the data from refs 83 and 86. As can be seen the trend is reasonably linear, but improvement can be achieved due to inclusion of the resonance energy correction.

Figure 1.10 generalizes the hydrogen abstraction step by iron(IV)-oxo oxidants using a valence bond curve crossing diagram. Thus, in the reactant state the system is in a reactant wave function (Ψ_r), which is an electronic configuration that correlates with an excited state in the iron(III)-hydroxo product complex (Ψ_p^*). Similarly, the product wave function (Ψ_p) for the iron(III)-hydroxo complex connects to an excited state in the reactant configuration (Ψ_r^*). These two wave functions cross and lead to an avoided crossing and a

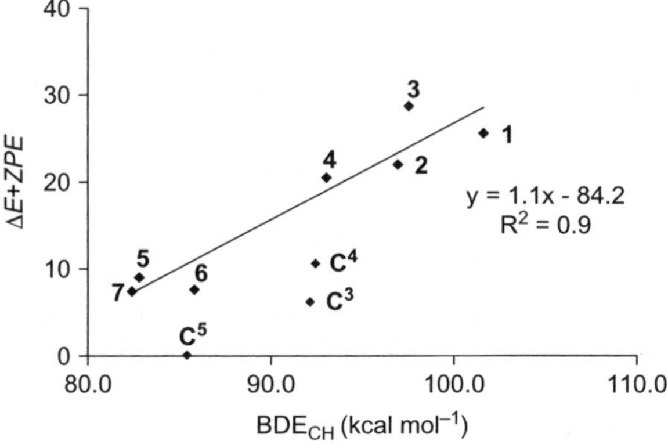

Figure 1.9 Correlation between hydrogen abstraction barrier height ($\Delta E + ZPE$) and BDE_{CH} value as calculated for the DFT model in Figure 1.5(b) with the following substrates: methane (1), ethane (2), C^1 position of propane (3), C^2 position of propane (4), propene (5), toluene, (6), ethylbenzene (7), C^3 position of proline (C^3), C^4 position of proline (C^4) and C^5 position of proline (C^5).

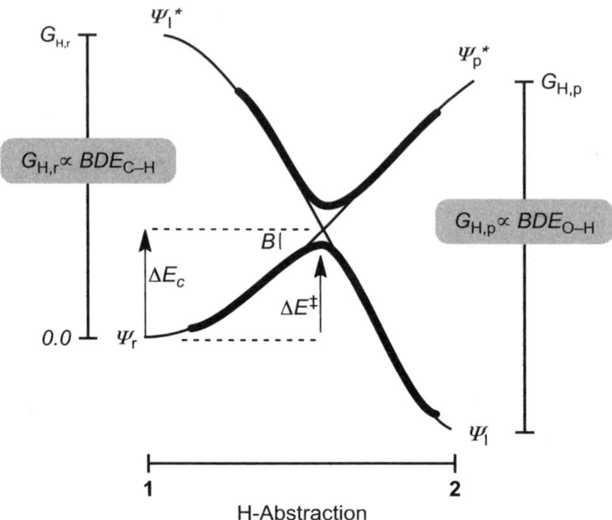

Figure 1.10 Valence bond curve crossing diagram for the hydrogen abstraction of substrates by iron(IV)-oxo oxidants.

transition state for the hydrogen atom abstraction. The curve crossing energy (ΔE_c) is located above the actual transition state energy ($\Delta E^{\ddagger}$) by the resonance energy (B). It was shown that the crossing point (ΔE_c) is proportional to a fraction (f) of the excitation energy in the reactants ($G_\mathrm{H,r}$), Equation (1.3), whereas for the reverse reaction process, *i.e.* from products to reactants, it is proportional to a fraction of the excitation energy in the products ($G_\mathrm{H,p}$). Therefore, the barrier height ($\Delta E^{\ddagger}$) for the hydrogen atom abstraction can be described by Equation (1.4). Studies using one oxidant and variable substrates showed that $G_\mathrm{H,r}$ correlates with BDE_CH, while work using one substrate and a range of iron(IV)-oxo oxidants gave a correlation of $G_\mathrm{H,p}$ with BDE_OH:[87]

$$\Delta E_\mathrm{c} = fG_\mathrm{H} \tag{1.3}$$

$$\Delta E^{\ddagger} = fG_\mathrm{H} - B \tag{1.4}$$

Thus, the reactant excitation energy, $G_\mathrm{H,r}$, is proportional to the singlet–triplet excitation energy in the C–H bond or the bond dissociation energy of the C–H bond (BDE_CH). On the other hand, the product excitation energy, from Ψ_I to Ψ_p^{*}, represents a singlet–triplet excitation in the O–H bond of the iron(III)-hydroxo complex that is formed or BDE_OH. These studies, therefore, generalized the reactivity patterns of iron(IV)-oxo complexes in enzymatic and biomimetic systems and show that the kinetics of the reaction of the ideal oxidant is dependent on the strength of the C–H bond of the substrate that is broken and the strength of the O–H bond that is formed. Moreover, the P4H studies (Figure 1.8) indicate that environmental perturbations, such as due to substrate binding can affect the kinetics of the reaction but not the thermodynamics.

1.2.4 α-Ketoglutarate Dependent Halogenases (αKDH)

Halogenation is a rare reaction mechanism in nature – nevertheless there are several enzymes that perform this task, such as haloperoxidases that contain a heme or vanadium dependent catalytic center.[37,89–91] A range of natural products contain halogens and some of these, such as vancomycin and chlortetracycline, have been shown to have therapeutic value.[89] Recently, a new class of halogenases was discovered that uses α-ketoglutarate and molecular oxygen on an iron(II) center to halogenate alkyl-groups in substrates.[92] Figure 1.11 displays the active site structure as taken from the 2FCT pdb file, which is a substrate-free form of the α-ketoglutarate dependent halogenase (αKDH) SyrB2 with chloride, Fe(II) and α-ketoglutarate (αKG) bound.[93]

Similar to the αKDD enzymes described above, the metal is anchored to the protein through interactions with two histidine side chains (of His_{116} and His_{235}). However, the αKDH have no 2-His/1-Asp ligand motif, since the carboxylic acid group is replaced by a chloride anion. The chloride anion is part of a hydrogen bonding network that includes several water molecules (identified with W in Figure 1.11) and resides in a hydrophobic binding pocket.[94] Similar to the structure of TauD shown above in Figure 1.3, αKG binds as a bidentate ligand and is involved in hydrogen bonding interactions with an arginine side chain (Arg_{254}). The active site is surrounded by a series of aromatic rings and apolar groups of which one of those (Phe_{104}) is highlighted in Figure 1.11. The last binding site of the metal is occupied by a water molecule that is part of a chain of water molecules just above the metal center

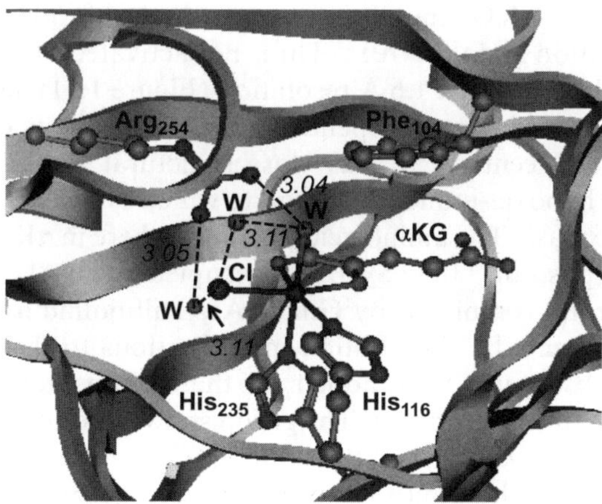

Figure 1.11 Active site structure of substrate bound αKDH as taken from the 2FCT pdb file with amino acids labeled as in the pdb file. Water molecules identified with W. Highlighted in italics are key hydrogen bond distances (Å) in the active site.

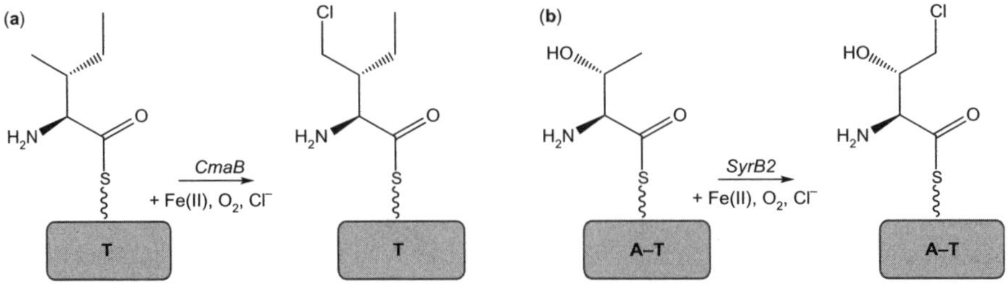

Scheme 1.3   Substrate chlorination by CmaB (left-hand side) and SyrB2 (right-hand side) enzymes.

surrounding the chloride and Arg_{254} groups. It may be anticipated that the substrate enters the pocket through this water channel and that molecular oxygen binds to the sixth coordination site of the metal.

The αKDH enzyme (CmaB) activates coronamic acid (CMA) or 1-amino-1-carboxy-2-ethylcyclopropane. The substrate derives from L-allo-Ile that is linked *via* a thioester group to the phosphopantetheinyl prosthetic group of 10 kDa carrier-protein domain T (CmaD). As shown in Scheme 1.3 (left-hand side) CmaB converts the Ile group into its chlorinated form on an Fe(II) center using αKG, chloride and molecular oxygen.[92] This is a key step in the biosynthesis of 1-amino-1-carboxy-2-ethylcyclopropane (CMA) a precursor molecule in the biosynthesis of coronatine. Thus, CmaB was the first non-heme iron enzyme identified with halogenation activity towards substrates.

A second enzyme in the category of αKDH enzymes is SyrB2, which as part of the syringomycin E biosynthesis regioselectively chlorinates the methyl group of L-Thr-*S*-SyrB1 on residue 9 of a 66 kDa protein containing an adenylation (A) and thiolation (T) domain, Scheme 1.3 (right-hand side).[95] The studies showed that αKG, molecular oxygen and chloride are essential for substrate chlorination and that free L-Thr is not activated by the enzyme. More recently, a crystal structure at 1.6 Å resolution (Figure 1.11) indeed proved it to belong to the α-ketoglutarate dependent enzymes with a chloride/bromide bound to the metal center.[93] It has close structural similarity with human factor-inhibiting hypoxia-inducible factor-1, which is a Fe(II)-α-ketoglutarate dependent hydroxylase. The amino acid position, where in αKDD the aspartate ligand is located (position 118 in SyrB2) is occupied by an Ala residue. Mutants where this Ala group is replaced by Glu or Asp still bound iron(II), but loss of all activity was observed.[93] In addition, modifications to the substrate carrier protein where substituents were added on the carbon atom adjacent to the halogenations site of the terminal L-Thr group showed decay rate constants and kinetic isotope effects (k_H/k_D) for the hydrogen abstraction reaction.[96] More-over, the formation of the iron(IV)-oxo species appears to be dependent on the substrate binding, where the native substrate gives its most efficient formation pattern. It has been proposed[92,93] that the αKDH enzymes undergo a catalytic cycle as depicted above in Figure 1.2, where molecular oxygen binding triggers decarboxylation of αKG to form succinate and an iron(IV)-oxo species. It has been anticipated that the iron(IV)-oxo species abstracts a hydrogen atom from the substrate and rebinds the chloride rather than hydroxo group to form halogenated products.

The first iron(IV)-oxo complex to be detected and characterized for αKDH enzymes was for the halogenase CytC3 that regioselectively halogenates an L-valine group attached *via* a thiolate linkage to CytC2 protein.[97] The addition of Fe(II) to a reaction mixture of substrate L-Aba-*S*-CytC2 and enzyme CytC3 resulted in the formation of an absorbance band at 520 nm. Subsequently, this reaction complex of CytC3–Fe(II)–αKG–Cl–L-Aba-*S*-CytC2 reacted with molecular oxygen to form an intermediate complex with an absorption band at 318 nm, a formation rate constant of $18 \, s^{-1}$ and decay rate constant of $0.20 \, s^{-1}$. This intermediate complex was found to react by hydrogen atom abstraction with a large kinetic isotope effect when one or more of the hydrogen atoms are replaced by deuterium. Subsequent Mössbauer spectroscopic experiments on these complexes revealed it to be a mixture of two stable intermediates: one with $\delta = 0.30$ mm s^{-1} and $\Delta E_Q = 1.09$ mm s^{-1} and the other with $\delta = 0.22$ mm s^{-1} and $\Delta E_Q = 0.70$ mm s^{-1}. The studies failed to identify these two Fe(IV) isomers. However, using bromide as a ligand rather than chloride an EXAFS spectrum was determined, which provided further evidence of a high-valent iron(IV)-oxo species.[98] Thus, bond distances of 1.60–1.62 Å for the Fe–O and 2.43 Å for the Fe–Br were measured.

After hydrogen atom abstraction by the iron(IV)-oxo species, halide rebound gives halogenated products. Interestingly, although an Cl–Fe(III)-hydroxo intermediate is proposed in the reaction mechanism (**F** in Figure 1.2) no hydroxylated by-products have been observed so far using the native substrate. This is surprising since the reaction of $Cr_2O_2Cl_2$ with toluene gave a mixture of 1-chloro-2-phenylethane (18% yield), phenacetaldehyde (32% yield) and 2-phenylethanol (5% yield) products.[99] Similarly, a non-heme iron(IV)-oxo complex with tris(2-pyridylmethyl)amine (TPA) and halide ligand gave significant amounts of alcohol by-products.[100] Clearly, the environment of the enzyme active site prevents formation of hydroxylated by-products. This is further supported by recent studies of Bollinger and coworkers[101] who used a substrate analogue where the L-Thr group that is linked to the SyrB1 carrier protein is replaced by L-norvaline. These studies showed a regioselectivity switch from substrate halogenations to hydroxylation, because L-Thr is a C_4 amino acid, whereas L-norvaline a C_5 amino acid. As a consequence, the L-norvaline group can penetrate deeper into the active site pocket and approaches the oxo group closer, while with L-Thr the distance is larger and hence in favor of halogenation.

Recently, DFT calculations on a αKDH model complex have proposed alternative mechanisms, where the lack of hydroxylated by-products is explained.[102,103] Figure 1.12 displays the DFT calculated reaction mechanism starting from an iron(IV)-oxo complex (**Re**) as taken from ref. 102, which appears in close lying septet and quintet spin states. It may very well be that these two spin states give rise to the two different Mössbauer signals described above but further studies are necessary to prove this. The initial reaction is a hydrogen atom

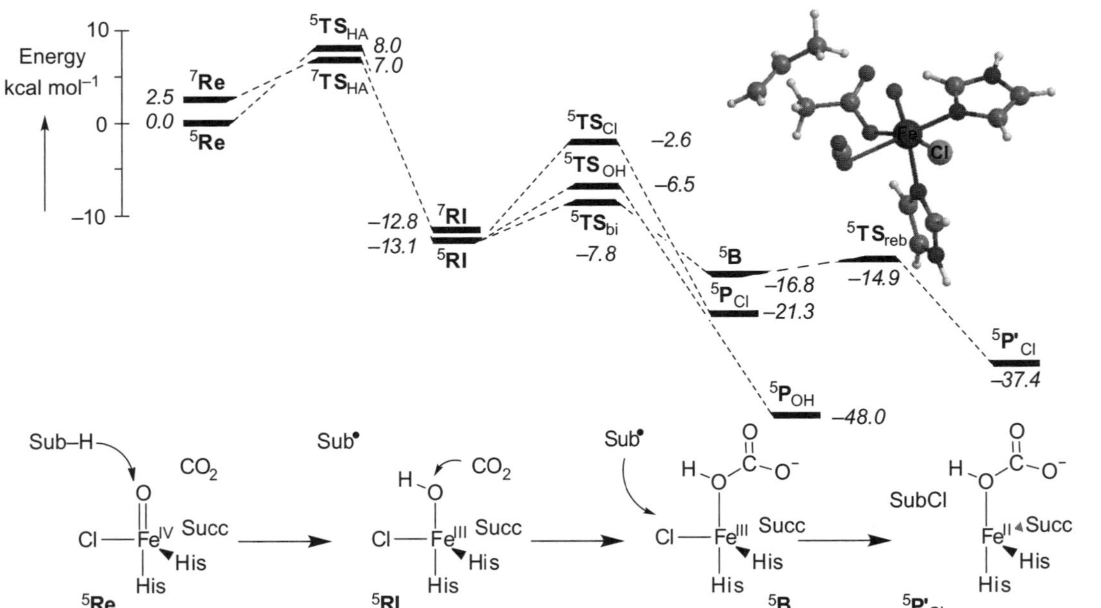

Figure 1.12 DFT calculated reaction mechanisms of substrate hydroxylation and chlorination of an αKDH model complex. All energies are in kcal mol^{-1} and contain zero-point and solvent ($\varepsilon = 5.7$) corrections.

abstraction of the substrate (SubH) to form an Cl–Fe(III)-hydroxo complex (**RI**). The barriers for this process are close in energy for the septet and quintet spin states and replacement of the transferrable hydrogen atom by deuterium gives an elevated KIE value in agreement with experiment. Moreover, the hydrogen abstraction barrier is the rate-determining step in the reaction mechanism. The complex **RI** is in an almost degenerate septet and quintet spin state with the same orbital occupation: $\pi^{*}_{x2-y2}{}^{1}\,\pi^{*}_{xz}{}^{1}\,\pi^{*}_{yz}{}^{1}\,\sigma^{*}_{xy}{}^{1}\,\sigma^{*}_{z2}{}^{1}\,\pi_{Sub}{}^{1}$, *i.e.* the metal d-block is singly occupied and either ferro- or antiferromagnetically coupled to a substrate radical (in π_{Sub}).

Three different reaction mechanisms were tested originating from ^{5}**RI**: (i) direct chlorination by attack of Sub$^{\bullet}$on the chloride ligand *via* barrier ^{5}**TS**$_{Cl}$, (ii) hydroxylation by attack of Sub$^{\bullet}$on the hydroxyl ligand *via* barrier ^{5}**TS**$_{OH}$ and (iii) bicarbonate formation by attack of the CO_2 group on the hydroxyl ligand *via* barrier ^{5}**TS**$_{bi}$. Thus, thermodynamically hydroxylation is the favorable process that gives products with the largest exothermicity (-48.0 kcal mol^{-1}) and as a consequence also the hydroxylation barrier is well below that of direct chlorination. This would explain the large amount of hydroxylated by-products observed in biomimetic reaction models.[100] The alternative reaction leading to bicarbonate gives a slightly lower barrier than that for direct hydroxylation and may explain the need of the enzyme to shuttle the hydroxo group away or shield it from the substrate to prevent hydroxylation. Further studies with advanced models as well as QM/MM are under way in our group at the moment to find the environmental effects of these various reaction components and to establish whether CO_2 trapping is a viable reaction pathway in αKGDH enzymes.

A series of spectroscopic studies on αKDH *vis-à-vis* TauD gave insight into the binding of Fe(II), whereby it was proposed that αKG may be involved in Fe(II) binding in αKDH because the two His ligands may be too weak to provide sufficient bonding to the metal.[104] Moreover, αKG binding weakens the remaining Fe–H$_2$O bond as well. Furthermore, based on the features of the active site pocket the authors concluded that hydrogen abstraction takes place from a substrate at a relatively large distance, and rebound barriers are dependent on the strength of the Fe–Cl *versus* Fe–OH bonds. Alternatively, OH trapping by proton donation in the active site may lead to Fe–H$_2$O and promote the chlorination process.[103,104]

1.3 Cysteine Dioxygenase (CDO)

A classical example of an enzyme where a combination of theoretical and experimental studies was necessary to characterize the catalytic cycle is cysteine dioxygenase (CDO), which is a non-heme iron-containing enzyme that catalyzes the conversion of L-cysteine into L-cysteine sulfinic acid (Scheme 1.4) by molecular oxygen. Studies with isotopically labeled O_2 proved that both oxygen atoms are incorporated in cysteine sulfinic acid products.[105] Thus, experimental studies initially failed to stabilize and characterize short-lived oxygen-bound intermediates in the catalytic cycle, so that several potential catalytic cycles

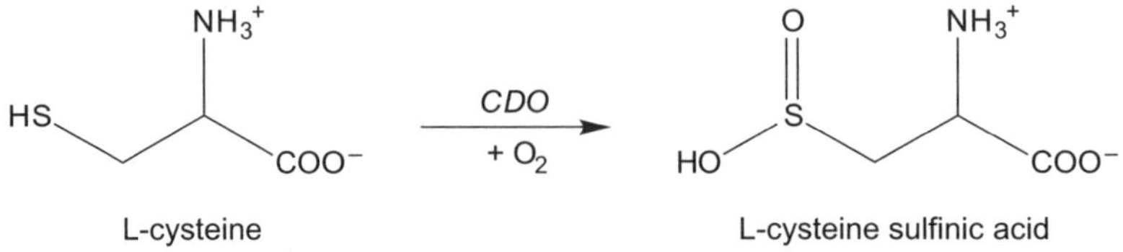

L-cysteine

CDO
+ O₂

L-cysteine sulfinic acid

Scheme 1.4   Conversion of cysteine into cysteine sulfinic acid by CDO enzymes.

were proposed.[106–108] Subsequent DFT studies gave further understanding of its mechanism and the oxygen activation process.[109,110]

CDO is an important enzyme in mammalian physiology that regulates the cysteine concentration in the body. Thus, cysteine is a non-essential amino acid that is synthesized in the body from methionine. However, in high concentrations cysteine is toxic and also precipitates to give cystine stones[111] so that a series of enzymes are involved in its metabolism. The first step in the biodegradation of cysteine is performed by CDO enzymes, where a dioxygenation of L-cysteine gives L-cysteine sulfinic acid products. Elevated levels of cysteine in the body have been correlated with neurological diseases, including motor neuron disease, Alzheimer's and Parkinson's. Therefore, understanding the mechanism of substrate dioxygenation by CDO enzymes is of importance to human health and as a consequence has attracted a lot of scientific research over the past years.

Structurally, CDO has a transition metal containing active site where an iron atom is bound to the protein backbone *via* three histidine linkages of His$_{86}$, His$_{88}$ and His$_{140}$ (Figure 1.13).[112] The remaining three ligand sites are occupied by substrate (cysteine) that binds as a bidentate ligand, while the final ligand position is empty in Figure 1.13 but will be occupied by molecular oxygen later on in the catalytic cycle. Electron paramagnetic resonance (EPR) studies on NO bound CDO provided compelling evidence that substrate binds prior to molecular oxygen to the metal center.[113] Figure 1.13 shows a representation of the active site structure as taken from the 2IC1 pdb file, which is a substrate bound human isoform of CDO.[112] Crystal structures of human CDO closely resemble those obtained of mouse and rat CDO.[114,115] The active site of the enzyme is located in a deep cavity on the protein surface. The substrate (cysteine) binds as a bidentate ligand with the thiolate group *trans* to His$_{88}$ and the amino group *trans* to His$_{140}$. The carboxylic acid group of cysteine is locked in a series of hydrogen bonding interactions, as shown in Figure 1.13, *via* a salt bridge with Arg$_{60}$ as well as stabilizing hydrogen bonds from the imidazole side chain of His$_{155}$ and the phenol groups of Tyr$_{157}$ and Tyr$_{58}$. Consequently, CDO enzymes are highly substrate specific and will not bind and dioxygenase substrates with considerably different features.[116] Studies using a set of structurally related cysteine analogues showed that a charged side group similar to the carboxylic acid group in cysteine is required for optimal substrate/inhibitor binding. However, none of the analogues led to product formation, not even D-cysteine. Interestingly, although α-ketoglutarate functions as a cofactor in many dioxygenase enzymes it is an inhibitor of CDO enzymes.[116]

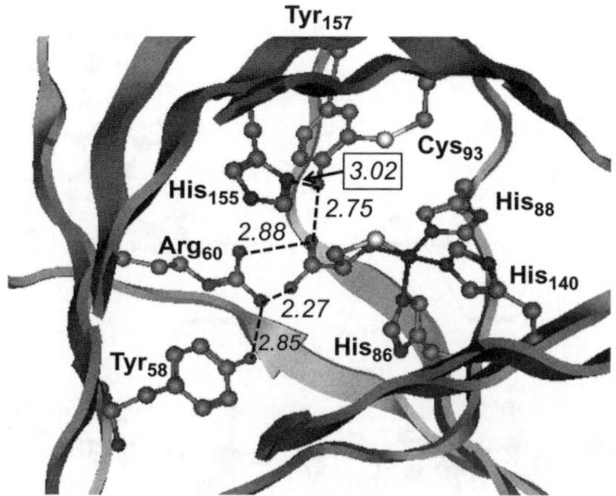

Figure 1.13 Active site structure of substrate bound CDO as taken from the 2IC1 pdb file with amino acids labeled as in the pdb file. Highlighted in italics are key hydrogen bond distances (Å) of the substrate in the active site.

Another interesting feature of the active site of CDO is the covalent linkage between Tyr_{157} and Cys_{93}, which is a unique feature the enzyme shares with galactose oxidase.[117] Recent, EPR studies on galactose oxidase assigned this crosslink as a Tyr-Cys• radical with significant unpaired spin density on the phenoxyl oxygen and sulfur atoms due to a possible singly occupied π orbital on the aromatic ring. Further studies on galactose oxidase showed that the thiolate linkage of the Tyr-Cys bound groups influences the bond dissociation energy of the phenolate O–H group and consequently its proton affinity.[118] In CDO enzymes the full function of this Tyr-Cys crosslink has yet to be established and no proof of radical character has been found so far, but biogenesis, electrochemical and mass spectrometric studies provided evidence that Fe^{2+} and O_2 are needed to make this crosslink.[119,120] However, absence of the cofactor retained most of the activity of CDO, so that it does not seem to be essential for biocatalysis. Nevertheless, formation of the crosslink is correlated with substrate binding and catalytic turnover, although with structurally analogues substrates such as cysteamine and β-mercaptoethanol no formation of a Tyr-Cys linkage was found.[119]

Since experimental studies failed to characterize intermediates in the catalytic cycle of CDO after dioxygen binding, a DFT study was performed[109] to establish the key factors in the later stages of the catalytic cycle (Figure 1.14). The model contained the iron with its direct ligands and an extensive part of the hydrogen bonded active site network as well as the Tyr_{157}-Cys_{93} cofactor. Although the dioxygen bound complex (**A**) has close lying singlet, triplet and quintet spin states the actual dioxygen activation mechanism takes place on the quintet spin state surface only. Thus, the dioxygen activation is initiated with a spin-state crossing from the singlet to the quintet spin state surface, which has an S–O bond formation transition state of about 10.0 kcal mol^{-1} and leads to a

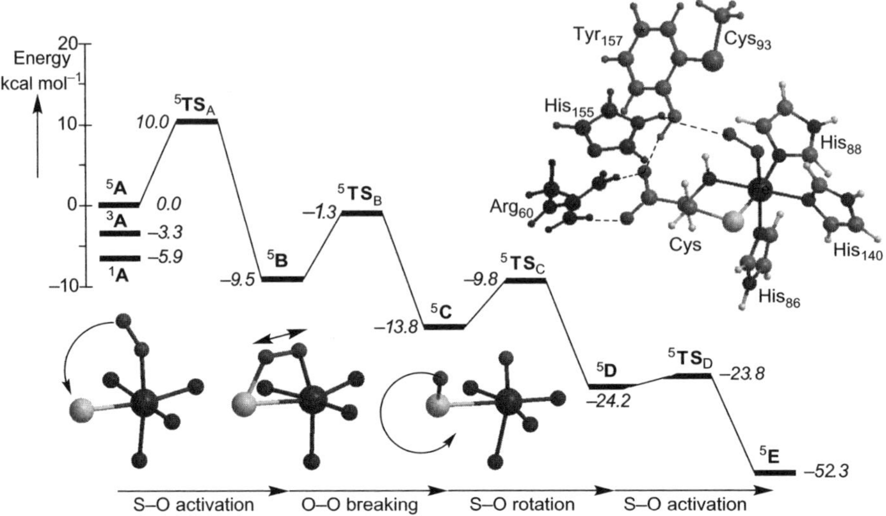

Figure 1.14   DFT calculated dioxygen activation mechanism of CDO active site model system. Data are taken from ref. 109. Spin states are indicated with a superscript and energies are gas-phase energies with zero-point energy included.

ring structure (**B**), where the dioxygen moiety bridges the sulfur and iron atoms. This is the rate-determining step in the reaction mechanism and explains why no dioxygen bound intermediates have been experimentally detected so far. Formation of the ring structure weakens the O–O bond from 1.338 to 1.513 Å, which is broken in the next step of the mechanism to form cysteine sulfoxide (**C**) intermediate. In this particular structure the sulfur atom is shielded from the second oxygen atom, so that an internal rotation of the sulfoxide group (to form **D**) is required prior to the second oxygenation step leading to cysteine sulfinic acid products (**E**). As follows from the energy diagram in Figure 1.14 each intermediate is lower in energy than its precursor and the same ordering is found for the transition states. Subsequent studies of active site mutants were performed where one of the three histidine ligands of the iron was replaced by a carboxylic acid group to create a 2-His/1-Asp bound motif, *i.e.* H86D, H88D and H140D mutants.[110] These studies showed that the 3-His ligand motif in CDO is essential for optimal dioxygenation of the substrate. Thus, a carboxylic acid group located *trans* to the sulfur atom of cysteinate weakens the Fe–S bond and thereby leads to loss of this bond after the first oxygen transfer process. In contrast, a carboxylic acid group *trans* to the dioxygen group increases the barriers of oxygen transfer processes due to a pull-effect of electrons.

This DFT established mechanism was disputed by recent crystallographic studies on rat CDO that detected a persulfenate bound structure (Figure 1.15).[121] Since, there are quite some geometric differences between the substrate bound structure of Figure 1.13 and the cysteine persulfenate structure in Figure 1.15 it is unclear whether the persulfenate system is an actual intermediate in the

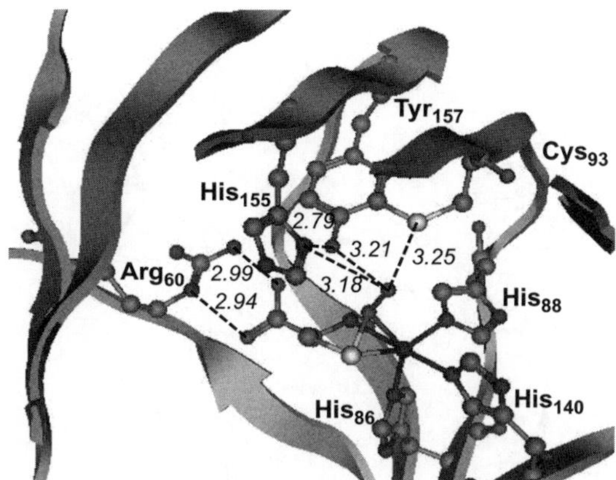

Figure 1.15 Active site structure of cysteine persulfenate bound CDO as taken from the 3ELN pdb file with amino acids labeled as in the pdb file. Highlighted in italics are key hydrogen bond distances (Å) of the substrate in the active site.

catalytic cycle or a dead-end by-product. Thus, close inspection of Figures 1.13 and 1.15 reveals critical differences in the substrate hydrogen bonding channel. In particular, the salt-bridge between the carboxylic acid group of cysteine with the side chain of Arg_{60} is twisted in such a fashion that the two terminal NH_2 groups of Arg_{60} are not both involved in hydrogen bonding, but only one of these. Moreover, the Tyr_{157}-Cys_{93} pair has also relocated to a slightly different position in the active site and is closely involved in hydrogen bonding interactions with the dioxygen moiety. Based on this crystal structure an alternative oxygen activation mechanism was suggested whereby the tyrosinate hydrogen bond guides the oxygen transfer process to cysteine.

Thus, to establish whether this cysteine persulfenate is an intermediate in the catalytic cycle of CDO enzymes, a combined quantum mechanics/molecular mechanics (QM/MM) study on the oxygen activation mechanism of CDO enzymes was performed.[122] This work focused on the energetic differences of the two mechanisms displayed in Scheme 1.5. Mechanism I is the one that was found using the DFT model complex and given by the potential energy surface in Figure 1.14. The QM/MM calculations reproduce most features of Mechanism I investigated with the small model complex mentioned above. Indeed, the potential energy profile and optimized geometries are similar for key steps in the catalytic cycle, although some differences due to the inclusion of the protein environment can be identified. Using both methods (DFT and QM/MM) the anchor point of the calculations (**A**) is a singlet spin ferric-dioxygen bound complex that reacts with substrate *via* an initial and rate determining S–O bond formation step *via* **TS$_A$**. The rate determining barrier involves a spin state crossing from the singlet to the quintet spin state prior to formation of the ring structure (**B**). In the later stages of the catalytic mechanism the differences between the QM/MM and DFT results are larger although they do not affect

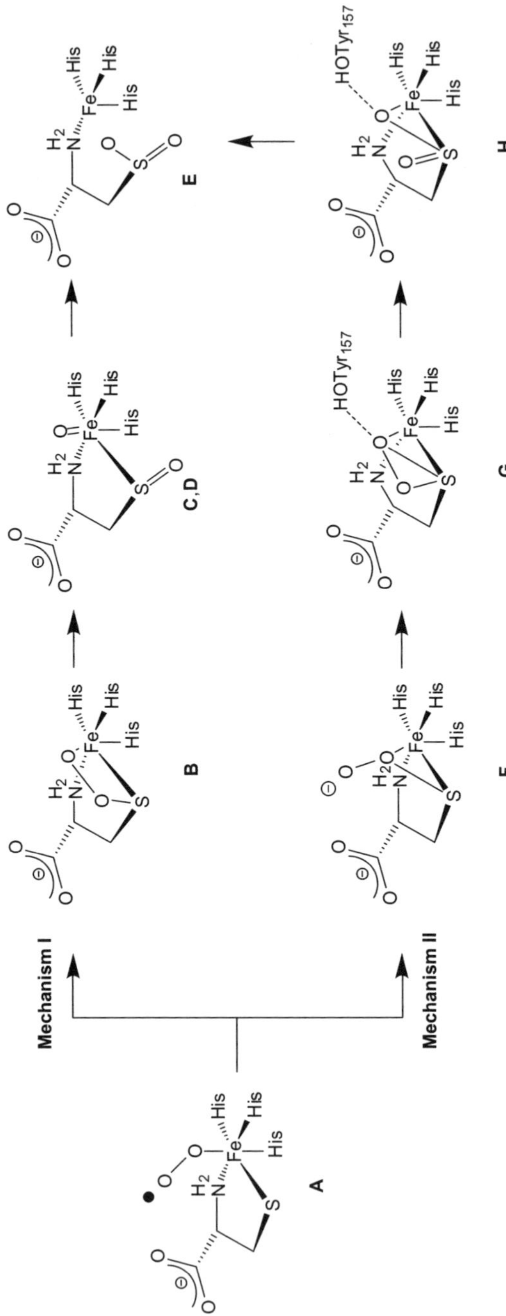

Scheme 1.5 Possible mechanisms (I and II) for the dioxygenation of cysteine by CDO enzymes.

the conclusions drawn in earlier work. Using small DFT model complexes the iron(IV)-oxo species (**C** and **D**) had a quintet spin ground state, while the other spin state surfaces were found to be much higher in energy. With QM/MM the ground state of **C** and **D** is the triplet spin state in agreement with biomimetic iron(IV)-oxo complexes. Despite these differences, the essential features of the reaction mechanism were confirmed with QM/MM.

Subsequently, Mechanism II in Scheme 1.5 was tested whereby the distal oxygen atom of the iron(III)-superoxo group forms a bond with the thiolate group of cysteinate. Our calculations give very high barriers for this process (>40 kcal mol^{-1}) for the singlet and triplet spin state mechanisms. Therefore, the alternative mechanism (II) is unlikely. In summary, QM/MM and DFT model calculations on the catalytic cycle of CDO enzymes support Mechanism I as the most likely mechanism for cysteine dioxygenation by CDO enzymes.

1.4 Isopenicillin *N* Synthase (IPNS)

Isopenicillin *N* synthase (IPNS) is a common mononuclear non-heme enzyme found in fungi and bacteria. It utilizes the linear tripeptide δ-(L-α-amino-adipoyl)-L-cysteinyl-D-valine (ACV), which is the direct precursor of all penicillins and cephalosporins in nature through the enzymatic conversion into bicyclic isopenicillin *N* (IPN) by the enzyme isopenicillin *N* synthase (IPNS) (Scheme 1.6).[123]

The enzyme IPNS utilizes molecular oxygen on a non-heme iron center, but in contrast to all dioxygenases described above does not incorporate oxygen atoms into the substrate. Instead, a four-electron oxidation, whereby four hydrogen atoms are abstracted from the substrate to form two water molecules is catalyzed. Similarly to CDO, IPNS also does not require a cofactor such as α-ketoglutarate to provide electrons for the reaction. Figure 1.16 displays the active site of IPNS with substrate and NO bound as taken from the 1BLZ pdb file.[124]

The metal (iron) is bound to the protein backbone *via* a typical 2-His/1-Asp ligand motif similar to TauD and other non-heme dioxygenases. Substrate ACV binds as a monodentate ligand to iron with a thiolate linkage. The remaining two ligand sites of the metal in the 1BLZ pdb file are occupied by a water molecule and NO. In the enzyme this NO site will be occupied by molecular oxygen instead, but so far no oxygen-bound intermediates have been crystallized.

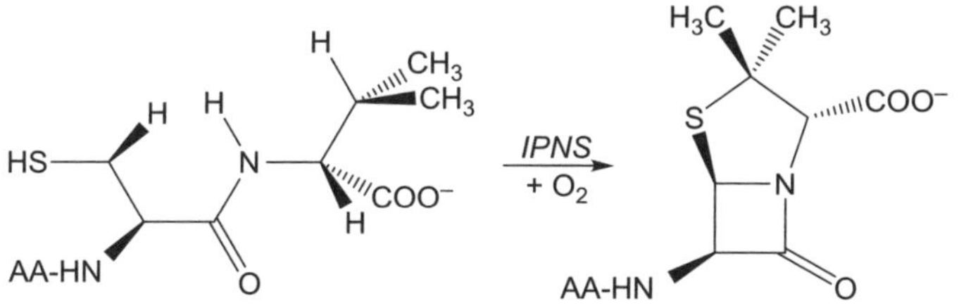

Scheme 1.6 Conversion of ACV into IPN by IPNS.

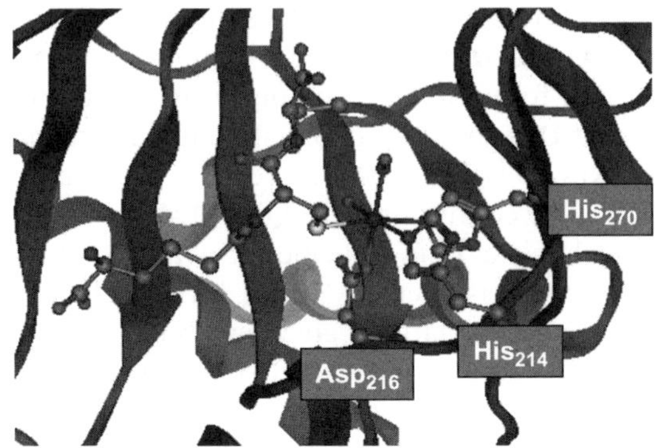

Figure 1.16 Active site structure of substrate bound IPNS as taken from the 1BLZ pdb file with amino acids labeled as in the pdb file.

QM/MM calculations on dioxygen binding to the active center of IPNS, as compared to small model DFT studies, showed that the protein environment stabilizes this process by about 8–10 kcal mol^{-1}, so that studies on small molecule complexes are inappropriate in this case.[125] A combination of side-on and end-on dioxygen geometries with either triplet, quintet or septet spin state is found within a window of $\Delta G = 4.1$ kcal mol^{-1}.

Subsequent studies using small active site models calculated with DFT and QM/MM provided insight into the dehydrogenation and ring-closure processes that convert ACV into IPN.[126,127] The calculated potential energy profile with structures of some of the transition state structures is shown in Figure 1.17. Thus, the dioxygen-bound complex (**1**) is a ferric-superoxo with a nearby ACV substrate. This was confirmed with a combination of spectroscopic studies on the substrate bound iron-NO complex in comparison with DFT modeling.[128] Although the electronic ground state of **1** is the septet spin state, actually the lowest hydrogen abstraction barrier ($^5\text{TS}_{\text{HA},1}$) from Cys-β-C–H is on the quintet spin state so that a fast spin state crossing from the septet to the quintet will precede the hydrogen atom transfer. However, this process does not lead to Fe(III)-hydroperoxo but to Fe(II)-hydroperoxo (**2**) due to an extra electron transfer from the substrate to the metal during the process. Experimentally, a free energy of activation of 16 kcal mol^{-1} was found,[129] while the corresponding activation barrier calculated for QM/MM is 14.6 kcal mol^{-1}. The hydroperoxo-iron complex, similar to studies on heme enzymes and biomimetics, is a weak oxidant unable to abstract the second hydrogen atom.[130,131] Instead, DFT provided an alternative mechanism whereby the water ligand of iron donates a proton to the terminal oxygen atom of the hydroperoxo group to form a hydroxo-iron(IV)-oxo complex (**3**). In this complex the quintet spin state is the ground state with the triplet and septet higher in energy by 5.2 and 10.3 kcal mol^{-1}. This spin state ordering and relative energies are very similar to those obtained for the iron(IV)-oxo species of TauD using either DFT or QM/MM calculations.[62,63] The second hydrogen abstraction from the valine

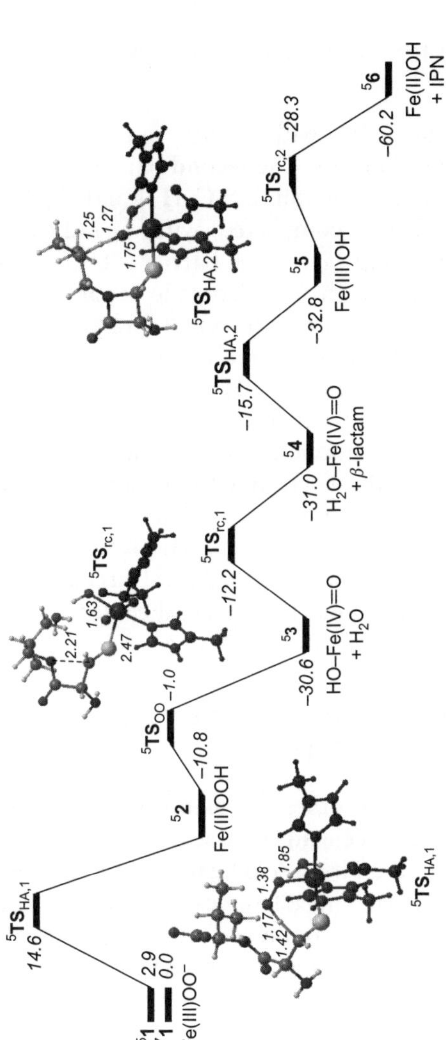

Figure 1.17 Potential energy profile of ACV activation by an Fe(III)OO• complex of IPNS. Energies are in kcal mol⁻¹ relative to the dioxygen bound complex in the septet spin state. Bond lengths are given in ångströms.

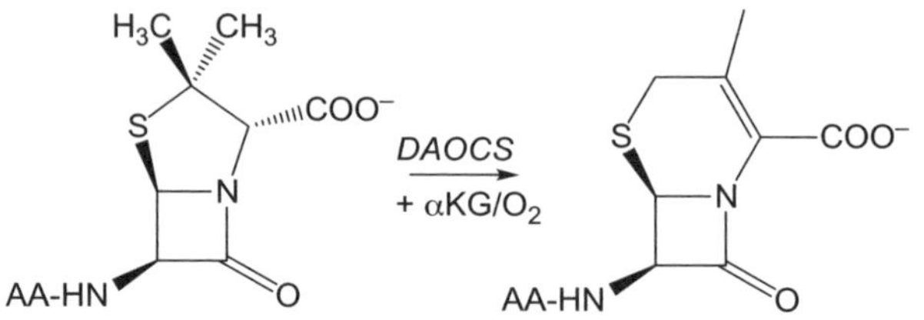

Scheme 1.7   Conversion of IPN into DAOC by DAOCS.

N–H group happens simultaneously with the first ring closure *via* barrier $^5TS_{rc}$ to form structure **4** in the mechanism. The second part of the mechanism starts with hydrogen abstraction of the valine-β-C–H bond *via* barrier $^5TS_{HA,2}$ leading to an iron(III)-hydroxo complexed with a radical on valine-β-C atom (**5**). A final ring-closure step *via* a small barrier ($^5TS_{rc,2}$) of 4.5 kcal mol^{-1} leads to IPN product and an iron(II)-hydroxo complex (**6**). Thus, the mechanism shown in Figure 1.17 implicates several rate-determining barriers, which is consistent with KIE studies that identified different effects upon labeling the cysteine β-C–H or valine β-C–H hydrogen atoms.[132]

IPN products are released and converted into deacetoxycephalosporin C (DAOC, Scheme 1.7) by the mononuclear non-heme iron-containing enzyme deacetoxycephalosporin-C synthase (DAOCS), which is an α-ketoglutarate dependent dioxygenase with close similarities to TauD.[133] The crystal structure provided evidence of an iron active site with a 2-His/1-Asp motif. Moreover, its catalytic cycle seems to resemble that described for TauD above since succinate and CO_2 products were found bound to the metal center.

1.5   1-Aminocyclopropane-1-carboxylic Acid Oxidase (ACCO)

The final step in the ethylene biosynthesis in plants is performed by the enzyme 1-aminocyclopropane-1-carboxylic acid oxidase (ACCO), which is a non-heme iron-containing enzyme with a 2-His/1-Asp ligand motif. Ethylene functions as a plant hormone that participates in all stages of plant growth and development, including responses to environmental stress and fruit ripening. The enzyme converts 1-aminocyclopropane-1-carboxylic acid into ethylene, CO_2 and HCN products on an iron center that utilizes molecular oxygen and ascorbate cofactor.[134] Thus, similarly to IPNS described above, ACCO also does not incorporate the oxygen atoms of O_2 into products, but releases them as water molecules. Although, the function of ascorbate is not completely clear studies showed that it is not essential for oxygen consumption.[135] On the other hand, bicarbonate is essential for product formation. EPR/ENDOR studies on NO-bound substrate complex provided evidence that substrate needs to bind prior to molecular oxygen and that bicarbonate and ascorbate are not directly ligated to the metal.[136] Figure 1.18 shows the active site geometry of ACCO as

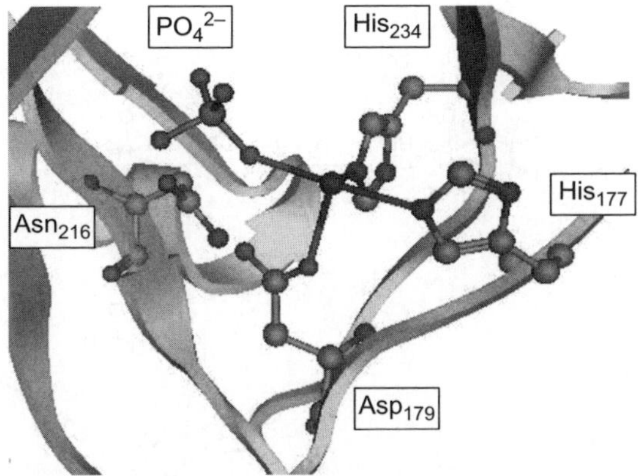

Figure 1.18 Active site structure of phosphate-bound ACCO as taken from the 1WA6 pdb file with amino acids labeled as in the pdb file.

taken from the phosphate bound crystal structure.[137] Thus, ACCO is a non-heme iron enzyme, where the metal is bound to two histidine side chains (of His$_{177}$ and His$_{234}$) and the carboxylic acid group of Asp$_{179}$ in a typical 2-His/1-Asp ligand motif. In the crystal structure, a phosphate molecule is bound to the metal, but substrate ACC should bind there as a bidentate ligand, similarly to αKG in TauD enzymes.

The catalytic cycle of ACCO starts from a resting state structure where the metal is bound to the 2-His/1-Asp ligand motif, while all other ligand sites are occupied with water molecules (**A** in Figure 1.19). Substrate (ACC) binding displaces two of these ligand waters and binds with the amino and carboxylic acid groups *trans* to the two histidine ligands (**B**). The rest of the mechanism is proposed based on DFT calculations on a model of the active site of ACCO, since no experimental data has characterized structures after dioxygen binding. Thus, molecular oxygen binds as an end-on superoxo complex (**C**) and abstracts a hydrogen atom from the amino group to give an iron(III)-hydroperoxo complex with a radical on the amino group (**D**), although its spin density is only 0.41.

From the Fe(III)-hydroperoxo complex **D** the mechanism can diverge in two directions, namely, (i) formation of a high-valent iron(IV)-oxo species (**E**) by proton abstraction from bicarbonate or (ii) ring-opening of the cyclopropane ring and rearrangement of the radical (**F**). In a subsequent step both mechanisms converge again to the iron(IV)-oxo species with a ring-opened ACC bound (**G**). The ordering of these two mechanisms was found to depend on the methods used to calculate the energies and whether zero-point and/or solvation corrections were included, but the conversion of either **D** into **E** or **D** into **F** is the rate-determining step in the reaction cycle. The processes that start with **G** are very fast and proceed almost barrierless: first a proton transfer from the amino group *via* a water molecule by bicarbonate gives structure **H** followed by proton transfer from bicarbonate to the iron(IV)-oxo species to give an

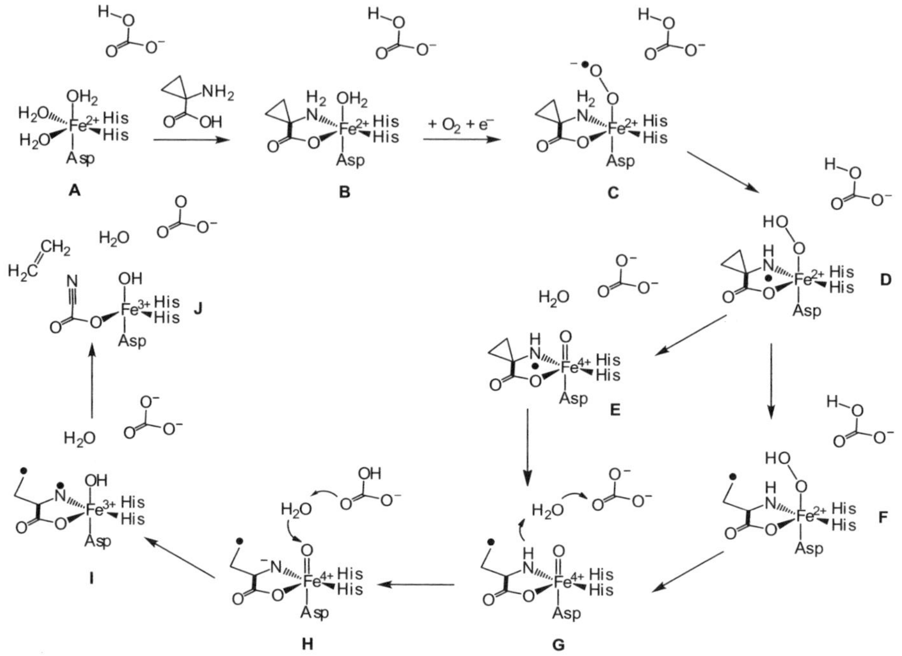

Figure 1.19 Catalytic cycle of ACCO as derived from DFT calculations on model complexes.

iron(III)-hydroxo complex (**I**). In this complex the C–C bond is considerably weakened and hence acetylene is released with high exothermicity, complex **J**. Recent experimental studies using steady-state kinetics, solvent kinetic isotope effects and competitive oxygen kinetic isotope effects provided further insight into the ACCO mechanism after dioxygen binding.[138] A large solvent KIE of 5.0 was observed, which implies that protons are relayed in the rate-determining step of oxygen activation. Moreover a large ^{18}O KIE provides evidence of an iron(IV)-oxo intermediate. These studies support the mechanism proposed by DFT calculations in Figure 1.19.

1.6 Rieske Dioxygenases

Metabolism of aromatic rings through dihydroxylation is a challenging process since the large resonance energy of the aromatic ring has to be broken in the process. Nevertheless, there are bacteria that contain enzymes, called Rieske dioxygenases or aromatic dihydroxylating dioxygenases, which perform this task efficiently.[5,139–141] As such these enzymes have important environmental and biocatalytic functions and much research has been devoted to this class of enzymes. Rieske dioxygenases are described in more detail in Chapter 2; therefore, in this chapter we give only a brief overview of the key developments in experimental and computational research of these dioxygenases.

Rieske dioxygenases contain a non-heme iron center and utilize molecular oxygen to react with aromatic substrates *via cis*-dihydroxylation. In contrast to the other mononuclear iron dioxygenases described above the Rieske dioxygenases use two externally provided electrons from a nicotinamide adenine nucleotide (NADH) cofactor that shuttles electrons *via* a flavoprotein reductase to the Rieske ferredoxin [2Fe:2S] cluster that is located nearby the active center.

Figure 1.20 displays the active site structure of naphthalene-1,2-dioxygenase (NDO), a typical Rieske dioxygenase, as taken from the 1O7M pdb file.[142] The enzyme has three domains: a 36 kDa reductase domain (not shown in Figure 1.20), a short 14 kDa region containing the Rieske ferredoxin cofactor and a 210 kDa oxygenase domain that includes the catalytic center. In the crystal structure the ferredoxin iron atoms are separated from the mononuclear iron center by about 44 Å. However, the distance between the two centers in adjacent subunits of the enzyme is only 12 Å and bridged by an aspartate carboxylic acid group. Indeed mutagenesis of this aspartic acid group in phthalate dioxygenase led to decreased activity.[143,144]

The ferredoxin is a [2Fe:2S] cluster that is linked to the protein *via* two sulfur linkages of cysteinate residues and two histidine side chains. The oxygenase domain has a buried active site with the metal bound to the protein with the 2-His/1-Asp ligand motif common to many mononuclear iron dioxygenases. In contrast to other dioxygenases described above, however, the carboxylic acid group of Asp_{362} binds as a bidentate ligand. Further hydrogen bonding interactions from the amide group of Asn_{201} with the distal ligand are in place. Near the metal center the aromatic substrate binds. EPR studies using NO-bound iron complexes confirmed the binding of substrate within 4–5 Å from the active site.[145] Further EPR/ENDOR studies using deuterated substrate indicated that the substrate moves away from the iron by approximately 0.5 Å upon reduction of the Rieske metal center.[146] Interestingly, molecular oxygen binds in a side-on fashion rather than end-on in the structure shown in

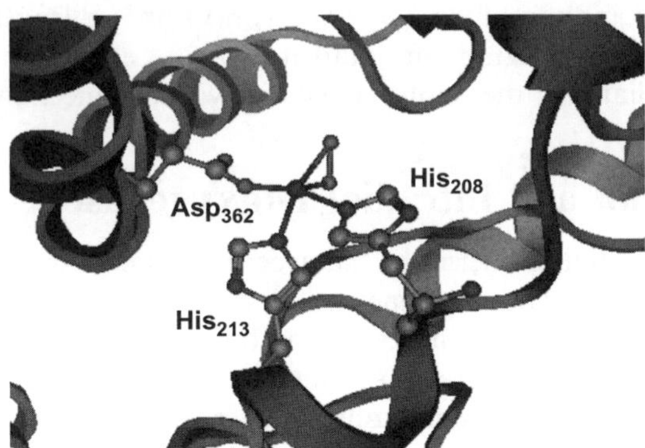

Figure 1.20 Active site structure of naphthalene dioxygenase as taken from the 1O7M pdb file with amino acids labeled as in the pdb file.

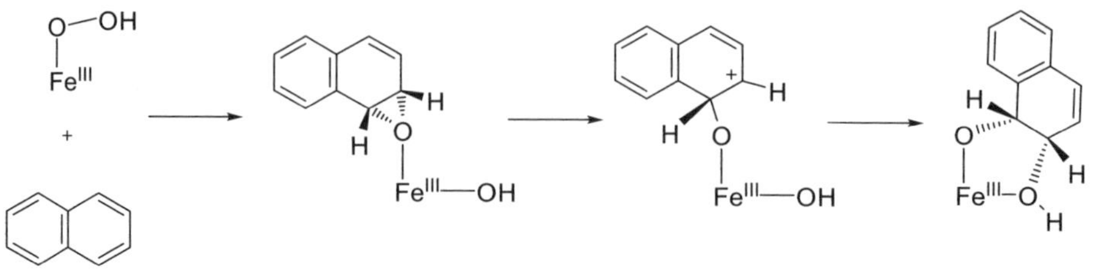

Scheme 1.8   Mechanism of *cis*-dihydroxylation of naphthalene by the iron-hydro-peroxo intermediate of NDO.

Figure 1.20, which contrasts an end-on bound NO crystal structure as well as biomimetic dioxygen bound structures.[147,148] Addition of hydrogen peroxide to the oxygenase domain led to *cis*-hydroxylation products efficiently without the need for additional electrons.[149] Moreover, use of $H_2{}^{18}O_2$ led to isotopically labeled products indicating that both oxygen atoms in the product originate from hydrogen peroxide.

DFT studies on the oxygen activation step in the catalytic cycle of NDO were performed and several possible mechanism tested.[38,150,151] In agreement with the crystal structure shown in Figure 1.20, dioxygen binds in a side-on fashion as $[Fe^{II}O_2]^{2+}$. One-electron reduction of this complex gives a Fe(II)-superoxo complex in an end-on configuration that is hydrogen bonded to a nearby water molecule. It has been suggested that the Fe(II)-superoxo abstracts a proton from a nearby residue to form the ferrous-hydroperoxo complex, which is the active oxidant in NDO enzymes (Scheme 1.8). The computational studies of Siegbahn and coworkers implicated a low-energy mechanism *via* an epoxide intermediate. Thus, naphthalene attacks the distal oxygen atom of the Fe(III)-hydroperoxo moiety to form an epoxide bound to Fe-hydroxo in a concerted reaction step. In a subsequent step the epoxide ring opens with the formation of a naphthalene oxide cation radical followed by attack of the hydroxo group on the cation site to give *cis*-hydroxy-naphthalene. These studies found no evidence of the participation of iron(IV)-oxo, iron(V)-oxo or iron(IV)-oxo-hydroxo intermediates in the reaction mechanism. So far, experimental evidence of epoxide intermediates in the naphthalene dioxygenase mechanism is missing.

1.7   Extradiol and Intradiol Dioxygenases

Extradiol and intradiol dioxygenases catalyze the aromatic ring opening of catechol compounds.[139–141] These enzymes will be discussed in more detail in Chapter 2, therefore we only give a short introduction. Intradiol dioxygenases utilize a non-heme iron(III) center to form (*E,E*)-muconic acid products, while extradiol dioxygenases contain a non-heme iron(II) center to yield 2-hydroxy-6-ketohexa-2,4-dienoic acid products. Homoprotocatechuate 2,3-dioxygenase (2,3-HPCD) from *Brevibacterium fuscum* is an extradiol dioxygenase for which several dioxygen bound intermediates have been characterized by crystallography studies.[152,153]

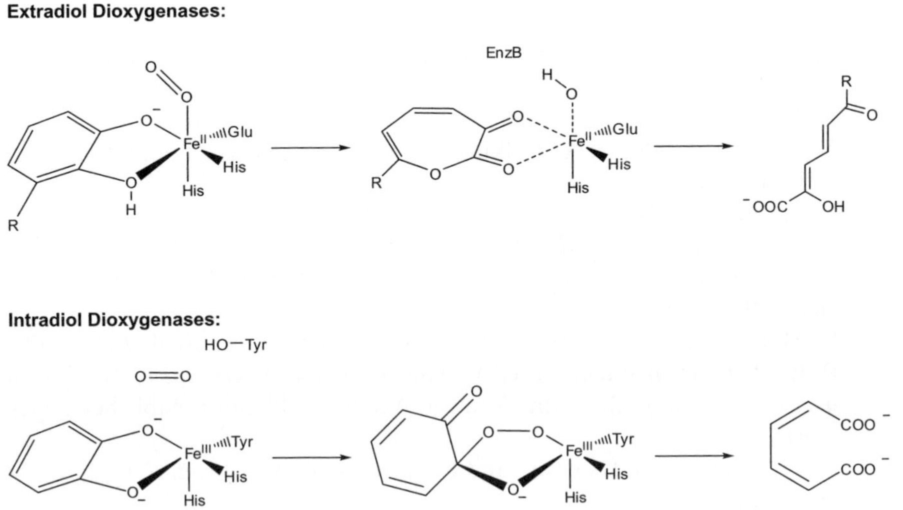

Extradiol Dioxygenases:

Intradiol Dioxygenases:

Scheme 1.9 Mechanism of arene dihydroxylation by extradiol and intradiol dioxygenases.

Scheme 1.9 displays the catalytic mechanism of arene dihydroxylation by extradiol dioxygenases (top) and intradiol dioxygenases (bottom). The active site of extradiol dioxygenase contains a 2His/1Glu iron binding motif, where the substrate binds as a bidentate monoanion ligand. Molecular oxygen binds to the iron center and using enzymatic hydrogen bonding interactions a bicyclic ring-structure is formed that after alkenyl transfer leads to lactone products. By contrast, in the intradiol dioxygenases the active center contains an unusual 2His/1-Tyr iron binding site, where the commonly found carboxylic acid amino acid (Asp/Glu) is replaced by a tyrosinate ligand. Here the substrate binds as a semiquinone and molecular oxygen is not covalently bound to the iron center. The reaction starts with a C–O bond formation followed by acyl transfer to form the muconic anhydride product.

1.8   Conclusion

Non-heme iron-containing dioxygenases are versatile enzymes that catalyze a large range of chemical reactions. In the past ten years significant progress has been made in the understanding of their reactivity and mechanism through a combination of experimental and computational techniques. These studies show the complementary nature of the techniques and give useful insights into these fascinating enzymes. This chapter has reviewed a range of different mononuclear non-heme iron dioxygenases with different structural characteristics (2His/1Asp ligand system *versus* 3His and 2His/1Tyr ligand systems) and the reactivity patterns *versus* substrates.

References

1. M. Sono, M. P. Roach, E. D. Coulter and J. H. Dawson, *Chem. Rev.*, 1996, **96**, 2841.
2. E. I. Solomon, T. C. Brunold, M. I. Davis, J. N. Kemsley, S.-K. Lee, N. Lehnert, F. Neese, A. J. Skulan, Y.-S. Yang and J. Zhou, *Chem. Rev.*, 2000, **100**, 235.
3. F. P. Guengerich, *Chem. Res. Toxicol.*, 2001, **14**, 611.
4. J. T. Groves, *Proc. Natl. Acad. Sci. USA*, 2003, **100**, 3569.
5. M. Costas, M. P. Mehn, M. P. Jensen and L. Que Jr., *Chem. Rev.*, 2004, **104**, 939.
6. B. Meunier, S. P. de Visser and S. Shaik, *Chem. Rev.*, 2004, **104**, 3947.
7. P. R. Ortiz de Montellano (ed.) *Cytochrome P450: Structure, Mechanism and Biochemistry*, 3rd edn, Kluwer Academic/Plenum Publishers, New York, 2005.
8. A. W. Munro, H. M. Girvan and K. J. McLean, *Nat. Prod. Rep.*, 2007, **24**, 585.
9. P. C. A. Bruijnincx, G. van Koten and R. J. M. Klein Gebbink, *Chem. Soc. Rev.*, 2008, **37**, 2716.
10. E. G. Kovaleva and J. D. Lipscomb, *Nat. Chem. Biol.*, 2008, **4**, 186.
11. E. L. Hegg and L. Que Jr., *Eur. J. Biochem.*, 1997, **250**, 625.
12. L. Que Jr., *Nat. Struct. Biol.*, 2000, **7**, 182.
13. X. Shan and L. Que Jr., *J. Inorg. Biochem.*, 2006, **100**, 421.
14. L. Que Jr., *Acc. Chem. Res.*, 2007, **40**, 493.
15. W. Nam, *Acc. Chem. Res.*, 2007, **40**, 522.
16. S. V. Kryatov, E. V. Rybak-Akimova and S. Schindler, *Chem. Rev.*, 2005, **105**, 2175.
17. I. J. Clifton, M. A. McDonough, D. Ehrismann, N. J. Kershaw, N. Granatino and C. J. Schofield, *J. Inorg. Biochem.*, 2006, **100**, 644.
18. M. L. Neidig and E. I. Solomon, *Chem. Commun.*, 2005, 5843.
19. T. D. H. Bugg, *Curr. Opin. Chem. Biol.*, 2001, **5**, 550.
20. M. J. Ryle and R. P. Hausinger, *Curr. Opin. Chem. Biol.*, 2002, **6**, 193.
21. C. Krebs, D. G. Fujimori, C. T. Walsh and J. M. Bollinger Jr., *Acc. Chem. Res.*, 2007, **40**, 484.
22. C. J. Schofield and Z. Zhang, *Curr. Opin. Chem. Biol.*, 1999, **9**, 722.
23. T. D. H. Bugg, *Tetrahedron*, 2003, **59**, 7075.
24. C. Loenarz and C. J. Schofield, *Nat. Chem. Biol.*, 2008, **4**, 152.
25. O. W. Choroba, D. H. Williams and J. B. Spencer, *J. Am. Chem. Soc.*, 2000, **122**, 5389.
26. L. J. Higgins, F. Yan, P. Liu, H.-W. Liu and C. L. Drennan, *Nature*, 2005, **437**, 838.
27. M. J. Bodner, R. M. Phelan, M. F. Freeman, R. Li and C. A. Townsend, *J. Am. Chem. Soc.*, 2010, **132**, 12.
28. Y. Mishina, E. M. Duguid and C. He, *Chem. Rev.*, 2006, **106**, 215.
29. P. J. O'Brien, *Chem. Rev.*, 2006, **106**, 720.

30. J. M. Simmons, T. A. Müller and R. P. Hausinger, *Dalton Trans.*, 2008, 5132.
31. R. K. Bruick and S. L. McKnight, *Science*, 2001, **294**, 1337.
32. E. Berra, E. Benizri, A. Ginouvès, V. Volmat, D. Roux and J. Pouysségur, *EMBO J.*, 2003, **22**, 4082.
33. C. M. West, H. van der Wel and Z. A. Wang, *Development*, 2007, **134**, 3349.
34. K. A. Lee, J. D. Lynd, S. O'Reilly, M. Kiupel, J. J. McCormick and J. J. LaPres, *Mol. Cancer Res.*, 2008, **6**, 829.
35. A. Seifert, D. M. Katschinski, S. Tonack, B. Fischer and A. N. Santos, *Chem. Res. Toxicol.*, 2008, **21**, 341.
36. R. Chowdhury, A. Hardy and C. J. Schofield, *Chem. Soc. Rev.*, 2008, **37**, 1308.
37. L. C. Blasiak and C. L. Drennan, *Acc. Chem. Res.*, 2009, **42**, 147.
38. A. Bassan, T. Borowski and P. E. M. Siegbahn, *Dalton Trans.*, 2004, 3153.
39. J. M. Bollinger Jr., J. C. Price, L. M. Hoffart, E. W. Barr and C. Krebs, *Eur. J. Inorg. Chem.*, 2005, 4245.
40. S. P. de Visser, *Coord. Chem. Rev.*, 2009, **253**, 754.
41. R. B. Muthukumaran, P. K. Grzyska, R. P. Hausinger and J. McCracken, *Biochemistry*, 2007, **46**, 5951.
42. J. C. Price, E. W. Barr, B. Tirupati, J. M. Bollinger Jr. and C. Krebs, *Biochemistry*, 2003, **42**, 7497.
43. D. A. Proshlyakov, T. F. Henshaw, G. R. Monterosso, M. J. Ryle and R. P. Hausinger, *J. Am. Chem. Soc.*, 2004, **126**, 1022.
44. P. J. Riggs-Gelasco, J. C. Price, R. B. Guyer, J. H. Brehm, E. W. Barr, J. M. Bollinger Jr. and C. Krebs, *J. Am. Chem. Soc.*, 2004, **126**, 8108.
45. J. M. Elkins, M. J. Ryle, I. J. Clifton, J. C. Dunning Hotopp, J. S. Lloyd, N. I. Burzlaff, J. E. Baldwin, R. P. Hausinger and P. L. Roach, *Biochemistry*, 2002, **41**, 5185.
46. J. R. O'Brien, D. J. Schuller, V. S. Yang, B. D. Dillard and W. N. Lanzilotta, *Biochemistry*, 2003, **42**, 5547.
47. P. K. Grzyska, M. J. Ryle, G. R. Monterosso, J. Liu, D. P. Ballou and R. P. Hausinger, *Biochemistry*, 2005, **44**, 3845.
48. K. P. McCusker and J. P. Klinman, *Proc. Natl. Acad. Sci. USA*, 2009, **106**, 19791.
49. K. P. McCusker and J. P. Klinman, *J. Am. Chem. Soc.*, 2010, **132**, 5114.
50. M. J. Ryle, R. Padmakumar and R. P. Hausinger, *Biochemistry*, 1999, **38**, 15278.
51. E. Eichhorn, J. R. van der Ploeg, M. A. Kertesz and T. Leisinger, *J. Biol. Chem.*, 1997, **272**, 23031.
52. D. Fujimoto and N. Tamiya, *Biochem. J.*, 1962, **84**, 333.
53. M. J. Ryle, A. Liu, R. B. Muthukumaran, R. Y. N. Ho, K. D. Koehntop, J. McCracken, L. Que Jr. and R. P. Hausinger, *Biochemistry*, 2003, **42**, 1854.
54. K. D. Koehntop, S. Marimanikkuppam, M. J. Ryle, R. P. Hausinger and L. Que Jr., *J. Biol. Inorg. Chem.*, 2006, **11**, 63.

55. J. C. Price, E. W. Barr, T. E. Glass, C. Krebs and J. M. Bollinger Jr., *J. Am. Chem. Soc.*, 2003, **125**, 13008.
56. J. C. Price, E. W. Barr, L. M. Hoffart, C. Krebs and J. M. Bollinger Jr., *Biochemistry*, 2005, **44**, 8138.
57. P. K. Grzyska, E. H. Appelman, R. P. Hausinger and D. A. Proshlyakov, *Proc. Natl. Acad. Sci. USA*, 2010, **107**, 3982.
58. T. Borowski, A. Bassan and P. E. M. Siegbahn, *Chem. Eur. J.*, 2004, **10**, 1031.
59. I. A. Topol, A. V. Nemukhin, K. Salnikow, R. E. Cachau, Y. G. Abashkin, K. S. Kasprzak and S. K. Burt, *J. Phys. Chem. A*, 2006, **110**, 4223.
60. S. P. de Visser, *Chem. Commun.*, 2007, 171.
61. L. M. Mirica, K. P. McCusker, J. W. Munos, H.-w. Liu and J. P. Klinman, *J. Am. Chem. Soc.*, 2008, **130**, 8122.
62. E. Godfrey, C. S. Porro and S. P. de Visser, *J. Phys. Chem. A*, 2008, **112**, 2464.
63. S. P. de Visser, *J. Am. Chem. Soc.*, 2006, **128**, 9813.
64. S. Sinnecker, N. Svensen, E. W. Barr, S. Ye, J. M. Bollinger Jr., F. Neese and C. Krebs, *J. Am. Chem. Soc.*, 2007, **129**, 6168.
65. A. V. Nemukhin, I. A. Topol, R. E. Cachau and S. K. Burt, *Theor. Chem. Acc.*, 2006, **115**, 348.
66. S. C. Trewick, T. F. Henshaw, R. P. Hausinger, T. Lindahl and B. Sedgwick, *Nature*, 2002, **419**, 174.
67. P. Ø. Falnes, R. F. Johansen and E. Seeberg, *Nature*, 2002, **419**, 178.
68. C. Yi, C.-G. Yang and C. He, *Acc. Chem. Res.*, 2010, **42**, 519.
69. Y. Mishina and C. He, *J. Inorg. Biochem.*, 2006, **100**, 670.
70. T. Duncan, S. C. Trewick, P. Koivisto, P. A. Bates, T. Lindahl and B. Sedgwick, *Proc. Natl. Acad. Sci. USA*, 2002, **99**, 16660.
71. B. Yu, W. C. Edstrom, J. Benach, Y. Hamuro, P. C. Weber, B. R. Gibney and J. F. Hunt, *Nature*, 2006, **439**, 879.
72. P. A. Aas, M. Otterlei, P. Ø. Falnes, C. B. Vågbø, F. Skorpen, M. Akbari, O. Sundheim, M. Bjørås, G. Slupphaug, E. Seeberg and H. E. Krokan, *Nature*, 2003, **421**, 859.
73. A. D. Winter and A. P. Page, *Mol. Cell Biol.*, 2000, **20**, 4084.
74. K. L. Gorres, R. Edupuganti, G. R. Krow and R. T. Raines, *Biochemistry*, 2008, **47**, 9447.
75. D. W. Lee, S. Rajagopalan, A. Siddiq, R. Gwiazda, L. Yang, M. F. Beal, R. R. Ratan and J. K. Andersen, *J. Biol. Chem.*, 2009, **284**, 29065.
76. L. M. Hoffart, E. W. Barr, R. B. Guyer, J. M. Bollinger Jr. and C. Krebs, *Proc. Natl. Acad. Sci. USA*, 2006, **103**, 14738.
77. M. Wu, H.-S. Moon, T. P. Begley, J. Myllyharju and K. I. Kivirikko, *J. Am. Chem. Soc.*, 1999, **121**, 587.
78. M. K. Koski, R. Hieta, C. Böllner, K. I. Kivirikko, J. Myllyharju and R. K. Wierenga, *J. Biol. Chem.*, 2007, **282**, 37112.
79. M. A. McDonough, L. A. McNeill, M. Tilliet, C. A. Papamicaël, Q.-Y. Chen, B. Banerji, K. S. Hewitson and C. J. Schofield, *J. Am. Chem. Soc.*, 2005, **127**, 7680.

80. M. B. Pappalardi, J. D. Martin, Y. Jiang, M. C. Burns, H. Zhao, T. Ho, S. Sweitzer, L. Lor, B. Schwartz, K. Duffy, R. Gontarek, P. J. Tummino, R. A. Copeland and L. Luo, *Biochemistry*, 2008, **47**, 11165.

81. M. A. Culpepper, E. E. Scott and J. Limburg, *Biochemistry*, 2010, **49**, 124.

82. M. K. Koski, R. Hieta, M. Hirsila, A. Ronka, J. Myllyharju and R. K. Wierenga, *J. Biol. Chem.*, 2009, **284**, 25290.

83. B. Karamzadeh, D. Kumar, G. N. Sastry and S. P. de Visser, *J. Phys. Chem. A*, 2010, **114**, 11324.

84. S. P. de Visser, D. Kumar, S. Cohen, R. Shacham and S. Shaik, *J. Am. Chem. Soc.*, 2004, **126**, 8362.

85. S. Shaik, D. Kumar and S. P. de Visser, *J. Am. Chem. Soc.*, 2008, **130**, 10128.

86. R. Latifi, M. Bagherzadeh and S. P. de Visser, *Chem. Eur. J.*, 2009, **15**, 6651.

87. S. P. de Visser, *J. Am. Chem. Soc.*, 2010, **132**, 1087.

88. D. Kumar, B. Karamzadeh, G. N. Sastry and S. P. de Visser, *J. Am. Chem. Soc.*, 2010, **132**, 7656.

89. F. H. Vaillancourt, E. Yeh, D. A. Vosburg, S. Garneau-Tsodikova and C. T. Walsh, *Chem. Rev.*, 2006, **106**, 3364.

90. D. G. Fujimori and C. T. Walsh, *Curr. Opin. Chem. Biol.*, 2007, **11**, 553.

91. A. Butler and M. Sandy, *Nature*, 2009, **460**, 848.

92. F. H. Vaillancourt, E. Yeh, D. A. Vosburg, S. E. O'Connor and C. T. Walsh, *Nature*, 2005, **436**, 1191.

93. L. C. Blasiak, F. H. Vaillancourt, C. T. Walsh and C. L. Drennan, *Nature*, 2006, **440**, 368.

94. C. Wong, D. G. Fujimori, C. T. Walsh and C. L. Drennan, *J. Am. Chem. Soc.*, 2009, **131**, 4872.

95. F. H. Vaillancourt, J. Yin and C. T. Walsh, *Proc. Natl. Acad. Sci. USA*, 2005, **102**, 10111.

96. M. L. Matthews, C. M. Krest, E. W. Barr, F. H. Vaillancourt, C. T. Walsh, M. T. Green, C. Krebs and J. M. Bollinger Jr., *Biochemistry*, 2009, **48**, 4331.

97. D. P. Galonić, E. W. Barr, C. T. Walsh, J. M. Bollinger and C. Krebs, *Nat. Chem. Biol.*, 2007, **3**, 113.

98. D. G. Fujimori, E. W. Barr, M. L. Matthews, G. M. Koch, J. R. Yonce, C. T. Walsh, J. M. Bollinger Jr., C. Krebs and P. J. Riggs-Gelasco, *J. Am. Chem. Soc.*, 2007, **129**, 13408.

99. J. M. Mayer, *Acc. Chem. Res.*, 1998, **31**, 441.

100. T. Kojima, R. A. Leising, S. Yan and L. Que Jr., *J. Am. Chem. Soc.*, 1993, **115**, 11328.

101. M. L. Matthews, C. S. Neumann, L. A. Miles, T. L. Grove, S. J. Booker, C. Krebs, C. T. Walsh and J. M. Bollinger Jr., *Proc. Natl. Acad. Sci. USA*, 2009, **106**, 17723.

102. S. P. de Visser and R. Latifi, *J. Phys. Chem. B*, 2009, **113**, 12.

103. S. Pandian, M. A. Vincent, I. H. Hillier and N. A. Burton, *Dalton Trans.*, 2009, 6201.

104. M. L. Neidig, C. D. Brown, K. M. Light, D. G. Fujimori, E. M. Nolan, J. C. Price, E. W. Barr, J. M. Bollinger Jr., C. Krebs, C. T. Walsh and E. I. Solomon, *J. Am. Chem. Soc.*, 2007, **129**, 14224.
105. J. B. Lombardini, T. P. Singer and P. D. Boyer, *J. Biol. Chem.*, 1969, **244**, 1172.
106. M. H. Stipanuk, *Annu. Rev. Nutr.*, 2004, **24**, 539.
107. G. D. Straganz and B. Nidetzky, *ChemBioChem.*, 2006, **7**, 1536.
108. C. A. Joseph and M. J. Maroney, *Chem. Commun.*, 2007, 3338.
109. S. Aluri and S. P. de Visser, *J. Am. Chem. Soc.*, 2007, **129**, 14846.
110. S. P. de Visser and G. D. Straganz, *J. Phys. Chem. A*, 2009, **113**, 1835.
111. C. L. Weinstein, R. H. Haschemeyer and O. W. Griffin, *J. Biol. Chem.*, 1988, **263**, 16568.
112. S. Ye, X. Wu, L. Wei, D. Tang, P. Sun, M. Bartlam and Z. Rao, *J. Biol. Chem.*, 2007, **282**, 3391.
113. B. S. Pierce, J. D. Gardner, L. J. Bailey, T. C. Brunold and B. G. Fox, *Biochemistry*, 2007, **46**, 8569.
114. C. R. Simmons, Q. Liu, Q. Huang, Q. Hao, T. P. Begley, P. A. Karplus and M. H. Stipanuk, *J. Biol. Chem.*, 2006, **281**, 18723.
115. J. G. McCoy, L. J. Bailey, E. Bitto, C. A. Bingman, D. J. Aceti, B. G. Fox and G. N. Phillips Jr., *Proc. Natl. Acad. Sci. USA*, 2006, **103**, 3084.
116. S. C. Chai, J. R. Bruyere and M. J. Maroney, *J. Biol. Chem.*, 2006, **281**, 15774.
117. Y.-K. Lee, M. M. Whittaker and J. W. Whittaker, *Biochemistry*, 2008, **47**, 6637.
118. R. Amorati, F. Catarzi, S. Menichetti, G. F. Pedulli and C. Viglianisi, *J. Am. Chem. Soc.*, 2008, **130**, 237.
119. J. E. Dominy Jr., J. Hwang, S. Guo, L. L. Hirschberger, S. Zhang and M. H. Stipanuk, *J. Biol. Chem.*, 2008, **283**, 12188.
120. T. Kleffmann, S. A. K. Jongkees, G. Fairweather, S. M. Wilbanks and G. N. L. Jameson, *J. Biol. Inorg. Chem.*, 2010, **14**, 913.
121. C. R. Simmons, K. Krishnamoorthy, S. L. Granett, D. J. Schuller, J. E. Dominy Jr., T. P. Begley, M. H. Stipanuk and P. A. Karplus, *Biochemistry*, 2008, **47**, 11390.
122. D. Kumar, W. Thiel and S. P. de Visser, *J. Am. Chem. Soc.*, 2011, **133**, 3869.
123. J. E. Baldwin and M. Bradley, *Chem. Rev.*, 1990, **90**, 1079.
124. P. L. Roach, I. J. Clifton, C. M. H. Hensgens, N. Shibata, C. J. Schofield, J. Hajdu and J. E. Baldwin, *Nature*, 1997, **387**, 827.
125. M. Lundberg and K. Morokuma, *J. Phys. Chem. B*, 2007, **111**, 9380.
126. M. Lundberg, P. E. M. Siegbahn and K. Morokuma, *Biochemistry*, 2008, **47**, 1031.
127. M. Lundberg, T. Kawatsu, T. Vreven, M. J. Frisch and K. Morokuma, *J. Chem. Theory Comput.*, 2009, **5**, 222.
128. C. D. Brown, M. L. Neidig, M. B. Neibergall, J. D. Lipscomb and E. I. Solomon, *J. Am. Chem. Soc.*, 2007, **129**, 7427.
129. A. Kriauciunas, C. A. Frolik, T. C. Hassell, P. L. Skatrud, M. G. Johnson, N. L. Holbrook and V. J. Chen, *J. Biol. Chem.*, 1991, **266**, 11779.

130. F. Ogliaro, S. P. de Visser, S. Cohen, P. K. Sharma and S. Shaik, *J. Am. Chem. Soc.*, 2002, **124**, 2806.
131. M. J. Park, J. Lee, Y. Suh, J. Kim and W. Nam, *J. Am. Chem. Soc.*, 2006, **128**, 2630.
132. J. E. Baldwin, R. M. Adlington, S. E. Moroney, L. D. Field and H.-H. Ting, *J. Chem. Soc., Chem. Commun.*, 1984, 984.
133. H.-J. Lee, M. D. Lloyd, K. Harlos, I. J. Clifton, J. E. Baldwin and C. J. Schofield, *J. Mol. Biol.*, 2001, **308**, 937.
134. M. C. Pirrung, *Acc. Chem. Res.*, 1999, **32**, 711.
135. A. M. Rocklin, K. Kato, H.-W. Liu, L. Que Jr. and J. D. Lipscomb, *J. Biol. Inorg. Chem.*, 2004, **9**, 171.
136. D. L. Tierney, A. M. Rocklin, J. D. Lipscomb, L. Que Jr. and B. M. Hoffman, *J. Am. Chem. Soc.*, 2005, **127**, 7005.
137. Z. Zhang, J.-S. Ren, I. J. Clifton and C. J. Schofield, *Chem. Biol.*, 2004, **11**, 1383.
138. L. M. Mirica and J. P. Klinman, *Proc. Natl. Acad. Sci. USA*, 2008, **105**, 1814.
139. D. R. Boyd and T. D. H. Bugg, *Org. Biomol. Chem.*, 2006, **4**, 181.
140. E. G. Kovaleva, M. B. Neibergall, S. Chakrabarty and J. D. Lipscomb, *Acc. Chem. Res.*, 2007, **40**, 475.
141. T. D. H. Bugg and S. Ramaswamy, *Curr. Opin. Chem. Biol.*, 2008, **12**, 134.
142. A. Karlsson, J. V. Parales, R. E. Parales, D. T. Gibson, H. Eklund and S. Ramaswamy, *Science*, 2003, **299**, 1039.
143. A. Pinto, M. Tarasev and D. P. Ballou, *Biochemistry*, 2006, **45**, 9032.
144. M. Tarasev, A. Pinto, D. Kim, S. J. Elliott and D. P. Ballou, *Biochemistry*, 2006, **45**, 10208.
145. T.-C. Yang, M. D. Wolfe, M. B. Neibergall, Y. Mekmouche, J. D. Lipscomb and B. M. Hoffman, *J. Am. Chem. Soc.*, 2003, **125**, 2034.
146. T. C. Yang, M. D. Wolfe, M. B. Neibergall, Y. Mekmouche, J. D. Lipscomb and B. M. Hoffman, *J. Am. Chem. Soc.*, 2003, **125**, 7056.
147. A. Karlsson, J. V. Parales, R. E. Parales, D. T. Gibson, H. Eklund and S. Ramaswamy, *J. Biol. Inorg. Chem.*, 2005, **10**, 483.
148. J. Annaraj, J. Cho, Y.-M. Lee, S. Y. Kim, R. Latifi, S. P. de Visser and W. Nam, *Angew. Chem. Int. Ed.*, 2009, **48**, 4150.
149. M. D. Wolfe and J. D. Lipscomb, *J. Biol. Chem.*, 2003, **278**, 829.
150. A. Bassan, M. R. A. Blomberg and P. E. M. Siegbahn, *J. Biol. Inorg. Chem.*, 2004, **9**, 439.
151. A. Bassan, M. R. A. Blomberg, T. Borowski and P. E. M. Siegbahn, *J. Phys. Chem. B*, 2004, **108**, 13031.
152. S. L. Groce and J. D. Lipscomb, *Biochemistry*, 2005, **44**, 7175.
153. E. G. Kovaleva and J. D. Lipscomb, *Science*, 2007, **316**, 453.

Non-heme Iron-Dependent Dioxygenases: Mechanism and Structure

TIMOTHY D. H. BUGG

Department of Chemistry, University of Warwick, Coventry CV4 7AL, United Kingdom

2.1 Introduction

Unlike the heme-dependent P450 monoxygenases, non-heme iron-dependent enzymes are able to catalyse dioxygenation reactions, in which both atoms of oxygen from dioxygen are incorporated into the product. There is a wide range of dioxygenase-catalysed reactions, which have been reviewed.[1,2] A number of these enzymes are involved in bacterial aromatic degradation pathways,[3,4] but dioxygenases are also involved in oxidative cleavage of carotenoids in plants and mammals, and specific reactions in natural product biosynthesis.[4] In this chapter, I will discuss enzymes involved in three general classes of reaction: oxidative C–C cleavage reactions (Section 2.2); reactions forming hydroperoxides and endoperoxides (Section 2.3); and dioxygenases catalysing hydroxylation and dihydroxylation reactions (Section 2.4). The α-ketoglutarate dependent dioxygenases will be discussed in much more detail in Chapter 3, and cysteine dioxygenase is specifically described in Chapter 1. In each case, I will describe what is known about the active site structure and reaction mechanism of each enzyme family.

Iron-Containing Enzymes: Versatile Catalysts of Hydroxylation Reactions in Nature
Edited by Sam P de Visser and Devesh Kumar
© Royal Society of Chemistry 2011
Published by the Royal Society of Chemistry, www.rsc.org

In most cases, the active catalytic species is a mononuclear iron(II) centre, ligated by three or four active site amino acid ligands. One common structural motif, that is found in many enzymes in this class, is the His-His-Asp/Glu motif,[5] in which the iron(II) centre is coordinated by two histidine residues and one aspartic acid or glutamic acid residue, in a facial tridentate ligand set. This coordination state allows the iron(II) centre to coordinate dioxygen and the organic substrate by three further coordination sites, offering many possible geometries for oxidative reactions, compared with the P450 enzymes, in which the heme iron centre has only one vacant coordination site. This may help to explain the wide range of non-heme iron-dependent oxygenase reactions.

2.2 Dioxygenases Catalysing Oxidative C–C Cleavage Reactions

2.2.1 Intradiol Catechol Dioxygenases

The catechol dioxygenases were discovered through the pioneering work of Osamu Hayaishi, who established that catechol, an intermediate on bacterial aromatic degradation pathways, could be oxidatively cleaved by two classes of enzymes: intradiol dioxygenases that require iron(III) for activity and extradiol dioxygenases that require iron(II) for activity.[6] Hayaishi was able to demonstrate in 1955, using $^{18}O_2$ labelling experiments, that catechol 1,2-dioxygenase from *Pseudomonas* incorporated two atoms of oxygen from dioxygen into the reaction products, which is consistent with a mechanism involving a four-membered dioxetane intermediate.[7]

The X-ray crystal structure of protocatechuate 3,4-dioxygenase (3,4-PCD) from *Pseudomonas putida* was solved by Ohlendorf *et al.* in 1988.[8] The mononuclear iron(III) cofactor is ligated by four amino acid side chains: His_{460}, His_{462}, Tyr_{408} and Tyr_{447} (Figure 2.1a). A fifth water ligand completes a trigonal bipyramidal structure. The two tyrosinate ligands give the enzyme its characteristic deep red colour, caused by ligand-to-metal charge transfer interactions, giving Raman vibrations at 1254 and 1266 cm^{-1}.[9] The active site of catechol 1,2-dioxygenase (1,2-CTD) from *Acinetobacter* sp. ADP1, which is sequence-related to 3,4-PCD, contains a very similar arrangement of iron(III) ligands: Tyr-200 and His-226 are the axial ligands, and Tyr_{164}, His_{224} and water molecules are the equatorial ligands.[10]

Structures of 3,4-PCD with bound catechol substrates reveal that, upon substrate binding, the axial tyrosine ligand Tyr_{447} and equatorial water ligand are displaced by substrate, to form a bidentate substrate complex.[11,12] Tyr_{447} swings away from the iron(III) cofactor to leave a penta-coordinate iron(III) centre with approximately octahedral geometry. Reaction of the iron(III)-catechol complex with dioxygen has been proposed to occur *via* a substrate activation mechanism, in which one electron transfer occurs from catecholate substrate to the iron(III) cofactor to form an iron(II)-semiquinone, which is able to react directly with dioxygen to form a hydroperoxide intermediate.

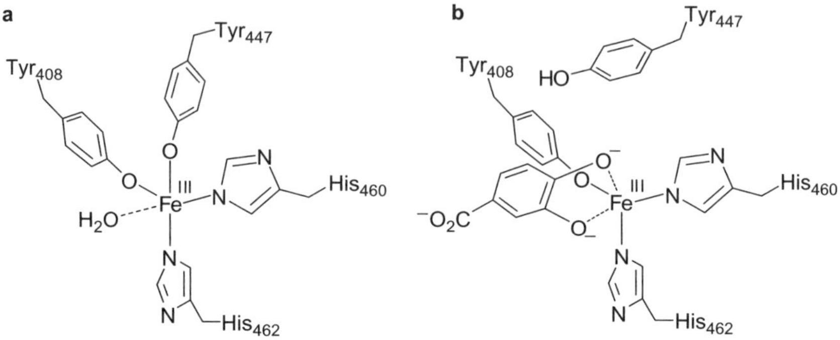

Figure 2.1 Active site of *Pseudomonas putida* protocatechuate 3,4-dioxygenase in the absence (a) and presence (b) of substrate.

Evidence in support of this mechanism comes from the studies of Que and coworkers on model complexes such as iron(III) TPA, TPA = tris-(2-pyridylmethyl)amine, in which the most reactive model complexes showed the most iron(II)-semiquinone character.[13,14] Studies of the 3,4-PCD reaction using magnetic circular dichroism (MCD) spectroscopy and electronic structure calculations by Solomon and coworkers have indicated that a highly covalent iron(III)-semiquinone complex interacts with dioxygen through a strong π interaction, with simultaneous interaction with the iron centre.[15] By replacement of the equatorial ligand Tyr$_{408}$, it was found that this Tyr$_{408}$ is essential for stabilizing the iron(III) centre for reaction with dioxygen.[15]

Subsequent reaction of the hydroperoxide intermediate is believed to occur *via* a Criegee rearrangement, with migration of the adjacent acyl group to yield muconic anhydride as an intermediate, which then undergoes hydrolysis to give the product muconic acid. $^{18}O_2$ labelling studies on catechol 1,2-dioxygenase from *Pseudomonas arvilla* have shown that the intradiol cleavage products contain 99% incorporation of a single atom of ^{18}O, and 74% incorporation of a second atom of ^{18}O, with 24% incorporation of only one atom of ^{18}O.[16] These data are not consistent with a dioxetane intermediate, but are consistent with a Criegee rearrangement to give an anhydride intermediate, followed by the partial exchange of the iron(III) ^{18}O-hydroxide with solvent water.[16] The role of Tyr$_{447}$ is thought to be as a base that deprotonates the second phenolic hydroxyl group of the substrate to form a catechol dianion. The proposed mechanism is shown in Scheme 2.1. Recent studies on *Acinetobacter* sp. catechol 1,2-dioxygenase using mechanistic probes have provided evidence in favour of a mechanism involving O–O bond homolysis, rather than a direct Criegee rearrangement.[17] Incubation of enzyme with stable peracid 3-chloroperoxybenzoic acid afforded 2-chlorophenol and 3-chlorophenol, consistent with a mechanism involving O–O bond homolysis, followed by decarboxylation. Incubation of enzyme with a stable hydroperoxide analogue 2-hydroperoxy-2-methyl-chlorohexanone led to the formation of an acyclic product 5,6-diketoheptan-1-ol, which is also consistent with a mechanism

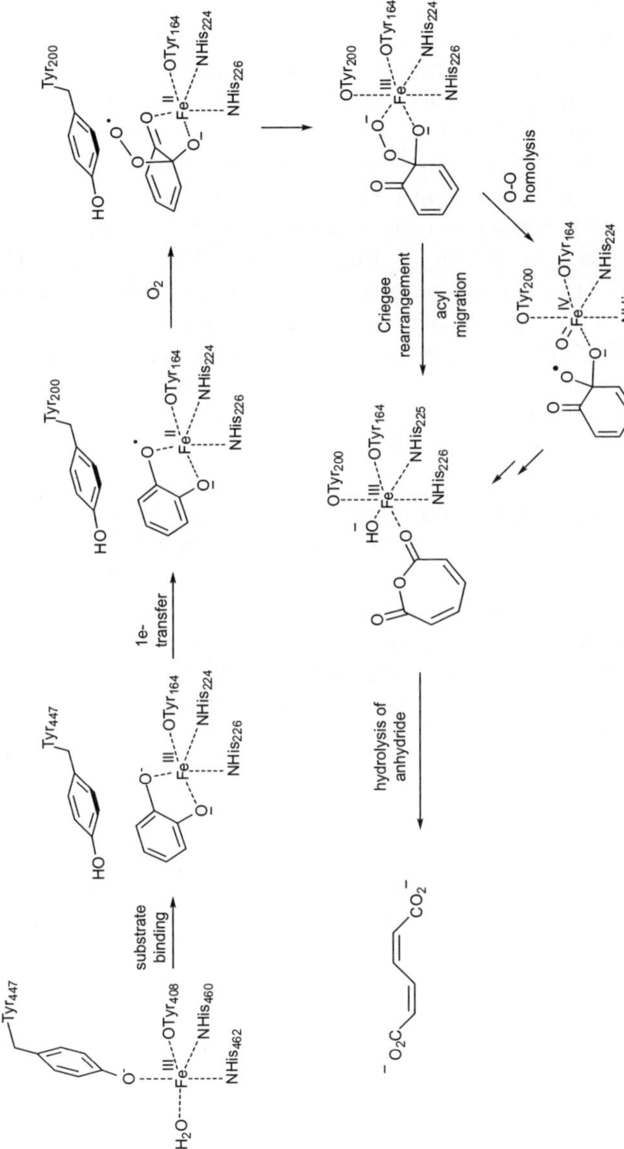

Scheme 2.1　Proposed catalytic mechanism for intradiol catechol dioxygenases (numbering for 3,4-PCD).

involving O–O bond homolysis.[17] A possible catalytic mechanism involving O–O bond homolysis is illustrated in Scheme 2.1.

2.2.2 Extradiol Catechol Dioxygenases

The extradiol catechol dioxygenases catalyse the oxidative cleavage of the aryl C–C bond adjacent to the two phenolic hydroxyl groups, to give a 2-hydroxymuconaldehyde product, using iron(II) as a cofactor. It was shown by Hayaishi that catechol 2,3-dioxygenase from *Pseudomonas arvilla* incorporated two atoms of ^{18}O from $^{18}O_2$ into the product.[18] The active site of 2,3-dihydroxybiphenyl 1,2-dioxygenase (BphC) from *Burkholderia xenovorans* LB400 contains a mononuclear iron(II) centre ligated by three amino acid side chains: His_{146}, His_{210} and Glu_{260} (Figure 2.2).[19] The structure of catechol 2,3-dioxygenase from *Pseudomonas putida mt-2*, an α_4 tetramer, is very similar to that of BphC, and the iron(II) cofactor is bound by His_{153}, His_{214} and Glu_{265}.[20] Protocatechuate 4,5-dioxygenase from *Sphingomonas paucimobilis* SYK-6, an $\alpha_2\beta_2$ tetramer, has no sequence similarity to BphC, yet the active site iron(II) ligands are very similar: the metal centre is coordinated by His_{12}, His_{61} and Glu_{242} (Figure 2.2).[21] The structure of 3-hydroxyanthranilate 3,4-dioxygenase from *Ralstonia metallidurans*, solved in the presence of bound NO, contains an iron(II) cofactor bound by a 2-His,1-Glu motif; however, the identity of outer sphere active site residues is different to that of the above enzymes, with Arg_{47} positioned close to bound NO, and Glu_{110} being the only active site acid–base residue.[22]

EPR spectroscopic studies by Lipscomb and coworkers of the NO complex of the protocatechuate 4,5-dioxygenase showed that the iron(II) cofactor binds both catecholic hydroxyl groups, and binds NO.[23] The reactive form of the bound catechol substrate is the catecholate monoanion, shown in the case of dioxygenase BphC using UV–visible and resonance Raman spectroscopy.[24] There are examples of transition metal semiquinone-superoxide complexes,[25]

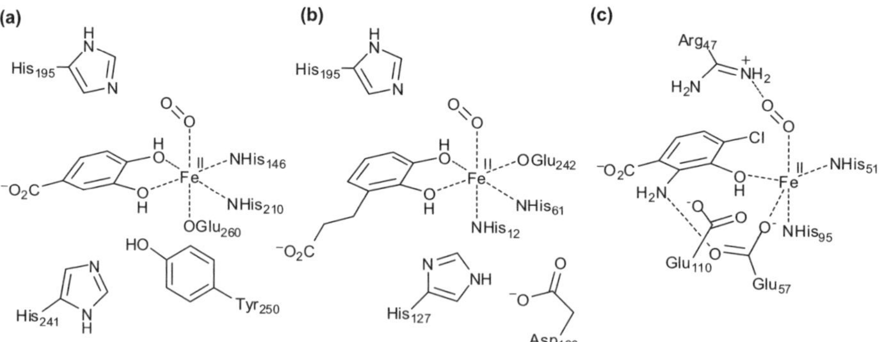

Figure 2.2 Active sites of (a) *Burkholderia xenovorans* BphC, (b) *Sphingomonas paucimobilis* protocatechuate 4,5-dioxygenase (LigB) and (c) *Ralstonia metallidurans* 3-hydroxyanthranilate 4,5-dioxygenase.

therefore, the iron(II) cofactor is thought to utilize dioxygen in the form of superoxide, and activate the catechol substrate as its semiquinone radical. Evidence for a semiquinone radical intermediate has been obtained for 2,3-dihydroxyphenylpropionate 1,2-dioxygenase (MhpB) from *Escherichia coli*, using substrate analogues containing cyclopropyl radical traps, in which processing of *trans-* and *cis-* substituted cyclopropyl analogues by MhpB was found to proceed with isomerization of the cyclopropyl ring substituents, *via* a transient iron(II)-semiquinone-superoxide intermediate.[26]

C–O bond formation between the two activated substrates would then give a proximal hydroperoxide intermediate, which has been mimicked by "carba" analogues in which the –OOH functional group was replaced by $–CH_2OH$.[27] Enzyme inhibition was only observed in analogues in which the hydroxymethyl substituent is positioned in an axial orientation with respect to the cyclo-hexanone ring, indicating a preference for a particular conformation of the bound hydroperoxide.[27] The proximal hydroperoxide intermediate has been observed directly in the crystal structure of homoprotocatechuate 2,3-dioxy-genase (2,3-HPCD) from *Brevibacterium fuscum*, in the presence of the slow substrate 4-nitrocatechol.[28] Three different intermediates were observed in the tetrameric crystal: the ternary complex of bound substrate and dioxygen, the proximal hydroperoxide intermediate, and the cleaved extradiol product. The hydroperoxide group is positioned axially with respect to the cyclohexadiene ring of the substrate, which is consistent with predictions based upon stereo-electronic arguments.[29]

The proximal hydroperoxide is then believed to undergo Criegee rearrangement to give a seven-membered α-keto-lactone intermediate, which is hydrolysed by iron(II)-hydroxide to give the extradiol product. Evidence for the lactone inter-mediate has come from ^{18}O labelling studies carried out on *E. coli* 2,3-dihydroxy-phenylpropionate 1,2-dioxygenase (MhpB), which revealed that after reaction in $H_2^{18}O$ the carboxylate group of the product was labelled to the extent of 30%, which is consistent with the formation of an α-keto lactone intermediate, and exchange of iron(II) hydroxide with solvent ^{18}O-labelled water.[30] The enzyme was also found to catalyse the hydrolysis of a seven-membered lactone analogue.[30] Scheme 2.2 shows the extradiol cleavage reaction mechanism.

The intradiol and extradiol catechol dioxygenase mechanisms both proceed through the same proximal hydroperoxide intermediate, but then undergo different Criegee rearrangements: intradiol cleavage occurs *via* 1,2-acyl migration to give an anhydride, whereas extradiol cleavage occurs *via* 1,2-alkenyl migration to give a lactone. More recent studies have shown that the extradiol *versus* intradiol reaction specificity can be altered by amino acid replacements.[31,33] In *Brevibacterium fuscum* 2,3-HPCD, mutant H200F con-verts 2,3-dihydroxybenzoic acid into the intradiol product.[31] In *Escherichia coli* MhpB, point mutants generated by directed evolution were found to generate 5–15% intradiol product.[32] These observations imply that iron(II) is able to support both extradiol and intradiol cleavage.

It has been proposed that acid–base catalysis is an important factor in extradiol cleavage.[33] A biomimetic model reaction for extradiol cleavage was

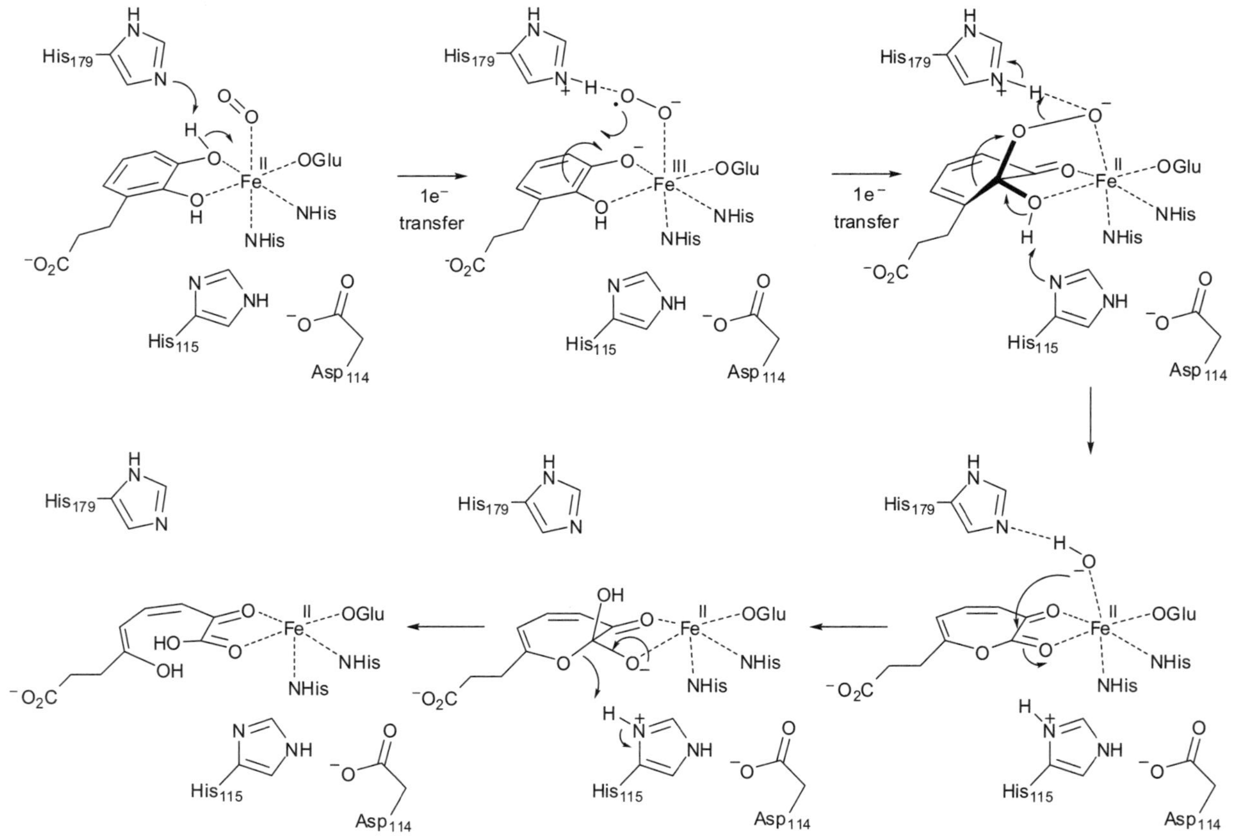

Scheme 2.2 Proposed catalytic mechanism for extradiol catechol dioxygenases (numbering for *Escherichia coli* MhpB).

reported by Lin *et al.*, which uses $FeCl_2$, 1,4,7-triazanonane and pyridine. In this reaction process pyridine has two roles: it first acts as a base to generate the catecholate monoanion; the pyridinium cation then acts as a proton donor to assist the Criegee rearrangement.[33] These observations are consistent with the presence of two acid–base groups in the outer sphere of the extradiol catechol dioxygenases. Site-directed mutagenesis of His_{179} and His_{115} in *E. coli* MhpB has shown that both residues are essential for activity.[34] His_{179} is proposed to act as the active site base, whereas His_{115}, possessing an abnormally high pK_a of 8.0, acts as a proton donor for the final lactone hydrolysis step.[34] A crystal structure of *Pseudomonas* KKS102 BphC complexed with substrate and NO showed that His_{194} was positioned close to NO, which occupies a sixth coordination site at the iron(II) centre, and therefore His_{194} was proposed to act as a base to deprotonate the substrate at C^3.[35] However, replacement of His_{200} in *Brevibacterium fuscum* 2,3-HPCD by Ala, Gln or Asn gives mutant enzymes that retain 30–40% activity;[36] therefore, it appears that the precise roles of active site acid–base residues in the two families of extradiol dioxygenases are somewhat different. Nevertheless, acid–base catalysis *via* outer-sphere active site residues appears to be a significant factor in extradiol cleavage catalysis. Although iron(II) is normally utilized as the metal ion cofactor in the extradiol dioxygenases, MndD from *Arthrobacter globiformis* contains manganese(II) at its active site.[37] A similar catalytic mechanism can be envisaged for Mn^{2+}, although the redox potential for the Mn^{3+}/Mn^{2+} redox couple is much higher (+1.6 V). The homoprotocatechuate 2,3-dioxygenases from *Arthrobacter globiformis* and *Brevibacterium fuscum* show 83% sequence identity, and have very similar active site structures, yet contain manganese(II) and iron(II) at their respective active sites.[38]

2.2.3 Carotenoid Cleavage Dioxygenases

A family of non-heme iron-dependent dioxygenases has been identified since 2000 that catalyse oxidative cleavage reactions on carotenoid substrates, in plants, microorganisms and mammals. The first members of this family to be identified were β-carotene 15,15′-dioxygenase, which catalyses the oxidative cleavage of β-carotene in mammals to form the visual pigment retinal,[39] and 9-*cis*-epoxycarotenoid dioxygenase (NCED), which catalyses the oxidative cleavage of the epoxy-carotenoid 9-*cis*-neoxanthin in plants, a key step in the biosynthesis of plant growth regulator abscisic acid.[40] A substantial family of enzymes with sequence similarity to NCED have been identified, which are believed to catalyse oxidative C–C cleavage reactions on a range of carotenoid substrates, to produce apocarotenoid cleavage products, several of which have biological activities as signalling molecules.[41,42] The NCED gene product has been shown to catalyse oxidative cleavage of 9-*cis*-neoxanthin.[43] Plant CCD1 cleaves several carotenoid substrates at the 9,10 or 9′,10′ positions, to release apocarotenoids such as β-ionone, a component of flower fragrance[44] (Scheme 2.3).

The *max3* (CCD7) and *max4* (CCD8) genes have been implicated in the shoot branching response in plants, and are believed to catalyse two steps in the biosynthesis of a plant hormone which controls shoot branching.[45,46] This hormone was identified in 2008 to be a member of the strigolactone family of natural products.[47,48]

There is a crystal structure of one member of the CCD family, a *Synechocystis* 15,15′-dioxygenase enzyme.[49] The non-heme iron(II) cofactor is ligated by four histidine residues, a different coordination state to other non-heme iron-dependent dioxygenases. The iron(II) centre is positioned 5–6 Å from the bound carotenoid substrate, which occupies a hydrophobic channel lined with several aromatic amino acid side chains.[49] The distance from the iron(II) cofactor to the polyene chain of the substrate suggests that dioxygen binds between the metal cofactor and the substrate.

The iron(II) cofactor could activate dioxygen and the carotenoid substrate *via* one-electron transfers, in a similar way to the extradiol catechol dioxygenases, to give a hydroperoxide intermediate, which would contain a carbocation (Scheme 2.4). In the later stages of the mechanism, either dioxetane formation or Criegee rearrangement steps are possible. Isotope incorporation studies on the mammalian β-carotene dioxygenase enzyme using an asymmetric substrate revealed approximately 50 atom% incorporation of oxygen into each aldehyde product from each source, which could be explained by an epoxide intermediate,[50] or could be explained by the exchange of an iron(II)-hydroxide intermediate with solvent water. The catalytic mechanism has also been examined using computational methods.[51]

2.2.4 Oxidative Cleavage of Aliphatic Substrates

A small group of non-heme iron-dependent dioxygenases catalyses the oxidative cleavage of aliphatic C–C bonds. Dioxygenase ARD from *Klebsiella pneumoniae* catalyses the oxidative cleavage of aci-reductone to 2-keto-4-thiomethyl-butyrate and formic acid, as part of the methionine salvage pathway (Scheme 2.5a).[52] Remarkably, the same polypeptide can utilize a Ni^{2+} cofactor (ARD′) to convert the same substrate into 3-thiomethyl-propionate, formic acid and CO.[53] The two reactions can be explained by the formation of a common hydroperoxide intermediate, followed either (in ARD′) by nucleophilic attack upon C-2 to form a dioxetane or (in ARD) by nucleophilic attack upon C-3 to form a five-membered endo-peroxide.[54] The two oxidative cleavages can also be observed non-enzymatically, *via* base-catalysed auto-oxidation.[54] The choice of metal ion cofactor therefore appears to dictate the choice of reactivity of the hydroperoxide intermediate. The structure of nickel(II)-containing ARD has been solved by NMR spectroscopy, and the mononuclear nickel(II) centre is ligated by four protein ligands: His_{96}, His_{98}, Glu_{102} and His_{140}.[55]

Acetylacetone dioxygenase from *Acinetobacter johnsonii* catalyses the oxidative cleavage of acetylacetone to give pyruvaldehyde and acetate. The purified enzyme requires both iron(II) and zinc(II).[56] Kinetic studies using a series of

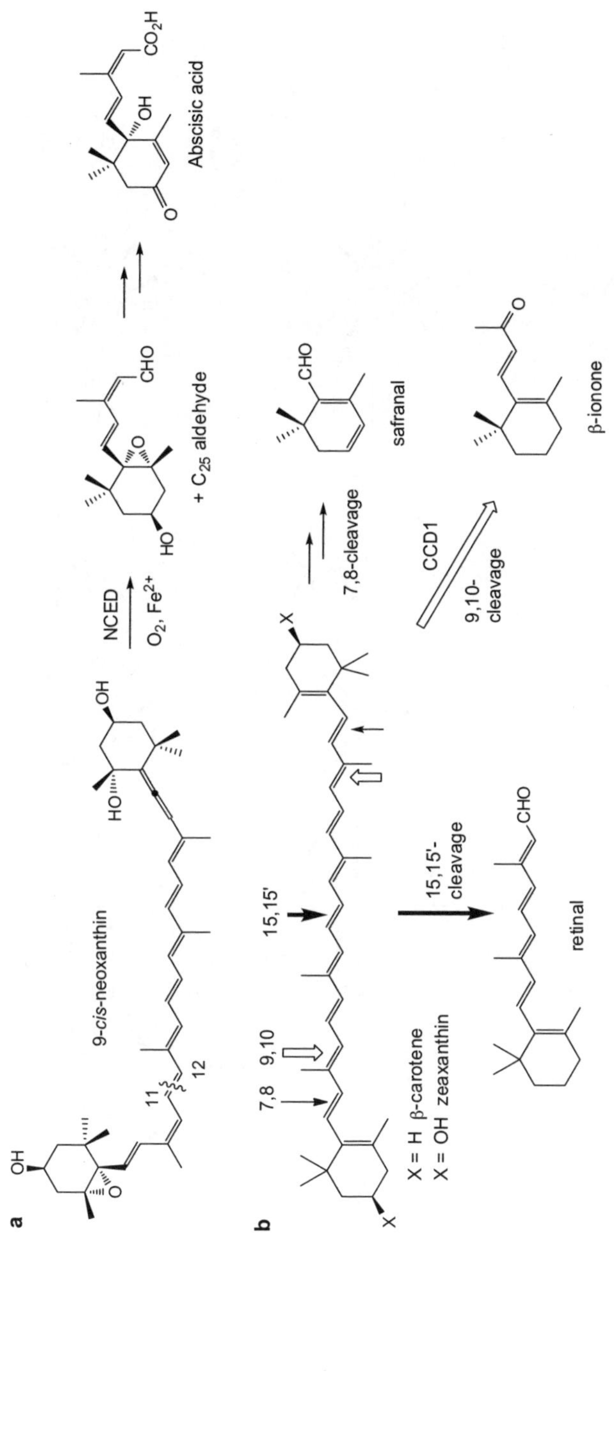

Scheme 2.3 Oxidative cleavage reactions catalysed by carotenoid cleavage dioxygenases.

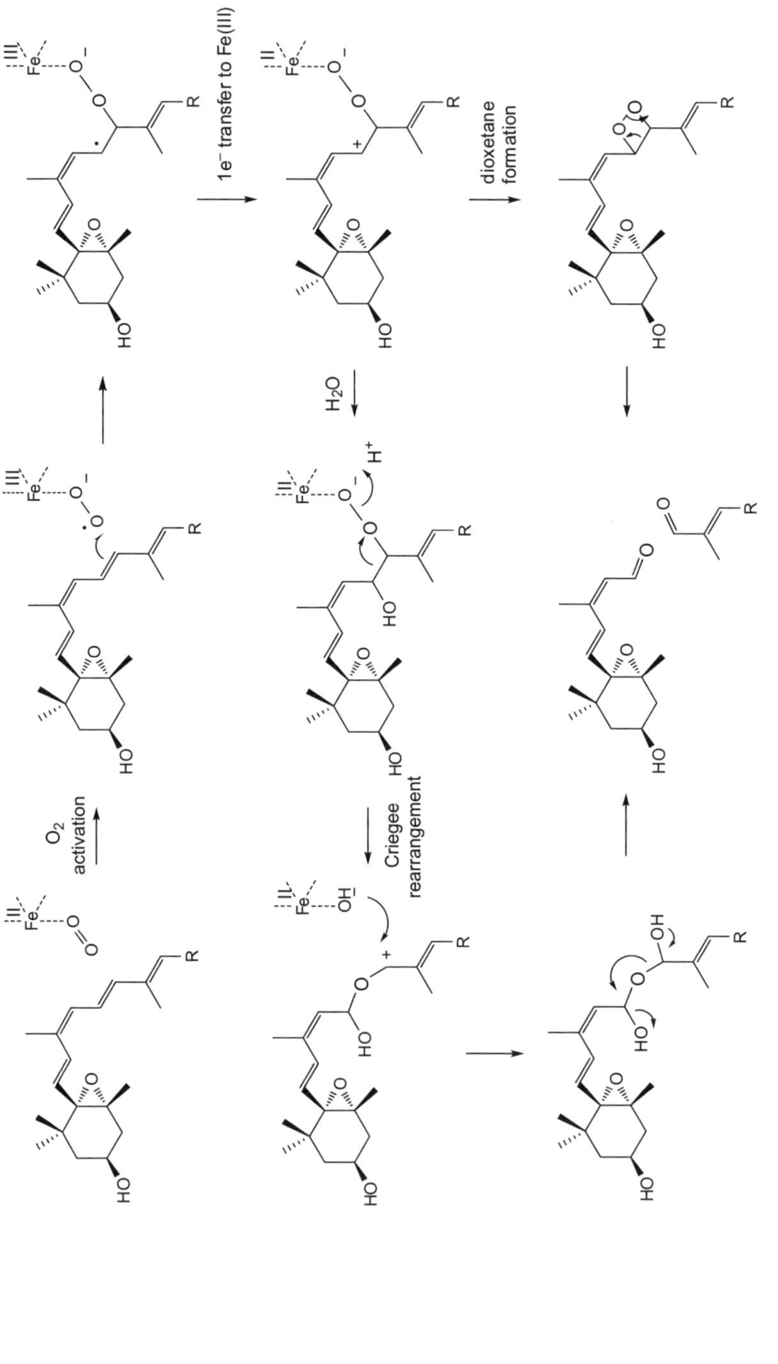

Scheme 2.4 Possible catalytic mechanisms for NCED-catalysed reaction.

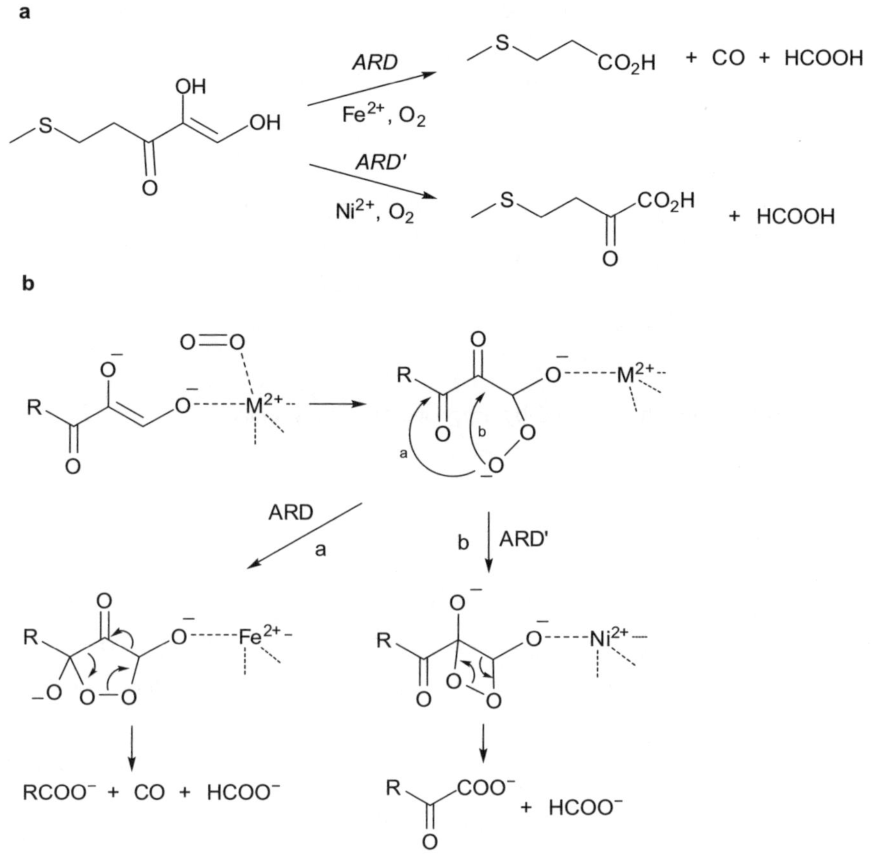

Scheme 2.5 Reactions (a) and possible mechanisms (b) for dioxygenases ARD and ARD'.

substrate analogues containing electron-donating and electron-withdrawing substituents has provided evidence in favour of a catalytic mechanism involving a dioxetane intermediate (Scheme 2.6).[57] The active site iron(II) cofactor is ligated by three histidine residues: His_{62}, His_{64} and His_{104}.[58] A related enzyme capable of oxidative cleavage of 1,3-diketone substrates has also been found in *Burkholderia xenovorans*.[59]

An unusual oxidative C–C bond cleavage reaction has been discovered on the biosynthetic pathway to phosphinothricin.[60] Hydroxyethylphosphonate dioxygenase (HEPD) catalyses the oxidation of 2-hydroxyethylphosphonate to hydroxymethylphosphonate and formic acid. The enzyme requires iron(II) for activity, and the crystal structure of the Cd(II) complex shows a mononuclear metal binding site, ligated by His_{129}, His_{182} and Glu_{176}, a similar facial tridentate ligand set to other non-heme iron-dependent dioxygenases. The enzyme catalyses the incorporation of two atoms of ^{18}O from $^{18}O_2$, one into the

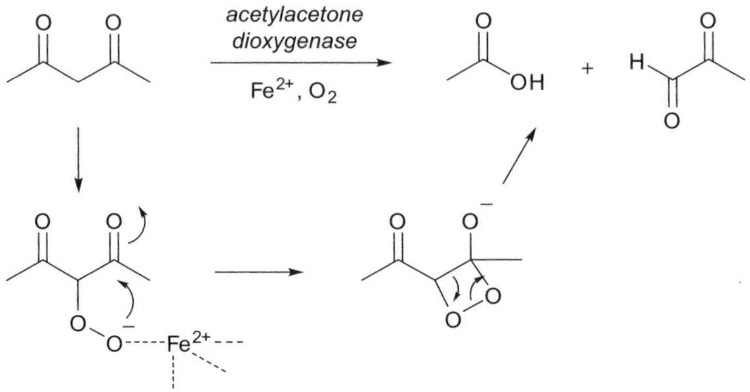

Scheme 2.6   Reaction catalysed by acetylacetone dioxygenase.

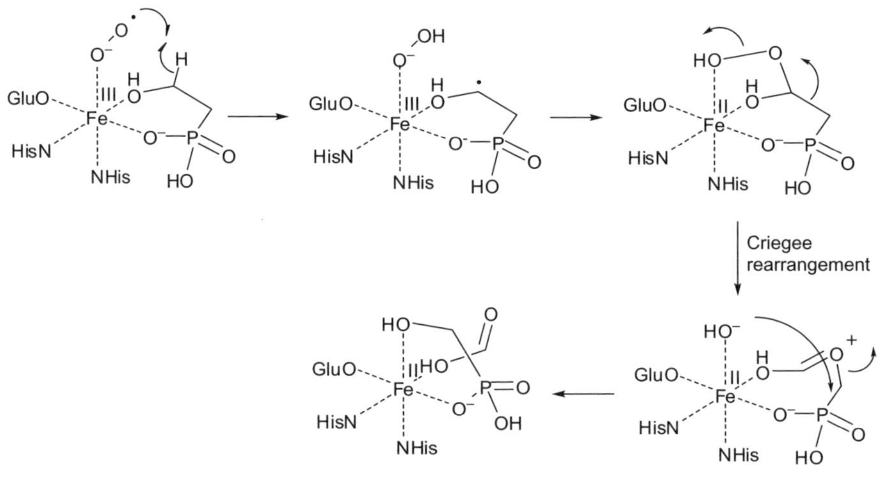

Scheme 2.7   Catalytic mechanism for hydroxyethylphosphonate dioxygenase.

hydroxyl group of the product, and one into formic acid.[60] Hydroxymethylphosphonate is then slowly converted by **HEPD** to yield phosphate and formate.[61] Both transformations can be rationalized by a mechanism in which a substrate hydroperoxide is formed, presumably *via* hydrogen atom abstraction by superoxide followed by recombination, and the hydroperoxide then undergoes a Criegee rearrangement (Scheme 2.7).

A related mechanism is thought to operate in the oxidative cleavage of *myo*-inositol, catalysed by *myo*-inositol dioxygenase, a binuclear iron-containing dioxygenase.[62] Hydrogen atom abstraction is believed to occur *via* a bound superoxide species, followed by recombination to form a substrate hydroperoxide, which undergoes a Criegee rearrangement to yield the product D-glucuronate (Scheme 2.8).[3] The crystal structure of human *myo*-inositol

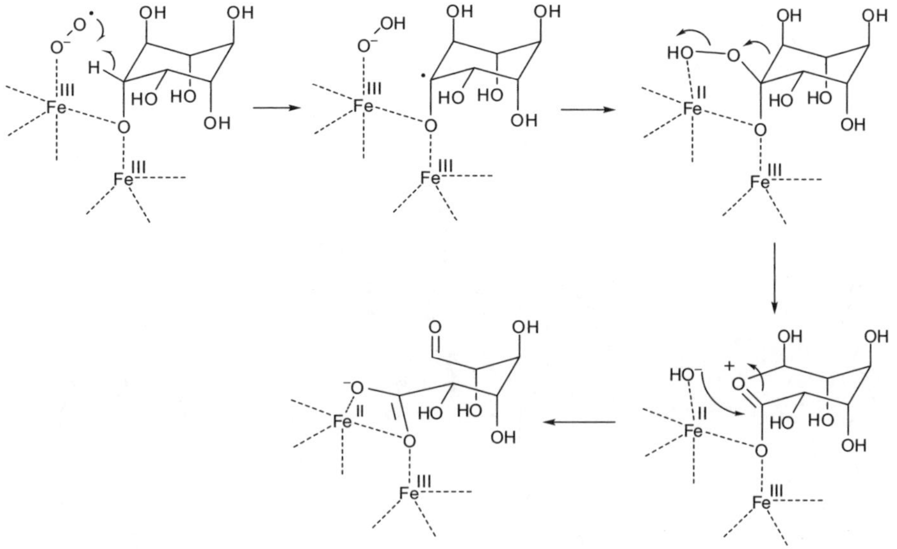

Scheme 2.8 Catalytic mechanism for *myo*-inositol dioxygenase.

dioxygenase has been determined, with *myo*-inosose bound to the diiron cluster.[63] The occurrence of Criegee rearrangements in the reaction mechanisms of these enzymes is closely reminiscent of the catechol dioxygenase mechanisms.

2.3 Dioxygenases Catalysing Formation of Peroxides: Lipoxygenases

Lipoxygenases are non-heme iron-dependent dioxygenases that catalyse the reaction of polyunsaturated fatty acids with dioxygen, to form lipid hydroperoxides. In plants the common substrate for lipoxygenase is linoleic acid, whereas in mammals the most common substrate is arachidonic acid, which is converted *via* its hydroperoxide into the leukotrienes (Scheme 2.9). A related process in mammals is the conversion of arachidonic acid and dioxygen into prostaglandin G_2, catalysed by cyclo-oxygenase, which does not require an iron cofactor, and is the molecular target for aspirin.[2]

Lipoxygenases contain a mononuclear iron cofactor, which is ligated by three histidine residues, the amide side chain of an asparagine residue and the carboxylate of the C-terminal amino acid.[64–66] For catalysis, the iron(II) centre is oxidized to a yellow iron(III)-hydroxide species.[67,68] There is evidence to support the formation of a substrate radical intermediate, *via* hydrogen atom abstraction, and one-electron transfer to iron(III),[69,70] although an alternative organo-iron species has also been proposed.[71] Reaction of the substrate radical with dioxygen then forms a hydroperoxy radical, which is reduced by iron(II)

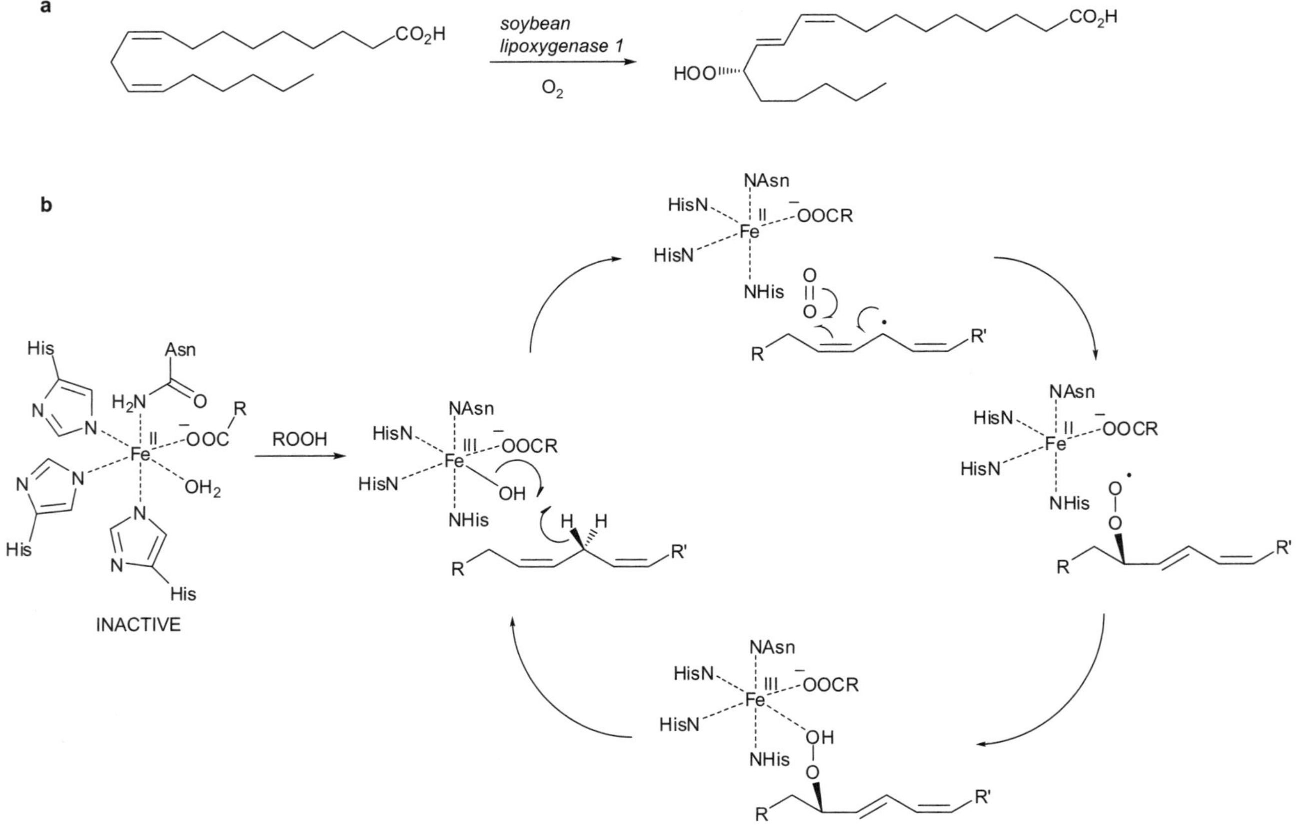

Scheme 2.9 Lipoxygenase-catalysed reaction (a) and catalytic mechanism (b).

to form the hydroperoxide product. Hydrogen atom abstraction is the rate-determining step for these enzymes, which exhibit extremely high kinetic isotope effects (KIEs) on k_{cat} (KIE $= k_H/k_D = 30$–80).[72–74] Hydrogen tunnelling has been invoked to rationalize these very large kinetic isotope effects, which are temperature independent.[75]

2.4 Dioxygenases Catalysing Hydroxylation Reactions

2.4.1 α-Ketoglutarate-Dependent Dioxygenases

A further class of non-heme iron-dependent dioxygenases catalyse mono-hydroxylation reactions and desaturation reactions on a range of amino acid, small molecule, polypeptide and nucleic acid substrates, utilizing the co-substrate α-ketoglutarate, which is converted into succinate and carbon dioxide. This family of enzymes will be discussed in much more detail in Chapter 3; here only a short summary is presented on their catalytic mechanism.

Prolyl hydroxylase was identified in 1967 as an α-ketoglutarate-dependent dioxygenase.[76] This enzyme catalyses the hydroxylation of prolyl residues in the structural protein collagen to 4-hydroxy-prolyl residues (Scheme 2.10). One oxygen atom is incorporated from dioxygen into the hydroxyprolyl product, and one into succinate.[76]

Several crystal structures have been solved for this class of enzymes, and in each case the mononuclear iron(II) centre is coordinated by a 2-His,1-Glu/Asp motif that is also observed in other non-heme iron-dependent enzymes.[77,78] The catalytic mechanism is believed to proceed *via* formation of an iron(III)–superoxide complex, followed by attack of superoxide upon the ketone carbonyl group of α-ketoglutarate. Decarboxylation of the resulting hydroperoxide intermediate, with cleavage of the O–O bond, then generates succinate, carbon dioxide and an iron(IV)-oxo intermediate (Scheme 2.11). The iron(IV)-oxo species then carries out hydroxylation of the substrate, probably *via* hydrogen atom abstraction to form a substrate radical intermediate.

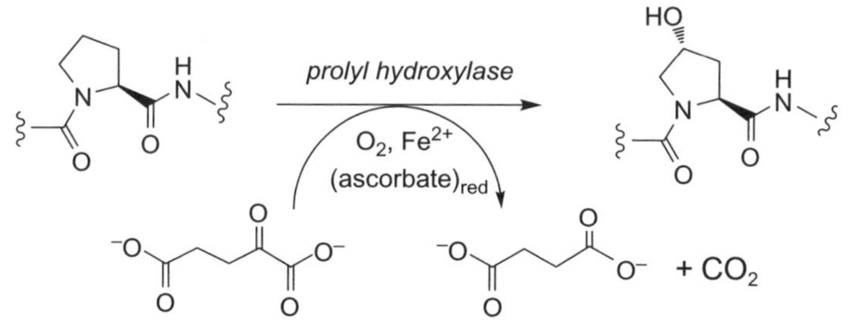

Scheme 2.10   Reaction catalysed by prolyl hydroxylase.

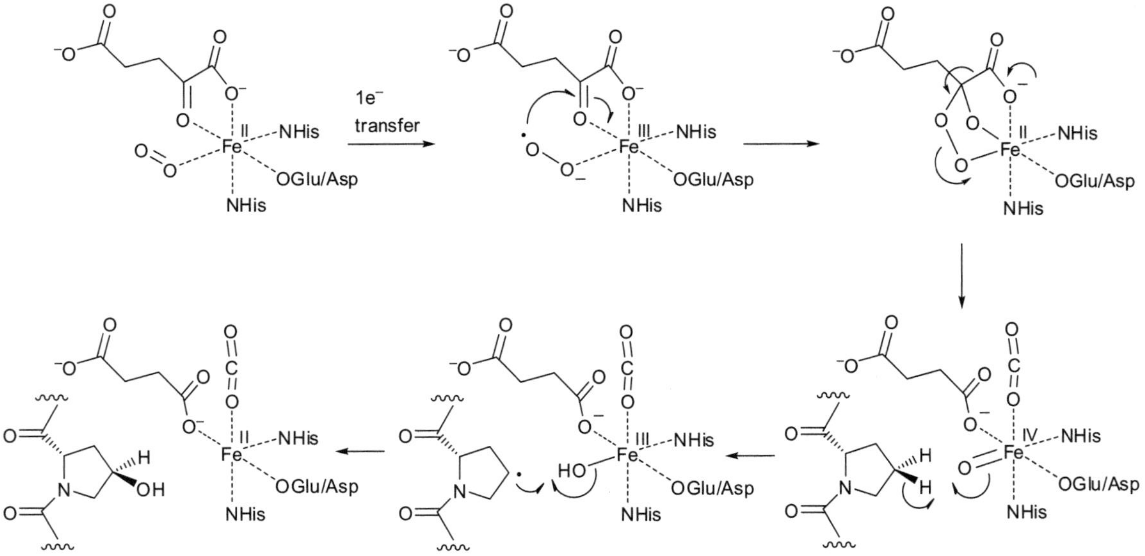

Scheme 2.11 Catalytic mechanism for α-ketoglutarate-dependent dioxygenases.

Direct spectroscopic evidence has been obtained for the existence of iron(IV)-oxo intermediates: in oxygenase TauD using Raman spectroscopy, where a band at 859 cm^{-1} corresponding to Fe(IV)=O was observed, which was shifted in the presence of ^{18}O;[79] and in prolyl hydroxylase, where an Fe(IV)=O intermediate was characterized kinetically and spectroscopically, using Mössbauer spectroscopy.[80] ^{18}O incorporation studies have been carried out on several enzymes of this family, and partial exchange of the iron(IV)-oxo intermediate with the oxygen atom of water is implicated in deacetoxy/deacetylcephalosporin C synthase,[81] *p*-hydrophenylpyruvate hydroxylase (which utilizes an α-keto acid group in the substrate for oxidative decarboxylation, rather than an α-ketoglutarate co-substrate),[82] α-ketoisocaproate oxygenase[83] and lysyl hydroxylase,[84] but not in prolyl hydroxylase.[85] Evidence for a substrate radical intermediate has been provided by the mechanism-based inactivation of prolyl hydroxylase by a substrate analogue containing a labile N–O bond adjacent to the site of hydroxylation.[86]

2.4.2 Arene (Rieske) Dioxygenases

A remarkable class of mononuclear non-heme iron-dependent enzymes catalyses the *cis*-dihydroxylation of arene substrates, to generate *cis*-dihydro-diol products. This reaction is found as the initial step of several aromatic degradation pathways in soil bacteria such as *Pseudomonas*. The best studied enzyme in this family is naphthalene dioxygenase, which consists of three components, forming an electron transfer chain (Figure 2.3): an NADH-dependent flavoprotein reductase,[87] a ferredoxin containing two [2Fe:2S] Rieske iron-sulfur clusters,[88] and an $\alpha_3\beta_3$ Rieske oxygenase containing both a [2Fe:2S] Rieske iron-sulfur cluster and a mononuclear iron(II) centre in the enzyme active site.[89,90]

The crystal structure of the terminal $\alpha_3\beta_3$ dioxygenase component of naphthalene dioxygenase was solved in 1998.[91] The active site of the α subunit contains a mononuclear iron centre, coordinated by His$_{208}$, His$_{213}$ and Asp$_{362}$.[91] Also in the α subunit is a [2Fe:2S] Rieske iron-sulfur cluster,

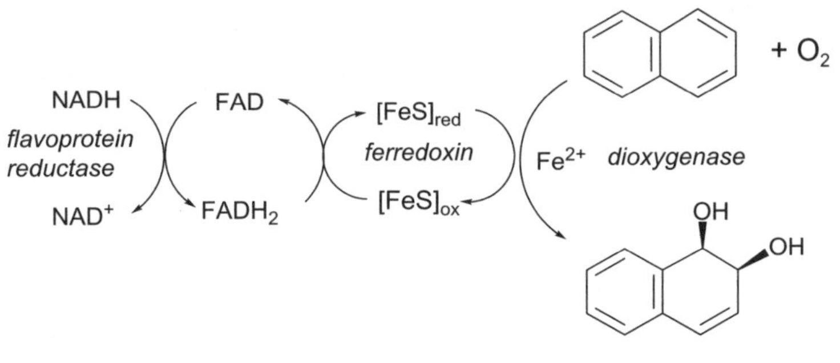

Figure 2.3 Electron transport chain in naphthalene 1,2-dioxygenase.

which is positioned 43 Å away from the mononuclear iron(II) centre within the same α subunit – too far to allow electron transfer within the same protein subunit.[91] However, the mononuclear Fe centre is positioned 12 Å away from the [2Fe:2S] cluster of a neighbouring α subunit, implying that electron transfer from the Rieske cluster to the active site takes place between sub-units (Figure 2.4).[91] The electron transfer pathway from the [2Fe:2S] cluster to the active site iron(II) centre passes through Asp₂₀₅, whose replacement results in loss of enzyme activity.[92] A further structure revealed electron density for an indole hydroperoxide, ligated to the iron(II) centre, with the indole ring positioned at about 4 Å from the iron(II) centre.[93] The presence of indole was believed to arise from the presence of L-tryptophan in the growth media.[93]

Low-temperature crystallographic studies have provided remarkable structures of ternary complexes with substrates (naphthalene or indene) and dioxygen.[94] Dioxygen was found to be bound side-on to the iron(II) centre, with Fe–O distances of 1.8 and 2.0 Å, and with an O–O distance of 1.4 Å. The aryl substrate was positioned 5–6 Å from the iron(II) centre, distal to the bound oxygen, suggesting that an activated oxygen species reacts with the aryl substrate.

Despite the detailed structural data for this enzyme, there are relatively few mechanistic studies for this unusual transformation. The original mechanistic proposal for dihydroxylation involved the formation and two-electron reduction of a dioxetane intermediate,[95] which now seems unlikely, given the presence of iron-sulfur clusters in the enzyme, which mediate one-electron

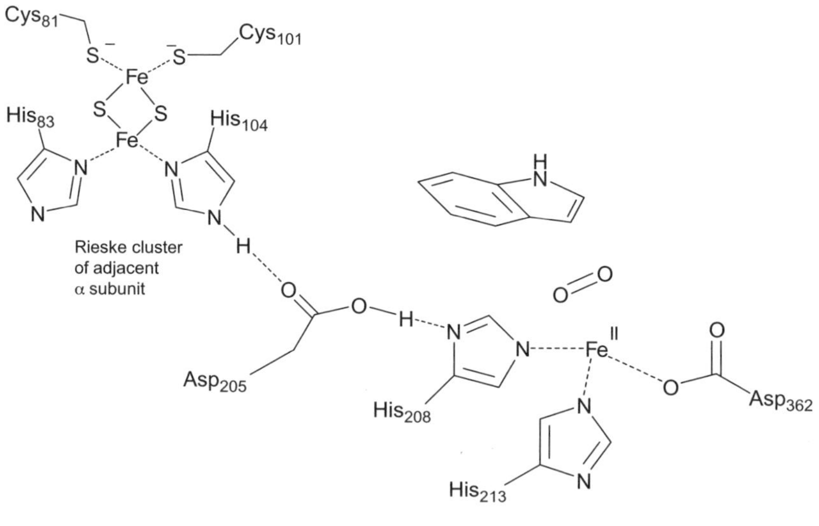

Figure 2.4 Active site of naphthalene 1,2-dioxygenase with bound dioxygen, showing the electron transfer pathway from the Rieske cluster of an adjacent protein subunit.

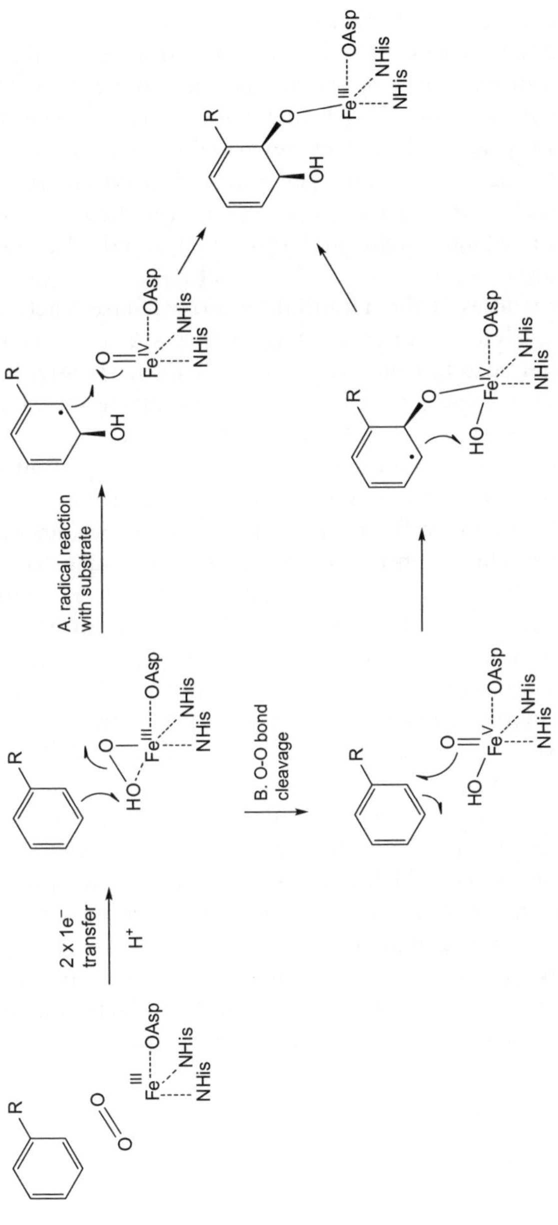

Scheme 2.12   Possible mechanisms for arene (Rieske) dioxygenases.

transfers. Toluene dioxygenase and naphthalene dioxygenase both exhibit monoxygenase activity in certain cases, using unnatural substrates, suggesting that dihydroxylation is stepwise, and not concerted.[96,97] The observation of monoxygenase activity suggests, by analogy with heme and non-heme mono-xygenases, the possible involvement of iron-oxo intermediates in catalysis. It has been observed that processing of benzene by naphthalene dioxygenase leads to the production of hydrogen peroxide, *via* uncoupling of oxygen activation from substrate hydroxylation.[98] Furthermore, fully reduced benzoate 1,2-dioxygenase is able to utilize hydrogen peroxide to form the *cis*-diol product.[99] These observations suggest that dioxygen is activated *via* superoxide, which can be further reduced to an iron(III) hydroperoxy intermediate, which might be the active oxidant, or which might undergo O–O bond cleavage to form a $Fe^V(=O)$-OH species.

Single turnover studies of the naphthalene dioxygenase reaction have shown that the catalytic cycle commences with active site non-heme cofactor as iron(III), which is reduced to iron(II) by electron transfer from a Rieske [2Fe:2S] cluster, and a further one-electron transfer occurs during the catalytic cycle.[100] Wolfe *et al.* have suggested that O–O bond cleavage could occur first, to give an O=Fe(v)-OH intermediate that could carry out dihydroxylation, similar to the dihydroxylation of alkenes by $NaIO_4$ or OsO_4, either *via* a concerted or step-wise mechanism (mechanism B, Scheme 2.12).[100] This mechanism is consistent with model studies using iron(III) complexes that are able to oxidize alkene substrates using hydrogen peroxide as oxidant, to form a mixture of *cis*-diol and epoxide products,[101–103] and one example of an iron(III) complex that is able to oxidize an arene substrate.[104] However, computational studies appear to disfavour the formation of an iron(v) oxo species.[105] The alternative step-wise mechanism involves reaction of an iron(III) hydroperoxy intermediate with the aryl substrate, with homolytic cleavage of the O–O bond, to form an arene radical, followed by delivery of the second oxygen to the same face of the substrate (mechanism A, Scheme 2.12). Studies using naphthalene dioxygenase have shown that ring opening of radical trap norcarene occurs, to give 60–70% of a ring-opened product, which is consistent with a substrate radical inter-mediate.[106] Either mechanism A or mechanism B could potentially involve transient arene radical intermediates.

The arene dioxygenases have been used extensively for whole-cell bio-transformations of arene substrates into *cis*-diol products, which have proved to be valuable starting materials for asymmetric synthesis.[107]

2.5 Conclusion and Summary

In summary, non-heme iron-dependent dioxygenases catalyse a diverse range of unusual oxidative C–C cleavage reactions, peroxidation and hydroxylation reactions. The ability of these enzymes to activate molecular oxygen, and control the reactivity of the activated oxygen species with arene or alkene substrates, permits a remarkable range of reactions that are very difficult to perform non-enzymatically.

References

1. L. Que Jr. and R. Y. N. Ho, *Chem. Rev.*, 1996, **96**, 2607.
2. T. D. H. Bugg, *Tetrahedron*, 2003, **59**, 7075.
3. T. D. H. Bugg and C. J. Winfield, *Nat. Prod. Rep.*, 1998, **15**, 513.
4. T. D. H. Bugg, in *Comprehensive Natural Products Chemistry II: Chemistry and Biology*, series eds. L. Mander and H.-W. Liu, Elsevier Science, Oxford, 2010, vol. 8, ed: C. P. Whitman, pp. 583–623.
5. E. L. Hegg and L. Que Jr., *Eur. J. Biochem.*, 1997, **250**, 625.
6. O. Hayaishi, *Bacteriol. Rev.*, 1966, **30**, 720.
7. O. Hayaishi, M. Katagiri and S. Rothberg, *J. Am. Chem. Soc.*, 1955, **77**, 5450.
8. D. H. Ohlendorf, J. D. Lipscomb and P. C. Weber, *Nature*, 1988, **336**, 403.
9. J. W. Pyrz, A. L. Roe, L. J. Stern and L. Que Jr., *J. Am. Chem. Soc.*, 1985, **107**, 614.
10. M. W. Vetting and D. H. Ohlendorf, *Structure*, **8**, 429.
11. A. M. Orville, N. Elango, J. D. Lipscomb and D. H. Ohlendorf, *Biochemistry*, 1997, **36**, 10039.
12. A. M. Orville, J. D. Lipscomb and D. H. Ohlendorf, *Biochemistry*, 1997, **36**, 10052.
13. D. D. Cox and L. Que Jr., *J. Am. Chem. Soc.*, 1988, **110**, 8085.
14. H. G. Jang, D. D. Cox and L. Que Jr., *J. Am. Chem. Soc.*, 1991, **113**, 9200.
15. M. Y. M. Pau, M. I. Davis, A. M. Orville, J. D. Lipscomb and E. I. Solomon, *J. Am. Chem. Soc.*, 2007, **129**, 1944.
16. R. J. Mayer and L. Que Jr., *J. Biol. Chem.*, 1984, **259**, 13056.
17. M. Xin and T. D. H. Bugg, *J. Am. Chem. Soc.*, 2008, **130**, 10422.
18. Y. Kojima, N. Itada and O. Hayaishi, *J. Biol. Chem.*, 1961, **236**, 2223.
19. S. Han, L. D. Eltis, K. N. Timmis, S. W. Muchmore and J. T. Bolin, *Science*, 1995, **270**, 976.
20. A. Kita, S. I. Kita, I. Fujisawa, K. Inaka, T. Ishida, K. Horiike, M. Nozaki and K. Miki, *Structure*, 1998, **7**, 25.
21. K. Sugimoto, T. Senda, H. Aoshima, E. Masai, M. Fukuda and Y. Mitsui, *Structure*, 1999, **7**, 953.
22. Y. Zhang, K. L. Colabroy, T. P. Begley and S. E. Ealick, *Biochemistry*, 2005, **44**, 7632.
23. D. M. Arciero and J. D. Lipscomb, *J. Biol .Chem.*, 1986, **261**, 2170.
24. F. H. Vaillancourt, C. J. Barbosa, T. G Spiro, J. T. Bolin, M. W. Blades, R. F. B. Turner and L. D. Eltis, *J. Am. Chem. Soc.*, 2002, **124**, 2485.
25. C. Bianchini, F. Frediani, F. Laschi, A. Meli, F. Vizza and P. Zanello, *Inorg. Chem.*, 1990, **29**, 3402.
26. E. L. Spence, G. J. Langley and T. D. H. Bugg, *J. Am. Chem. Soc.*, 1996, **118**, 8336.
27. C. J. Winfield, Z. Al-Mahrizy, M. Gravestock and T. D. H. Bugg, *J. Chem. Soc., Perkin Trans.*, 2000, **1**, 3277.
28. E. G. Kovaleva and J. D. Lipscomb, *Science*, 2007, **316**, 453.

29. T. D. H. Bugg and G. Lin, *Chem. Commun.*, 2001, 941.
30. J. Sanvoisin, G. J. Langley and T. D. H. Bugg, *J. Am. Chem. Soc.*, 1995, **117**, 7836.
31. S. L. Groce and J. D. Lipscomb, *J. Am. Chem. Soc.*, 2003, **125**, 11780.
32. J. Schlosrich, K. L. Eley, P. J. Crowley and T. D. H. Bugg, *ChemBiochem*, 2006, **7**, 1899.
33. G. Lin, G. Reid and T. D. H. Bugg, *J. Am. Chem. Soc.*, 2001, **123**, 5030.
34. S. Mendel, A. Arndt and T. D. H. Bugg, *Biochemistry*, 2004, **43**, 13390.
35. N. Sato, Y. Uragami, T. Nishizaki, Y. Takahashi, G. Sazaki, K. Sugimoto, T. Nonaka, E. Masai, M. Fukuda and T. Senda, *J. Mol. Biol.*, 2002, **321**, 621.
36. S. L. Groce and J. D. Lipscomb, *Biochemistry*, 2005, **44**, 7175.
37. A. K. Whiting, Y. R. Boldt, M. P. Hendrich, L. P. Wackett and L. Que Jr., *Biochemistry*, 1996, **35**, 160.
38. M. W. Vetting, L. P. Wackett, L. Que Jr., J. D. Lipscomb and D. H. Ohlendorf, *J. Bacteriol.*, 2004, **186**, 1945.
39. A. Wyss, G. Wirtz, W. D. Woggon, R. Brugger, M. Wyss, A. Friedlein, H. Bachmann and W. Hunziker, *Biochem. Biophys. Res. Commun.*, 2000, **271**, 334.
40. S. H. Schwartz, B. C. Tan, D. A. Gage, J. A. Zeevaart and D. R. McCarty, *Science*, 1997, **276**, 1872.
41. G. Giuliano, S. Al-Babili and J. von Lintig, *Trends Plant Sci.*, 2003, **8**, 145.
42. M. E. Auldridge, D. R. McCarty and H. J. Klee, *Curr. Opin. Plant Biol.*, 2006, **9**, 315.
43. B. C. Tan, S. H. Schwartz, J. A. Zeevaart and D. R. McCarty, *Proc. Natl. Acad. Sci. USA*, 1997, **94**, 12235.
44. S. H. Schwartz, X. Qin and J. A. Zeevaart, *J. Biol. Chem.*, 2001, **276**, 25208.
45. K. Sorefan, J. Booker, K. Haurogne, M. Goussot, K. Baibridge, E. Foo, S. Chatfield, S. Ward, C. Beveridge, C. Rameau and O. Leyser, *Gene Dev.*, 2003, **17**, 1467.
46. J. Booker, M. Auldridge, S. Wills, H. Klee and O. Leyser, *Curr. Biol.*, 2004, **14**, 1.
47. V. Gomez-Roldan, S. Fermas, P. B. Brewer, V. Puech-Pages, E. A. Dun, J. P. Pillot, F. Letisse, R. Matusova, S. Danoun, J. C. Portais, H. Bouwmeester, G. Becard, C. A. Beveridge, C. Rameau and S. F. Rochange, *Nature*, 2008, **455**, 189.
48. M. Umehara, A. Hanada, S. Yoshida, K. Akiyama, T. Arite, N. Takeda-Kamiya, H. Magome, Y. Kamiya, K. Shirasu, K. Yoneyama, J. Kyozuka and S. Yamaguchi, *Nature*, 2008, **455**, 195.
49. D. P. Kloer, S. Ruch, S. Al-Babili, P. Beyer and G. E. Schulz, *Science*, 2005, **308**, 267.
50. W. D. Woggon, *Pure Appl. Chem.*, 2002, **74**, 1397.
51. T. Borowski, M. R. A. Blomberg and P. E. M. Siegbahn, *Chem. Eur. J.*, 2008, **14**, 2264.

52. J. W. Wray and R. H. Abeles, *J. Biol. Chem.*, 1995, **270**, 3147.
53. Y. Dai, P. C. Wensink and R. H. Abeles, *J. Biol. Chem.*, 1999, **274**, 1193.
54. Y. Dai, T. C. Pochapsky and R. H. Abeles, *Biochemistry*, 2001, **40**, 6379.
55. T. C. Pochapsky, S. S. Pochapsky, T. Ju, H. Mo, F. Al-Mjeni and M. Maroney, *Nat. Struct. Biol.*, 2002, **9**, 966.
56. G. D. Straganz, L. Brecker, H. J. Weber, W. Steiner and D. W. Ribbons, *Biochem. Biophys. Res. Commun.*, 2002, **297**, 232.
57. G. D. Straganz and B. Nidetzky, *J. Am. Chem. Soc.*, 2005, **127**, 12306.
58. S. Leitgeb, G. D. Straganz and B. Nidetzky, *Biochem. J.*, 2009, **418**, 403.
59. S. Leitgeb, G. D. Straganz and B. Nidetzky, *FEBS J.*, 2009, **276**, 5983.
60. R. M. Cicchillo, H. Zhang, J. A. V. Blodgett, J. T. Whitteck, G. Li, S. K. Nair, W. A. van der Donk and W. W. Metcalf, *Nature*, 2009, **459**, 871.
61. J. T. Whitteck, R. M. Cicchillo and W. A. van der Donk, *J. Am. Chem. Soc.*, 2009, **131**, 16225.
62. G. Xing, Y. Diao, L. M. Hoffart, E. W. Barr, K. S. Prabhu, R. J. Arner, C. C. Reddy, C. Krebs and J. M. Bollinger Jr., *Proc. Natl. Acad. Sci. USA*, 2006, **103**, 6130.
63. A.-G. Thorsell, C. Persson, N. Voevodskaya, R. D. Busam, M. Hammarström, S. Gräslund, A. Gräslund and B. M. Hallberg, *J. Biol. Chem.*, 2008, **283**, 15209.
64. J. C. Boyington, B. J. Gaffney and L. M. Amzel, *Science*, 1993, **260**, 1482.
65. W. Minor, J. Steczko, B. Stec, Z. Otwinowski, J. T. Bolin, R. Walter and B. Axelrod, *Biochemistry*, 1996, **35**, 10687.
66. S. A. Gillmor, A. Villasenor, R. Fletterick, E. Sigal and M. F. Browner, *Nat. Struct. Biol.*, 1997, **4**, 1003.
67. M. C. Feiters, R. Aasa, B. G. Malmstrom, G. A. Veldink and J. F. G. Vliegenthart, *Biochim. Biophys. Acta*, 1986, **873**, 182.
68. R. C. Scarrow, M. G. Trimitsis, C. P. Buck, G. N. Grove, R. A. Cowling and M. J. Nelson, *Biochemistry*, 1994, **33**, 15023.
69. J. J. M. C. De Groot, G. J. Garssen, J. F. G. Vliegenthart and J. Boldingh, *Biochim. Biophys. Acta*, 1973, **326**, 279.
70. M. J. Nelson and R. A. Cowling, *J. Am. Chem. Soc.*, 1990, **112**, 2820.
71. E. J. Corey and R. Nagata, *J. Am. Chem. Soc.*, 1987, **109**, 8107.
72. C. C. Hwang and C. B. Grissom, *J. Am. Chem. Soc.*, 1994, **116**, 795.
73. M. H. Glickman, J. S. Wiseman and J. P. Klinman, *J. Am. Chem. Soc.*, 1994, **116**, 793.
74. M. H. Glickman and J. P. Klinman, *Biochemistry*, 1995, **34**, 14077.
75. M. J. Knapp, K. Rickert and J. P. Klinman, *J. Am. Chem. Soc.*, 2002, **124**, 3865.
76. J. J. Hutton, A. L. Tappel and S. Udenfriend, *Arch. Biochem. Biophys.*, 1967, **118**, 231.
77. C. J. Schofield and Z. Zhang, *Curr. Opin. Struct. Biol.*, 1999, **9**, 722.
78. M. Ryle and R. P. Hausinger, *Curr. Opin. Chem. Biol.*, 2002, **6**, 193.
79. D. A. Proshlyakov, T. F. Henshaw, G. R. Monterosso, M. J. Ryle and R. P. Hausinger, *J. Am. Chem. Soc.*, 2004, **126**, 1022.

80. L. M. Hoffart, E. W. Barr, R. B. Guyer, J. M. Bollinger and C. Krebs, *Proc. Natl. Acad. Sci. USA*, 2006, **103**, 14738.
81. J. E. Baldwin, R. W. Adlington, N. P. Crouch, I. A. C. Pereira, R. T. Aplin and C. Robinson, *J. Chem. Soc. Chem. Commun.*, 1993, 105.
82. B. Lindblad, G. Lindstedt and S. Lindstedt, *J. Am. Chem. Soc.*, 1970, **92**, 7446.
83. P. J. Sabourin and L. L. Bieber, *J. Biol. Chem.*, 1982, **257**, 7468.
84. Y. Kikuchi, Y. Suzuki and N. Tamiya, *Biochem J.*, 1983, **213**, 507.
85. M. Wu, T. P. Begley, J. Myllyharju and K. I. Kivirikko, *Bioorg. Chem.*, 2000, **28**, 261.
86. M. Wu, H. S. Moon, T. P. Begley, J. Myllyharju and K. I. Kivirikko, *J. Am. Chem. Soc.*, 1999, **121**, 587.
87. B. E. Haigler and D. T. Gibson, *J. Bacteriol.*, 1990, **172**, 457.
88. B. E. Haigler and D. T. Gibson, *J. Bacteriol.*, 1990, **172**, 465.
89. B. D. Ensley and D. T. Gibson, *J. Bacteriol.*, 1983, **155**, 505.
90. W. C. Suen and D. T. Gibson, *J. Bacteriol.*, 1993, **175**, 5877.
91. B. Kauppi, K. Lee, E. Carredano, R. E. Parales, D. T. Gibson, H. Eklund and S. Ramaswamy, *Structure*, 1998, **6**, 571.
92. R. E. Parales, J. V. Parales and D. T. Gibson, *J. Bacteriol.*, 1999, **181**, 1831.
93. E. Carredano, A. Karlsson, B. Kauppi, D. Choudhury, R. E. Parales, J. V. Parales, K. Lee, D. T. Gibson, H. Eklund and S. Ramaswamy, *J. Mol. Biol.*, 2000, **296**, 701.
94. A. Karlsson, J. V. Parales, R. E. Parales, D. T. Gibson, H. Eklund and S. Ramaswamy, *Science*, 2003, **299**, 1039.
95. D. T. Gibson, G. E. Cardini, F. C. Maseles and R. E. Kallio, *Biochemistry*, 1970, **9**, 1631.
96. L. P. Wackett, L. D. Kwart and D. T. Gibson, *Biochemistry*, 1988, **27**, 1360.
97. S. M. Resnick, D. S. Torok, K. Lee, J. M. Brand and D. T. Gibson, *Appl. Environ. Microbiol.*, 1994, **60**, 3323.
98. K. Lee, *J. Bacteriol.*, 1999, **181**, 2719.
99. M. B. Neibergall, A. Stubna, Y. Mekmouche, E. Münck and J. D. Lipscomb, *Biochemistry*, 2007, **46**, 8004.
100. M. D. Wolfe, J. V. Parales, D. T. Gibson and J. D. Lipscomb, *J. Biol. Chem.*, 2001, **276**, 1945.
101. K. Chen and L. Que Jr., *Angew. Chem., Intl. Ed.*, 1999, **38**, 2227.
102. M. Costas, A. K. Tipton, K. Chen, D. H. Jo and L. Que Jr., *J. Am. Chem. Soc.*, 2001, **123**, 6722.
103. K. Chen, M. Costas, J. Kim, A. K. Tipton and L. Que Jr., *J. Am. Chem. Soc.*, 2002, **124**, 3025.
104. Y. Feng, C. Y. Ke, G. Q. Xue and L. Que Jr., *Chem. Commun.*, 2009, 50.
105. A. Bassan, M. R. A. Blomberg and P. E. M. Siegbahn, *J. Biol. Inorg. Chem.*, 2004, **9**, 439.
106. S. Chakravarty, R. N. Austin, D. Deng, J. T. Groves and J. D. Lipscomb, *J. Am. Chem. Soc.*, 2007, **129**, 3514.
107. D. R. Boyd and T. D. H. Bugg, *Org. Biomol. Chem.*, 2006, **4**, 181.

Transient Iron Species in the Catalytic Mechanism of the Archetypal α-Ketoglutarate-Dependent Dioxygenase, TauD

DENIS A. PROSHLYAKOV* AND
ROBERT P. HAUSINGER*

Department of Chemistry, Department of Biochemistry and Molecular
Biology, and Department of Microbiology and Molecular Genetics,
Michigan State University, East Lansing, MI 48824, United States of
America

3.1 Introduction

Iron(II)- and α-ketoglutarate (αKG)-dependent dioxygenases encompass a large family of oxidative enzymes that use variations to a common mechanism to hydroxylate structural proteins, regulate the hypoxic response, modulate the level of histone methylation, repair alkylated DNA/RNA, synthesize certain antibiotics, flavonoids, gibberellins, and other metabolites, and degrade selected lipids, herbicides, nucleotides, and a wide range of other molecules.[1-4] Nearly 30 years ago, Hanauske-Abel and Günzler proposed a theoretical mechanism of catalysis for prolyl 4-hydroxylase, a collagen-modifying representative of this family enzyme, in which αKG decomposes to CO_2 and succinate as an Fe(IV)-oxo intermediate forms, with the latter species abstracting a hydrogen atom from substrate with subsequent recombination to yield the product.[5] Investigation of this reaction mechanism is most advanced using the

Iron-Containing Enzymes: Versatile Catalysts of Hydroxylation Reactions in Nature
Edited by Sam P de Visser and Devesh Kumar
© Royal Society of Chemistry 2011
Published by the Royal Society of Chemistry, www.rsc.org

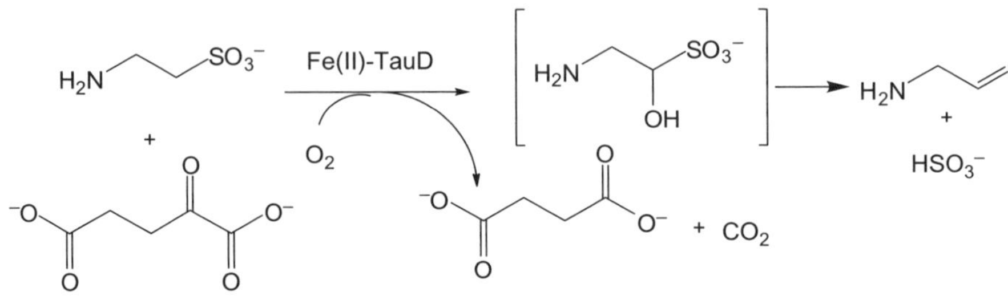

Scheme 3.1 Overall reaction catalyzed by TauD enzymes.

taurine (2-aminoethanesulfonic acid)-degrading enzyme TauD (taurine hydroxylase or taurine:αKG dioxygenase) of *Escherichia coli*. After a brief introduction to the enzyme we describe in subsequent sections evidence related to each step of its mechanism.

Depletion of sulfur from the medium of *E. coli* cells leads to increased transcription of *tauABCD*, which encodes both an ATP-binding cassette-type taurine transport system and the TauD enzyme.[6] TauD oxidatively decomposes αKG while hydroxylating the C^1 carbon of taurine to create an unstable intermediate that spontaneously decomposes to aminoacetaldehyde and sulfite, which serves as a nutritional source of sulfur (Scheme 3.1).[7]

TauD homologues are present in many other bacteria and fungi, where they are presumed to catalyze the same reaction. A closely related sulfate:αKG dioxygenase (AtsK) has been studied from *Pseudomonas putida* S-313,[8] indicating that some TauD-like proteins may in fact utilize sulfates rather than sulfonates. In addition, a broad-specificity sulfonate:αKG dioxygenase has been characterized from *Saccharomyces cerevisiae* and shown to prefer several other sulfonates over taurine.[9]

3.2 Structure of the TauD Active Site

Crystal structures have been determined for TauD in the presence of Fe(II), αKG, and taurine under anaerobic conditions (PDB 1GY9 and 1OS7) and for TauD apoprotein with bound taurine (PDB 1OTJ).[10,11] In the PDB files 1OS7 and 1OTJ, one protomer of the four subunits per unit cell lacks taurine, thus providing structures of species with Fe(II) plus αKG and the apoprotein. As for other members of the Fe(II)/αKG dioxygenases, TauD exhibits a double-stranded β-helix fold (Figure 3.1), alternatively called a cupin or β-jellyroll structure, with Fe(II) coordinated by a two-His-1-carboxylate facial triad motif.[12,13] The TauD metal-coordinating residues are His_{99}, Asp_{101}, and His_{255}. The αKG binds to Fe(II) in a bi-dentate fashion *via* its C^1 carboxylate and C^2 keto group. Binding of αKG is stabilized further by an interaction between the C^1 carboxylate and Arg_{270}, as well as an interaction of the C^5 carboxylate with Arg_{266} *via* a salt bridge and hydrogen bonding to Thr_{126}. Arg_{270} along with His_{70} and the backbone amide of Val_{102} bind the sulfonate group of taurine, with the substrate amine group participating in a hydrogen bond with the

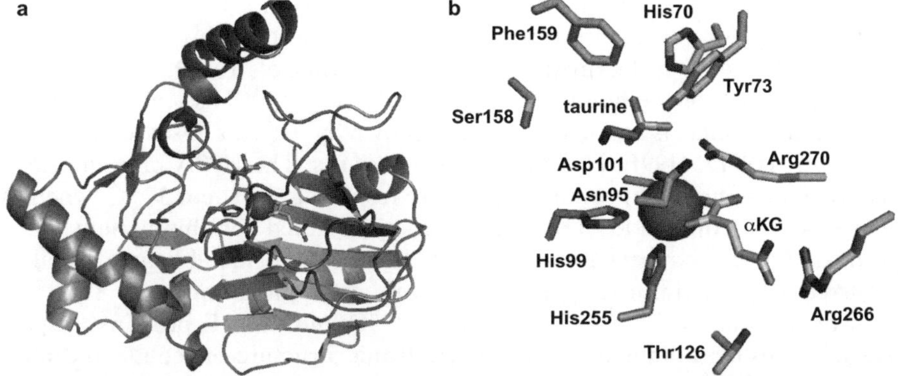

Figure 3.1 TauD structure. (a) Depiction of the overall fold of the TauD protomer with the double-stranded β-helix strands shown as ribbons. (b) Illustration of the active site Fe(II) (sphere), metal ligands, αKG, taurine, and selected other side chains.

amide oxygen of Asn_{95} and two water molecules. Notably, taurine does not coordinate the Fe(II).

3.2.1 Metal Binding to TauD Apoprotein

To avoid oxidative inactivation of TauD during purification and storage, EDTA is added during cell disruption and the enzyme is purified as the apoprotein. Subsequent binding of Fe(II) to the TauD apoprotein can be detected by quenching of $\sim 35\%$ of the endogenous protein fluorescence signal or by difference UV spectral changes at 298 nm.[14] Titration of apoprotein with the metal ion as monitored by either method provides an Fe(II) K_d of ~ 100 nM. TauD is active only with Fe(II), whereas Mg(II), Ca(II), Mn(II), Co(II), Cu(II), and Zn(II) inhibit the enzyme.[7] Of significance to those working with related enzymes that are purified by using His-tags, Co(II) and Ni(II) act as slow-binding competitive inhibitors of TauD by substituting for the active site metal ion.[15]

Iron(II)-bound TauD is colorless when prepared from protein generated in cells grown under anoxic conditions.[16] In other preparations of TauD, however, the addition of Fe(II) to anaerobic samples of apoprotein generates a weak and broad absorption band near 650 nm that is attributed to an Fe(III)-catecholate species.[17] This green-colored feature is formed by electron transfer from Fe(II) to an ortho-quinone at position 73, where the oxidized side chain is derived from intracellular enzyme self-hydroxylation during aerobic cell growth followed by oxidation of the catechol to the quinone state during further growth and protein purification. *In vitro* studies have demonstrated the conversion of Tyr_{73} into dihydroxy-Phe *via* the intermediacy of a tyrosyl radical.[18–20] Additional evidence for the presence of an ortho-quinone in a portion of the enzyme isolated from aerobically grown cells is derived from studies involving Cr(II) addition to anaerobic TauD apoprotein.[21] The Cr(II)-treated sample generates a diagnostic spectrum consistent with a Cr(III)-semiquinone, again likely arising from electron transfer from metal to ortho-quinone.

Substitutions of metal ligands in variant forms of TauD were used to examine the role of the binding motif in TauD activity.[17] His_{99}, located in the αKG binding plane, is a critical residue as shown by the near absence of activity for H99C, H99D, H99E, H99N, or H99Q variants. The H99A protein also is inactive when purified at room temperature from cells grown at 37 °C, whereas nearly 40% of the wild-type enzyme activity is noted for this variant when purified at 4 °C from cells grown at 16 °C.[22] The His_{255} metal ligand, distal to the likely oxygenation axis position, is more accommodating, with 81% and 33% of the wild-type k_{cat} observed for H255Q and H255E proteins, respectively, suggesting a limited role in the electronic structure of the metal center. Other His_{255} TauD variants, namely H255A, H255C, H255D, and H255N, are less than 1% active. Asp_{101} can be replaced by Glu (D101E) with more than 20% retention of k_{cat}, but the non-carboxylic variants D101A, D101C, D101H, D101N, or D101Q proteins are nearly inactive (with the most active variant, D101Q, displaying $\sim 1\%$ of the k_{cat} as for wild-type protein), highlighting the role of a carboxylate in the facial triad.

3.2.2 Substrate Binding to TauD

The addition of αKG to samples of TauD and Fe(II) leads to the formation of a lilac color with an absorbance maximum at 530 nm ($\varepsilon_{530} = 140\text{--}240$ M^{-1} cm^{-1}).[16,23] Exploiting this chromophore, αKG was titrated into Fe(II)-TauD to obtain an αKG K_d of 270 ± 100 μM. Significantly, stopped-flow UV–vis studies reveal that the rate of chromophore formation is independent of the concentration of αKG or Fe(II)-TauD. This result is consistent with αKG binding to the protein, followed by a conformational change that gives rise to the absorption. A likely scenario involves monodentate binding of the αKG C^1 carboxylate to TauD-bound Fe(II), followed by chelate formation that is associated with the color (Scheme 3.2). Circular dichroism (CD) and magnetic circular dichroism (MCD) studies of another family member, clavaminate synthase, identified an analogous chromophore in that protein as arising from low-lying metal-to-ligand charge-transfer (MLCT) transitions associated with the bidentate binding mode of αKG.[24]

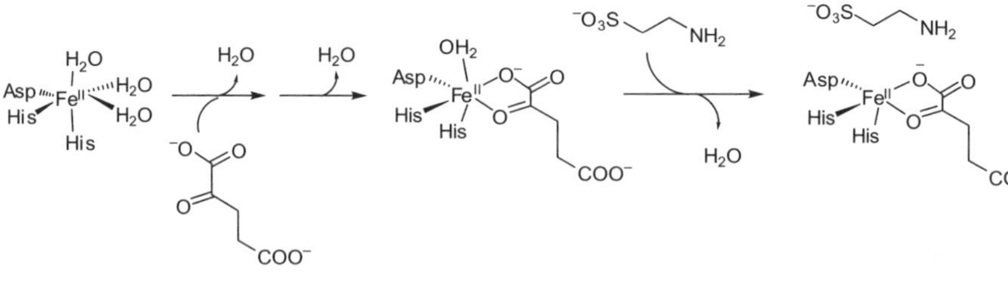

Scheme 3.2 Sequential binding of substrates to the active site of TauD.

Taurine binds to TauD apoprotein with a K_d of 175 ± 10 µM according to results from difference UV spectroscopy.[14] It is reasonable to assume that taurine binds similarly to Fe(II)-TauD; however, the taurine K_d is unknown for protein in the presence of Fe(II). The crystal structure reveals that taurine fills a channel leading into the active site and this substrate thus is likely to bind after Fe(II) and αKG.[10,11]

The addition of taurine to αKG-Fe(II)-TauD perturbs the visible spectrum with a shift in absorbance maximum to 520 nm, a small enhancement of intensity ($\varepsilon_{520} = 180$–270 M^{-1} cm^{-1}), and improved resolution of three distinct MLCT transitions.[16,23] Resonance Raman spectroscopy reveals a 10 cm^{-1} upshift in two features associated with the chelate mode for taurine-αKG-Fe(II)-TauD compared to αKG-Fe(II)-TauD.[25] These results are consistent with the addition of taurine causing a shift from six-coordinate to five-coordinate geometry (Scheme 3.2). CD and MCD spectroscopic comparison of these TauD species confirms the reduction in coordination number,[26] consistent with the probable loss of a metal-bound water ligand upon taurine binding. Density functional theory (DFT) calculations show that dissociation of the water ligand may be facilitated by rotation of the Asp_{101} carboxylate following taurine binding, which leads to the loss of hydrogen bonding between the carboxylate oxygen and bound water, with destabilization of the latter by 8–9 kcal mol^{-1}.[26]

Mutational studies have provided additional insights into binding of the primary substrate. His_{70} and Arg_{270} are critical in binding the sulfonate group of taurine (Figure 3.1), since substitution by Ala at either position abolishes enzyme activity while still allowing Fe(II) and αKG to bind, as demonstrated by the resulting MLCT transitions.[23] Development of taurine-specific UV changes also is abolished in the H70A mutant.[14] By contrast, substitution of the taurine amine-binding Asn_{95} residue by Ala or Asp reduces k_{cat} to 7% and 14% of the wild-type value, respectively, but leads to 25-fold increases in K_m,[23] suggesting that the residue contributes to both taurine binding and proper positioning of the substrate for catalysis. Likewise, substitution of Tyr_{73} with Ile or Ser leads to small reductions in k_{cat} and four- or eleven-fold increases in K_m, respectively;[23] this residue lies near the taurine-binding site and undergoes self-hydroxylation in the absence of substrate.[18,19] By contrast, Ala substitution of Ser_{158}, which is close to the taurine amine functionality, results in modest changes including a 50% reduction in k_{cat} and two-fold increase in K_m.[23] Phe_{159}, on the distal side of taurine when referenced to the metallocenter, has been substituted with Leu, Val, and Ala.[27,28] These variants exhibit reduced rates (8.2-, 62-, and 77-fold, respectively) of substrate binding consistent with the hydrophobic bulk at this position enhancing the productive positioning of taurine.

3.2.3 Characterization of the NO-Bound Quaternary Complex

The reaction of taurine-αKG-Fe(II)-TauD with O_2 is fast and proceeds without build-up of stable O_2 intermediates that can be studied in detail. Treatment of taurine-αKG-Fe(II)-TauD with NO, a surrogate for O_2, results in a stable paramagnetic ($S = {}^3/_2$) species that was studied by electron spin echo envelope modulation (ESEEM) spectroscopy.[29] Major ESEEM contributions from ^{1}H and ^{14}N modulations can be eliminated by obtaining a ratio between deuterated and protonated taurine data, which accentuates the modulations from ^{2}H. ESEEM analysis was carried out by using field positions of $g=4$ to $g=2$, allowing calculation of ^{2}H-to-Fe(II) distances along with estimation of the geometry with respect to the Fe–NO magnetic axis for the C^1 and C^2 deuterons of taurine. For example, a deuteron on C^1 is 3.4 ± 0.2 Å from the metal ion and lies on a cone that makes a $20 \pm 3°$ angle to the g_z axis. Similarly, a deuteron on C^2 is 4.5 ± 0.2 Å from Fe at an angle of $63 \pm 5°$ from the Fe–NO axis. Together, these measurements compare very favorably with the distances and geometry of the taurine C^1 and C^2 protons from the metal center as derived from the crystal structure.[10,11]

Cryoreduction of the NO adduct of taurine-αKG-Fe(II)-TauD leads to a species believed to be isoelectronic with the proposed O_2 adduct of the enzyme.[30] Computational analysis of the starting quartet species confirms its identity as high-spin Fe(III) $(S = {}^5/_2)$ antiferromagnetically coupled to NO$^-$ $(S = 1)$, with the bent NO having lowest energy positions over the carbonyl group of αKG or the carboxylate side chain. The one-electron reduced form of this species was generated by low-temperature γ-irradiation of the enzyme. Using the convention of Enemark and Feltham,[31] this converts the {FeNO}^{7} species into an {FeNO}^{8} species which, when examined by Mössbauer spectroscopy, gives rise to a quadrupole doublet at small magnetic field, but a split spectrum in large magnetic fields. Additional studies suggest this to be a triplet paramagnetic species best described as a quintet high-spin Fe(II) $(S = 2)$ antiferromagnetically coupled to a triplet NO$^-$ $(S = 1)$. In contrast to the reactivity observed with the {FeO$_2$}^{8} species that is presumed to form transiently in the enzyme, the {FeNO}^{8} species converts into a quintet {FeHNO}^{8} species upon warming to 193 K.

3.3   The Fe(IV)-oxo Species

In 1982, the reaction mechanism of Fe(II)/αKG hydroxylases was hypothesized to proceed *via* a highly oxidized Fe(IV)-oxo species similar to heme-based oxygenases.[5] Recent studies of TauD provide direct evidence for an Fe(IV)-oxo reaction intermediate, which is the first such species in a mononuclear non-heme iron oxygenase. This key species has since been observed in a prolyl 4-hydroxylase,[32] two αKG-dependent halogenases,[33,34] and tetrahydropterin-dependent tyrosine hydroxylase.[35] The following subsections describe the experiments that identified the Fe(IV)-oxo species in TauD, the electronic configuration of this species, its ability to abstract a hydrogen atom from substrate taurine, and the thermodynamics of this process.

3.3.1   Experimental Detection of Fe(IV)-oxo

Taurine-induced displacement of water from the TauD metallocenter serves as a trigger to create an O_2 binding site. In this manner, O_2 reactivity is reduced until

Scheme 3.3 Oxygen activation steps in TauD.

both the αKG and taurine substrates are in place. Nevertheless, the αKG-Fe(II)-TauD species does exhibit slow reactivity toward O_2 leading to enzyme side chain hydroxylation involving Tyr$_{73}$, as mentioned above,[18–20] and several more distant Trp residues.[10] These aberrant reactions are avoided by the presence of taurine. The quaternary complex rapidly reacts with O_2 (with a second-order rate constant greater than 1.5×10^5 M^{-1} s^{-1} at 5 °C)[36] and, on the basis of model chemistry,[37] generally is considered to undergo electron transfer to yield an Fe(III)-superoxo species that has not been directly observed (denoted by brackets in Scheme 3.3). Attack of the superoxide on the carbonyl carbon of αKG is inferred to yield a peroxohemiketal bicyclic complex. An ^{18}O kinetic isotope effect of 1.0102 ± 0.0002 suggests that formation of this intermediate is the rate-limiting step in oxygen activation.[38] The peroxohemiketal complex decomposes to yield CO_2, bound succinate, and an Fe(IV)-oxo species (Scheme 3.3).

The first evidence for the Fe(IV)-oxo species was the detection of a transient absorption at 318 nm when taurine-αKG-Fe(II)-TauD was mixed with oxygenated buffer in stopped-flow experiments.[36] At 5 °C, this species develops within $\sim$25 ms and decays by 600 ms. Rapid freeze-quench approaches combined with Mössbauer spectroscopy were used to demonstrate the integer spin nature of this species, with $S \geq 2$. The absence of an EPR signal for this sample is consistent with $S = 2$, and the finding that cryoreduction gives rise to a high-spin Fe(III) state identifies the intermediate as a high-spin Fe(IV) species. Extended X-ray absorption fine structure spectroscopic analysis of the sample obtained by rapid freeze-quench reveals a short Fe–O distance (1.62 Å), assigned to an Fe(IV)-oxo species.[39] This assignment was demonstrated convincingly by continuous-flow, cryogenic, isotope-difference Raman spectroscopy that reveals Fe–O vibrations at 825/788 ($^{16}O_2$/$^{18}O_2$) cm^{-1} in agreement with known Fe(IV)-oxo species.[40] Final assignment of the Fe(IV)-oxo species was obtained by using the mixed isotope $^{16}O^{18}O$, where the Raman vibrations demonstrate the presence of a single bound O atom.[41]

Significantly, the apparent first-order rate constant (k_1) for formation of the Fe(IV)-oxo species (Scheme 3.3) is linearly dependent on the O_2 concentration.[23,42] This result is consistent with the intermediate arising from a second-order reaction between the quaternary complex and oxygen, and suggests that intervening species are not formed in appreciable amounts under these conditions. Of great importance, chemical quench-flow experiments demonstrate that CO_2 formation is coincident with Fe(IV)-oxo formation, thus supporting concerted cleavage of the peroxo O–O bond and the C^1–C^2 αKG bond at this step.[42] DFT calculations support prompt dissociation of CO_2 from the complex following O–O bond scission.[43] Deuteration of substrate does not affect k_1, which is consistent with the Fe(IV)-oxo formation step being separate from substrate C–H cleavage.[44] Of the three metal-ligand mutants that exhibit significant rates, k_1 nearly is unaffected for H255E and H255Q variants, but it is only $\sim 10\%$ of that for wild-type protein in the D101E variant.[17] Several active site variants (Y73I, Y73S, W98I, S158A, Y256F) exhibit k_1 values of the same order of magnitude as the wild-type protein; however, substitutions of Asn_{95} (N95A or N95D) fail to generate a detectable Fe(IV)-oxo species. Similarly, k_1 is too small to detect when using the alternative substrates MOPS or pentanesulfonic acid.[23] These results suggest that the rate of Fe(IV)-oxo formation is sensitive to the identity of metal ligands in the same plane as αKG along with the proper positioning of the substrate.

3.3.2 Electronic Configuration of the Fe(IV)-oxo Species

Extensive computational efforts on the TauD Fe(IV)-oxo species, aligned with experimental data, reveal significant details of its electronic configuration and underlying structural properties. Such analysis often is carried out in the context of heme systems, particularly cytochromes P450. The key difference between the two enzyme classes stems from the binding geometry of a heme, which is effectively confined by the porphyrin macrocycle to a fairly symmetrical square bipyramidal configuration, whereas the structure of the non-heme iron centers does not impose such limitations. Density functional theory (DFT) analysis of the Fe(IV)-oxo species in TauD shows that a high-spin six-coordinated complex provides the best fit with the Mössbauer data.[22] Several variations of this complex with close energies (all in $S = 2$ states) are possible, including either a distorted octahedral geometry with a combination of bidentate/monodentate binding of carboxylates or a trigonal bipyramidal complex with two monodentate carboxylate ligands in the equatorial plane relative to the z-axis defined by the Fe–O bond. Neither the high-spin five-coordinated square pyramidal complex nor any of the low-spin ($S = 1$) complexes, also square pyramidal but with an equatorial oxo group, are consistent with Mössbauer data.[22] DFT analysis by de Visser[45] leads to similar conclusions favoring a six-coordinated complex with the metal ligands being a monodentate or bidentate Asp_{101} carboxylate in the presence and absence of CO_2, respectively. The energy difference between the two configurations is small and Asp_{101} universally assumes an equatorial bidentate configuration in the ensuing steps.[45–47]

The geometry of the Fe(IV)-oxo species affects the spin state configuration of the ground state. The lowest energy configuration is defined as the $\pi^*_{xz}{}^1 \pi^*_{yz}{}^1 \pi^*_{x^2-y^2}{}^1 \sigma^*_{xy}{}^1 \sigma^*_{z^2}{}^0$ quintet state,[22,43,45,48] using the P450 orbital notation.[49] The septet and triplet spin states are 7.2 and 15.8 kcal mol^{-1} greater in energy, respectively, relative to the quintet state.

The high-spin state configuration in TauD may be stabilized by the geometry of the complex due to equatorial ligation by the Asp_{101} carboxylate and succinate, mostly involving the metal σ^*_{xy} orbital.[46] Spectroscopic and computational studies on pentadentate model complexes with nitrogen ligands show a reversed order of energy levels where the triplet state is favored over the quintet state by $7\,kcal\,mol^{-1}$,[50] making them similar to heme complexes. Spin ordering in non-heme complexes is most sensitive to the ligand in the *trans* position relative to the Fe–O group; its influence is mediated *via* metal 3d orbitals.[47,50] Increasing electron donation from the ligands lowers the energy of the quintet state of the Fe(IV)-oxo relative to the triplet state[50] and additional effects are observed in the ensuing transition states. The presence of two electrons in the antibonding $\pi^*_{x^2-y^2}$ and σ^*_{xy} orbitals destabilizes the Fe(IV)-oxo species in TauD and increases its reactivity relative to heme Compound I, where both electrons are located in the non-bonding $\delta_{x^2-y^2}$ orbital.[45]

The length of the Fe–O bond in the quintet spin state of TauD (calculated at 1.65 Å) is comparable to many other non-heme and heme models and enzymes, but this bond is estimated to be significantly longer in the triplet state.[45,51–53] Spin density in both the triplet and quintet states is defined predominantly by the singly occupied d_{xz} and d_{yz} metal orbitals interacting with the p_x and p_y oxygen orbitals, which are antibonding with respect to the Fe–oxo bond.[22,46] Spin density analysis shows an uneven distribution with approximately $-\frac{2}{3}$ and $+3$ spin found on the oxygen and the metal, respectively, with small variations between studies.[22,43,46] The Fe–O bond in the septet spin state is formally a single bond (1.95 Å) due to a $\pi \to \sigma^*$ excitation, and spin density on the oxygen is approximately double that in other states.[43,45]

The spin density of TauD models is confined mostly to the metal-oxo group with little delocalization onto other metal ligands. Although the protonation state of the metal ligands does not affect spin density on either the oxygen or metal, calculations show that both His_{99} and His_{255} must be neutral (protonated) to yield an electronic configuration in agreement with the Mössbauer data.[46] This spin density pattern differs from that found in heme systems where the spin is delocalized over both the porphyrin and the axial residue ligand.[54,55] Hydrogen bonding interactions of the proximal residue in hemes play a major role in their electronic configuration and reactivity *via* a so-called "push-pull" mechanism.[56] In contrast, the effect of the local environment on the complex geometry, bond lengths, and electronic structure of the TauD Fe(IV)-oxo complex is small.[46] For example, inclusion of the protein environment in quantum mechanics/molecular mechanics (QM/MM) and DFT calculations amounts to only a few $kcal\,mol^{-1}$ of stabilization and no clear dipole or electrostatic interactions are found. While the amine and a sulfonate oxygen atom of taurine form a hydrogen bond in calculations, that bond may be attributed to the lack of an H-bond acceptor in the computational model. The limited effect of imidazole ligands on the structure and properties of the Fe(IV)-oxo complex found in calculations is in agreement with the moderate effect observed when using the H99A TauD variant.[22] In this case a water molecule occupies the vacant position of the equatorial imidazole ligand to the metal and, while some heterogeneity was observed in the Fe(II) form of this variant, the properties of its Fe(IV)-oxo intermediate resemble those of the wild-type enzyme.

3.3.3 Hydrogen Atom Abstraction by Fe(IV)-oxo

The Fe(IV)-oxo species decays with a first-order rate constant k_2 (*e.g.*, see Scheme 3.4) that is dramatically reduced by substrate deuteration.[44] The estimated k_H/k_D of 28–50 provides compelling evidence that this intermediate catalyzes hydrogen atom transfer (HAT) from the substrate. Such a strong kinetic $^1H/^2H$ effect is corroborated by the temporal profiles of the 821/788 cm^{-1} species observed by transient Raman spectroscopy.[41] In addition, the spectral sensitivity of this Raman vibration to substrate $^1H/^2H$ substitution reveals direct interactions between the oxo group and the substrate proton positioned for HAT. This proximity is in good agreement with the Fe-taurine distance observed by crystallography[10,11] or obtained from ESEEM measurements.[29]

The rate constant of Fe(IV)-oxo decay is perturbed in several mutant proteins. For example, k_2 of the S158A variant is less than half that of wild-type protein, leading to an enhanced accumulation and a prolonged lifetime of the Fe(IV)-oxo species.[23] This result is interpreted in terms of Ser_{158} helping to orient the taurine appropriately for catalysis. In a similar manner, the k_2 values measured for the F159L and F159V proteins are significantly smaller than the native enzyme.[27,28] The large aromatic side chain is suggested to reduce the internuclear distance between taurine and the Fe(IV)-oxo intermediate, thus promoting hydrogen atom transfer, whereas the smaller side chains increase the distance and reduce the transfer rate. In contrast to these mutant results, the concentration of O_2 does not affect k_2.[23,42]

Computational studies have provided details of HAT in TauD and highlighted mechanistic differences in reactivity from Compound I of heme enzymes and pentadentate non-heme model compounds.[45,57,58] The hydroxylation of substrates by TauD models generally is expected to proceed *via* a classical HAT mechanism (Scheme 3.4).[45,47] Similar to the ground state of the Fe(IV)-oxo species, the quintet transition state is favored energetically over the triplet and septet states.[50] Consequently, DFT calculations on TauD models predict the reaction to proceed *via* a single-state quintet reactivity pathway with the formation of an Fe(III)-OH species and a substrate radical. The electron is transferred from the substrate into an unoccupied $\sigma^*_{z^2}$ orbital that is antibonding in respect to the Fe–O bond. Formation of a spin sextet in the complete, singly occupied metal 3d orbitals leads to stabilization of this state and lowering of the barrier. Although the height of the HAT barrier is similar between the quintet and septet pathways, the quintet pathway is significantly lower in energy through the rest of the reaction and crossover is not expected to play a major role.[45,47,50] In non-heme pentadentate

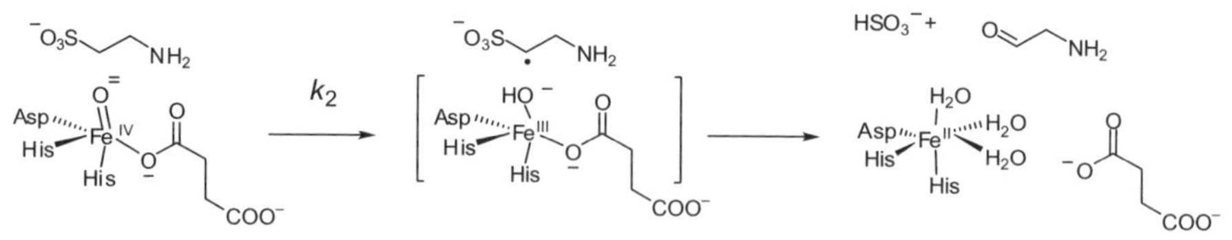

Scheme 3.4 Classical mechanism for substrate hydroxylation by TauD.

models, the two-electron oxo transfer to triphenylphosphine also follows a single-state quintet path, while HAT follows a two-states reactivity pathway with a triplet to quintet transition upon C–H bond activation similar to heme systems.[50]

The HAT barrier and the length of the O–H bond correlate with the strength of the C–H bond. The resulting O–H bond is shorter ($d_{O-H} \leq 1.2$ Å) and the reaction barriers are higher for substrates with a strong C–H bond, while substrates with weaker C–H bonds yield a lower reaction barrier and more symmetrical O–H–C bonds ($d_{O-H} > 1.3$ Å).[47] Hydrogen bonding in the active site has little effect on either spin state separation or height of barriers. In all calculated transition states the carboxylate group of either Asp_{101} or succinate shows bidentate binding to the metal and CO_2 is lost.[45,47] However, no clear transition states are found in QM/MM calculations of taurine oxygenation.[46]

3.3.4 Thermodynamics of Hydrogen Atom Abstraction by Fe(IV)-oxo

The ability of a metal-oxo species to abstract a hydrogen atom, or its HAT capacity, is determined by the relative bond dissociation energies (BDE) of the newly formed O–H bond of the metal-OH complex *versus* that of the C–H bond of the substrate. Bordwell's thermodynamic cycle (*e.g.*, Scheme 3.5) can be used to calculate the BDE of a metal-oxo center from the one-electron redox potential, $E_{1/2}$, of the oxidized complex and the proton affinity, or pK_a, of the reduced complex,[59] as brilliantly illustrated by Mayer in a study on a series of organic substrates.[60] Assuming equal entropies of reactants and products, the following equation can be written for reactions at ambient temperature by converting the pK_a and $E_{1/2}$ into standard free energy:

$$BDE_{O-H}(\text{kcal mol}^{-1}) = 23.06\, E_{1/2}\left([Me-O]^{(n)}/[Me-O]^{(n-1)}\right)(V)$$
$$+ 1.37\, pK_a\left([Me-O]^{(n-1)}/[Me-OH]^{(n)}\right) + C \qquad (3.1)$$

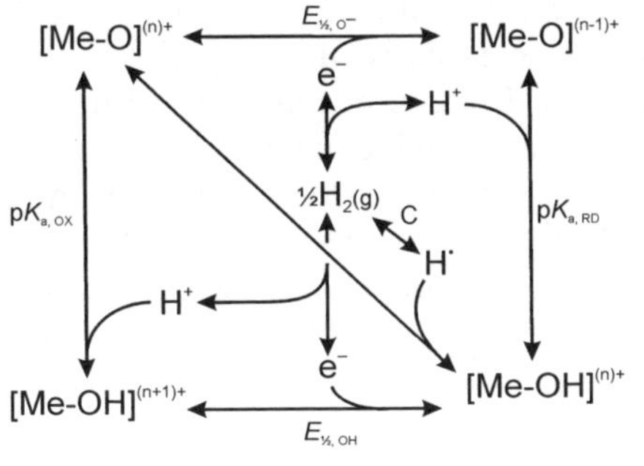

Scheme 3.5   Bordwell thermodynamic cycle for hydrogen atom abstraction reactions.

where the constant C represents the bond dissociation and solvation energy of hydrogen and equals 57 ± 2 kcal mol^{-1}.[60] This correlation finds further support in model work on pentadentate iron-oxo model complexes,[61] homologous heme models,[62] as well as DFT calculations.[47]

Green[53] examined the protonation states of several iron-oxo and iron-hydroxo complexes from the perspective of Badger's rule.[63] From the correlation between the bond length and the force constant, he concludes that Fe(IV)-oxo groups in most representative heme proteins with a proximal imidazole ligand are acidic ($pK_a < 4$). Increased electron donation from the proximal cysteine ligand, in contrast, is cited as a primary cause of an unusually high proton affinity of (porphyrin)Fe(IV)–OH in Compound II of chloroperoxidase, which allows its precursor (porphyrin)$^{\bullet+}$Fe(IV)=O, Compound I, to oxidize halides.[64] Indeed, non-heme complexes with more electron-donating ligands also are more reactive toward C–H and O–H bonds.[50] A DFT study on these complexes shows that the charge transferred onto the terminal oxo group increases from 0.27 to 0.88 as the electron donation increases from the weak axial ligand CH_3CN to an electron-rich RS^- ligand, correlating with the HAT capacity.[50]

The O–H BDE in model non-heme iron-oxo complexes is surprisingly unaffected by increasing the electron donation from ligands.[50] Kinetic trends with increasing electron donation are attributed to the decrease in the triplet–quintet spin-state gap, but the $E_{1/2}$ values observed for Fe(IV)-oxo complexes are unexpectedly small for their HAT capacity. While there is a possibility that the $E_{1/2}$ values are underestimated due to kinetic inhibition,[50,65] the results raise the question of the relative contribution of the electron and proton affinities to the BDE. According to Equation (3.1), a lower $E_{1/2}$ can be offset by a larger pK_a and *vice versa* without a significant effect on the BDE. Excess negative charge on the terminal oxo group is expected to decrease the electron affinity and increase the proton affinity *via* the redox Bohr effect, which is fully consistent with the decreased redox potentials of Fe(IV)–oxo and the elevated pK_a in sulfur-ligated complexes. This interplay between pK_a and $E_{1/2}$ can rationalize the ability of cytochromes P450 to catalyze HAT with moderate $E_{1/2}$ to avoid side reactions associated with long-range protein oxidation by Compounds I and II. In hemes, $E_{1/2}$ can be reduced further by charge delocalization over the porphyrin macrocycle, which is not available in non-heme systems. A net more negative charge on the cysteine-ligated heme in cytochromes P450 relative to imidazole-ligated peroxidases or TauD models is likely to further decrease the $E_{1/2}$ and raise the pK_a of oxygenic ligands.[41,56]

The electronic structure of the TauD Fe(IV)-oxo species shows a minimal effect of metal ligands with negligible electron donation, solidly siding it with peroxidases rather than with cytochromes P450. Nevertheless, its BDE (98.3 kcal mol^{-1}) is significantly greater than that of heme enzymes (87.4–88.7 kcal mol^{-1}), which is rationalized as the result of its five-coordinate geometry (*versus* six-coordinate in hemes) and its single-state high-spin reactivity.[45] It can be seen from Equation (3.1) that a greater pK_a and/or $E_{1/2}$ must contribute to the increased BDE in TauD. While neither value is currently known, they can be estimated from available data. In the absence of substrate, the Fe(II)-TauD-αKG complex slowly reacts with O_2. This non-productive reaction leads to extensive self-hydroxylation of the protein, including Tyr$_{73}$ and several Trp residues.[10,18,20] While Tyr$_{73}$ is the closest among those residues to the metal center, based on crystallographic data it is not likely to approach the Fe(IV)-oxo group close enough to allow direct HAT reaction. Therefore, oxidation is likely to proceed *via* one-electron oxidation of Tyr followed by its deprotonation to yield the observed neutral tyrosyl radical.[18] One-electron oxidation potentials of Tyr and Trp in water at neutral pH are 0.93 and 1.02 V *versus* NHE, respectively.[66] Potentials and proton affinities of individual residues can be greatly affected by the local protein environment; however, the promiscuity of oxidation observed for TauD in the absence of substrate and the relative openness of the TauD structure to proton equilibrium with the bulk solution suggest that the $E_{1/2}$ for the Fe(IV)-oxo species in TauD is at least comparable to that of imidazole-ligated heme Fe(IV)-oxo compounds and is likely to be above 1 V.[67,68] Taking the potential of the TauD Fe(IV)-oxo species to be equal to that of horseradish peroxidase Compound I (HRP-I) as a low limit estimate, the entire increase in BDE of O–H in TauD relative to HRP would be attributed to the change in its pK_a with a maximal increase of 7 pH units relative to HRP Compound II. The pK_a of the latter species was shown to be less than 4,[53] leading to an upper limit for the pK_a of Fe(III)–OH in TauD of 11.

It is likely that $E_{1/2}$ of the Fe(IV)-oxo species in TauD is significantly greater than that of HRP-I on the basis of the long-range and wide-spread protein oxidation found for TauD in the absence of taurine along with the intrinsic water-mediated inhibition that controls the metallocenter reactivity with O_2. For example, the pH-dependent oxidation of Tyr or Trp residues is observed in the active site of the Compound I–like $\boldsymbol{P_m}$ species of cytochrome c oxidase, but not in the subsequent Compound II-like $\boldsymbol{F}$ species.[69,70] Both of these species have very similar Fe(IV)-oxo structures, with $E_{1/2}$ of 1.2 and 1.1 V, respectively.[68,71] Again using HRP-I as a reference and accounting for the 10 kcal mol^{-1} stronger O-H bond in TauD, a 200 mV increase in $E_{1/2}$ would allow for a smaller pK_a increase of 3.4 units, which reduces the upper limit of the pK_a value of the Fe(III)–OH from 11 to 7.4.

Importantly, the properties of the non-heme Fe(III)–O(H) species should be analyzed relative to Compound II in heme systems (Figure 3.2), since these two species contribute to the BDE according to Equation (3.1). A deceptively more intuitive analogy between non-heme Fe(III)–OH and ferric aqua/hydroxy hemes would be incorrect despite their structural similarity. Ferric hemes are a full oxidizing equivalent lower than the corresponding Compound II species and the pK_as of the ferric–aqua complexes are generally assumed to be significantly above the attainable level for enzymatic systems. Recently, however, this assumption has been challenged based on comparative analysis of a series of synthetic tetradentate and porphyrin complexes in comparison to the corresponding biological complexes,[72–74] which suggests that the pK_a of ferric ligands may be significantly smaller than previously assumed.[41]

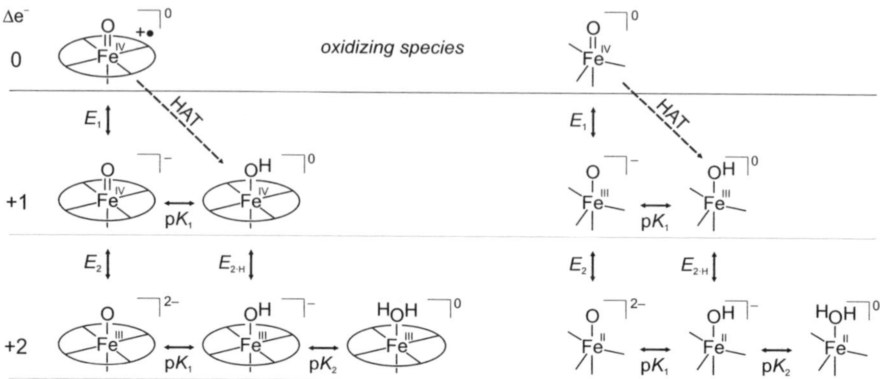

Figure 3.2 Reduction and protonation states in heme and non-heme Fe centers relative to the catalytic, hydrogen atom abstracting species. The first electron added to heme systems (E_1) reduces the porphyrin π-cation radical in Compound I (left, top) to form the neutral heme-containing Compound II (left, middle), which maintains the iron in the Fe(IV) state. The corresponding transition in non-heme enzymes (right-hand side) involves reduction of the metal to Fe(III). In both cases, addition of a second electron leads to metal reduction. The electron count and net charges are shown relative to the highly oxidized state.

3.4 Fe(III)-O(H) Species and Oxygen Transfer

The transient Fe(III)–O(H) species lies at another critical point in the reaction sequence and its structure has major mechanistic implications for both the preceding HAT, involving the Fe(IV)=O species as described above, and for mediating transfer of the oxygen atom to the substrate (Scheme 3.4).

DFT calculations show that upon oxygen rebinding in the quintet spin state, the second antiferromagnetically-coupled (down spin) substrate electron is transferred to the $\pi^*_{x^2-y^2}$ orbital, which becomes doubly occupied.[47] In the process of oxygen transfer, the π^* orbitals of the Fe–O species become atomic 3d orbitals of the metal as the overlap with the oxygen orbitals is lost.[45] The energy barrier for the rebinding step is determined by the difference between the ionization potential of the substrate and the electron affinity (EA) of Fe(III)–OH. The small EA of the non-heme Fe(III)–OH complex in TauD (37.9 kcal mol^{-1}) results in a higher rebinding barrier when compared to the corresponding step in heme systems ($EA = 69.0$ kcal mol^{-1}) for a given substrate.[47] A small barrier of 3.5 kcal mol^{-1} was found for oxygen transfer to propene, increasing to 17.7 kcal mol^{-1} for propane.[45,47] Thus, the Fe(III)–OH species in the non-heme oxygenation pathway of most alkanes is predicted to have a significant lifetime, sufficient for substrate rearrangement and, possibly, for spectroscopic detection.

Somewhat surprisingly, the QM/MM study on hydroxylation of taurine (as opposed to the DFT analysis for hydroxylation of alkanes and alkenes) shows concerted behavior where products are formed immediately following the transition state in the HAT step.[46] Overall, the exothermicity (56.8 kcal mol^{-1}) and the initial barrier (6.7 kcal mol^{-1}) for the taurine reaction are similar to those for propene (52.7 and 5.4 kcal mol^{-1}, respectively). Not only are no intervening species found, as expected for a stepwise mechanism, but efforts to purposely generate a Fe(III)–OH species and a taurine radical fail. Considering the small overall effect of the protein found in QM/MM calculations in comparison to the DFT studies, it is likely that the lack of intermediates is due to the nature of the substrate, not the protein environment.

The lack of a stable Fe(III)–OH species is consistent with the freeze-quench Mössbauer and stopped-flow UV/visible spectroscopic studies, which identify an Fe(II) intermediate following the Fe(IV)-oxo species.[36] Follow-up investigations found that this Fe(II)-TauD complex was distinct from the binary, ternary, and quaternary species,[42] and is likely the product complex whose release was the rate-determining step in the overall reaction.

In recently reported cryogenic continuous-flow Raman studies, however, evidence is described for two new oxygen-isotope sensitive intermediates whose temporal profiles trail the Fe(IV)–oxo intermediates.[40,41] The earlier of the two species exhibits a $^{16}O/^{18}O$ vibrational mode at 578/555 cm^{-1}. The second new species exhibits a vibrational mode at 815/787 cm^{-1} that overlaps the Fe(IV)-oxo features, but with a significantly narrower $^{16}O/^{18}O$ isotopic shift. Kinetic profiles of both of these Raman species show pronounced sensitivity to substrate deuteration supporting their catalytic relevance. Resolution of these species supports a predicted stepwise mechanism, while the small Raman intensities and short lifetimes suggest that the energy barriers are relatively low. The vibrational properties of the new species raise questions about their structural assignment, and efforts were made to identify these intermediates.

The frequency and $^{16}O/^{18}O$ shift of the 578 cm^{-1} species is close to the Fe–O stretching frequency of either Fe–O$_2$ or Fe–O(H) complexes. The lack of intermediate frequencies with $^{16}O^{18}O$ mixed-isotope oxygen unambiguously rules out a diatomic oxygen structure for this species, leaving a monooxy Fe–O(H) structure as the only possible assignment. Significantly, while similar in frequency to calculated frequencies of several Fe(III)–O(H) or Fe(IV)–OH species,[53] the 578 cm^{-1} species shows no sensitivity to substrate or solvent $^{1}H/^{2}H$ substitution. The absence of a covalent proton cannot be unambiguously proven by the lack $^{1}H/^{2}H$ isotope sensitivity alone, since an isotope-sensitive shift can be reduced in strongly bent ligands or even reversed by hydrogen bonds.[75–79] However, such special cases are well understood and the crystal structure of the Fe(II)-TauD-αKG-taurine complex shows no steric or hydrogen bonding interaction that could distort the Fe–O–H angle. Particularly significant is that the Asp$_{101}$ carboxylate is rotated away from the aqueous ligand and cannot form a strong hydrogen bond.[26] Several computational studies show that the carboxylate assumes an equatorial bidentate configuration in the Fe(IV)-oxo species or following HAT; this configuration prevents hydrogen bonding to the axial water.[22,45,47] Even the extra water coordinated to the metal in H99A variant does not show significant interactions with the carboxylate.[22] Furthermore, computation shows that a favorable spin configuration in TauD facilitates an axial approach of the substrate C–H bond relative to the Fe(IV)-oxo group.[45] Electron transfer into the $\sigma^*_{z^2}$ orbital results in a nearly linear geometry of Fe(III)–O–H group, which must result in the largest $^1H/^2H$ isotope sensitivity of the Fe–OH stretching mode. Taken together, the calculations on TauD models give no indication of the special interactions that would render the Fe(III)–OH insensitive to deuteration, thus providing indirect support for assigning the 578 cm^{-1} species to a deprotonated Fe–O structure.[41]

These results clearly contradict the potential assignment of the 578 cm^{-1} species to either the Fe(III)–OH or the Fe(IV)–OH species. Rather, the 578/555 cm^{-1} mode is most consistent with the presence of a deprotonated Fe(III)–O$^-$ or Fe(IV)–O$^-$ species. DFT calculations show that the Fe(IV)–oxo bond in the septet spin state is significantly elongated and the formal bond order is single due to the extra electron in the antibonding orbital.[43] Since the quintet and septet states are close in energy, it is conceivable that both the 821 and 578 cm^{-1} vibrations are Fe(IV)–oxo stretching modes on the quintet and septet pathways, respectively. A longer lifetime of the 578 cm^{-1} species with deuterated substrate would support HAT by this species. This possibility would imply that spin crossing does not readily occur, based on the difference in temporal profiles of the two species, and this situation could be tested by resolving the early phases of the reaction. It does not, however, provide a rational interpretation for the unusual 815 cm^{-1} species.

An alternative assignment of the 578 cm^{-1} mode involves an unusual deprotonated Fe(III)–O$^-$ species and requires modification of the classical hydroxyl radical rebinding mechanism (Scheme 3.6).[41] In this interpretation, the formation of an Fe(III)–OH species upon HAT from taurine to Fe(IV)–oxo is followed immediately by proton transfer to a nearby base yielding a deprotonated Fe(III)–O$^-$ species. So far, such a species has been observed

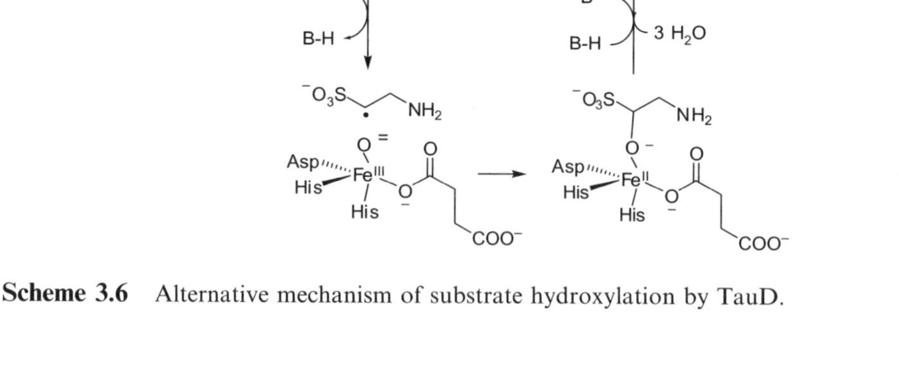

Scheme 3.6 Alternative mechanism of substrate hydroxylation by TauD.

experimentally only upon cryogenic reduction of the Fe(IV)–oxo species in hemes.[80] The identity of the proton acceptor is not currently known in TauD, but it must have a sufficiently large pK_a to compete with the Fe(III)–OH for the proton. Comparative analysis of the proton affinity of the Fe(III)–OH presented above suggests that the basicity of many protein side chains, such as phenols or amines, may be sufficient for this purpose, at least in transient conditions.

The presence of a putative Fe(III)–O⁻ in close proximity to a substrate radical may lead to the formation of a transient Fe–O–C (metal-alkoxo) intermediate (Scheme 3.6). Such a species could give rise to the unusual isotopic shift at $815\,cm^{-1}$ observed in the transient Raman study, although its accurate characterization is complicated due to spectral overlap with the Fe(IV)–oxo species.[41] The $^{16}O/^{18}O$ isotopic shift of the $815\,cm^{-1}$ species is significantly smaller than expected for the Fe–O diatomic oscillator and suggests involvement of relatively light atoms, such as in a C–O bond. In fact, a similar vibrational signature is observed in an isotopically labeled Zn^{2+}-isopropoxide complex as the simplest model of a possible intermediate in TauD. By contrast, isotopic shifts observed for models of other potential targets of oxygen labeling in the active site of TauD (a carboxylic acid, its metal salts, or a secondary alcohol) are significantly different from that of the $815\,cm^{-1}$ species and those assignments were dismissed.[41]

Despite the lack of evidence for the presence of structural protons in the 578 and $815\,cm^{-1}$ species, taurine deuteration increases their lifetime. Although less pronounced than that for the Fe(IV)–oxo species, such a kinetic isotope effect strongly implies the involvement of a substrate proton at the corresponding steps in the reaction. If the $578\,cm^{-1}$ species is an Fe(IV)–oxo on the septet pathway, its kinetic sensitivity is expected. For a pathway involving deprotonation of a quintet Fe(III)–OH, a kinetic effect is interpreted in terms of reusing the substrate proton for charge neutralization upon protonation of the product sulfite group at the last step of the reaction.[41]

3.5 Conclusions

By combining steady-state and transient kinetic approaches with a diverse array of spectroscopic methods along with structural and computational studies a rather detailed picture of the TauD reaction mechanism has emerged. The TauD–Fe(II) holoenzyme binds αKG and taurine, priming the metal center for oxygen binding by opening a coordination site (Scheme 3.2). Reaction of this activated complex with oxygen likely proceeds *via* Fe(III)–superoxo and Fe(IV)–peroxohemiketal species, the latter of which decomposes to form CO_2 and an Fe(IV)–oxo complex (Scheme 3.3). HAT from the substrate to the Fe(IV)–oxo species is expected to yield an Fe(III)–OH and a substrate centered radical. In the classic mechanism, these species are transformed to products by hydroxyl radical transfer to the substrate, followed by spontaneous decomposition of the resulting hydroxytaurine to aminoacetaldehyde and sulfite (Scheme 3.4). An alternative, more complex mechanism is postulated on the basis of continuous-flow Raman studies that show two additional oxygen isotope-sensitive species. These intermediates are tentatively assigned to Fe(III)–oxo and Fe(II)–alkoxo species (Scheme 3.6). As described in the text of this chapter, multiple questions remain to be answered in the mechanism of substrate oxygenation by TauD. Additional studies are needed to examine the validity of the intermediates described here for TauD and to explore the universality of these species in the diverse range of other family members.

Acknowledgements

TauD research in the authors' laboratories was supported by the National Institutes of Health. (GM070544 to D.A.P. and GM063584 to R.P.H.)

References

1. R. P. Hausinger, *Crit. Rev. Biochem. Mol. Biol.*, 2004, **39**, 21.
2. V. M. Purpero and G. R. Moran, *J. Biol. Inorg. Chem.*, 2007, **12**, 587.
3. C. Loenarz and C. J. Schofield, *Trends Biochem. Sci.*, 2011, **36**, 7.
4. J. M. Simmons, T. A. Müller and R. P. Hausinger, *Dalton Trans.*, 2008, **38**, 5132.
5. H. M. Hanauske-Abel and V. Günzler, *J. Theor. Biol.*, 1982, **94**, 421.
6. J. R. Van Der Ploeg, M. A. Weiss, E. Saller, H. Nashimoto, N. Saito, M. A. Kertesz and T. Leisinger, *J. Bacteriol.*, 1996, **178**, 5438.
7. E. Eichhorn, J. R. van der Ploeg, M. A. Kertesz and T. Leisinger, *J. Biol. Chem.*, 1997, **272**, 23031.
8. A. Kahnert and M. A. Kertesz, *J. Biol. Chem.*, 2000, **275**, 31661.
9. D. A. Hogan, T. A. Auchtung and R. P. Hausinger, *J. Bacteriol.*, 1999, **181**, 5876.
10. J. M. Elkins, M. J. Ryle, I. J. Clifton, J. C. Dunning Hotopp, J. S. Lloyd, N. I. Burzlaff, J. E. Baldwin, R. P. Hausinger and P. L. Roach, *Biochemistry*, 2002, **41**, 5185.
11. J. R. O'Brien, D. J. Schuller, V. S. Yang, B. D. Dillard and W. N. Lanzilotta, *Biochemistry*, 2003, **42**, 5547.
12. E. L. Hegg and L. Que Jr., *Eur. J. Biochem.*, 1997, **250**, 625.
13. K. D. Koehntop, J. P. Emerson and L. Que Jr., *J. Biol. Inorg. Chem.*, 2005, **10**, 87.
14. P. K. Grzyska, R. P. Hausinger and D. A. Proshlyakov, *Anal. Biochem.*, 2010, **399**, 64.
15. E. Kalliri, P. K. Grzyska and R. P. Hausinger, *Biochem. Biophys. Res. Commun.*, 2005, **338**, 191.
16. M. J. Ryle, R. Padmakumar and R. P. Hausinger, *Biochemistry*, 1999, **38**, 15278.
17. P. K. Grzyska, T. A. Müller, M. G. Campbell and R. P. Hausinger, *J. Inorg. Biochem.*, 2007, **101**, 797.

18. M. J. Ryle, A. Liu, R. B. Muthukumaran, R. Y. N. Ho, K. D. Koehntop, J. McCracken, L. Que Jr. and R. P. Hausinger, *Biochemistry*, 2003, **42**, 1854.
19. M. J. Ryle, K. D. Koehntop, A. Liu, L. Que Jr. and R. P. Hausinger, *Proc. Natl. Acad. Sci. USA*, 2003, **100**, 379.
20. K. D. Koehntop, S. Marimanikkuppam, M. J. Ryle, R. P. Hausinger and L. Que Jr., *J. Biol. Inorg. Chem.*, 2006, **11**, 63.
21. P. K. Grzyska and R. P. Hausinger, *Inorg. Chem.*, 2007, **46**, 10087.
22. S. Sinnecker, N. Svensen, E. W. Barr, S. Ye, J. M. Bollinger Jr., F. Neese and C. Krebs, *J. Am. Chem. Soc.*, 2007, **129**, 6168.
23. P. K. Grzyska, M. J. Ryle, G. R. Monterosso, J. Liu, D. P. Ballou and R. P. Hausinger, *Biochemistry*, 2005, **44**, 3845.
24. E. G. Pavel, J. Zhou, R. W. Busby, M. Gunsior, C. A. Townsend and E. I. Solomon, *J. Am. Chem. Soc.*, 1998, **120**, 743.
25. R. Y. N. Ho, M. P. Mehn, E. L. Hegg, A. Liu, M. A. Ryle, R. P. Hausinger and L. Que Jr., *J. Am. Chem. Soc.*, 2001, **123**, 5022.
26. M. L. Neidig, C. D. Brown, K. M. Light, D. G. Fujimori, E. M. Nolan, J. C. Price, E. W. Barr, J. M. Bollinger Jr., C. Krebs, C. T. Walsh and E. I. Solomon, *J. Am. Chem. Soc.*, 2007, **129**, 14224.
27. K. P. McCusker and J. P. Klinman, *Proc. Natl. Acad. Sci. USA*, 2009, **106**, 19791.
28. K. P. McCusker and J. P. Klinman, *J. Am. Chem. Soc.*, 2010, **132**, 5114.
29. R. B. Muthakumaran, P. K. Grzyska, R. P. Hausinger and J. McCracken, *Biochemistry*, 2007, **46**, 5951.
30. S. Ye, J. C. Price, E. W. Barr, M. T. Green, J. M. Bollinger Jr., C. Krebs and F. Neese, *J. Am. Chem. Soc.*, 2010, **132**, 4739.
31. J. H. Enemark and R. D. Feltham, *Coord. Chem. Rev.*, 1974, **13**, 339.
32. L. M. Hoffart, E. W. Barr, R. B. Guyer, J. M. Bollinger Jr. and C. Krebs, *Proc. Natl. Acad. Sci. USA*, 2006, **103**, 14738.
33. D. P. Galonic, E. W. Barr, C. T. Walsh, J. M. Bollinger Jr. and C. Krebs, *Nat. Chem. Biol.*, 2007, **3**, 113.
34. M. M. Matthews, C. Krest, E. W. Barr, F. Vaillancourt, C. T. Walsh, M. Green, C. Krebs and J. M. Bollinger Jr., *Biochemistry*, 2009, **48**, 4331.
35. B. E. Eser, E. W. Barr, P. A. Frantom, L. Saleh, J. M. Bollinger Jr., C. Krebs and P. F. Fitzpatrick, *J. Am. Chem. Soc.*, 2007, **129**, 11334.
36. J. C. Price, E. W. Barr, B. Tirupati, J. M. Bollinger Jr. and C. Krebs, *Biochemistry*, 2003, **42**, 7497.
37. A. Mukherjee, M. A. Cranswick, M. Chakrabarti, T. K. Paine, K. Fujisawa, E. Münck and L. Que Jr., *Inorg. Chem.*, 2010, **49**, 3618.
38. L. M. Mirica, K. P. McCusker, J. W. Munos, H.-W. Liu and J. P. Klinman, *J. Am. Chem. Soc.*, 2008, **130**, 8122.
39. P. J. Riggs-Gelasco, J. C. Price, R. B. Guyer, J. H. Brehm, E. W. Barr, J. M. Bollinger Jr. and C. Krebs, *J. Am. Chem. Soc.*, 2004, **126**, 8108.
40. D. A. Proshlyakov, T. F. Henshaw, G. R. Monterosso, M. J. Ryle and R. P. Hausinger, *J. Am. Chem. Soc.*, 2004, **126**, 1022.

41. P. K. Grzyska, E. H. Appelman, R. P. Hausinger and D. A. Proshlyakov, *Proc. Natl. Acad. Sci. USA*, 2010, **107**, 3982.
42. J. C. Price, E. W. Barr, L. M. Hoffart, C. Krebs and J. M. Bollinger Jr., *Biochemistry*, 2005, **44**, 8138.
43. T. Borowski, A. Bassan and P. E. M. Siegbahn, *Chem. Eur. J.*, 2004, **10**, 1031.
44. J. C. Price, E. W. Barr, T. E. Glass, C. Krebs and J. M. Bollinger Jr., *J. Am. Chem. Soc.*, 2003, **125**, 13008.
45. S. P. de Visser, *J. Am. Chem. Soc.*, 2006, **128**, 9813.
46. E. Godfrey, C. S. Porro and S. P. de Visser, *J. Phys. Chem. A*, 2008, **112**, 2464.
47. R. Latifi, M. Bagherzadeh and S. P. de Visser, *Chem. Eur. J.*, 2009, **15**, 6651.
48. A. Bassan, T. Borowski and P. E. M. Siegbahn, *Dalton Trans.*, 2004, 3153.
49. S. Shaik, D. Kumar, S. P. de Visser, A. Altun and W. Thiel, *Chem. Rev.*, 2005, **105**, 2279.
50. C. V. Sastri, J. Lee, K. Oh, Y. J. Lee, J. Lee, T. A. Jackson, K. Ray, H. Hirao, W. Shin, J. A. Halfen, J. Kim, L. Que Jr., S. Shaik and W. Nam, *Proc. Natl. Acad. Sci. USA*, 2007, **104**, 19181.
51. P. Andreoletti, A. Pernoud, G. Sainz, P. Gouet and H. M. Jouve, *Acta Crystallogr. D*, 2003, **59**, 2163.
52. G. I. Berglund, G. H. Carlsson, A. T. Smith, H. Szoke, A. Henriksen and J. Hajdu, *Nature*, 2002, **417**, 463.
53. M. T. Green, *J. Am. Chem. Soc.*, 2006, **128**, 1902.
54. J. C. Schöneboom, H. Lin, N. Reuter, W. Thiel, S. Cohen, F. Ogliaro and S. Shaik, *J. Am. Chem. Soc.*, 2002, **124**, 8142.
55. G. H. Loew and Z. S. Herman, *J. Am. Chem. Soc.*, 1980, **102**, 6173.
56. J. D. Dawson, *Science*, 1988, **240**, 433.
57. A. Decker and E. I. Solomon, *Curr. Opin. Chem. Biol.*, 2005, **9**, 152.
58. D. Kumar, H. Hirao, L. Que Jr. and S. Shaik, *J. Am. Chem. Soc.*, 2005, **127**, 8026.
59. F. G. Bordwell, J.-P. Cheng, G.-Z. Ji, A. V. Satish and X. Zhang, *J. Am. Chem. Soc.*, 1991, **113**, 9790.
60. J. M. Mayer, *Acc. Chem. Res.*, 1998, **31**, 441.
61. J. Kaizer, E. J. Klinker, N. Y. Oh, J.-U. Rohde, W. J. Song, A. Stubna, J. Kim, E. Münck, W. Nam and L. Que Jr., *J. Am. Chem. Soc.*, 2004, **126**, 472.
62. Y. J. Jeong, Y. Kang, A. R. Han, Y. M. Lee, H. Kotani, S. Fukuzumi and W. Nam, *Angew Chem., Int. Ed.*, 2008, **47**, 7321.
63. R. M. Badger, *J. Chem. Phys.*, 1935, **3**, 710.
64. M. T. Green, J. D. Dawson and H. B. Gray, *Science*, 2004, **304**, 1653.
65. L. D. Slep, A. Mijovilovich, W. Meyer-Klaucke, T. Weyhermuller, E. Bill, E. Bothe, F. Neese and K. Wieghardt, *J. Am. Chem. Soc.*, 2003, **125**, 15554.
66. A. Harriman, *J. Phys. Chem.*, 1987, **91**, 6102.

67. Y. Hayashi and I. Yamazaki, *J. Biol. Chem.*, 1979, **254**, 9101.
68. M. Wikstrom and J. E. Morgan, *J. Biol. Chem.*, 1992, **267**, 10266.
69. D. A. Proshlyakov, M. A. Pressler, C. DeMaso, J. F. Leykam, D. L. DeWitt and G. T. Babcock, *Science*, 2000, **290**, 1588.
70. D. A. Proshlyakov, *Biochim. Biophys. Acta*, 2004, **1655**, 282.
71. D. A. Proshlyakov, T. Ogura, K. ShinzawaItoh, S. Yoshikawa and T. Kitagawa, *Biochemistry*, 1996, **35**, 8580.
72. R. Gupta and A. S. Borovik, *J. Am. Chem. Soc.*, 2003, **125**, 13234.
73. T. H. Parsell, M.-Y. Yang and A. S. Borovik, *J. Am. Chem. Soc.*, 2009, **131**, 2762.
74. D. E. Lansky and D. P. Goldberg, *Inorg. Chem.*, 2006, **45**, 5119.
75. A. J. Sitter, C. M. Reczek and J. Terner, *J. Biol. Chem.*, 1985, **260**, 7515.
76. S. Hashimoto, J. Teraoka, T. Inubushi, T. Yonetani and T. Kitagawa, *J. Biol. Chem.*, 1986, **261**, 11110.
77. D. A. Proshlyakov, T. Ogura, K. Shinzawa-Itoh, S. Yoshikawa, E. H. Appelman and T. Kitagawa, *J. Biol. Chem.*, 1994, **269**, 29385.
78. A. Feis, M. P. Marzocchi, M. Paoli and G. Smulevich, *Biochemistry*, 1994, **33**, 4577.
79. S. Ogo, R. Yamahara, M. Roach, T. Suenobu, M. Aki, T. Ogura, T. Kitagawa, H. Masuda, S. Fukuzumi and Y. Watanabe, *Inorg. Chem.*, 2002, **41**, 5513.
80. R. Davydov, R. L. Osborne, S. H. Kim, J. D. Dawson and B. M. Hoffman, *Biochemistry*, 2008, **47**, 5147.

Density Functional Theory Studies on Non-heme Iron Enzymes

TOMASZ BOROWSKI*[a] AND PER E. M. SIEGBAHN*[b]

[a] Institute of Catalysis and Surface Chemistry, Polish Academy of Sciences, ul. Niezapominajek 8, 30-239, Kraków, Poland; [b] Department of Physics, Albanova, Department of Biochemistry and Biophysics, Arrhenius Laboratories, Stockholm University, S-106 91, Stockholm, Sweden

4.1　Introduction

A considerable number of enzymes utilize non-heme iron cofactors to activate molecular dioxygen and then catalyse a wide variety of important and chemically interesting reactions.[1,2] In this chapter we provide a synthesis of computational studies, performed mostly in our groups, which shed light on the catalytic mechanisms of these interesting metalloenzymes. In such studies several different mechanism and/or spin states are usually considered, and, based on calculated barrier heights, the most likely reaction scenario is suggested.[3] Since all the details concerning the mechanisms dismissed are available in the original papers, in this chapter we present only the most likely reaction paths. Moreover, instead of presenting full catalytic cycles of specific enzymes we split them into well-defined chemical steps that can be compared easily. As we will see below such an approach greatly aids identification of general catalytic strategies employed by non-heme iron enzymes.

Iron-Containing Enzymes: Versatile Catalysts of Hydroxylation Reactions in Nature
Edited by Sam P de Visser and Devesh Kumar
© Royal Society of Chemistry 2011
Published by the Royal Society of Chemistry, www.rsc.org

4.1.1 Reactions Catalysed by Non-heme Iron Enzymes and their Biological Significance

Catalytic metalloproteins that host in their active sites iron bound by means of oxygen and/or nitrogen ligands, such as the side chains of aspartate, glutamate, tyrosine and histidine, are customarily called non-heme iron enzymes (NIEs). This name itself emphasises differences between this and two other large groups of iron metalloproteins, which are heme proteins and iron-sulfur proteins. NIEs are present in all types of living organisms (bacteria, plants and animals) where they take part in a plethora of important processes by catalysing various types of oxidative transformations (Scheme 4.1). The catalytic reactions of NIEs

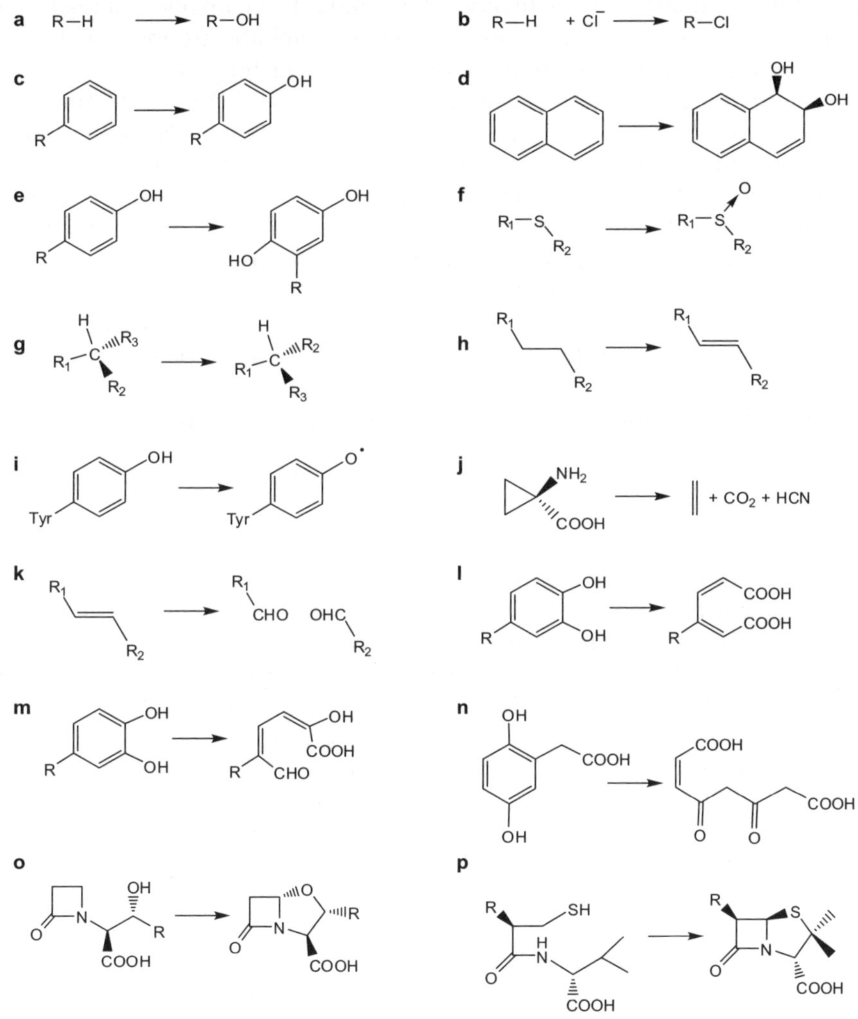

Scheme 4.1 Chemical transformations catalyzed by non-heme iron enzymes.

range from relatively simple aliphatic hydroxylation (Scheme 4.1a) or sulfoxidation (Scheme 4.1f) to synthetically unprecedented ring closure reactions (Scheme 4.1o and p).[4,5]

Hydroxylation of aliphatic groups (Scheme 4.1a) is probably the most common reaction catalysed by NIEs, and, notably, the hydroxylated C–H bond can come from the primary, secondary or tertiary aliphatic group. Methane, which is the aliphatic hydrocarbon with the strongest C–H bond, is transformed into methanol by soluble methane monooxygenase (sMMO) produced by methanotrophic bacteria.[6] This is the first step to metabolize methane as the only source of carbon and energy for these species. Other NIEs catalyse hydroxylation of aliphatic groups in larger organic substrates. Thus, methyl groups attached to nitrogen are hydroxylated by DNA repairing (demethylating) enzymes and histone demethylases,[7,8] the latter being involved in regulation of genes expression. Hydroxylation of secondary aliphatic groups is performed by, for example, proline hydroxylases taking part in biosynthesis of collagen and cellular oxygen sensing,[9] clavaminic acid synthase (CAS) – a trifunctional enzyme involved in biosynthesis of the useful antibiotic clavulanic acid,[10] and taurine/α-ketoglutarate dependent dioxygenase (TauD),[11] which catalyses cleavage of taurine concomitant with release of sulfate. Hydroxylation at the more reactive benzylic and tertiary carbons is catalysed by phenylalanine hydroxylase (PAH),[12] hydroxymandelate synthase (HMS) and 4-hydroxyphenylpyruvate dioxygenase (4-HPPD).[13,14]

Chlorination, or more generally halogenation, of aliphatic carbon atoms (Scheme 4.1b) is a relatively newly discovered function of NIEs, encountered, for example, in the biosynthesis of the antifungal antibiotic syringomycin E.[15]

Oxygenation of aromatic rings without their cleavage can be achieved by NIEs in three different ways. First, a single hydroxyl group can be introduced into the ring (Scheme 4.1c), as exemplified by PAH. This type of reaction is a crucial step in the biosynthesis of neurotransmitters such as dopamine, adrenaline, noradrenaline and serotonin. Second, the aromatic ring can be *cis*-dihydroxylated in a single enzymatic step (Scheme 4.1d), a reaction performed by naphthalene dioxygenase (NDO).[16] This is the first step in biodegradation of naphthalene. Finally, hydroxylation of the aromatic ring can be coupled to a 1,2-shift of an alkyl substituent (Scheme 4.1e), which is a unique reaction catalysed by 4-HPPD, an enzyme involved in tyrosine catabolism.[14]

Besides oxygenation reactions, which involve incorporation of an oxygen atom into the substrate, NIEs catalyse several biologically important oxidation processes. For example, inversion of configuration at the chiral centre (Scheme 4.1g) is coupled to desaturation (Scheme 4.1h) in the enzymatic reaction of carbapenem synthase (CarC),[17] an ultimate enzyme of the biosynthetic route leading to the simplest carbapenem antibiotic. Oxidation of a tyrosine residue to a tyrosyl radical (Scheme 4.1i) is one of the initial steps in the catalytic cycle of ribonucleotide reductase (RNR),[18] an enzyme that synthesizes deoxyribonucleotides. Another interesting example of an oxidation reaction is formation of ethylene – the signalling molecule in plants (Scheme 4.1j), which is catalysed by 1-aminocyclopropane-1-carboxylic acid oxidase (ACCO).[19]

Oxygenation reactions coupled to C–C bond cleavage constitute another group of reactions performed by NIEs (Scheme 4.1k–n). Splitting long polyene chains is, for example, (Scheme 4.1k) an ultimate step in the synthesis of vitamin A,[20] opening of the homogentisate ring (Scheme 4.1n) is a step in tyrosine catabolism,[21] whereas *intra-* (Scheme 4.1l) or *extradiol* (Scheme 4.1m) type cleavages of catechol rings are pivotal processes in degradation of aromatic compounds (bioremediation).[22]

The most spectacular processes catalysed by NIEs are ring closure reactions (Scheme 4.1o and p), which are important steps in the biosynthesis of antibiotics. The closure of the oxazolidine ring (Scheme 4.1o) is catalysed by CAS,[10] whereas the β-lactam and thiazolidine rings are formed in a single enzymatic reaction catalysed by isopenicillin *N* synthase (IPNS).[23] These are two- and four-electron oxidation reactions, respectively, taking place without incorporation of any oxygen atom into the organic substrate.

It is important to note here that the survey of the reactions catalysed by NIEs presented in Scheme 4.1 is by no means exhaustive. We have chosen only these types of the reactions for which we performed computational studies. Nevertheless, the selection is rather representative, as an interested reader can check against many interesting published reviews.[1,2,4,5,24,25]

4.1.2 Iron Binding Sites

Iron cofactors of NIEs are composed of one or two Fe(II) or Fe(III) ions coordinated by oxygen and nitrogen-based ligands such as side chains of aspartate, glutamate, tyrosine and/or histidine.[2] Since such a composition of the first coordination sphere amino acids produces a weak ligand field, the ground state electronic configuration of the Fe ions becomes high-spin, which for the mononuclear cofactors leads to high-spin states, and for the coupled iron dimers to ferro- or antiferro-magnetically coupled spin states. The metal cofactors of NIEs have a remarkable degree of similarity (Figure 4.1). At a first glance it is apparent from Figure 4.1 that the key feature distinguishing NIEs from heme enzymes is the number of coordination sites available for binding of external ligands.

Thus, at least two coordination sites of the structures shown in Figure 4.1 are not occupied by proteinogenous groups, and in the resting state conformations these are either filled with solvent derived ligands (H_2O or OH^-) or left vacant. This feature has mechanistic relevance, as in many cases besides dioxygen also the organic substrate or co-substrate needs to bind to the metal during the catalytic cycle.

Within the mononuclear NIEs there exists a group of apparently unrelated enzymes featuring a common cofactor structure, with the so-called 2-His-1-carboxylate facial triad (Figure 4.1a).[26,27] More specifically, in this particular cofactor one face of the Fe(II) coordination octahedron is occupied by two imidazole groups of histidine and a single carboxylic acid group of an aspartate or glutamate. Notably, the other face of the octahedron is available for binding of reactants. So far, four families and two individual NIEs enzymes

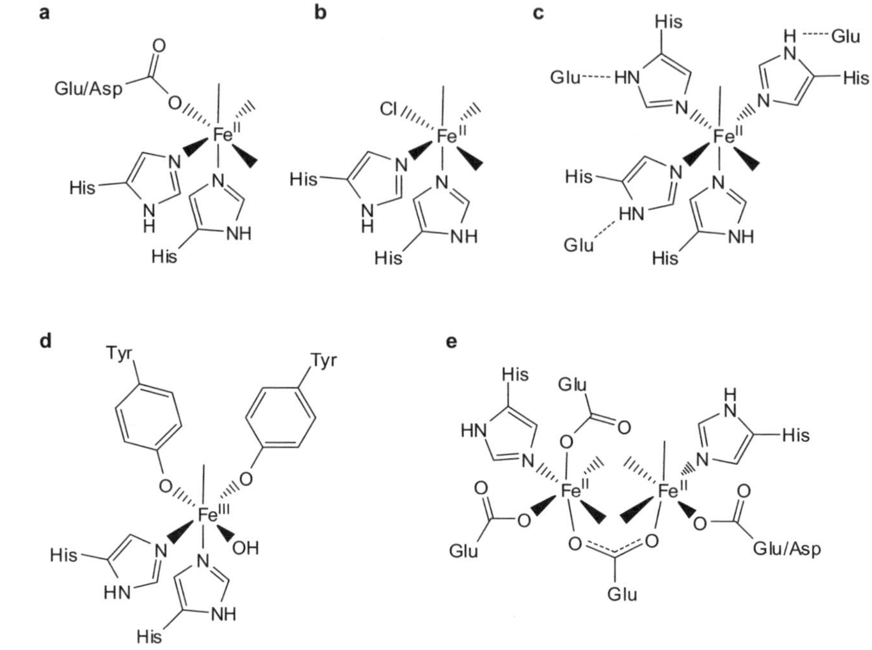

Figure 4.1 Composition of the first coordination shell of cofactors in non-heme iron enzymes.

containing this type of iron cofactor have been identified, namely, the α-ketoacid dependent enzymes (αKDE),[28] pterin dependent hydroxylases (PDH), extradiol dioxygenases and Rieske dioxygenases and ACCO and IPNS. The first two families utilize a cofactor/cosubstrate (α-ketoacid or pterin), which provides two electrons during the activation of dioxygen. The oxoferryl species ($Fe^{IV}=O$), which was observed with spectroscopic methods for αKDE,[29–31] is responsible for various chemical transformations, such as those presented in Scheme 4.1(a), (c), (e), (f), (g), (h), (j) and (o). With respect to the metal coordination, α-ketoacid binds to Fe(II) as a bidentate ligand, whereas pterin is placed in the vicinity of the metal cofactor. ACCO and IPNS that catalyse the transformations depicted in Scheme 4.1(j) and (p), respectively, are similar to these two families in this respect, whereby they also produce an oxoferryl species during their catalytic cycle. ACC binds to Fe(II) as a bidentate ligand, whereas the IPNS substrate provides one iron ligand that is bound *via* the thiol group. The other two families with a 2-His-1-carboxylate motif are Rieske dioxygenases and extradiol dioxygenases, both of which are isolated from soil bacteria degrading aromatic compounds. Rieske dioxygenases,[32] exemplified by NDO, initiate the catabolism of aromatic compounds by transforming them into *cis*-dihydrodiols (Scheme 4.1d), and the two electrons required to complete the reaction between dioxygen and the aromatic compound are provided by the external reductant. Since naphthalene lacks functional groups capable of Fe binding, it is located in the active site in the proximity of the iron cofactor. In the case of extradiol dioxygenases,[33] which catalyse the transformation depicted in Scheme 4.1(m), the organic substrate, *i.e.* catechol, binds to Fe(II) as a bidentate ligand and the third coordination site is available for O_2.

A ligand arrangement very similar to the 2-His-1-carboxylate motif was found in α-ketoglutarate dependent halogenases (αKDH), where the place of glutamate or aspartate is taken by the chloro ligand (Figure 4.1b).[34] These enzymes catalyse halogenation of aliphatic groups (Scheme 4.1b) and, similarly to other αKDE, they bind α-ketoglutarate to Fe^{II} and produce the highly reactive oxoferryl species as a key intermediate.

A somewhat unusual cofactor (Figure 4.1c) was found for apocarotenoid oxygenase (ACO),[35] which is a representative enzyme cleaving long polyene chains. In this case there is no negatively charged proteinogenous ligand in the first coordination sphere, and four coordination sites are occupied by histidine amino acids. However, in the second shell there are three glutamate amino acids that form hydrogen bonds with the first shell histidine residues. This class of enzymes cleave the polyene chain into two aldehydes (Scheme 4.1k) and, as in NDO, the hydrophobic substrate is located very close to the iron complex.

In the Fe(III)–binding active site of intradiol dioxygenases (Figure 4.1d) two coordination sites are occupied by tyrosyl anions and two by imidazole groups of histidine groups.[36] The interesting feature of this cofactor is that binding of the substrate, *i.e.* catechol, is coupled to dissociation of one of the tyrosinate groups from the metal. This dissociating tyrosinate residue acts together with the solvent derived OH ligand as a proton acceptor and deprotonates the substrate, and, thus, the dianionic form of the catechol binds as a bidentate ligand to Fe(III). The sixth coordination site of the metal is available for binding of a reduced form of O_2, which will elicit the intradiol ring cleavage reaction (Scheme 4.1l).

In the binuclear NIEs (Figure 4.1e) each ferrous ion is coordinated by one side chain of histidine and two or three oxygen atoms from the carboxylic acid groups of glutamates or aspartates.[37,38] The two shared coordination sites located between the Fe(II) ions are used to bind and cleave dioxygen, and the resulting oxoferryl species performs either methane hydroxylation (Scheme 4.1a) or tyrosine oxidation (Scheme 4.1i).

4.2 Computational Methods

The reaction pathways were investigated by means of hybrid DFT with the B3LYP exchange-correlation functional.[39–41] Two quantum chemical programs were used, *i.e.* Jaguar and Gaussian.[42,43] Active site models were constructed employing available experimental structures of the enzymes under investigation. Amino acids from the metal first and second coordination sphere were included in the form of suitable smaller molecules, and, for example, histidine was replaced with imidazole or 4-methylimidazole, and aspartate and glutamate with acetate, formate or propionate. The rigidity of the protein matrix was usually incorporated into the model by freezing coordinates of some terminal atoms according to the available X-ray data. This procedure is generally very useful,[44] and especially when the model contains molecules that are not anchored in the first coordination sphere of the metal. Once the model had been set up, the potential energy surface(s) (PES), *i.e.* energy of the system as a function of nuclear coordinates, was explored to examine different *a priori* plausible reaction mechanisms. As a result, transition states (TSs) and intermediates lying along different reaction paths were localized, and based on their energies the most likely mechanism(s) was identified.

Molecular Hessians were computed for TSs to confirm that the structure has one imaginary frequency corresponding to the normal mode associated with the reaction coordinate. Standard double zeta basis sets were employed for geometry optimizations and for Hessian calculations. However, the final energies of the optimized structures were calculated with larger basis sets of triple zeta quality and including polarization functions. Solvent effects due to the remaining part of the protein were computed with the Poisson–Boltzmann model where the solvent is a macroscopic continuum with the dielectric constant $\varepsilon = 4$.[45,46] Computation of solvent effects was performed with the same basis set as for the geometry optimization.

4.3   Dioxygen Binding and Generation of Peroxo Intermediates

Molecular dioxygen in its triplet ground electronic state is relatively inert with respect to organic compounds for two reasons. First, O_2 is a weak one electron oxidant and, as a result, the spin allowed reactions $R–H + O_2 \rightarrow R^{\bullet} + HO_2^{\bullet}$ or $R + O_2 \rightarrow R^{+\bullet} + O_2^{-\bullet}$ are very endothermic and thus unlikely at room temperature. Second, even though the two electron reduction of O_2 is thermodynamically favourable, such a reaction, leading to the peroxo species: $^1\{R–H\} + {}^3O_2 \rightarrow {}^1\{R–OOH\}$, is spin-forbidden because the spin state changes from triplet for the reactants to singlet for the product. One of the roles of the iron cofactors in NIEs is to facilitate the two-electron reduction of O_2, which is achieved by providing an easily accessible electron for the initial reduction of O_2 to $O_2^{-\bullet}$ and/or allowing for an efficient change of the spin state, *i.e.* an intersystem crossing. It seems appropriate to note here that in our computational studies we have never observed a single Fe(II) ion acting as a two-electron donor during O_2 binding leading to a peroxo species, and, thus, we consider the $Fe^{IV}–OO–R$ intermediate to be very unlikely. The detailed mechanisms for O_2 binding and formation of the peroxo species are dependent on the (co-)substrate, and they are described in the following subsections.

4.3.1 O_2 Binding with Oxidation of Fe(II)

For the enzymes whose substrates or co-substrates are difficult to oxidize, the first step of the enzymatic reaction, taking place after substrate binding, is one-electron reduction of O_2 (Scheme 4.2a–c).

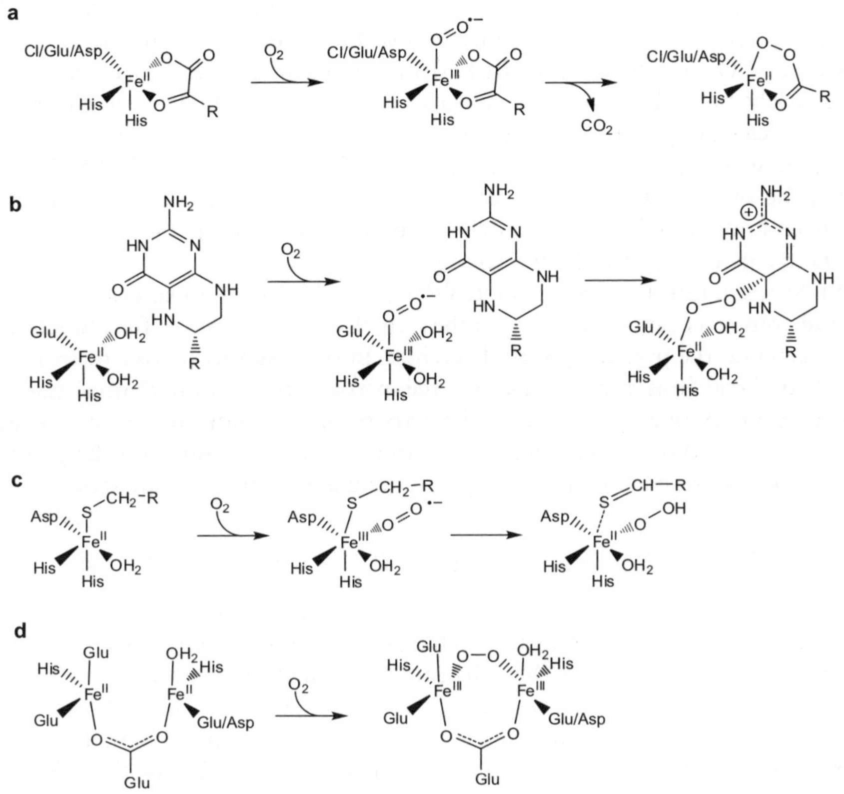

Scheme 4.2 Mechanisms for O_2 binding and formation of peroxo species – first electron provided by Fe(II).

The resulting intermediate features the superoxo radical anion coordinated to Fe(III), as was found for αKDE (Scheme 4.2a),[47] PDH (Scheme 4.2b)[48] and IPNS (Scheme 4.2c).[49,50] The ground electronic state of this species is usually a spin septet with the spin of the unpaired electron of the superoxide arranged parallel to those of the high-spin ferric ion. Changing the relative orientation of spins on Fe(III) and $O_2^{-\bullet}$, from parallel to antiparallel, leads to the quintet spin state, which lies 3–6 kcal mol^{-1} above the ground state. Such septet to quintet intersystem crossing is certainly facilitated by a high value of the spin–orbit coupling constant of iron, and the fact that a minimum energy crossing point is located very near the quintet reactant, both in terms of energy and geometry.[51]

The septet → quintet intersystem crossing is of mechanistic relevance as the next step, which leads to the peroxo species, has the lowest activation barrier on the quintet PES. The gain in activation energy more than compensates the energy required for the septet to quintet excitation. The preference for the quintet spin state is easily understood if one knows that the quintet (high-spin Fe^{II}) is the ground state for the peroxo species, whereas singlet, triplet and septet states lie at considerably higher energies.

Going into slightly more detail, in αKDE (Scheme 4.2a) an attack of the superoxo group on the carbonyl carbon of the ketoacid is concerted with the release of CO_2, which leads to the Fe(II)-peroxo acid species.[45,52] In the case of PDH (Scheme 4.2b), the two electrons necessary for reduction of the Fe(III)–$O_2^{-\bullet}$ species are provided by the pterin, which is transformed into an organic peroxide with a positive charge in the ring.[46] In the reaction of IPNS (Scheme 4.2c), the two electrons come from the thiol group of the substrate, which is oxidized to thioaldehyde with a delivery of a proton from the methylene group to the oxygen atom distal with respect to the iron.[48]

Finally, the diiron cofactors of RNR and sMMO are special cases, since the two ferrous ions can easily provide the two electrons necessary for reduction of dioxygen into the peroxo species. Therefore, in the resulting peroxo intermediate the two ferric ions are bridged by the peroxo group bound in a *cis*-μ-1,2 coordination geometry.[53] Notably, the peroxo species is not capable of activating a C–H bond[54] and, consequently, in enzymes with such a function the peroxide must be transformed into an oxoferryl intermediate (as discussed in Section 4.4.1).

4.3.2 O_2 Binding with Oxidation of the Organic Substrate

When a substrate is a good one-electron donor, because the resulting radical is stabilized by resonance, for example, the binding of dioxygen is then coupled to a one-electron oxidation of the substrate (Scheme 4.3).

For the Fe(II)-dependent enzymes (Scheme 4.3a, c, d) the ground state of the ternary enzyme (Fe)–O_2–substrate complex is usually a spin septet, as with the mononuclear NIEs from Scheme 4.2. However, the coupling of the spins now extends to three centres: high-spin Fe(II), superoxide anion and the organic radical. In the septet ground state the spins of all unpaired electrons are aligned parallel, whereas in the reactive quintet state, which is found at the most 5.5 kcal mol^{-1} higher in energy, the spins of the unpaired electrons on $O_2^{-\bullet}$ and the organic radical are opposite. Such an arrangement of spins is best suited for the next reaction step where the superoxide and the organic radical recombine to form organic peroxide interacting with a high-spin Fe(II).

For extradiol dioxygenases (Scheme 4.3a) and homogentisate dioxygenase (HGD, Scheme 4.3c), the attack on the ring and formation of the peroxo bridge triggers proton transfer to the peroxide oxygen that is proximal to iron.[55–57] In both cases it is the second shell histidine amino acid that plays the role of proton donor or mediator. Interestingly, also in the case of Fe(III)-dependent intradiol dioxygenases (Scheme 4.3d) O_2 binds to iron in the first place.[58] It is possible because when O_2 binds it oxidizes catecholate to a semiquinone radical anion, and as a result $O_2^{-\bullet}$ binds to Fe(III). The ground state of this adduct is spin quartet where the spins of the unpaired electrons on the high-spin Fe(III) centre are antiparallel to those on the superoxo and semiquinone radicals. Changing the spin directions on the latter two radicals, from the same to opposite, leads to a sextet excited state, which lies only 7 kcal mol^{-1} higher in energy. Importantly, in this sextet state the two spins on the radicals are arranged appropriately for developing an O–C bond of the peroxo intermediate, whose ground state is also a sextet (high-spin Fe^{III}).

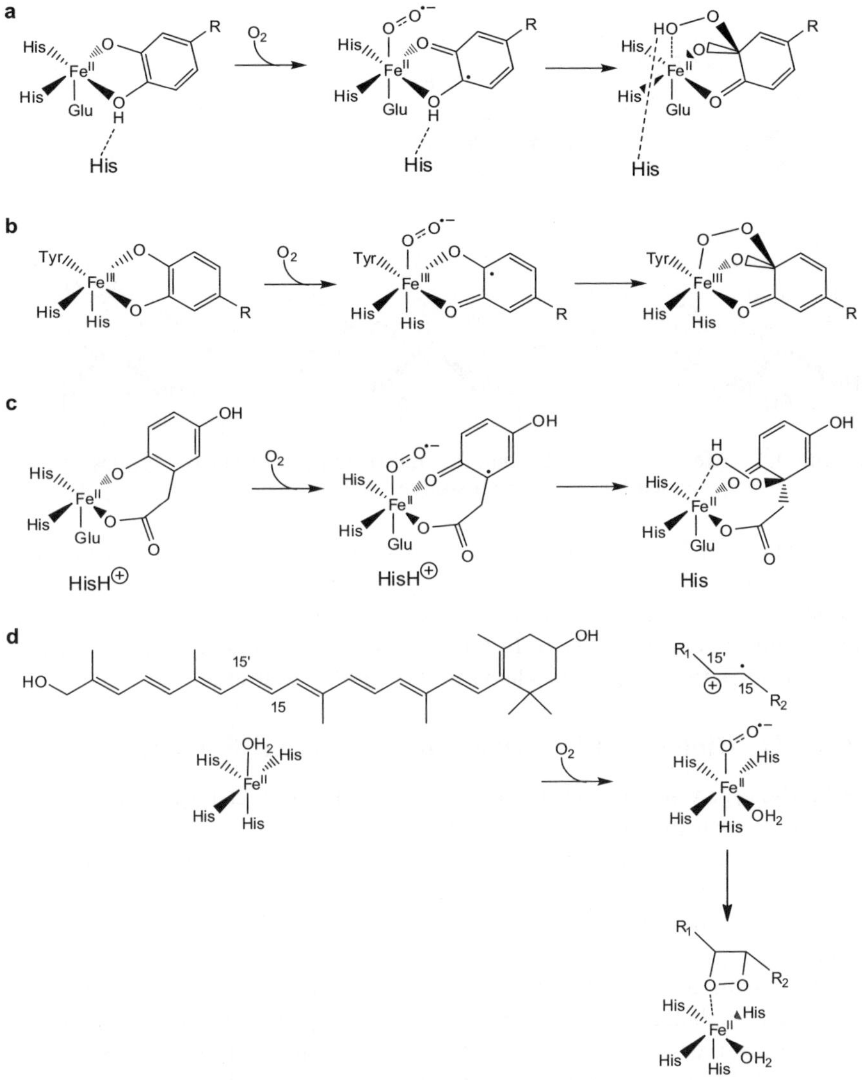

Scheme 4.3 Mechanisms for O_2 binding and formation of peroxo species – first electron provided by the organic substrate.

The reaction of ACO (Scheme 4.3d) differs slightly, as the recombination of the two radicals does not lead to a peroxo bridge between iron and the organic substrate, but instead a dioxetane ring is formed, which only weakly interacts with the ferrous ion.[59] This different reactivity has its origin in the fact that once the first C–O bond is established the adjacent carbon of the substrate has a carbocation character, and, thus, it easily forms a bond with the second peroxo atom. On the other hand, in the reactions of dioxygenases cleaving aromatic

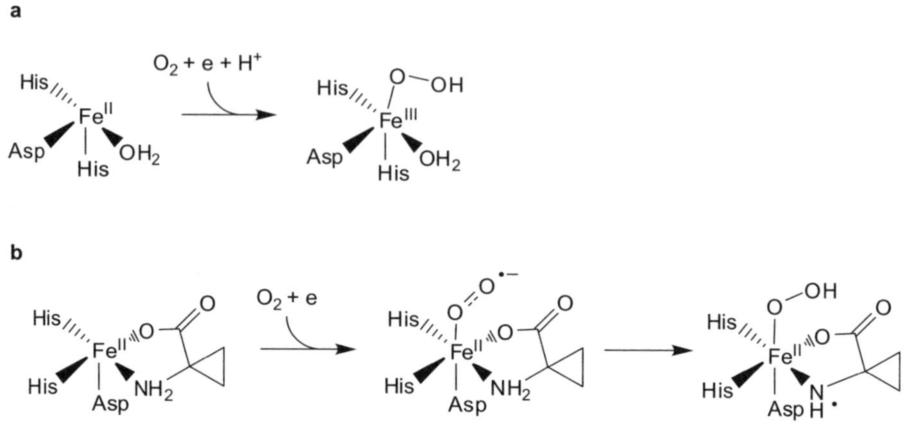

a

b

Scheme 4.4 Mechanisms for O_2 binding and formation of peroxo species – electron provided by the external source.

rings (Scheme 4.3a–c), formation of a dioxetane-like intermediate requires an attack of the peroxo atom on the carbonyl group which is conjugated with two double bonds in the ring. For this reason, formation of a dioxetane-like intermediate is energetically unfavourable, and such a species does not lie on the reaction paths of the ring cleaving dioxygenases studied.

4.3.3 O_2 Binding with Oxidation of External Reductants

Rieske dioxygenases, represented by NDO, and ACCO require one external electron to form the peroxo species (Scheme 4.4).

The requirement of the external electron most probably comes from the fact that the substrates, *i.e.* naphthalene and ACC, cannot be oxidized by Fe^{III}–$O_2^{-\bullet}$ and instead require more powerful oxidants such as Fe^{III}–OOH and Fe^{II}–$O_2^{-\bullet}$, respectively.

For the iron cofactor of NDO the electron is provided by the Rieske iron-sulfur cluster placed in the protein structure very near the catalytic centre,[60] whereas in the reaction of ACCO the role of an external reductant is played by ascorbate.[61] In both cases the resulting intermediates in their electronic ground states possess iron ions in high-spin configurations, which means a sextet state for the Fe^{III}–OOH complex in NDO and Fe^{II}–$O_2^{-\bullet}$ in ACCO. In the latter system, four unpaired electrons of iron couple ferromagnetically with the unpaired electron of the superoxide. Fe^{II}–$O_2^{-\bullet}$ is an oxidant powerful enough to abstract a hydrogen atom from the amino group of ACC (Scheme 4.4b), which is the ultimate step leading to the peroxo intermediate.

4.4 Strategies for O–O Bond Cleavage

From the preceding subsections it follows that the peroxo species are formed in the catalytic cycles of all NIEs. This is a manifestation of the chemical preference of dioxygen to be fully reduced in two two-electron steps and, accordingly, it suggests that the next reaction step will be the final two-electron reduction of the peroxo intermediates yielding a species with fully reduced [O(2–)] oxygen. Indeed, this is what was found for the NIEs studied, yet, the O–O bond cleavage can proceed according to three different scenarios: (i) heterolytic cleavage, leading to an oxoferryl species (Fe^{IV}=O), which is most common, (ii) homolytic cleavage affording alkoxyl radical intermediates (R–O$^{\bullet}$) and (iii) a single step heterolysis of O–O coupled with a two-electron oxidation of the organic substrate. These three general mechanisms are presented in the following three subsections.

4.4.1 Heterolytic O–O Bond Cleavage Leading to Fe(IV)=O

Many classes of NIEs produce reactive oxoferryl species that are subsequently used to oxidize chemically inert substrates. In most such systems, the O–O bond cleavage leading to Fe(IV)=O is a heterolytic process, which implies that both oxygen atoms of the peroxo group are reduced to the O^{2-} state. The two electrons required for this reaction are provided by the Fe ion(s), which is oxidized to the ferryl state (Fe^{IV}, Scheme 4.5), and the process can usually be divided into two steps: the first and more difficult step is connected with the stretch of the O–O bond to a distance of ~ 2.10 Å and transfer of the first electron to the O–O moiety. The transfer of the second electron follows and it is an easy process connected with the development of a second covalent bond by the distal oxygen.

For mononuclear NIEs with a peroxo structure featuring Fe^{II} (Scheme 4.5a–d) the O–O bond cleavage requires overcoming a barrier of 2–10 kcal mol^{-1}. In αKDE (Scheme 4.5a) the cleavage of the O–O bond in the peroxo acid/high-spin Fe^{II} complex starts with delivery of the β electron from Fe^{II} to the elongating peroxo bond, and finishes with a transfer of the second, in this case α, electron from iron to the peroxide.[45,50] During this second step the oxygen atom distal with respect to the metal develops a second covalent bond with the adjacent carbon, *i.e.* it forms a carboxylic acid group. The resulting oxoferryl intermediate has a high-spin (quintet) ground state, which was determined with computational and experimental methods.[29,45] Importantly, a comparative QM–QM/MM study demonstrated the relative energies of spin states for the oxoferryl intermediate are insensitive to presence of the protein surrounding of the active site,[62] and thus the description of the reaction path obtained with active site QM-only models is most likely correct.

A very similar reaction scenario was found for PDE,[46] IPNS[48] and ACCO (Scheme 4.5b–d),[58] with the minor difference that the distal oxygen develops a bond with a proton delivered by an aqua ligand (PAH, Scheme 4.5b), an amido group of the substrate (IPNS, Scheme 4.5c) or bicarbonate present in the active site (ACCO, Scheme 4.5d).

Concerning the structure of the oxoferryl species, α-ketoglutarate dependent halogenases (Scheme 4.5a Cl-form) are an interesting case because two distinctive forms of the oxoferryl intermediate were detected.[63] The identity of the

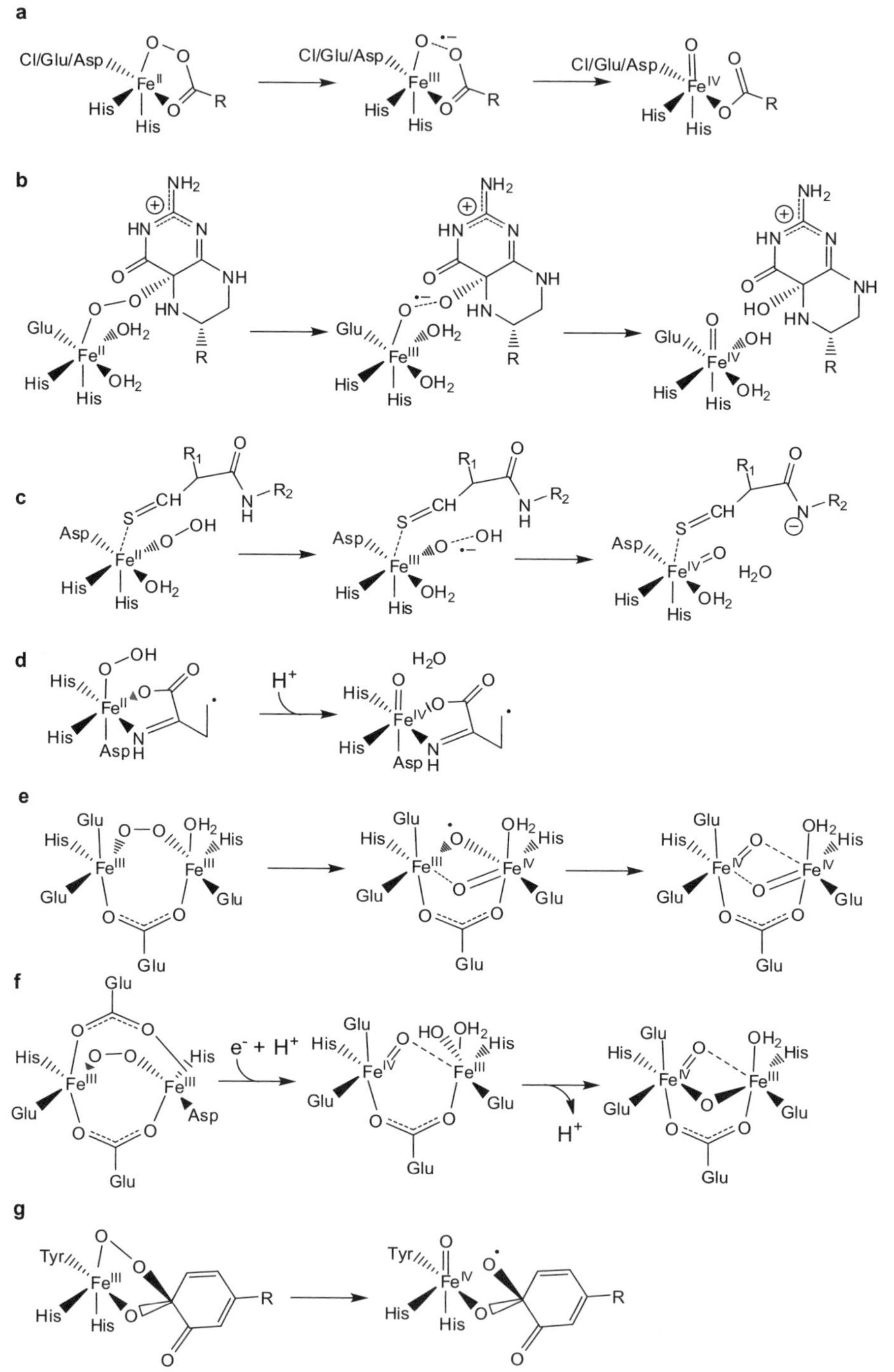

Scheme 4.5 Mechanisms for O–O bond splitting – (heterolytic) cleavage leading to Fe(IV)=O.

two species was successfully elucidated by means of DFT calculations.[64] More specifically, from the computational results it follows that the two forms of the complex differ in the positions of the oxo and chloro ligands. The two forms obtained by interchanging the positions of the chloro and oxo ligands are almost isoenergetic, and can transform into one another with a barrier significantly lower than that for the C–H cleavage step, and have distinctive Mössbauer spectra. Notably, these three features are consistent with the experimental data available for the oxoferryl intermediate of αKDH.[65]

The two diiron enzymes studied, sMMO (Scheme 4.5e) and RNR (Scheme 4.5f), have slightly different O–O cleavage mechanisms.[66] In sMMO (Scheme 4.5e) the two electrons required in the heterolytic process are provided by two ferric ions, which yields the di(oxoferryl) intermediate capable of hydroxylating methane. On the other hand, in RNR there is a tryptophan residue placed in the vicinity of the iron cofactor, and it is a source of an electron delivered during the O–O cleavage step. This electron transfer reduces the barrier for the cleavage of the peroxide bond, so that the reaction rate is two orders of magnitude faster than in sMMO. The resulting reactive intermediate is a mixed-valence diiron species.

The peroxo intermediate formed during the catalytic reaction of intradiol dioxygenases may undergo a homolytic cleavage leading to the oxoferryl/R–O$^{\bullet}$ species (Scheme 4.5g) (T. Borowski *et al.*, unpublished results). This is an unusual reaction for two reasons. First, homolysis of the O–O bond is usually catalysed by Fe^{II} (*vide infra*) and not by Fe^{III} as in this case. Second, the oxoferryl species produced in the reactions depicted in Scheme 4.5(a)–(f) are energetically more stable, or only by a few kcal mol^{-1} less stable, than the preceding peroxo species, whereas the Fe^{IV}=O intermediate of intradiol dioxygenases was computed to lie 18.5 kcal mol^{-1} above the peroxo intermediate. The substantial endothermicity of this reaction is compensated only in the following step, when the oxyl group is incorporated into the ring and the radical becomes stabilized by the resonance. Finally, interestingly, such cleavage of the peroxo intermediate of the intradiol dioxygenase takes place on the quartet PES, which means that in the reactant there is one β electron on the Fe 3d orbitals, in analogy to high-spin ferrous peroxo intermediates in the reactions presented in Scheme 4.5(a)–(d).

4.4.2 Homolytic O–O Bond Cleavage Leading to R–O$^{\bullet}$

Several classes of NIEs catalyse C–C bond cleavage reactions, and in their catalytic cycles alkoxyl radical (R–O$^{\bullet}$) intermediates are usually encountered as a reactive species ultimately responsible for splitting a ring or a chain (Scheme 4.6).

In all cases studied, one peroxo oxygen atom is in contact with the iron ion, and as the O–O bond cleaves this oxygen develops a bond with the ferric ion. One electron is necessary for the homolysis, as it reduces the iron-bound oxygen to the O^{2-} redox state, and this electron is provided either by the ferrous ion (Scheme 4.6a–c) or by the tyrosinate, which is oxidized to tyrosyl radical (Scheme 4.6d).

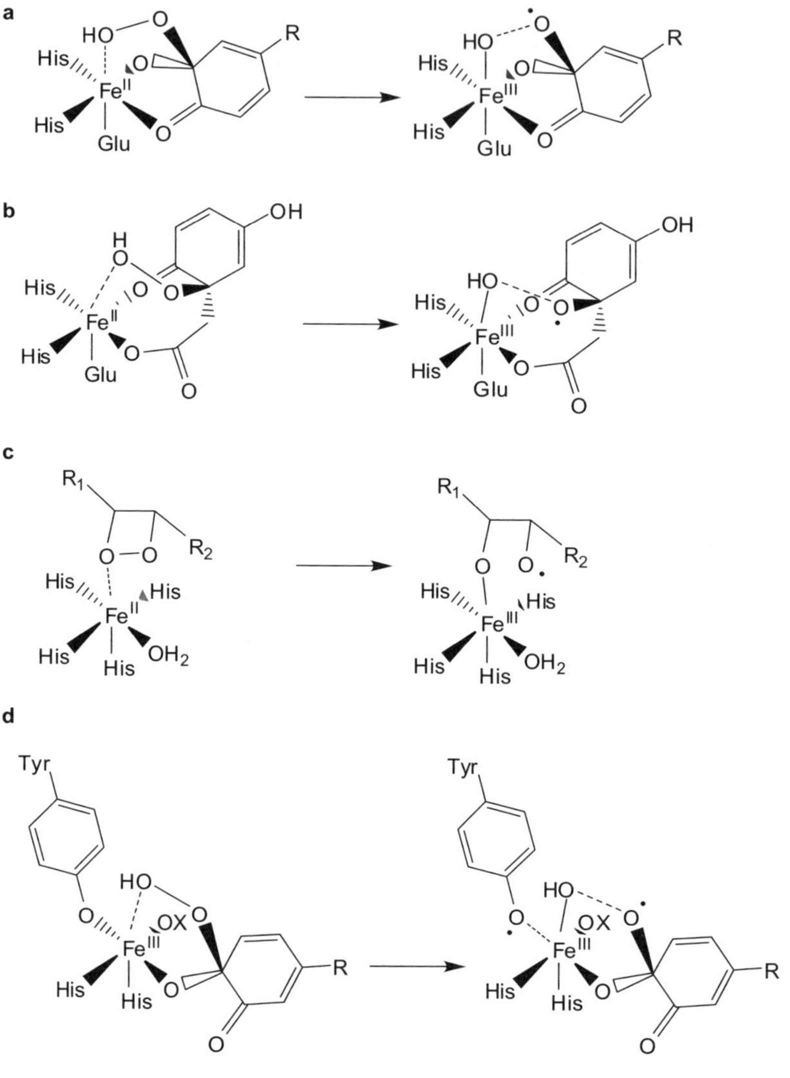

Scheme 4.6   Mechanisms for O–O bond splitting – homolytic cleavage leading to R–O$^•$.

The peroxo intermediates of extradiol dioxygenases (Scheme 4.6a) and HGD (Scheme 4.6b) feature a protonated oxygen proximal with respect to the high-spin ferrous ion.[52–54] The splitting of the O–O bond is concomitant with the transfer of the β electron to the OH and, thus, the unpaired electron of the oxyl radical also has β spin. However, in the Fe^{III}–OH/R–O$^•$ intermediate the O–OH bond is not fully cleaved, as can be recognized from the O---O distance of around 2.2 Å, and its NBO bond order of 0.18, both of which indicate a presence of a partial bond. The orientation of this partial O–O bond with respect to the ring determines the site of the attack in the following step (*vide infra*).

The cleavage of the O–O bond in the dioxetane intermediate of ACO (Scheme 4.6c) is a similar process with the exception that the O–O bond is cleaved completely, probably due to the strain of the four-membered ring.[56]

As was already mentioned, the active site in intradiol dioxygenases has a peculiar feature as one of the protein ligands, tyrosine, dissociates from the Fe(III) centre during substrate binding.[34] The coordination site from which this tyrosine dissociated becomes again available once the peroxo intermediate is formed and it attains a reactive conformation. Thus, is was suggested that at this stage of the catalytic cycle the tyrosine can return to the coordination sphere of iron, where it binds to Fe(III) as a tyrosinate (XO in Scheme 4.6d) and its proton is bound to the peroxo group.[55] Then, the cleavage of the O–OH group proceeds similarly as in the enzymes with Fe^{II} (Scheme 4.6a–c) with the exception that the β electron is taken from the tyrosinate ligand. The barrier heights calculated for this step range from 5 kcal mol^{-1} for ACO (Scheme 4.6c) to 13 kcal mol^{-1} for intradiol dioxygenases (Scheme 4.6d).

4.4.3　Heterolytic O–O Bond Cleavage in Fe(III)–OOH

In the case of two NIEs, namely, NDO (representative Rieske dioxygenase) and intradiol dioxygenases, it was found that the O–O bond can be cleaved heterolytically without the redox involvement of iron, as the two electrons necessary for reduction of peroxide are provided by the substrate (Scheme 4.7).

The side-on bound hydroperoxo group of NDO attacks the double bond of naphthalene with its deprotonated atom (Scheme 4.7a). The optimized TS

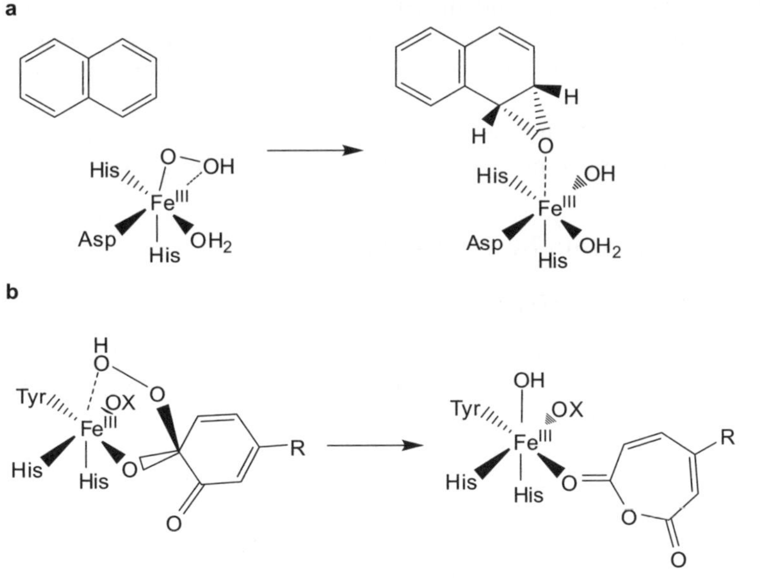

Scheme 4.7　Mechanisms for O–O bond splitting – heterolytic cleavage without iron participation.

indicates that this is a concerted step, where the O–OH bond is cleaved concomitantly with the formation of two C–O bonds, leading to the production of an epoxide.[67] The calculated activation energy of this step amounts to 17.5 kcal mol^{-1}. In the following steps, involving low barriers, the epoxide converts *via* cleavage of one of the C–O bonds into a carbocation, which subsequently reacts with the OH ligand to yield the final product – *cis*-diol.

The homolytic cleavage of the O–OH bond in intradiol dioxygenases, which was discussed in the preceding subsection, is characterized by an activation barrier similar to that found for a heterolytic process (Scheme 4.7b). This latter is an example of a Criegee rearrangement where the acyl group migrates to the peroxo atom distal with respect to the Fe(III) ion.[55] Such a mechanism was suggested in the literature, yet only when the computational study was performed did it became apparent that such a heterolytic reaction requires that the two bonds cleaved in this step, *i.e.* C–C and O–O, have to be placed in one plane. This coplanarity guarantees a proper overlap between the O–O σ* and the C–C σ orbitals, which are the two major orbitals active in this heterolytic step. The subsequent step proceeds with a low activation barrier, and is a nucleophilic attack of the OH ligand on the muconic anhydride, which leads to the final product *via* opening of the seven-membered ring.

4.5 Reactions of the High-Valent Intermediates

The reactive oxoferryl species, usually afforded by a heterolytic O–O bond cleavage, has a mechanistic role to play, namely, a two-electron oxidation of the organic substrate. The detailed mechanism of this reaction depends on the identity of the substrate, and in general it can be either oxygenation of the organic compound, *i.e.* incorporation of an oxygen atom into the substrate, or a two-electron oxidation without oxygen transfer. Within the latter category a special case is ring closure reactions, which are encountered in the biosynthetic routes leading to β-lactam antibiotics. Several examples of oxygenation and oxidation reactions are presented in the following three subsections.

4.5.1 Oxygenation by Fe(IV)=O

4.5.1.1 *Aliphatic Hydroxylation*

Hydroxylation of aliphatic hydrocarbon groups is probably the most common reaction of NIEs with an oxoferryl species and, thus, it is also the reaction most intensely studied with theoretical methods.[68–70] The reaction follows the rebound paradigm formulated by Groves,[71] which means that the cleavage of the C–H bond and the C–OH bond formation take place in two consecutive steps (Scheme 4.8a).

In the mononuclear NIEs the reaction proceeds on the quintet PES, and in the first step the C–H bond is cleaved with a proton and an electron transferred to the oxo ligand and the iron ion, respectively. In the radical intermediate the ferrous cofactor assumes a high-spin configuration, where the unpaired

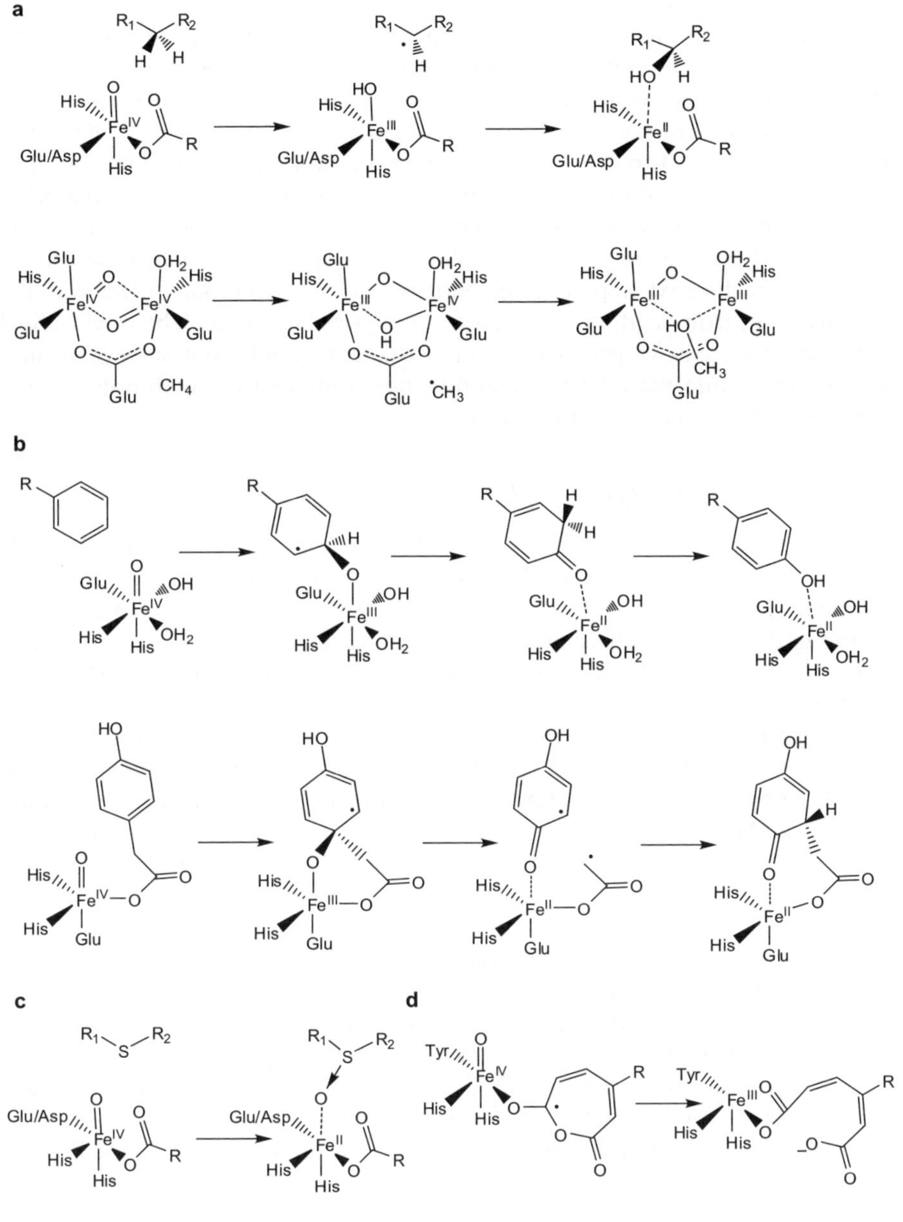

Scheme 4.8 Oxygenation reactions performed by Fe(IV)=O.

electron of the organic radical is antiferromagnetically coupled with those located on the iron. In the second step, which involves a significantly lower barrier than the C–H cleavage, the C–OH bond develops simultaneously with a transfer of the β electron from the carbon to iron, yielding a high-spin ferrous complex with a weakly bound alcohol product. Hydroxylation of methane performed by sMMO, which has been one of the subjects most studied by theory the past decade,[72–74] follows the same two-step mechanism, but the presence of two iron ions leads to several possible spin states and electromers. After the initial formation of symmetric peroxide, dioxygen is cleaved homolytically. Only one of the iron atoms is redox active in the symmetric diamond-shaped TS of this cleavage, changing the oxidation state from III to IV. The barrier is about 15 kcal mol^{-1}. The subsequent electron transfer from the other iron atom to oxygen is without barrier and the product has the oxidation state Fe$_2$(IV,IV) with antiferromagnetic coupling of the spins of the two irons. At an early stage of the next step, an electron is transferred from one of the bridging oxygens to an iron, leading to a Fe$_2$(III,IV)-oxyl radical state. The oxyl radical then abstracts the hydrogen atom from methane, with a total barrier of around 14 kcal mol^{-1}, and the rebound step follows spontaneously, with only a small energy barrier associated with this step.

4.5.1.2 Aromatic Hydroxylation

Oxygenation of aromatic rings follows a different mechanism because the C–H bond in the aromatic ring is difficult to cleave, while the π-electrons are easily available (Scheme 4.8b). Thus, in the first step the oxoferryl group attacks a carbon in the ring, which changes hybridization from sp^2 to sp^3 in a so-called σ-complex.[75] This first step is usually coupled with one-electron oxidation of the ring to a corresponding radical. The hydrogen bound to the attacked carbon can be easily dispatched in the second step, as the oxygen bound to this carbon forms a keto group. This step, known as a NIH-shift, involves migration of the hydrogen to the neighbouring carbon coupled with the second one-electron oxidation of the ring. The final step of the reaction is a keto–enol tautomerization, which can take place either in the active site or in solution, *i.e.* once the products have been released.

A different mechanism for a ring rearrangement was found for 4-HPPD, shown in the lower scheme of Scheme 4.8(b).[50] Thus, instead of a single step NIH-shift, the migration of the carboxymethyl substituent is proposed to follow a two-step radical mechanism. Accordingly, once the radical σ-complex is formed, the C–C bond cleaves homolytically and the ring is one-electron oxidized to a semiquinone radical. The unpaired electrons of the two organic radicals have opposite spins, which greatly facilitates recombination of the two, *i.e.* establishing the new C–C bond. The keto form of the product, thus formed, easily tautomerizes to the phenol, either in the active site or when released to solution.

4.5.1.3 Sulfoxidation

Mechanistically, oxygenation of thioethers to sulfoxides is the simplest reaction catalysed by the oxoferryl species (Scheme 4.8c). Since quintet spin is the ground state for both reactant and product, the two-electron oxidation of sulfur is a single step reaction on the quintet PES. The barrier calculated for such a reaction in the active site of 4-HPPD is 14 kcal mol^{-1}.[65]

4.5.1.4 Ring Oxygenation

The oxoferryl group which may be formed in the catalytic cycle of intradiol dioxygenases attacks the seven-membered ring of the muconic anhydride radical (Scheme 4.8d).[62] This is an ultimate step in the ring cleavage, as the product of this reaction is a Fe(III) complex with acyclic muconic acid. The reaction proceeds on the sextet PES, which implies that in the oxoferryl intermediate the unpaired electron of the organic radical is anti-ferromagnetically coupled with those on Fe(IV).

4.5.2 Oxidation by Fe(IV)=O

4.5.2.1 Generation of a Tyrosyl Radical

The reactive oxoferryl species in RNR, also known as compound X, is responsible for generation of a tyrosyl radical (Scheme 4.9a). This tyrosyl radical then communicates with another subunit of the protein to create a cysteinyl radical that is responsible for transformation of ribonucleotides into deoxyribonucleotides.

A contact between the iron cofactor and the tyrosine residue is made by a single water molecule, which forms hydrogen bonds with the first-shell

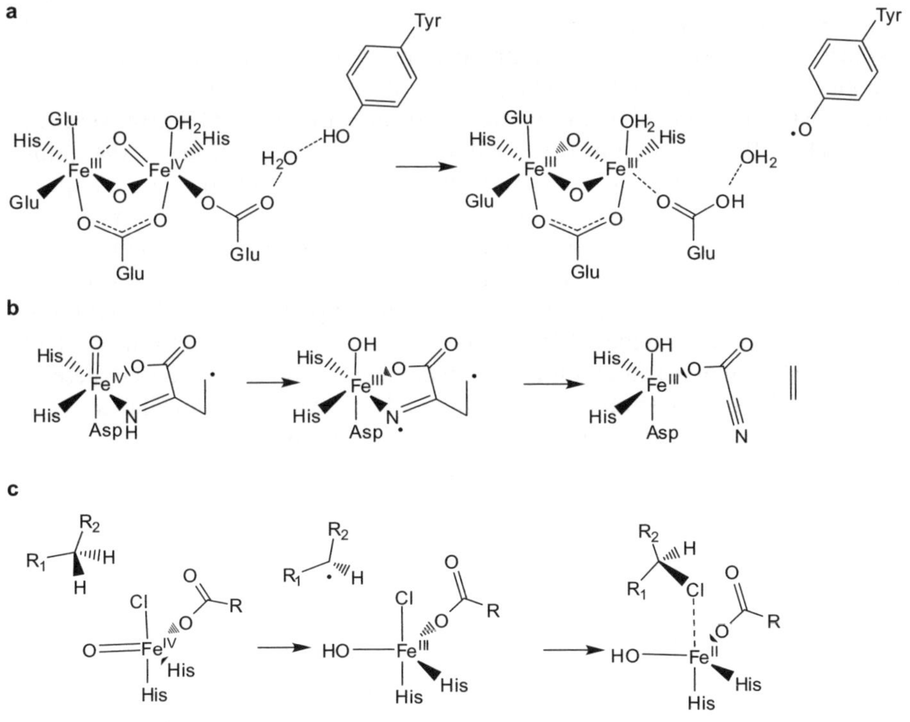

Scheme 4.9 Oxidation reactions performed by Fe(IV)=O.

glutamate and the phenolic group of tyrosine. This hydrogen bond bridge enables a proton transfer between the tyrosine and the glutamate, while an electron is accepted by the ferryl ion of compound X,[76] with a calculated barrier of 11 kcal mol^{-1}. As usual, in the transition state, neither the proton that is transferred nor the bridging water molecule has any noticeable spin density. This type of proton coupled electron transfer (PCET) is still termed hydrogen atom transfer (HAT) since the proton and electron have the same donor and go to the same acceptor complex.

4.5.2.2 Ethylene Formation

The role of the oxoferryl group produced in the catalytic reaction of ACCO is to oxidize the acyclic radical intermediate to ethylene and cyanoformate (Scheme 4.9b).[58] The electron transfer and subsequent barrierless cleavage of the C–C bond are triggered by a proton transfer from the imido nitrogen to the oxo ligand, which is mediated by two second shell residues: water and bicarbonate. As for the majority of mononuclear NIEs, this reaction proceeds on the PES where the iron ion is in a high-spin configuration. The calculated barrier connected with this process is very low, *ca.* 3 kcal mol^{-1}, which indicates an efficient reaction.

4.5.2.3 Chlorination by Cl-Fe(IV)=O

The oxidative halogenation of aliphatic groups, which is performed by the oxoferryl intermediate of αKDH, proceeds in an analogous way as the two-step rebound mechanism of hydroxylation, yet with the exception that it is the chloro ligand that is rebounded by the aliphatic radical, and not the OH group (Scheme 4.9c). Thus, the first step in the reaction is C–H bond cleavage on the quintet PES, leading to a high-spin Cl–Fe(III)–OH form of the cofactor and an aliphatic radical with β spin. In the subsequent step the chloro ligand, which is positioned substantially closer to the organic radical than the OH group, recombines with R$^{\bullet}$ forming the chlorinated product.[60] Importantly, the configuration of the oxoferryl form of the cofactor that guarantees the preferential chlorination, *i.e.* the one shown in Scheme 4.9(c), is not the one achieved directly in oxidative decarboxylation of α-ketoglutarate. Instead, a fast equilibrium between these two forms of Cl–Fe^{IV}=O establishes, and the C–H cleavage reaction has a slightly lower barrier for the configuration favouring chlorination.

Other authors proposed alternative mechanisms whereby they explain selective chlorination by αKDH, such as blocking the reactivity of the Fe^{III}–bound OH ligand by its reaction with CO_2 or a proton,[77,78] or coupling the C–H cleavage with the rebound step.[79]

4.5.2.4 Coupled Epimerization and Desaturation

Oxidation reactions performed by the oxoferryl species of NIEs involved in biosynthesis of antibiotics belong to the most intricate ones and they usually have no precedence in synthetic organic chemistry (Scheme 4.10).

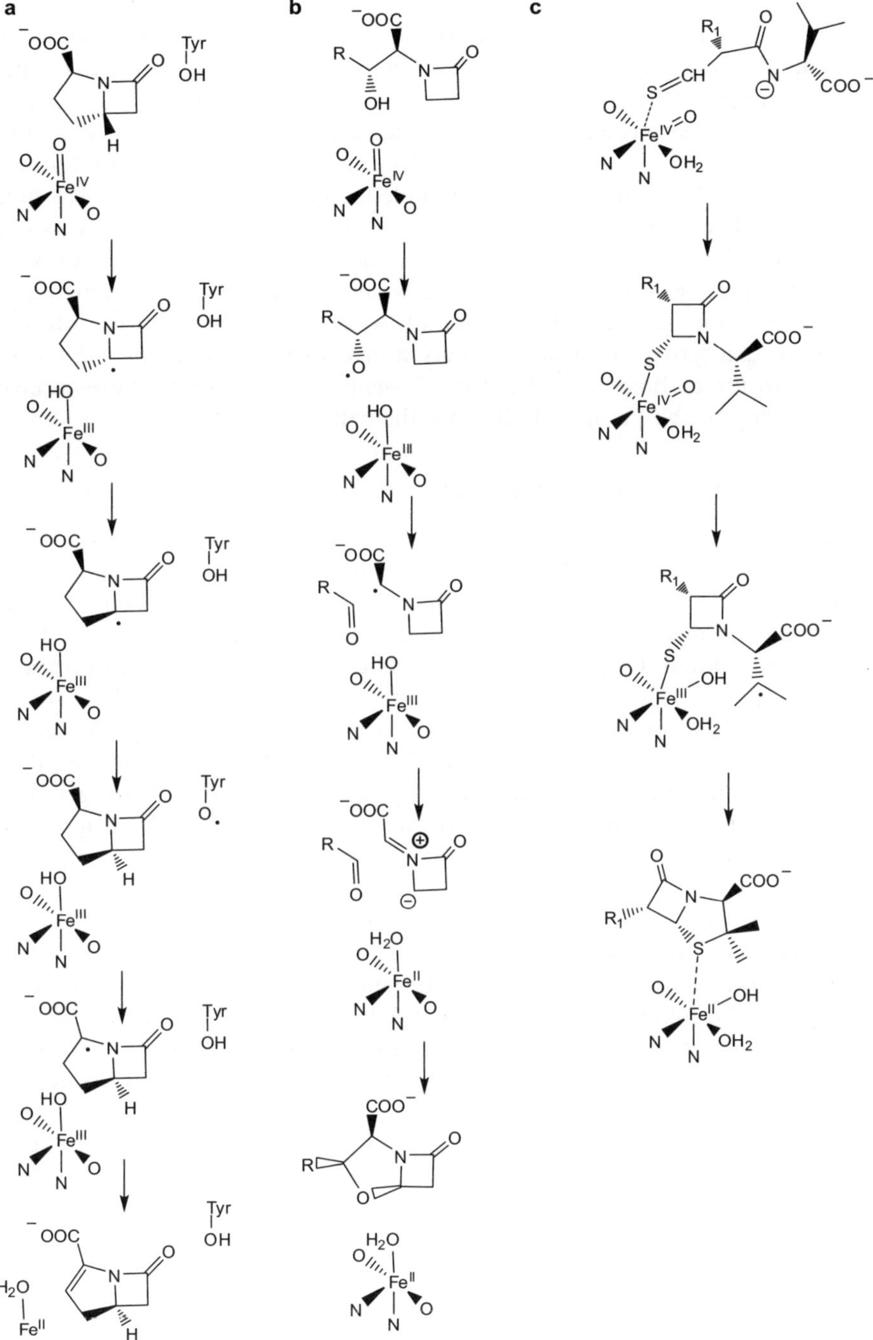

Scheme 4.10 Oxidation reactions performed by Fe(IV)=O – epimerization/desaturation and ring-closure reactions.

The reaction catalyzed by CarC in the biosynthesis of carbapenem antibiotics (Scheme 4.10a) involves inversion of the stereochemistry at a chiral carbon and subsequent desaturation of the five-membered ring. As found in the computational studies,[80,81] the two reactions are coupled by means of a reductant, most likely a tyrosine residue, which provides a hydrogen atom at the end of the epimerization and then initiates the desaturation by abstracting the first hydrogen from the ring. Thus, the coupled reactions start with C–H bond cleavage by Fe(IV)=O, which leads to the tertiary radical that inverts its configuration with a very low barrier. The tyrosine residue, located on the opposite face of the bicyclic substrate radical, provides a hydrogen atom reducing the ring to the epimer of the starting material. At this point the desaturation stage begins as the tyrosyl radical abstracts a hydrogen atom from the carbon adjacent to the carboxylic acid group. Transfer of the second hydrogen atom from the ring to the Fe(III)–OH finishes the catalytic reaction.

4.5.2.5 *Ring Closure Reactions*

In the catalytic reactions of CAS (Scheme 4.10b) and IPNS (Scheme 4.10c), the oxidative power of the oxoferryl species is harnessed to form strained bicyclic β-lactams, which are subsequently transformed into clinically useful antibiotics.

The reaction mechanism for cyclization by CAS, which was suggested based on a computational study, starts off with a hydrogen atom abstraction from the hydroxyl group.[82] This quite unusual reaction involves a reasonable barrier of 16 kcal mol^{-1} and produces an alkoxyl radical, which very easily decomposes into a more stable carbon-based radical and an aldehyde. In the following step, the β-lactam radical fragment is one-electron oxidized by hydrogen atom abstraction from the ring; a step that affords azomethine ylide. Finally, the ylide and aldehyde fragments undergo 1,3-dipolar cycloaddition to form the bicyclic product.

The two fused rings of isopenicillin-*N* are formed in the active site of IPNS in two consecutive steps of the same catalytic cycle.[47,48] Thus, the β-lactam ring forms when the deprotonated amido nitrogen, which delivered a proton to the leaving OH during the preceding O–O cleavage step, attacks the thioaldehyde carbon. Closure of the five-membered thiazolidine ring is initiated by Fe(IV)=O abstracting a hydrogen atom from the tertiary carbon of the substrate. The resulting aliphatic radical attacks the sulfur atom concertedly with an electron transfer to the metal cofactor. As a result, the ring is closed while the metal achieves its resting ferrous state.

4.5.3 Reactions of R-O·

A group of NIEs catalysing cleavage of the C–C bond in unsaturated compounds usually produces an alkoxyl radical intermediate, which is responsible for initiating the C–C bond breakage (Scheme 4.11).

In extradiol dioxygenases (Scheme 4.11a) the O-radical attacks the extradiol carbon of the ring, forming in this way a radical arene oxide.[52,53] Rupture of

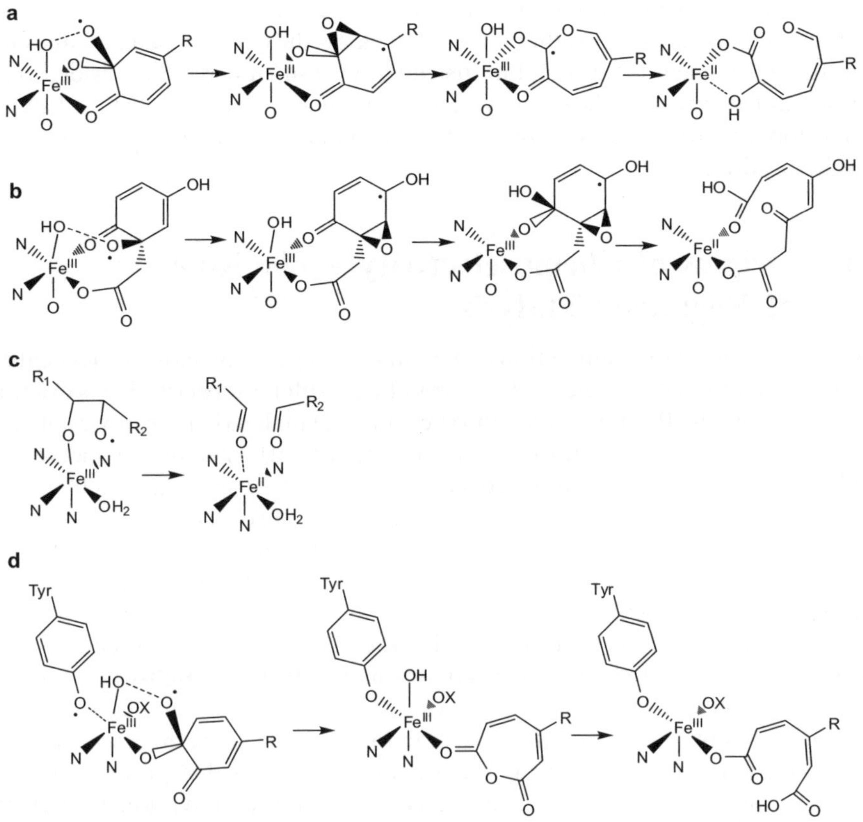

Scheme 4.11 Reactions of the alkoxyl radical species.

the C–C bond in the oxirane ring yields a seven-membered lactone radical, which subsequently recombines with the OH ligand, a process coupled to one-electron oxidation of the ring and its opening to the final acyclic product.

The reaction of HGD (Scheme 4.11b) starts off in a similar way, with an attack of the O-radical on the ring.[54] However, due to the differences in the structure of the substrate, the oxirane ring is considerably more resistant to C–C bond cleavage, and thus the next step is an attack of the OH ligand on the carbonyl carbon. Then, the oxirane ring re-opens and a C–C bond cleavage, coupled to one-electron oxidation of the organic intermediate, yields the final acyclic product. This final step has a mechanism basically identical with the C–C cleavage in the catalytic reaction of ACO (Scheme 4.11c).

The alkoxyl radical species produced in the active site of ACO (Scheme 4.11c) has all atoms in the right place to form the final products and, thus, an easy C–C cleavage, concerted with one-electron oxidation of the organic intermediate, yields the aldehyde products.[56]

Evolution of the alkoxide radical produced in the catalytic cycle of intradiol dioxygenases (Scheme 4.11d) largely parallels that in extradiol enzymes

(Scheme 4.11a). Thus, in the first step the O-radical attacks the adjacent ring carbon, in this case the intradiol one.[55] Attack on the carbonyl carbon does not lead to a three-membered ring, but instead a seven-membered anhydride ring is formed as the tyrosyl radical is reduced to tyrosinate. The final chemical step is a nucleophilic attack of the OH ligand on the anhydride, which leads to the acyclic product.

4.6 Origins of Chemoselectivity – The Role of Negative Catalysis

As seen in the above subsections, the catalytic cycles of most NIEs feature reactive intermediates, either in the form of an oxoferryl species ($Fe^{IV}=O$) or as alkoxide radicals ($R-O^{\bullet}$). Taking into account the high oxidative power of such intermediates, implying that the substrates could be attacked by them at several different sites, a valid question concerns the origin of enzyme specificity. In other words, how does the enzyme guide the reaction so that only one reaction channel, out of many possible ones, is taken? Theoretical methods are best suited for pursuing this question, since they allow for testing various mechanistic hypotheses without altering the system.

At this point it is useful to emphasize the principal difference between the enzymes producing reactive intermediates like radicals or high-valent species and other, more "classical", enzymes catalysing, for example, hydrolytic reactions. In the latter case, the same reaction with the same mechanism as the one taking place in the enzyme active site can be envisioned to proceed in water solution, and the role of the enzyme is to lower the activation barrier for this reaction, usually *via* electrostatic effects. For the former group, however, the reaction mechanisms are tightly dependent on the cofactors, and the enzymatic reactions therefore have no well-defined solution equivalents. For example, the reaction between methane and dioxygen takes completely different routes in the sMMO active site and in solution. Moreover, the presence of highly reactive species, produced with the help of cofactors, makes the accelerating role of the enzyme unnecessary, but instead the protein has to play a role of a guide that channels the reaction into an appropriate direction. This has been suggested, in the form of a "negative catalysis hypothesis",[83] in the sense that such guidance has a passive character, *i.e.* the enzyme does not accelerate the progress of the reaction in the appropriate channel, but instead hinders all other alternative processes. In the computational studies several examples were found supporting this hypothesis, which will be discussed below.

The cyclization reaction catalysed by CAS (Schemes 4.1o and 4.10b) requires that the oxidative power of the oxoferryl species is directed at two groups of the substrate: the hydroxyl group in the aliphatic chain and/or the methylene group in the β-lactam ring. The structure of the active site region in the CAS-$Fe^{IV}=O$---substrate complex is presented in Figure 4.2(a),[70] where the oxo group forms a hydrogen bond with the hydroxyl. Two ionized fragments of the

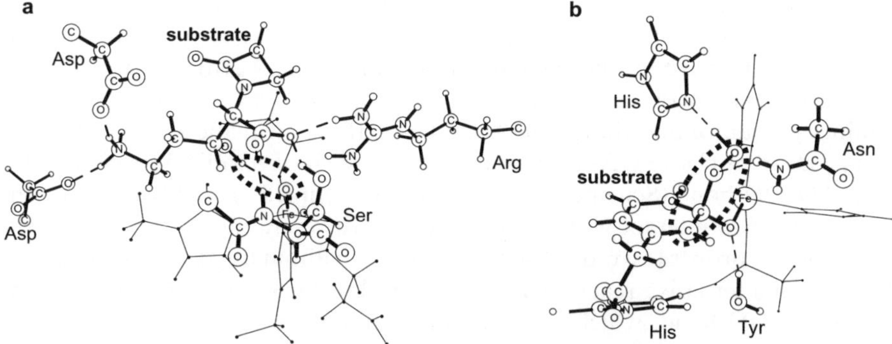

Figure 4.2 Active sites for: (a) CAS–Fe(IV)=O---substrate complex and (b) extradiol dioxygenase–peroxo intermediate. The first shell proteinogenous ligands are drawn with thin lines and circles.

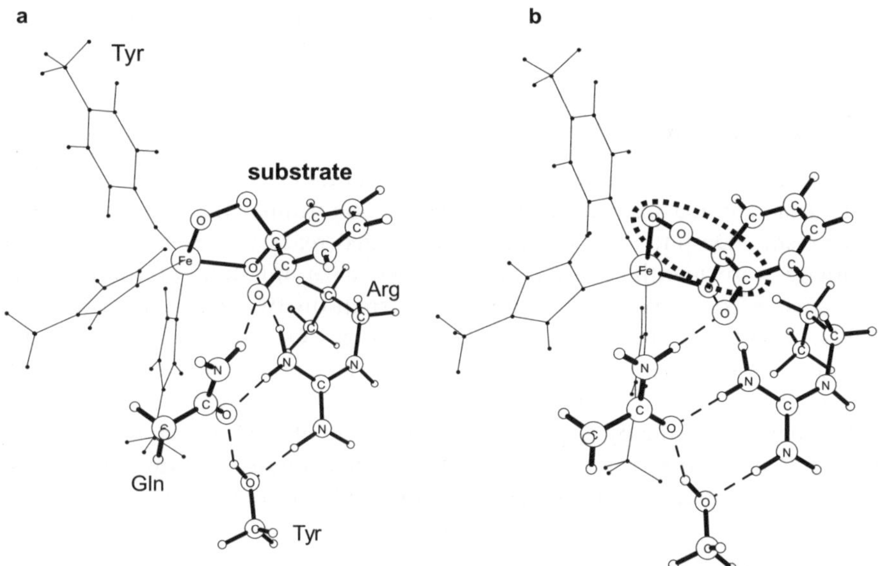

Figure 4.3 Structures of peroxo intermediate for intradiol dioxygenase: (a) initially formed and less stable conformation and (b) the more stable and reactive conformation. The first shell proteinogenous ligands are drawn with thin lines and circles.

substrate, *i.e.* the amino and carboxyl groups, make several hydrogen bonding contacts with the protein side chains (two Asp, Ser and Arg residues), and those interactions guarantee that (i) the oxo group is positioned close to the OH group and (ii) approach of Fe(IV)=O to other sites of the substrate would elicit energy increases.

Another example comes from the study on extradiol dioxygenases (Schemes 4.1m, 4.3b, 4.6a and 4.11a), where the alkoxyl radical is formed from the cleavage of the peroxo intermediate shown in Figure 4.2(b).[52,53] In this case hydrogen bonds between the organic hydroperoxide and several second shell residues guarantee that the O–OH bond lies in the same plane as the C–C bond that needs to be cleaved. As a consequence, in the alkoxyl radical species, with a partial O–OH bond (Scheme 4.11a), the singly occupied orbital lying along the O–O vector can interact with the π-orbital on the extradiol carbon. Thus, the attack on the proper (extradiol) site can develop without any additional barrier, whereas an alternative reaction with the intradiol carbon requires cleaving the partial O---OH bond, which costs several kcal mol^{-1}.

The final example comes from the study on intradiol dioxygenases (Schemes 4.11 and 4.3b).[55,62] In this case, the peroxo intermediate is first formed in a configuration not compatible with the intradiol-type cleavage (Figure 4.3a), yet, the hydrogen bonds with the second shell residues drive a conformational change. Thus, the structure presented in Figure 4.3(b) is more stable by *ca.* 5 kcal mol^{-1}, and it guarantees the proper arrangement of the bonds for an intradiol-type reaction, irrespective of the detailed mechanism for coupling the O–O and C–C cleavage (Schemes 4.5g, 4.6d or 4.7b).

4.7 Conclusions

In this chapter the results of computational studies on non-heme iron enzymes have been summarized with the emphasis on similarities of the catalytic mechanisms. Despite a wide range of catalyzed reactions (Scheme 4.1) many analogies were identified, and they allowed a formulation of three general mechanisms for dioxygen binding and formation of peroxo intermediates. Similarly, also for the next step, which involves O–O bond cleavage, a few general reaction schemes were found. Finally, concerning the specificity of the reactions it seems that the chemoselectivity of the catalytic process is achieved by the metalloenzyme by appropriate positioning of the reactive fragments. In this way alternative reaction channels become almost unavailable as they require substantial structural changes connected with energy penalties.

References

1. M. J. Ryle and R. P. Hausinger, *Curr. Opin. Chem. Biol.*, 2002, **6**, 193.
2. E. I. Solomon, T. C. Brunold, M. I. Davis, J. N. Kemsley, S.-K. Lee, N. Lehnert, F. Neese, A. J. Skulan, Y.-S. Yang and J. Zhou, *Chem. Rev.*, 2000, **100**, 235.
3. P. E. M. Siegbahn and T. Borowski, *Acc. Chem. Res.*, 2006, **39**, 729.
4. S. J. Lange and L. Que Jr., *Curr. Opin. Chem. Biol.*, 1998, **2**, 159.
5. R. P. Hausinger, *Crit. Rev. Biochem. Mol.*, 2004, **39**, 21.
6. M.-H. Baik, M. Newcomb, R. A. Friesner and S. J. Lippard, *Chem. Rev.*, 2003, **103**, 2385.
7. B. Sedgwick, *Nat. Rev. Mol. Cell Biol.*, 2004, **5**, 148.

8. Y. Tsukada, J. Fang, H. Erdjument-Bromage, E. Warren, C. H. Borchers, P. Tempst and Y. Zhang, *Nature*, 2006, **439**, 811.

9. R. K. Bruick and S. L. McKnight, *Science*, 2001, **294**, 1337.

10. Z. Zhang, J. Ren, D. K. Stammers, J. E. Baldwin, K. Harlos and C. J. Schofield, *Nat. Struct. Biol.*, 2000, **7**, 127.

11. E. Eichhorn, J. R. van der Ploeg, A. Kertesz and T. Leisinger, *J. Biol. Chem.*, 1997, **272**, 23031.

12. H. Erlandsen, F. Fusetti, A. Martinez, E. Hough, T. Flatmark and R. C. Stevens, *Nat. Struct. Biol.*, 1997, **4**, 995.

13. O. W. Choroba, D. H. Williams and J. B. Spencer, *J. Am. Chem. Soc.*, 2000, **122**, 5389.

14. G. R. Moran, *Arch. Biochem. Biophys.*, 2005, **433**, 117.

15. F. H. Vaillancourt, J. Yin and C. T. Walsh, *Proc. Natl. Acad. Sci. USA*, 2005, **102**, 10111.

16. A. Karlsson, J. V. Parales, R. E. Parales, D. T. Gibson, H. Eklund and S. Ramaswamy, *Science*, 2003, **299**, 1039.

17. A. Stapon, R. Li and C. A. Townsend, *J. Am. Chem. Soc.*, 2003, **125**, 15746.

18. J. Stubbe and W. A. van der Donk, *Chem. Rev.*, 1998, **98**, 705.

19. Z. Zhang, J.-S. Ren, I. J. Clifton and C. J. Schofield, *Chem. Biol.*, 2004, **11**, 1383.

20. A. Wyss, G. Wirtz, W.-D. Woggon, R. Brugger, M. Wyss, A. Friedlein, G. Riss, H. Bachmann and W. Hunziker, *Biochem. J.*, 2001, **354**, 521.

21. W. E. Knox and S. W. Edwards, *J. Biol. Chem.*, 1955, **216**, 479.

22. M. Costas, M. P. Mehn, M. P. Jensen and L. Que Jr., *Chem. Rev.*, 2004, **104**, 939.

23. J. E. Baldwin and E. Abraham, *Nat. Prod. Rep.*, 1988, **5**, 129.

24. C. J. Schofield and Z. Zhang, *Curr. Opin. Struct. Biol.*, 1999, **9**, 722.

25. T. D. H. Bugg, *Tetrahedron*, 2003, **59**, 7075.

26. E. L. Hegg and L. Que Jr., *Eur. J. Biochem.*, 1997, **250**, 625.

27. L. Que Jr, *Nat. Struct. Biol.*, 2000, **7**, 182.

28. A. G. Prescott and M. D. Lloyd, *Nat. Prod. Rep.*, 2000, **17**, 367.

29. J. C. Price, E. W. Barr, B. Tirupati, J. M. Bollinger Jr. and C. Krebs, *Biochemistry*, 2003, **42**, 7497.

30. P. J. Riggs-Gelasco, J. C. Price, R. B. Guyer, J. H. Brehm, E. W. Barr, J. M. Bollinger Jr. and C. Krebs, *J. Am. Chem. Soc.*, 2004, **126**, 8108.

31. D. A. Proshlyakov, T. F. Henshaw, G. R. Monterosso, M. J. Ryle and R. P. Hausinger, *J. Am. Chem. Soc.*, 2004, **126**, 1022.

32. D. T. Gibson and R. E. Parales, *Curr. Opin. Biotechnol.*, 2000, **11**, 235.

33. T. D. H. Bugg and C. Winfield, *Nat. Prod. Rep.*, 1998, **15**, 513.

34. L. C. Blasiak, F. H. Vaillancourt, C. T. Walsh and C. L. Drennan, *Nature*, 2006, **440**, 368.

35. D. P. Kloer, S. Ruch, S. Al-Babili, P. Beyer and G. E. Schulz, *Science*, 2005, **308**, 267.

36. C. K. Brown, M. W. Vetting, C. A. Earhart and D. H. Ohlendorf, *Annu. Rev. Microbiol.*, 2004, **58**, 555.

37. P. Nordlund and H. Eklund, *J. Mol. Biol.*, 1993, **232**, 123.
38. A. C. Rosenzweig, C. A. Frederic, S. J. Lippard and P. Nordlund, *Nature*, 1993, **366**, 537.
39. A. D. Becke, *J. Chem. Phys.*, 1993, **98**, 5648.
40. C. Lee, W. Yang and R. G. Parr, *Phys. Rev. B: Condens. Matter*, 1988, **37**, 785.
41. P. J. Stevens, F. J. Devlin, C. F. Chablowski and M. J. Frisch, *J. Phys. Chem.*, 1994, **98**, 11623.
42. *JAGUAR* 4.0, 2000, Schrödinger, Inc., Portland, Oregon.
43. M. J. Frisch, G. W. Trucks, H. B. Schlegel, G. E. Scuseria, M. A. Robb, J. R. Cheeseman, J. A. Montgomery Jr., T. Vreven, K. N. Kudin, J. C. Burant, J. M. Millam, S. S. Iyengar, J. Tomasi, V. Barone, B. Mennucci, M. Cossi, G. Scalmani, N. Rega, G. A. Petersson, H. Nakatsuji, M. Hada, M. Ehara, K. Toyota, R. Fukuda, J. Hasegawa, M. Ishida, T. Nakajima, Y. Honda, O. Kitao, H. Nakai, M. Klene, X. Li, J. E. Knox, H. P. Hratchian, J. B. Cross, C. Adamo, J. Jaramillo, R. Gomperts, R. E. Stratmann, O. Yazyev, A. J. Austin, R. Cammi, C. Pomelli, J. W. Ochterski, P. Y. Ayala, K. Morokuma, G. A. Voth, P. Salvador, J. J. Dannenberg, V. G. Zakrzewski, S. Dapprich, A. D. Daniels, M. C. Strain, O. Farkas, D. K. Malick, A. D. Rabuck, K. Raghavachari, J. B. Foresman, J. V. Ortiz, Q. Cui, A. G. Baboul, S. Clifford, J. Cioslowski, B. B. Stefanov, G. Liu, A. Liashenko, P. Piskorz, I. Komaromi, R. L. Martin, D. J. Fox, T. Keith, M. A. Al-Laham, C. Y. Peng, A. Nanayakkara, M. Challacombe, P. M. W. Gill, B. Johnson, W. Chen, M. W. Wong, C. Gonzalez and J. A. Pople, *GAUSSIAN03*, Gaussian Inc., Pittsburgh PA, 2003.
44. V. Pelmenschikov, M. R. A. Blomberg and P. E. M. Siegbahn, *J. Biol. Inorg. Chem.*, 2002, **7**, 284.
45. D. J. Tannor, B. Marten, R. Murphy, R. A. Friesner, D. Sitkoff, A. Nicholls, M. Ringnalda, W. A. Goddard III and B. Honig, *J. Am. Chem. Soc.*, 1994, **116**, 11875.
46. B. Marten, K. Kim, C. Cortis, R. A. Friesner, R. Murphy, M. Ringnalda, D. Sitkoff and B. Honig, *J. Phys. Chem.*, 1996, **100**, 11775.
47. T. Borowski, A. Bassan and P. E. M. Siegbahn, *Chem. Eur. J.*, 2004, **10**, 1031.
48. A. Bassan, M. R. A. Blomberg and P. E. M. Siegbahn, *Chem. Eur. J.*, 2003, **9**, 106.
49. M. Wirstam and P. E. M. Siegbahn, *J. Am. Chem. Soc.*, 2000, **122**, 8539.
50. M. Lundberg, P. E. M. Siegbahn and K. Morokuma, *Biochemistry*, 2008, **47**, 1031.
51. A. Bassan, T. Borowski and P. E. M. Siegbahn, *Dalton Trans.*, 2004, 3153.
52. T. Borowski, A. Bassan and P. E. M. Siegbahn, *Biochemistry*, 2004, **43**, 12331.
53. W.-G. Han, T. Liu, T. Lovell and L. Noodleman, *J. Am. Chem. Soc.*, 2005, **127**, 15778.
54. S. P. de Visser, *Chem. Commun.*, 2007, 171.

55. P. E. M. Siegbahn and F. Haeffner, *J. Am. Chem. Soc.*, 2004, **126**, 8919.
56. V. Georgiev, T. Borowski, M. R. A. Blomberg and P. E. M. Siegbahn, *J. Biol. Inorg. Chem.*, 2008, **13**, 929.
57. T. Borowski, V. Georgiev and P. E. M. Siegbahn, *J. Am. Chem. Soc.*, 2005, **127**, 17303.
58. T. Borowski and P. E. M. Siegbahn, *J. Am. Chem. Soc.*, 2006, **128**, 12941.
59. T. Borowski, M. R. A. Blomberg and P. E. M. Siegbahn, *Chem. Eur. J.*, 2008, **14**, 2264.
60. A. Bassan, M. R. A. Blomberg, T. Borowski and P. E. M. Siegbahn, *J. Phys. Chem. B*, 2004, **108**, 13031.
61. A. Bassan, T. Borowski, C. J. Schofield and P. E. M. Siegbahn, *Chem. Eur. J.*, 2006, **12**, 8835.
62. E. Godfrey, C. S. Porro and S. P. de Visser, *J. Phys. Chem. A.*, 2008, **112**, 2464.
63. D. P. Galonic, E. W. Barr, C. T. Walsh, J. M. Bollinger Jr. and C. Krebs, *Nat. Chem. Biol.*, 2006, **3**, 113.
64. T. Borowski, H. Noack, M. Radoń, K. Zych and P. E. M. Siegbahn, *J. Am. Chem. Soc.*, 2010, **132**, 12887.
65. M. L. Matthews, C. M. Krest, E. W. Barr, F. H. Vaillancourt, C. T. Walsh, M. T. Green, C. Krebs and J. M. Bollinger Jr., *Biochemistry*, 2009, **48**, 4331.
66. P. E. M. Siegbahn, *Chem. Phys. Lett.*, 2002, **351**, 311.
67. A. Bassan, M. R. A. Blomberg and P. E. M. Siegbahn, *J. Biol. Inorg. Chem.*, 2004, **9**, 439.
68. A. Bassan, M. R. A. Blomberg, T. Borowski and P. E. M. Siegbahn, *J. Inorg. Biochem.*, 2006, **100**, 727.
69. S. P. de Visser, *J. Am. Chem. Soc.*, 2006, **128**, 9813.
70. R. Latifi, M. Bagherzadeh and S. P. de Visser, *Chem. Eur. J.*, 2009, **15**, 6651.
71. J. T. Groves, G. A. McClusky, R. E. White and M. J. Coon, *Biochem. Biophys. Res. Commun.*, 1978, **81**, 154.
72. B. F. Gherman, M.-H. Baik, S. J. Lippard and R. A. Friesner, *J. Am. Chem. Soc.*, 2004, **126**, 2978.
73. P. E. M. Siegbahn, *J. Biol. Inorg. Chem.*, 2001, **6**, 27.
74. H. Basch, K. Mogi, D. G. Musaev and K. Morokuma, *J. Am. Chem. Soc.*, 1999, **121**, 7249.
75. A. Bassan, M. R. A. Blomberg and P. E. M. Siegbahn, *Chem. Eur. J.*, 2003, **9**, 4055.
76. P. E. M. Siegbahn, *Q. Rev. Biophys.*, 2003, **36**, 91.
77. S. P. de Visser and R. Latifi, *J. Phys. Chem. B*, 2009, **113**, 12.
78. S. Pandian, M. Vincent, I. Hillier and N. Burton, *Dalton Trans.*, 2009, 6201.
79. H. J. Kulik, L. C. Blasiak, N. Marzari and D. L. Drennan, *J. Am. Chem. Soc.*, 2009, **131**, 14426.

80. M. Topf, G. M. Sandala, D. M. Smith, C. J. Schofield, C. J. Easton and L. Radom, *J. Am. Chem. Soc.*, 2004, **126**, 9932.
81. T. Borowski, E. Broclawik, C. J. Schofield and P. E. M. Siegbahn, *J. Comput. Chem.*, 2006, **27**, 740.
82. T. Borowski, S. de Marothy, E. Broclawik, C. J. Schofield and P. E. M. Siegbahn, *Biochemistry*, 2007, **46**, 3682.
83. J. Rétey, *Angew. Chem.,Int. Ed.*, 1990, **29**, 355.

Theoretical Spectroscopies of Iron-Containing Enzymes and Biomimetics

SHENGFA YE, GEMMA J. CHRISTIAN, CAIYUN GENG AND FRANK NEESE[*]

Lehrstuhl für Theoretische Chemie, Universität Bonn, Wegelerstr. 12, D-53115 Bonn, Germany

5.1 Introduction

Spectroscopic methods play a prominent role in the study of iron-containing enzymes particularly for the study of reactive species, which are often short-lived and difficult to characterize *via* crystallography.[1,2] Spectroscopic techniques allow us to probe the oxidation states of metal ions in active sites, their electronic structures and local coordination geometries and gain insight into the reactivity of these important species. Mössbauer spectroscopy, electron paramagnetic resonance, electron nuclear double resonance, absorption spectroscopy, and nuclear resonance vibrational spectroscopy, particularly in combination with rapid freeze-quench techniques, are most successful for iron-containing enzymes, and will most likely also be key techniques in future studies. The interpretation of the spectroscopic parameters, however, is not always straightforward and in some cases is even ambiguous. More importantly, comparison with model systems and/or interpretation based on simple empirical models is not sufficient to extract all the information contained in experimental spectra. This process can be significantly facilitated by theoretical predictions of the corresponding spectroscopic parameters.

Iron-Containing Enzymes: Versatile Catalysts of Hydroxylation Reactions in Nature
Edited by Sam P de Visser and Devesh Kumar
© Royal Society of Chemistry 2011
Published by the Royal Society of Chemistry, www.rsc.org

Theoretical spectroscopy in combination with experiment is a powerful tool in the study of reactive intermediates. Spectroscopic parameters are much more sensitive indicators of the electronic and geometric structure than the total energy itself, and hence can be used to differentiate between proposed structural models of the reaction intermediate. At present density functional theory (DFT) based approaches are widely used to study these problems;[3] however, studies based purely on DFT can be complicated by the uncertainties of spin-state energetics for transition metal containing species with close lying spin states[4] and other shortcomings of DFT itself. If theoretical spectroscopy is able to reliably predict distinct spectroscopic properties for different spin ground states, it provides a powerful means of determining the ground state spin multiplicities without reliance on total energies of uncertain quality.

The second central theme is that spectroscopic measurements yield not only information about the geometric structure of a given active site but are also sensitive to the details of their electronic structures. Because electronic structure and reactivity are intimately linked, only spectroscopy provides an experimental means of studying the electronic structure contributions to the reactivity. Thus, the correct interpretation of electronic structure from a combination of theory and experiment is pivotal for the understanding of the reactivity of enzymes and, as an important long-term goal, also for the design of novel biomimetic catalysts. Successful reproduction of spectroscopic parameters can lend credence to the correct understanding of the electronic structure of reaction intermediates and greatly strengthen faith in the calculated reaction pathways.

One field in which theoretical spectroscopy has made an extremely useful contribution is the study of high-valent iron-oxo species in enzymatic and biomimetic systems. Iron-oxo species are a recurring theme in iron-containing O_2 activating enzymes and are proposed as key intermediates in the catalytic cycles of numerous iron-containing enzymes.[5,6]

This chapter aims to demonstrate how theoretical spectroscopy in combination with traditional quantum chemistry can be used to study reactive intermediates. The theory and applications of theoretical Mössbauer (MB), electron paramagnetic resonance (EPR), absorption spectra (ABS) and X-ray absorption spectroscopy (XAS), and nuclear resonance vibrational spectroscopy (NRVS) are described with a focus on the contribution that theoretical spectroscopy has made to the study of high-valent iron-oxo species in enzymatic and biomimetic systems.

5.2 Mössbauer Spectroscopy

Of all theoretical spectroscopies, the prediction of Mössbauer parameters is perhaps the most widely used in the study of iron-containing enzymes. Experimentally, Mössbauer spectroscopy is used extensively, often in combination with freeze-quench techniques, which make it possible to trap unstable or short-lived intermediates.[7,8]

^{57}Fe Mössbauer spectroscopy probes the transitions of the nucleus between the $I = \frac{1}{2}$ ground state and the $I = \frac{3}{2}$ excited state, which is 14.4 keV above the

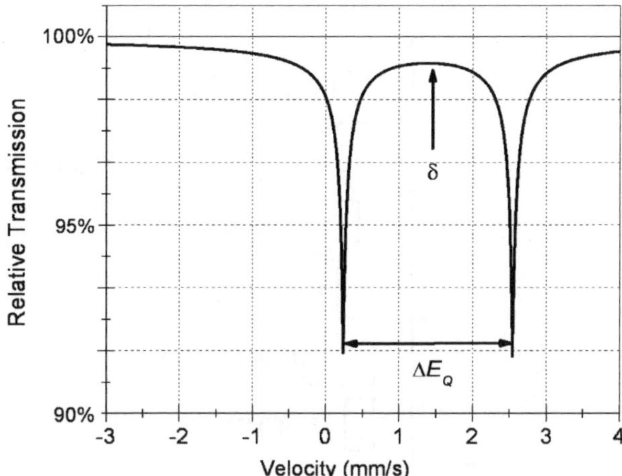

Figure 5.1 Typical Mössbauer spectrum showing both the isomer shift (δ) and quadrupole splitting (ΔE_Q).

former.[9] The important features of the Mössbauer spectrum are the isomer shift and the quadrupole splitting. Figure 5.1 shows an idealized spectrum. The isomer shift measures the shift in the energy of the γ-ray absorption relative to a standard, usually Fe foil.[10] The isomer shift is sensitive to the electron density at the nucleus, and indirectly probes changes in the bonding of the valence orbitals due to variations in covalency and 3d shielding. Thus, it can be used to probe oxidation and spin states, and the coordination environment of the iron.

The quadrupole splitting arises from the interaction of the nuclear quadrupole moment of the excited state with the electric field gradient (EFG) at the nucleus. The former is related to the non-spherical charge distribution in the excited state. As such it is extremely sensitive to the coordination environment and the geometry of the complex. Using a combination of isomer shifts and quadrupole splittings it is often possible to assign the oxidation and spin state of the iron centre and gain information about the coordination environment. However, while assignments are often unambiguous for high-spin Fe(II), there is significant overlap for other oxidation and spin states, which makes it more difficult to assign the spectra (Figure 5.2).[7] In these instances theoretical spectroscopy is extremely useful, for both evaluating proposed models with different oxidation states and also aiding understanding which factors influence the spectroscopic parameters.

5.2.1 Theoretical Prediction of Mössbauer Parameters

Both the isomer shift and quadrupole splitting can be successfully predicted using computational methods. DFT is usually the method of choice although correlated methods can also be used.[11–13]

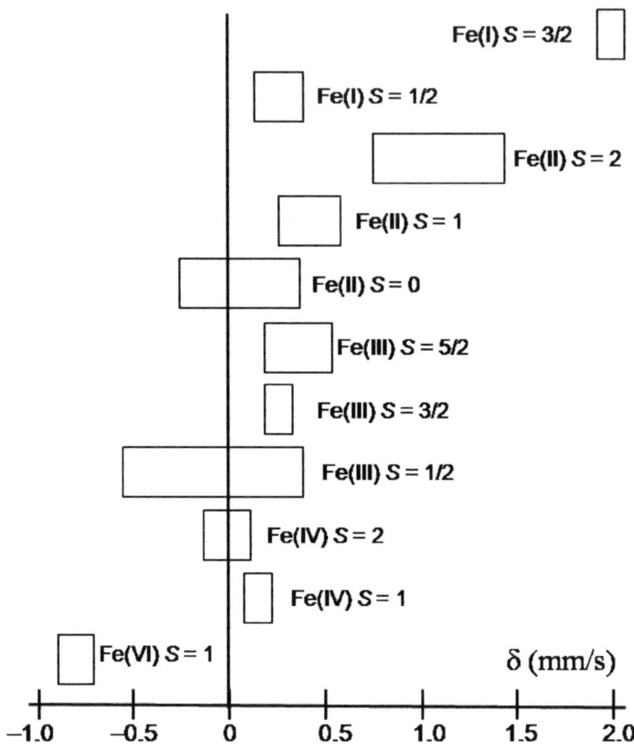

Figure 5.2   Ranges of isomer shifts observed in various oxidation and spin states of iron.

The isomer shift (δ) is directly related to the s electron density at the nucleus and can be calculated using the formula:

$$\delta = \alpha(\rho_0 - C) + \beta \tag{5.1}$$

Where α is a constant that depends on the change in the distribution of the nuclear charge upon absorption, and ρ_0 is the electron density at the nucleus. The constants α and β are usually determined *via* linear regression analysis of a plot of the experimental isomer shifts *versus* the theoretically calculated electron density for a series of iron compounds with various oxidation and spin states. The constant C is employed to improve the sensitivity of the linear fit, because the dominant part of the electron density arising from 1s-, 2s- and 3s-electrons is nearly a constant for all iron complexes irrespective of oxidation states, spin states and coordination environments.[14] Since the electron density depends on the functional and basis set employed, fitting must be carried out for each combination used.[15,16] Usually, an accuracy of better than 0.10 mm s^{-1} can be achieved for DFT with reasonably sized basis sets.[14,16–18] Theoretical studies have shown that the density at the nucleus is determined by a subtle interplay of the shielding and covalency of valence 3d electrons, and especially metal–ligand bond lengths. For discussions of how different factors affect the isomer shift see refs 15 and 19. Recently an elegant linear response theory has been proposed for predicting isomer shifts without empirical constants.[11–13] However, at this stage the errors can be as large as 0.6 mm s^{-1}.

The quadrupole splitting is proportional to the largest component of the EFG at the iron nucleus and can be calculated using the formula:

$$\Delta E_Q = \frac{1}{2} e Q V_{zz} \left(1 + \frac{\eta^2}{3} \right)^{\frac{1}{2}} \tag{5.2}$$

where e is the electrical charge of an electron, Q is the nuclear quadrupole moment (approx. 0.16 barn) of Fe. V_{xx}, V_{yy} and V_{zz} are components of the electric field gradient tensor and $\eta = |(V_{xx} - V_{yy})/V_{zz}|$ is the asymmetry parameter in a coordinate system chosen such that $|V_{zz}| \geq |V_{yy}| \geq |V_{xx}|$. The EFG tensors are usually determined from DFT calculations. Good agreement with experiment has been found for heme,[20] phthalocyanines,[21] corroles,[22] two and three coordinate iron complexes[23] and diiron complexes.[24] However, since the quadrupole splitting is very sensitive to the geometry the quality of the predictions depend on the size of the model and flexibility of the coordination sphere.[3] In our experience in cases where the coordination shell is rather flexible the accuracy of the predictions is sometimes limited to determining whether the quadrupole splitting is expected to be "large" (*e.g.* >1 mm s^{-1}) or "small" (*e.g.* <0.5 mm s^{-1}). Furthermore, significant effects of counterions can sometimes be observed in the study of highly charged model complexes.[25]

5.2.2 Examples from the Literature

One of the strengths of Mössbauer spectroscopy is its ability to distinguish between proposed models for key intermediates. This approach has been applied to the study of heme[26–29] and non-heme iron[30–32] enzymes and biomimetic complexes.[10] Noodleman *et al.*[33] as well as Bominaar and coworkers[19] have used predicted Mössbauer isomer shifts to probe the nature of the interstitial atom in the FeMo cofactor in nitrogenase. Prediction of isomer shifts and quadrupole splittings has been used extensively in the study of diiron enzymes, *e.g.* methane monoxygenase (sMMO)[34] and ribonucleotide reductase subunit R2 (RNR-R2),[16,17,35–38] along with other model complexes.[24,39]

5.2.2.1 *Taurine:α-ketoglutarate dioxygenase (TauD)*

An example of the strength of theoretical Mössbauer spectroscopy in differentiating between proposed models is the study of the proposed Fe(IV)-oxo intermediate (intermediate **J**) of taurine:αKG dioxygenase (TauD).[40] TauD belongs to the family of α-ketoglutarate (αKG) dependent enzymes that couple the activation of oxygen to substrate oxidation. In most cases the substrate is hydroxylated, but other outcomes such as ring closure and halogenation have also been observed. Intermediate **J** is proposed to abstract a hydrogen atom from the substrate, which is bound nearby, followed by rapid OH-rebound to give the product.[41]

Mössbauer and EPR studies have been used to identify intermediate J as an Fe(IV) species with the unusual $S = 2$ spin state but do not give much information about the possible coordination geometry. A series of models with different spin states ($S = 1$, 2) and coordination geometries were constructed, including distorted octahedral, trigonal bipyramidal and square pyramidal structures. Monodentate and bidentate binding modes of the carboxylate ligands were also considered. A representative computational model is shown in Figure 5.3 and the calculated Mössbauer parameters for different coordination geometries are shown in Table 5.1.

The calculated parameters for the intermediate spin state ($S = 1$) are incompatible with experiment. In particular the isomer shift is on average 0.19 mm s^{-1} lower than experiment. As shown in Table 5.1 the agreement with experiment of the isomer shift and quadrupole splittings calculated for the high-spin state models is very good with the exception of the square pyramidal structure. Therefore, based on these results we can conclude that intermediate **J** is high spin and has a ligand sphere with two histidine residues, an oxo group and two carboxylates (succinate and Asp$_{101}$) forming an overall distorted

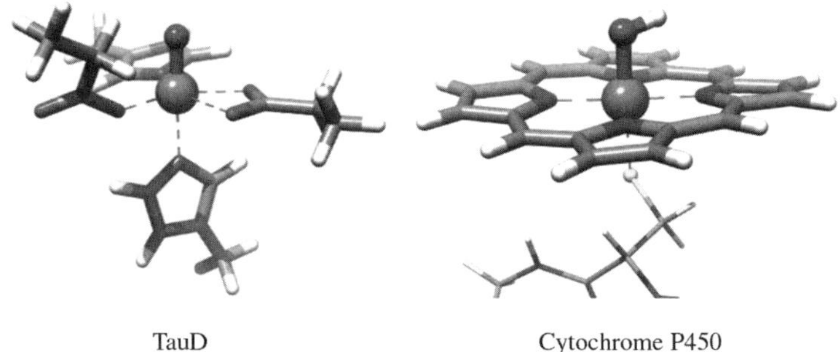

Figure 5.3   Left: computational model of TauD Intermediate **J** with a distorted octahedral geometry; right: protonated model of Compound II of cytochrome P450.

Table 5.1   Calculated Mössbauer parameters (mm s^{-1}) for the proposed high-spin models for TauD intermediate **J** compared to the experimental values.

Parameters	*Exp*	*Distorted octahedron*	*Trigonal bipyramid*	*Square pyramid*
δ	0.30	0.27	0.27	0.17
ΔE_Q	−0.90	−0.65	−0.82	−0.77

octahedral or trigonal bipyramidal geometry depending on whether Asp binds in a mono-dentate or bi-dentate manner. Notably, of the nine models investigated none could have been selected in favor of the others on the basis of total energy considerations.[40]

5.2.2.2 Cytochromes P450

Another excellent example of how theoretical Mössbauer spectroscopy can be used to distinguish between proposed models is given by the study of the ferryl complex compound II of cytochrome P450.[42] In the consensus mechanism for cytochrome P450, compound I is an Fe(IV)-oxo species that is coordinated to a porphyrin radical. It abstracts a hydrogen atom from the substrate to form a protonated ferryl species, similar to compound II of chloroperoxidase (CPO). Recent studies of CPO-II,[43,44] including theoretical Mössbauer studies,[45] have suggested that it is an Fe(IV)-OH species and, based on comparisons with CPO-II, it is likely that Compound II of cytochrome P450 (CYP 450) is also an Fe(IV)-OH species, a model of which is shown in Figure 5.3.

To explore the protonation state of the complex, Mössbauer parameters were calculated for two computational models of compound II of CYP 450_{BM3} and CYP 450_{CAM} with Fe(IV)=O and Fe(IV)-OH.[42] The calculated isomer shifts were not greatly affected by the difference in protonation but the quadrupole splitting was found to be more sensitive, with calculated values of 1.05 and 2.17 mm s^{-1} for the Fe(IV)=O and Fe(IV)-OH forms, respectively. The latter is in good agreement with the experimental value of the quadrupole splitting of 2.16 mm s^{-1}. Therefore these results demonstrate that Compound II is likely to be protonated.

5.3 Nuclear Resonance Vibrational Spectroscopy

The synchrotron-based vibrational technique known as nuclear resonance vibrational spectroscopy (NRVS) utilizes a high resolution (~ 1 meV) incident X-ray beam that is tuned to the nuclear transition energy of a Mössbauer active nucleus, *e.g.* ^{57}Fe. By scanning the incident beam relative to the Mössbauer resonance, sidebands that correspond to the nuclear transition with excitation or de-excitation of vibrational levels are observed. More importantly, NRVS is not subject to the selection rules of Raman and infrared spectroscopy, and thus can provide the complete set of vibrational bands corresponding to normal modes that only involve the motion of the iron atom.[46–52] The NRVS intensity is directly related to the vibrational frequencies and normal mode composition factors that characterize the extent of involvement of the resonant nucleus in a given normal mode.[53,54] The normal mode composition factors are determined by the molecular force field, and thus their values reflect the details of the electronic structure. Details of how to calculate NRVS spectra are somewhat complex and have been discussed elsewhere.[55]

5.3.1 Examples from the Literature

Recently, NRVS has been successfully applied in bioinorganic chemistry to heme proteins[47,56] and model complexes,[46,57–59] an iron-sulfur cluster,[60] the FeMo-cofactor in nitrogenase[61] and Fe(IV)-oxo model complexes.[62]

As an example, we discuss the NRVS spectrum of the high-valent [Fe(V)-nitrido cyclam-acetato]$^+$ complex[63] shown in Figure 5.4. This complex has a d^3-configuration, and hence a ground state with either $S = \frac{1}{2}$ or $S = \frac{3}{2}$ can be envisioned. The calculations predict quite different Fe–N stretching frequencies for these two spin multiplicities. NRVS provides a unique opportunity to experimentally measure this fingerprint. Using the normal mode composition factor and vibrational frequencies calculated by DFT methods, the NRVS spectrum can be fitted and simulated. The good agreement between the experimental and calculated spectra allows for a clear interpretation of the NRVS data. The isolated band at 864 cm^{-1} is unambiguously assigned to a Fe–N stretch and provides concrete experimental evidence that the Fe(V)-nitrido complex possesses a doublet ground state.

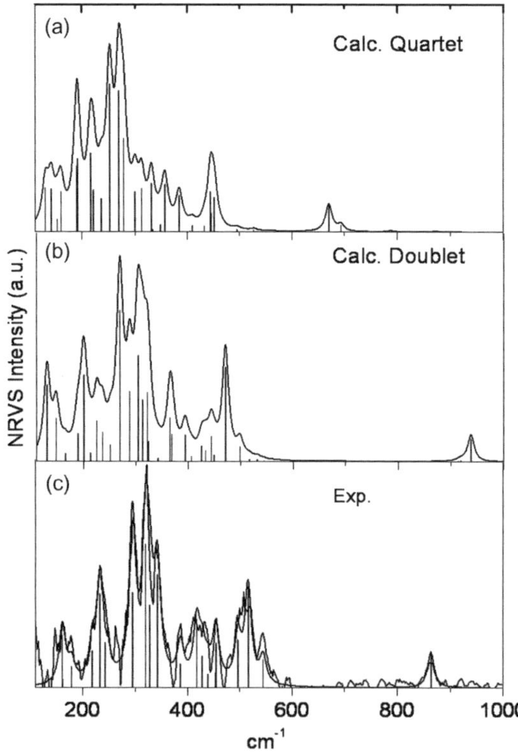

Figure 5.4 Calculated (BP86/TZVP level) NRVS spectra of quartet (a) and doublet (b) for [Fe(V)-nitrido cyclam-acetato]$^+$; (c) is the experimental smoothed (grey curve) and fitted (black) NRVS spectrum. Bar graphs represent the intensities of the individual vibrational transitions.

5.4 Electron Paramagnetic Resonance

Electron paramagnetic resonance (EPR) is one of the most successful spectroscopic methods for the study of iron-containing enzymes. This technique, in principle, is applicable to any chemical species with one or more unpaired electrons. In bioinorganic chemistry, it has the advantage of being selective for the active site, which often contains metal ions in an open-shell configuration, while the remaining part of the protein is "EPR silent". For example, the ferric peroxy anion and the ferric hydroperoxy intermediate of CYP 450_{cam} were identified by EPR, ^{1}H- and ^{14}N-ENDOR studies.[64]

5.4.1 Theoretical EPR Spectroscopy

5.4.1.1 g-Values

EPR features stem from the interaction of the magnetic field and the molecular magnetic dipole moment $-\beta(\boldsymbol{L} + g_e\boldsymbol{S})$, where $\boldsymbol{L}$ is the total angular momentum operator, $\boldsymbol{S}$ is the total spin angular momentum operator, β is Bohr's magneton and g_e is the free electron g-value (*ca.* 2.002319). Fundamentally, there are two kinds of sources that contribute to the magnetic dipole moment. The first part $(-\beta g_e\boldsymbol{S})$ is the isotropic magnetic dipole moment of the electron, which is an intrinsic property of elementary particles. It is a constant and therefore not of chemical interest. The second contribution comes from the orbital motion (orbital angular momentum) of the unpaired electrons $(-\beta\boldsymbol{L})$. In most cases the system under investigation has an orbital non-degenerate ground state where the orbital angular momentum is "quenched." A limited amount of orbital angular momentum can be restored by spin–orbit coupling (SOC) between the ground state and the various excited states of the system. The additional magnetic dipole moment can be calculated from the small electric currents that are generated by the mixing of excited states into the ground state *via* SOC. This mixing is necessarily anisotropic, and hence so is the corresponding contribution to the magnetic dipole moment. In the formalism of the spin Hamiltonian (SH) the extra angular momentum is introduced by modifying the g-value of the free electron. Because the electronic motion is anisotropic, the modified g-values are described by a 3×3 matrix ($\boldsymbol{g}$-matrix) and the chemical information is contained in the g-shift ($g - g_e$), where g represents any of the (positive) square root eigenvalues of the symmetric matrix $\boldsymbol{g}\boldsymbol{g}^{\mathrm{T}}$.

The g-matrix has been well studied and there are several implementations and applications available.[65–70] As explained in detail elsewhere, the g-matrix contains the four contributions:[3,68]

$$g^{kl} = g_e\delta_{kl} + \Delta g^{\mathrm{RMC}}\delta_{kl} + \Delta g_{kl}^{\mathrm{GC}} + \Delta g_{kl}^{\mathrm{OZ/SOC}} \tag{5.3}$$

$$\Delta g^{\mathrm{RMC}} = -\frac{\alpha^2}{S}\sum_{\mu,\nu} P_{\mu\nu}^{\alpha-\beta}\left\langle\varphi_\mu\left|-\tfrac{1}{2}\nabla^2\right|\varphi_\nu\right\rangle \tag{5.4}$$

$$\Delta g_{kl}^{\mathrm{GC}} = \frac{\alpha^2}{4S} \sum_{\mu,\nu} P_{\mu\nu}^{\alpha-\beta} \left\langle \varphi_\mu \middle| \sum_A Z_A r_A^{-3} [r_A - r_{A,k} r_l] \middle| \varphi_\nu \right\rangle \tag{5.5}$$

$$\Delta g_{kl}^{\mathrm{(OZ/SOC)}} = \frac{1}{2S} \sum_{\mu,\nu} \frac{\partial P_{\mu\nu}^{\alpha-\beta}}{\partial B_k} \left\langle \varphi_\mu \middle| \hat{Z}_l^{\mathrm{SOMF}} \middle| \varphi_\nu \right\rangle \tag{5.6}$$

Here, the first term is the isotropic free electron g-value. The second and third terms correspond to the relativistic mass and gauge corrections, respectively. The final term represents the contribution from the orbital angular momentum, which is usually dominant over the second and third parts. The term α is the fine structure constant, S the total spin of the ground state, $P_{\mu\nu}^{\alpha-\beta}$ the spin density matrix, Z_A the nuclear charge for atom A, and $\hat{Z}_l^{\mathrm{SOMF}}$ is the SOC coupling operator modeled by the spin–orbit mean field approximation (SOMF).[71–73] In recent years, extensive test calculations on some well-defined systems have been conducted. From the results obtained so far in the DFT field, two conclusions have been reached:[74–79]

First, very good accuracy has been obtained for organic and inorganic radicals, *i.e.* typically within 500 ppm of the experimental values. For transition metal complexes, errors of a factor of ~ 2 in the g-shifts are not uncommon. Second, g values are not sensitive to density functional. Different functionals predict similar g values although hybrid functionals such as B3LYP tend to give more accurate results for transition metal complexes than GGA functional such as BP86.

An *ab initio* methodology that can be extended to reasonably large molecules is available.[80] In this case, it is also possible to resort to a spin–orbit-coupled configuration interaction (CI) treatment that avoids a perturbative treatment of SOC.[81,82]

5.4.1.2 Zero-Field Splitting

If the system contains more than one unpaired electron, even in the absence of magnetic field, degeneracy of the M_s magnetic sublevels is lifted by zero-field splitting (ZFS). It is well known that, starting from a spin-free wave function and introducing relativistic effects using perturbation theory up to second order, ZFSs have two physical origins:[83,84] (i) the direct electron–electron spin–spin coupling (SSC) to first order in perturbation theory and (ii) the SOC to second order in perturbation theory. The SSC part is usually the leading contribution for the ZFSs of organic radicals.[85] By contrast, the SOC contribution has long been believed to dominate the ZFS of transition metal complexes.[86] Analogous to the g-tensor, the SOC mixes different excited states into the distinct ground M_s microstates and hence splits the M_s members of a given total S multiplet. However, a recent study on $Mn(acac)_3$ suggested that SOC does not always strongly dominate and that the SSC term must be taken into account to arrive at quantitative results.[87]

In a sum-over-states formulation the SOC contribution to the ZFS (***D***-tensor) contains three terms that arise from the excited states of the same total spin as the ground state as well as the excited states differing by one unit of spin-angular momentum, explicitly:[88]

$$D_{kl}^{SOC} = D_{kl}^{SOC(0)} + D_{kl}^{SOC(-1)} + D_{kl}^{SOC(+1)} \tag{5.7}$$

with:

$$D_{kl}^{SOC(0)} = -\frac{1}{S^2} \sum_{b(S_b=S)} \Delta_b^{-1} \left\langle 0SS \left| \sum_i h_i^{k,SOC} s_{i,0} \right| bSS \right\rangle \left\langle bSS \left| \sum_i h_i^{l,SOC} s_{i,0} \right| 0SS \right\rangle \tag{5.8}$$

$$D_{kl}^{SOC(-1)} = -\frac{1}{S(2S-1)} \sum_{b(S_b=S-1)} \Delta_b^{-1} \left\langle 0SS \left| \sum_i h_i^{k,SOC} s_{i,+1} \right| bS-1S-1 \right\rangle$$
$$\left\langle bS-1S-1 \left| \sum_i h_i^{l,SOC} s_{i,-1} \right| 0SS \right\rangle \tag{5.9}$$

$$D_{kl}^{SOC(+1)} = -\frac{1}{(S+1)(2S+1)} \sum_{b(S_b=S+1)} \Delta_b^{-1} \left\langle 0SS \left| \sum_i h_i^{k,SOC} s_{i,-1} \right| bS+1S+1 \right\rangle$$
$$\left\langle bS+1S+1 \left| \sum_i h_i^{l,SOC} s_{i,+1} \right| 0SS \right\rangle \tag{5.10}$$

Here, S is the total spin of the ground state and the sums indexed by b refer to the electronically excited states $|bSS>$ of the appropriate total spin.

In the case of non-degenerate ground states where the SH is applicable, the ZFS tensor can also be described by a 3×3 tensor (**D**-tensor). The **D**-tensor can be reduced to only two parameters by choosing its principal axes: $D = D_{zz} - (D_{xx} + D_{yy})/2$ (ZFS parameter) and $E = (D_{xx} - D_{yy})/2$ ($0 \le E/D \le \frac{1}{3}$ where E is referred to as the rhombicity and measures the deviation from axial symmetry).[88]

Recently, quasi-degenerate perturbation theory (QDPT), which is based on *ab initio* methods such as complete active space self-consistent field (CASSCF), or multireference configuration interaction (MRCI), has been used to calculate ZFSs.[89] The SOC contributions of ZFSs are computed by direct diagonalization of the SOC operator using the basis of a preselected number of roots of the spin-free Hamiltonian. This amounts to an infinite order treatment of the SOC in terms of perturbation theory. In parallel, response equations of **D**-tensor were derived,[90] which are easily implemented in single determinant methods such as Hartree–Fock (HF) and DFT. For transition metal complexes the QDPT method delivers much better results than DFT methods.[91,92] In some cases the DFT calculation even predicts the wrong sign of D.[91]

5.4.1.3   Hyperfine Coupling

Hyperfine coupling (HFC) arises from the interaction of the unpaired electrons with the magnetic dipole moments of nearby magnetic nuclei. This interaction results in the so-called hyperfine splitting in EPR spectra if the nucleus is where the unpaired electron is formally localized, and superhyperfine splitting if the nucleus is different. In a transition metal compound, the hyperfine splitting is associated with the metal center, and the superhyperfine splitting with ligand nuclei. There are three different physical mechanisms that contribute to HFC:[93] (i) the non-classical Fermi contact interaction that originates from the presence of unpaired spin density at the nucleus in question; (ii) the magnetic interaction between the electron dipole and the nuclear dipole moment; and (iii) the interaction of electronic angular momentum with the magnetic dipole moment of the nucleus. Because only s-electrons have non-vanishing probabilities at the nucleus, the spin density required for the Fermi contact term comes either from mixing of s-orbitals into singly occupied molecular orbitals (SOMOs) or from spin polarization of the core s-orbitals. The second mechanism may be considered to be a classical magnetic dipolar interaction between two magnetic moments with the trace being zero, and is proportional to the inverse third power of the distance between the nucleus and the unpaired electrons.[94] The contribution from the third mechanism may, in ligand field theory, be expressed as a function of g-shifts. Thus, it is usually negligible for ligand nuclei, but has considerable magnitude for transition metal ions.

The Fermi contact and magnetic dipole contributions are first order properties and can be calculated directly as expectation values over the ground state wave function. To properly describe the core polarization, the calculation of the Fermi contact term requires the use of basis sets that are more flexible in the core region. However, in the case of iron-containing complexes a significant underestimation of the Fermi contact contribution has been observed. Therefore, scaling factors have to be used to correct computed values of the Fermi contact contribution to obtain quantitative results.[14]

5.4.2   Examples from the Literature

EPR spectroscopy has played an essential role in the characterization of short-lived enzymatic intermediates. Pioneering work was carried out by Noodleman on Fe/S proteins,[95–98] and theoretical EPR has since been used to study FeNO-porphyrins,[99–102] non-heme metal-nitrosyl systems,[103–105] plastocyanin,[106] d^1 metal-porphyrins[100] and cytochrome P450[26] systems. The studies of FeNO-porphyrins present a plausible explanation for the occurrence of two different species observed experimentally.[99] In another study Fe(I) and Fe(III) porphyrins were treated with the ZORA method, which was able to deal with the orbitally nearly degenerate low-spin d^5 configuration.[101] More recently, the ZORA method was used in conjunction with state-of-the-art single-crystal EPR experiments to obtain insight into the structures of the reaction intermediates in Ni-hydrogenases.[107–109] For example, Foerster and coworkers combined experimental and relativistic DFT studies on geometrically optimized model structures for the active site of the [NiFe] hydrogenase.[109] Pierce and coworkers also used the combination of EPR spectroscopy and DFT computational methods in probing the proposed cysteine dioxygenase catalytic cycle.[110]

Recently, high-valent Fe(IV) species have been investigated with DFT and correlated *ab initio* methods to predict their EPR properties.[26,111] For calibration purposes, some small model complexes were studied, *e.g.* [Fe(O)(TMC)-(CH$_3$CN)]$^{2+}$ (TMC = tetramethylcyclam). In the classical ligand field treatment, a direct proportionality between the g-shift and the ZFS was proposed. In contrast to this long-held assumption, Schöneboom *et al.* predicted a g-tensor with small anisotropy and a very large ZFS, which is in good agreement with the recent experimental data.[112] The discrepancy may be rationalized by the fact that only excited states of the same total spin ($\Delta S = 0$) contribute to the g-shift while excited states with $\Delta S = 0$, ± 1 contribute to the $\boldsymbol{D}$-tensor.[88] The calculated g-tensors for [Fe(O)(TMC)(CH$_3$CN)]$^{2+}$ complex are 2.015 ($\parallel$ to Fe=O), 2.024, and 2.026 ($\perp$ to Fe=O). For $\boldsymbol{D}$-tensor calculations, the computations indicate that there are contributions from singlet, triplet and quintet states. Schöneboom and coworkers made semi-quantitative estimates of the contributions of the leading excited states to the ZFS parameter of the (FeO)$^{2+}$ unit which are presented in Table 5.2.

Using the ligand field arguments in ref. 88 one obtains the following equations:

$$D \cong \zeta_{\mathrm{Fe}}^2 \left[\frac{1}{3} \frac{\alpha_{xy}^2 \alpha_{x^2-y^2}^2}{\Delta[^5A_1(1b_2 \rightarrow 2b_1)]} + \frac{\alpha_{xz,yz}^4}{\Delta[^1A_1(3e \rightarrow 3e)]} \right.$$
$$\left. + \frac{1}{4} \left(\frac{\alpha_{xz,yz}^2 \alpha_{xy}^2}{\Delta[^3E(1b_2 \rightarrow 3e)]} + \frac{\alpha_{xz,yz}^2 \alpha_{x^2-y^2}^2}{\Delta[^3E_1(2e \rightarrow 1b_1)]} + \frac{3\alpha_{xz,yz}^2 \alpha_{z^2}^2}{\Delta[^3E(3e \rightarrow 4a_1)]} \right) \right] \quad (5.11)$$

These equations emphasize the important role of the low-lying singlet and quintet states that act in concert to produce the very large and positive D that is characteristic of (FeO)$^{2+}$ systems. The final value of $D = +22.5$ cm^{-1} in Table 5.2 is fairly realistic and comes close to the experimental result of $+26.95$ cm^{-1}.[112] A comparison of the individual contribution of all excited

Table 5.2 Individual contributions to the D value in the [Fe(IV)(O)-(TMC)(CH$_3$CN)]$^{2+}$ complex.

State^a	Energy (cm^{-1})	Contribution to D (cm^{-1})
$^5A_1(1b_2 \rightarrow 1b_2)$	2402	+14.7
$^1A_1(2e \rightarrow 2e)$	10 000	+2.9
$^3E(1b_2 \rightarrow 2e)$	8065	+2.9
$^3E(2e \rightarrow 1b_2)$	15 320	+1.0
$^3E(2e \rightarrow 2a_1)$	16 530	+0.9
Total		+22.5

^{a}See Section 5.5 for discussion of the excited states of [Fe(IV)(O)(TMC)(CH$_3$CN)]$^{2+}$.

states shows that the lowest lying quintet state gives the leading contribution. Thus, the result obviously depends very much on the position of the first quintet state.

As a good example we discuss the ^{57}Fe HFCs for the different spin states of ferryl complexes.[31] Calculations predict that the parallel component (along the Fe–O bond) of the HFC for the quintet Fe(IV)oxo is significantly larger than that of the triplet species. This may be interpreted by basic ligand field theory as follows:

$$A_{\parallel}(^{3}A_{2g}) \cong P_{Fe}\left\{ \frac{4\pi}{3}\rho(0) + \frac{2}{7}\langle r^{-3}\rangle_{3d}\alpha^{2}_{xz,yz} \right\} \tag{5.12}$$

$$A_{\perp}(^{3}A_{2g}) \cong P_{Fe}\left\{ \frac{4\pi}{3}\rho(0) + \frac{1}{7}\langle r^{-3}\rangle_{3d}\alpha^{2}_{xz,yz} \right\} \tag{5.13}$$

$$A_{\parallel}(^{5}A_{1g}) \cong P_{Fe}\left\{ \frac{4\pi}{6}\rho(0) + \frac{1}{7}\langle r^{-3}\rangle_{3d}\left[\alpha^{2}_{xz,yz} - \alpha^{2}_{xy} - \alpha^{2}_{x^{2}-y^{2}}\right] \right\} \tag{5.14}$$

$$A_{\perp}(^{5}A_{1g}) \cong P_{Fe}\left\{ \frac{4\pi}{6}\rho(0) + \frac{1}{14}\langle r^{-3}\rangle_{3d}\left[\alpha^{2}_{xz,yz} - \alpha^{2}_{xy} - \alpha^{2}_{x^{2}-y^{2}}\right] \right\} \tag{5.15}$$

As seen in Table 5.3, the larger HFC for the quintet state may come from two contributions: (i) the isotropic Fermi contact contribution and (ii) the dipolar contribution. Despite the fact that the $S=2$ state has a smaller prefactor for the isotropic term, the core polarization in the presence of four unpaired electrons of the high spin Fe(IV) center is much more effective than that of the intermediate spin Fe(IV) counterpart and, consequently, the isotropic ^{57}Fe-HFC is predicted to be roughly a factor of two greater in magnitude for $S=2$ compared to $S=1$. For the dipolar contribution the factor of two holds in the limit of isotropic covalency (all αs are equal), but it is evident that the anisotropic covalency will enhance the dipolar HFCs of the $S=2$ state as $\alpha^{2}_{xz,yz} < \alpha^{2}_{x^{2}-y^{2}} < \alpha^{2}_{xy}$. Since the core polarization will give rise to a negative spin density at the iron nucleus, it is therefore expected that for the triplet ground state of ferryl species the isotropic and dipolar contributions partially cancel each other along the Fe–O bond direction while they will reinforce each other

Table 5.3 Contribution to the ^{57}Fe hyperfine interaction as determined by spin polarized B3LYP calculations (MHz).

	$S=1$		$S=2$	
	$A_{\perp}$	$A_{\parallel}$	$A_{\perp}$	$A_{\parallel}$
Isotropic[a]	−20.9	−20.9	−36.7	−36.7
Dipolar	−7.6	+15.2	+6.5	−13.1
Spin–orbit	−1.6	+0.8	−0.6	−0.5
Total	−30.1	−4.9	−29.6	−50.3

[a]According to the results of ref. 14 the isotropic term was scaled by 1.8 to compensate for the intrinsic underestimation of the core polarization by DFT methods.

Table 5.4 Spectroscopic parameters of selected oxoiron(IV) complexes with varying ground state spin multiplicities.

	S	A_x	A_y	A_z
TauD-J[a]	2	−18.4	−17.6	−31.0
$[Fe(IV)(O)(TMG_3tren)]^{2+}$ [b]	2	−15.5	−14.8	−28.0
$[Fe(IV)(O)(TMC)(NCCH_3)]^{2+}$ [c]	1	−22.6	−18.3	−2.9

[a]Data from refs 40 and 113–115.
[b]$TMG_3tren=N[CH_2CH_2N=C(NMe_2)_2]$, data from ref. 116.
[c]Data from refs 117 and 118.

for the quintet ground state. The predicted signs of all HFC components match the expectations from ligand field theory. As shown in Table 5.3, irrespective of spin multiplicities, the calculated orbital contributions (SOC) are negligible, which is in agreement with the very small g-shifts of Fe(IV)oxo complexes. For the model complex $[Fe(IV)(O)(NH_3)_5]^{2+}$ the partial cancellation of the isotropic and dipolar contributions in the $S=1$ state leads to a total HFC value of −4.9, while the reinforcement of them gives a value of −50.3 in the $S=2$ state. Consequently, the magnitude of the HFC component parallel to the Fe–O bond relative to those perpendicular to the Fe–O bond is diagnostic for the spin state of the $(FeO)^{2+}$ core. This prediction is in excellent agreement with the available experimental data (Table 5.4).

5.5 Absorption Spectroscopy

Absorption spectroscopy probes the transitions between many particle electronic energy levels induced by electromagnetic radiation. For transition metal containing systems there are mainly five types of transitions that fall in the UV–visible range: (i) d–d transitions, (ii) ligand-to-metal charge transfer (LMCT), (iii) metal-to-ligand charge transfer (MLCT), (iv) intra-ligand and (v) ligand-to-ligand charge transfers. The energy of these transitions contains information about the electronic and geometric structure.

The spectroscopy of d–d transitions has been long interpreted using ligand field theory but despite its usefulness in interpretation, ligand field theory cannot be used to predict spectra. Semi-empirical approaches such as INDO/S have been very successful in the past[119–122] but more rigorous approaches are desirable.

5.5.1 Theoretical Prediction of Absorption Spectroscopy

There are several ways of predicting transition energies using theoretical methods. One of the simplest is the ΔSCF method with DFT,[123–129] otherwise known as the Slater transition state vertical self-consistent reaction field methodology,[130] where an electron is excited from the donor to the acceptor molecular orbital and the change in energy after convergence is taken as the excitation energy. Greater accuracy can be achieved with wave-function-based methods such as the complete active space SCF with second-order perturbation (CAS-PT2). Multireference configuration interaction (MRCI) also gives good results. The theory of these methods is complex and is outlined in some detail in ref. 131. While these methods give excellent accuracy, they are computationally demanding especially for transition metal systems and absorption spectra are more commonly calculated with time-dependent density functional theory (TD-DFT).

TD-DFT rests on the Runge–Gross theorem, which states that there exists a one to one mapping from an external potential to the electron density of a system, which has evolved from a certain initial state. This has the consequence that when the initial state and the density are known the potential can be determined and the time-dependent Kohn–Sham equation can be solved. In other words all properties of a system can be determined from the electron density (see, for example, ref. 132 for an introduction to the theory).

While standard DFT formulations are concerned with the ground state density and properties, excitation energies are probed in TD-DFT *via* linear response theory. In this approach the excitation energies are obtained from the way the system responds to a small time-dependent perturbation.[132]

TD-DFT often overestimates d–d transitions and underestimates charge transfer bands for transition metal systems.[3] A further limitation is that TD-DFT probes only one-electron excitations so it is not appropriate to treat systems where double excitations are important such as $[Ni^{II}(H_2O)_6]^{2+}$.[133] For more detailed discussions about theoretical absorption spectra see refs 3 and 133.

5.5.2   Examples from the Literature

TD-DFT and standard ground state DFT are often used in combination as tools to understand the electronic structure of complexes and to assign transitions in experimental spectra. One of the earliest TD-DFT studies of iron-containing systems was the study of the electronic structure and d–d transitions of the $[Fe^{II}(Cys)_4]^{2-}$ core of rubredoxins.[134] Since then, TD-DFT has been used to study the absorption spectra of a range of enzymes and biomimetic complexes, including intradiol dioxygenases,[135] phthalocyanine complexes,[136] high-valent Fe(IV)oxo species,[137] diazide Fe(III) model complexes for superoxide dismutase (SOD),[138] diiron reactive centers such as those found in intermediate X from ribonucleotide reductase subunit R2 (RNR-R2),[139] and models of the 2Fe subcluster of the Fe-only hydrogenases.[140]

The absorption spectra of RNR-R2 have also been studied using the ΔSCF method,[130] and Compound I of cytochrome P450 and models of SOD have been studied with semi-empirical INDO / S methods.[141–143] Several other model complexes have also been studied with these methods, *e.g.* [Fe(TPP)(Cl)] (TPP = tetraphenylporphyrin).[144] Along with TD-DFT studies, the MRCI method in the form of spectroscopy oriented CI (SORCI)[145] has been used for the study of Fe(IV)=O(L) (L = NH_3, H_2O) complexes[31] and Compound I of cytochrome P450.[26]

5.5.2.1   High-Valent Iron-oxo

The absorption spectrum of the high-valent iron model complex, $[Fe^{IV}(O)(NH_3)_4(H_2O)]^{2+}$ has been calculated by both TD-DFT and SORCI methods.[26,118,137] Figure 5.5 shows the key transitions and the calculated transition energies are included in Table 5.5.

Promotion of one electron from the $Fe\text{-}d_{xy}$ based $1b_2$ MO to the singly occupied degenerate $2e$-set, which are π-antibonding MOs involving the $Fe\text{-}d_{xz,yz}$ and $O\text{-}p_{x,y}$ orbitals, results in a triplet $^3E(1b_2 \rightarrow 2e)$ excited state. The single excitation of $1b_2 \rightarrow 1b_1$ $(Fe\text{-}d_{x2-y2})$ leads to a series of excited states due to different spin coupling schemes, which are of $^{3,5}A_1(1b_2 \rightarrow 1b_1)$, $^3A_2(1b_2 \rightarrow 1b_1)$, $^3B_1(1b_2 \rightarrow 1b_1)$ and $^3B_2(1b_2 \rightarrow 1b_1)$ symmetry, respectively. Similarly the single excitation of $1b_2 \rightarrow 2a_1$ $(Fe\text{-}d_{z2})$ gives rise to excited states of $^{3,5}B_1(1b_2 \rightarrow 2a_1)$, $^3A_1(1b_2 \rightarrow 2a_1)$, $^3A_2(1b_2 \rightarrow 2a_1)$ and $^3B_2(1b_2 \rightarrow 2a_1)$ symmetry.

TD-DFT is based on the linear response formalism starting from the single Kohn–Sham determinant of the ground state. Thus, it is not able to properly describe multiplet effects and spin-coupling, for which explicitly spin-coupled multireference methods such SORCI are required. As a consequence (Table 5.5), TD-DFT calculations yields only two roots for the excitations of $(1b_2 \rightarrow 1b_1)$ and $(1b_2 \rightarrow 2a_1)$, respectively, instead of five as required by spin-coupling.

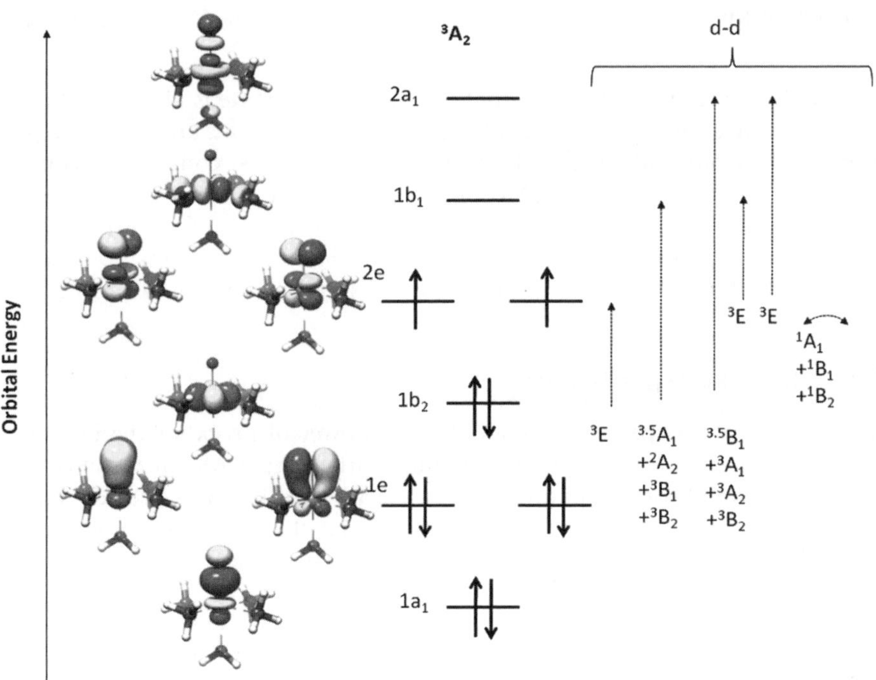

Figure 5.5   Metal d-based MOs and term symbols arising from single excitations for $[Fe(O)(NH_3)_4(H_2O)]^{2+}$. The indicated orbital occupation pattern corresponds to the $^3A_2$ ground state.

Table 5.5 Calculated d–d triplet → triplet vertical excitation energies (in eV relative to the lowest $^3A_2$ state) of $[Fe(O)(NH_3)_4(H_2O)]^{2+}$ using the SORCI and B3LYP methods. The calculations refer to vertical excitations energies at the B3LYP optimized geometry of the ground state. Calculated oscillator strengths ($\times 10^4$) are given in parentheses.[a]

State (transition)	SORCI	B3LYP	Exp[146]
$^3E(1b_2 \to 2e)$	1.00 (0.02)	1.95 (6.1)	1.3
$^3E(2e \to 1b_1)$	1.90 (20.9)	2.20 (51.2)	1.6
$^3E(2e \to 2a_1)$	2.05 (7.0)	2.61 (9.0)	2.1
$^3\Gamma(1b_2 \to 1b_1)$	1.32 (0.00)	1.79[b] (0.00)	~1.1
	1.37 (0.00)	2.53[b] (0.00)	–
	1.56 (0.00)	–	–
	1.60 (0.00)	–	–
	2.08 (0.00)	–	–
$^3\Gamma(1b_2 \to 2a_1)$	2.77 (0.00)	2.86[b] (0.00)	3.2
	2.84 (0.03)	4.20[b] (0.00)	–
	2.93 (0.16)	–	–
	3.20 (0.09)	–	–
	3.35 (0.01)	–	–

[a]Since our model does not have strict C_{4v} symmetry small splittings within formally doubly degenerate states arise (<0.05 eV). The transition energies in the table were consequently averaged and the oscillator strengths were added.
[b]These roots are of mixed symmetry since the spin unrestricted calculation is unable to give the correct multiplets.

The SORCI calculation predicts that the first excited state is about 1 eV and a series of d–d excited states extends to ~3.5 eV with the most intense peak originating from the $^3E(2e \to 1b_1)$ transition. This is consistent with the experimental results of $[Fe^{IV}O(TMC)(NCMe)]^{2+}$, which show a weak absorption band between 600 and 700 nm (1.7–2.0 eV) tailing off into the near-IR region. By contrast, the TD-DFT calculation estimates most transition energies to be too high (sometimes by more than 1 eV) and not balanced (Table 5.5).

5.6   X-Ray Spectroscopy

X-Ray absorption spectroscopy (XAS) is a powerful probe of the local geometric and electronic structure of a photoabsorber. An XAS edge results when a core electron absorbs a photon whose energy is equal to or greater than its bonding energy. The prominent feature of XAS is that the position of the edge is element specific. Therefore, it is a widely used spectroscopic technique in the field of (bio)inorganic chemistry and material science.[147]

The XAS spectra can be divided into three regions[148] (Figure 5.6): (i) the pre-edge region involving transitions from core-orbitals into valence orbitals, (ii) the edge region consisting of transitions from the core-level into high lying empty orbitals close to the continuum and (iii) transitions from the core-level into the continuum leading to the extended X-ray absorption fine structure (EXAFS). A detailed analysis of the pre-edge and edge regions provides

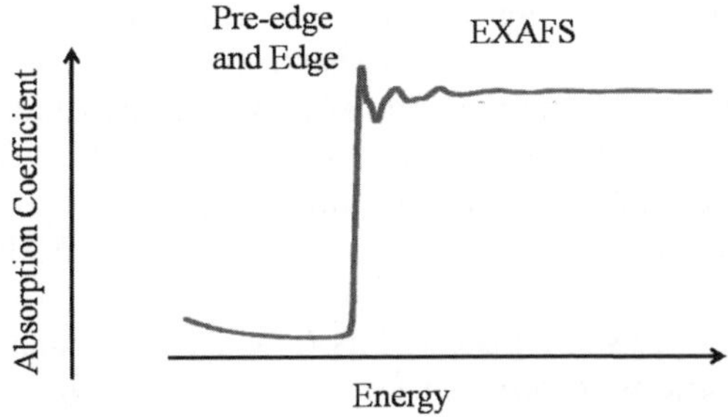

Figure 5.6 Schematic demonstration of an X-ray absorption spectrum with its spectroscopical domains.

information about the electronic structure of the absorbing atoms such as the physical oxidation state, the coordination geometry and the metal–ligand covalency. The EXAFS region is used to determine metal–ligand bond distances and coordination numbers.

5.6.1 Theoretical Prediction of Metal and Ligand *K*-Edge Spectra

Here we focus on the calculation of metal and ligand *K*-edge spectra. In fact, there is a considerable history of DFT-based calculations of pre-edge XAS spectra.[147] On a most elementary level, the intensity of the pre-edge peak from a metal 1s- into a metal d-based molecular orbital (MO) Ψ is given by:

$$I(1s \rightarrow \Psi) = a\alpha^2 \tag{5.16}$$

Where a is a proportionality factor and α is the MO coefficient of the metal in the electron acceptor Ψ. In the most straightforward approach, metal–ligand covalencies are determined by a population analysis of the ground-state DFT calculations and can be used to interpret the intensity distribution of the experimental XAS pre-edge peaks *via* a semiempirical calibration procedure. For such analyses standardized effective values for the radial transition moment integral must be defined. This approach has been thoroughly explored by Solomon and coworkers.[149–155] The second type of approach is based on the Slater transition state concept also used for the prediction of UV–vis spectra.[123–129] Using this method each transition needs to be calculated independently, and hence many individual SCF calculations have to be performed to construct the spectrum and the calculated excited states are not orthogonal. However, this method has the advantage that electronic relaxation effects in the course of core-electron excitation are properly treated.

There are several differences between calculations of core-to-valence transitions and those of valence-to-valence transitions:

1. Unlike valence-to-valence transitions in the UV/Vis spectral range, for core-to-valence transitions the long wavelength approximation does not

hold where the wavelength of the radiation in XAS is not necessarily larger than the size of the absorbers, especially for the metal *K*-edge region. Thus in XAS magnetic dipole and electric quadrupole transitions show comparable intensities with electric dipole transitions.

2. In XAS relativistic effects become very important because X-ray absorption generates a "hole" in the core where electrons approach relativistic velocities.

3. If one applies the same protocol as that used for calculating valence-to-valence transitions to XAS spectra, a very large number of roots would be required to cover the entire range of the spectrum. Hence, a method that focuses on the pre-edge region is required.

More recently, a simple and effective protocol for calculating pre-edge spectra based on DFT linear response theory has been proposed.[156–159] In these calculations, excitations are only allowed out of localized core-holes into the entire virtual space of MOs. Thus, in the standard TD-DFT treatment the excitations included fall into the *K*-edge region of the given absorber. The initial localization of the core-hole is consistent with the so-called sudden approximation.[160] The inclusion of scalar relativistic effects and large and flexible basis sets in the core region has not been found to improve correlation with experiment. This approach does not provide accurate absolute transition energies because the present day DFT potentials have the wrong asymptotic behavior not only in the long range but also close to the nucleus. However, relative transition energies for series of complexes or different transitions of the same species are usually well predicted with an error of a few tenths of an eV. Therefore, a constant shift can be applied to each absorber for a given DFT functional and basis set to yield quantitative transition energies. More importantly, the intensity distributions can be accurately reproduced by this simple approach and the overall calculations are very efficient and successful.[156–159]

5.6.2   Examples from the Literature

Solomon and coworkers carried out systematic studies of a series of synthetic iron complexes, using ligand field arguments to understand the energy splitting and intensity distribution.[161] Arrio and coworkers have calculated quadrupole transition intensity in iron minerals also using ligand field theory.[162]

Iron *K*-edge XAS has had an important impact on our understanding of high-valent iron intermediates. The XAS pre-edge feature is formally a 1s to 3d quadrupole allowed transition, which is much weaker than the dipole allowed 1s to 4p "main" edge transition. However, studies by Que,[163–165] Wieghardt[166,167] and coworkers have clearly shown that high-valent iron-oxo and -nitrido model complexes are characterized by pre-edges that have higher energy and greater intensity than well-known ferrous and ferric complexes. The increased energy may be attributed to the higher effective nuclear charge of the high-valent iron center. The higher intensity is due to the very short Fe-oxo/nitride bonds

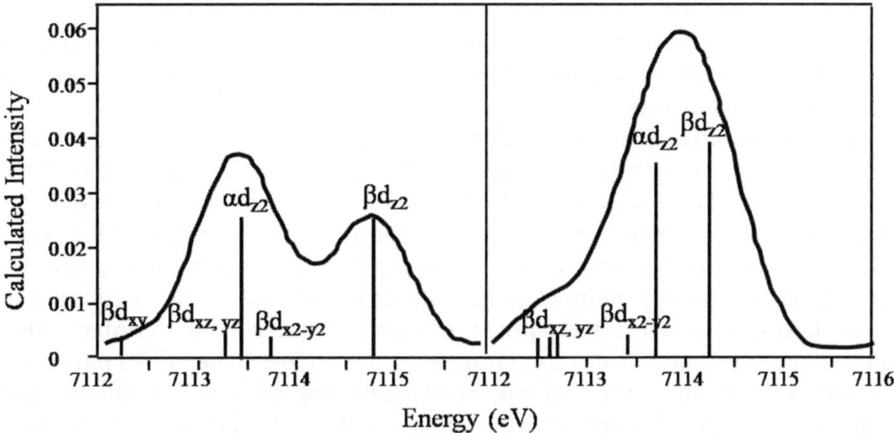

Figure 5.7　Comparison of the calculated pre-edge spectra for intermediate spin complex $[Fe(IV)(O)(NH_3)_4(OH)]^+$ (right) and the corresponding high-spin partner (left).

in these complexes, which leads to a $4p_z$-$3d_{z2}$ metal-orbital mixing and hence to "borrowing" of the intensity from the edge transition.[168–170]

Calculations on the model complex $[Fe^{IV}(O)(NH_3)_4(OH)]^+$ predict that there are mainly two peaks in the pre-edge spectra of high-spin Fe(IV)=O complexes ($S = 2$), while the spectrum of the corresponding triplet partner only shows one peak, as depicted in Figure 5.7. In fact, for both species pre-edges are dominated by transitions to the d_{z2} based orbitals. However, in the case of the high-spin state a greater degree of spin polarization occurs, which lowers the energy of the alpha d_{z2}-orbital and hence results in considerable splitting of the alpha and beta d_{z2}-orbital. This prediction was verified recently in the high-resolution XAS studies of the high-spin ferryl complex $[Fe^{IV}(O)(TMG_3tren)]^{2+}$.[116]

The method of X-ray emission spectroscopy (XES) is a complementary technique to XAS.[171] Kβ XES arises from the emission of photons by the decay of electrons after ionization of the metal 1s electron.[171] Only a few Kβ XES studies on transition metal complexes have been reported thus far.[172,173] One example is a systematic study of a series of ferrous and ferric complexes using Fe Kβ XES.[174] A DFT based approach was employed to compute valence-to-core Kβ XES spectra with surprisingly high success. The transition energies and intensities distribution were demonstrated to be sensitive to spin state, ligand identity, ligand ionization energy and metal–ligand bond distance. Good agreement between the experimental and calculated XES spectra was observed; thus, one may anticipate widespread application of XES to bioinorganic chemistry in the near future.

5.7　Conclusion

In the present chapter, the underlying physics of theoretical spectroscopy has been briefly reviewed. The focus is laid on how experimental and theoretical spectroscopy can be combined to gain more insight from the experimental spectra. As shown here theoretical spectroscopy has been applied to a varied set of problems of iron-containing systems from heme and non-heme enzymes, to a range of interesting model complexes.

Of the theoretical spectroscopies reviewed in this chapter, Mössbauer and EPR spectroscopy have been the most widely used to date. This is due to the relative ease and accuracy with which they can be calculated, and their selectivity with respect to the central metal ion in the protein environment. Thus they are often used to differentiate between proposed models for key reactive intermediates, and to explore electronic structures in detail. The calculation of absorption spectra is also widespread, usually with the purpose of interpreting ambiguous experimental spectra. However, accurate predictions – in particular for spin coupled sites – are generally challenging and frequently require the use of laborious multireference *ab initio* methods. The calculation of XAS, XES and NRVS spectroscopies is a growing field of investigation and we hope that in this chapter we have at least hinted at the possibilities that these techniques offer for the study of iron-containing enzymes.

References

1. L. Que Jr., *Physical Methods in Bioinorganic Chemistry*, University Science Books, Sausalito, CA, 2000.
2. E. I. Solomon and A. Lever, *Inorganic Electronic Structure and Spectroscopy*, John Wiley & Sons, Inc., New York, 2000.
3. F. Neese, *Coord. Chem. Rev.*, 2009, **253**, 526.
4. S. Ye and F. Neese, *Inorg. Chem.*, 2010, **49**, 772.
5. C. Krebs, D. G. Fujimori, C. T. Walsh and J. M. Bollinger Jr., *Acc. Chem. Res.*, 2007, **40**, 484.
6. L. Que Jr., *Acc. Chem. Res.*, 2007, **40**, 493.
7. E. Münck, in *Physical Methods in Bioinorganic Chemistry*, ed. L. Que Jr., University Science Books, Sausalito, 2000, p. 287.
8. C. Krebs and J. M. Bollinger Jr., *Photosynth. Res.*, 2009, **102**, 295.
9. P. Gütlich, E. Bill and A. X. Trautwein, *Mössbauer Spectroscopy and Transition Metal Chemistry: Fundamentals and Application*, Springer, Berlin, 2010.
10. J. F. Berry, S. DeBeer George and F. Neese, *Phys. Chem. Chem. Phys.*, 2008, **10**, 4361.
11. M. Filatov, *J. Chem. Phys.*, 2007, **127**, 084101.
12. M. Filatov, *Coord. Chem. Rev.*, 2009, **253**, 594.
13. R. Kurian and M. Filatov, *J. Chem. Theor. Comput.*, 2008, **4**, 278.
14. S. Sinnecker, L. D. Slep, E. Bill and F. Neese, *Inorg. Chem.*, 2005, **44**, 2245.
15. F. Neese, *Inorg. Chim. Acta*, 2002, **337**, 181.
16. T. Q. Liu, T. Lovell, W. G. Han and L. Noodleman, *Inorg. Chem.*, 2003, **42**, 5244.

17. W. G. Han, T. Q. Liu, T. Lovell and L. Noodleman, *J. Comput. Chem.*, 2006, **27**, 1292.
18. M. Römelt, S. Ye and F. Neese, *Inorg. Chem.*, 2009, **48**, 784.
19. V. Vrajmasu, E. Münck and E. L. Bominaar, *Inorg. Chem.*, 2003, **42**, 5974.
20. D. M. A. Smith, M. Dupuis, E. R. Vorpagel and T. P. Straatsma, *J. Am. Chem. Soc.*, 2003, **125**, 2711.
21. V. N. Nemykin, N. Kobayashi, V. Y. Chernii and V. K. Belsky, *Eur. J. Inorg. Chem.*, 2001, 733.
22. O. Zakharieva, V. Schünemann, M. Gerdan, S. Licoccia, S. Cai, F. A. Walker and A. X. Trautwein, *J. Am. Chem. Soc.*, 2002, **124**, 6636.
23. Y. Zhang, E. Oldfield, *J. Phys. Chem. B*, 2003, 107, 7180.
24. A. Chanda, F. T. de Oliveira, T. J. Collins, E. Münck and E. L. Bominaar, *Inorg. Chem.*, 2008, **47**, 9372.
25. K. Ray, A. Begum, T. Weyhermüller, S. Piligkos, J. van Slageren, F. Neese and K. Wieghardt, *J. Am. Chem. Soc.*, 2005, **127**, 4403.
26. J. C. Schöneboom, F. Neese and W. Thiel, *J. Am. Chem. Soc.*, 2005, **127**, 5840.
27. R. H. Havlin, N. Godbout, R. Salzmann, M. Wojdelski, W. Arnold, C. E. Schulz and E. Oldfield, *J. Am. Chem. Soc.*, 1998, **120**, 3144.
28. Y. Ling and Y. Zhang, *J. Am. Chem. Soc.*, 2009, **131**, 6386.
29. Y. Zhang and E. Oldfield, *J. Am. Chem. Soc.*, 2004, **126**, 4470.
30. Y. Zhang and E. Oldfield, *J. Am. Chem. Soc.*, 2004, **126**, 9494.
31. F. Neese, *J. Inorg. Biochem.*, 2006, **100**, 716.
32. S. Ye, J. C. Price, E. W. Barr, M. T. Green, J. M. Bollinger Jr., C. Krebs and F. Neese, *J. Am. Chem. Soc.*, 2010, **132**, 4739.
33. T. Lovell, J. Li, T. Q. Liu, D. A. Case and L. Noodleman, *J. Am. Chem. Soc.*, 2001, **123**, 12392.
34. T. Lovell, W. G. Han, T. Q. Liu and L. Noodleman, *J. Am. Chem. Soc.*, 2002, **124**, 5890.
35. W. G. Han, T. Lovell, T. Q. Liu and L. Noodleman, *Inorg. Chem.*, 2003, **42**, 2751.
36. W. G. Han, T. Lovell, T. Q. Liu and L. Noodleman, *Inorg. Chem.*, 2004, **43**, 613.
37. W. G. Han, T. Q. Liu, T. Lovell and L. Noodleman, *J. Am. Chem. Soc.*, 2005, **127**, 15778.
38. W. G. Han and L. Noodleman, *Dalton. Trans.*, 2009, 6045.
39. M. Martinho, G. Q. Xue, A. T. Fiedler, L. Que Jr., E. L. Bominaar and E. Münck, *J. Am. Chem. Soc.*, 2009, **131**, 5823.
40. S. Sinnecker, N. Svensen, E. W. Barr, S. Ye, J. M. Bollinger Jr., F. Neese and C. Krebs, *J. Am. Chem. Soc.*, 2007, **129**, 6168.
41. J. T. Groves and G. A. McClusky, *J. Am. Chem. Soc.*, 1976, **98**, 859.
42. R. K. Behan, L. M. Hoffart, K. L. Stone, C. Krebs and M. T. Green, *J. Am. Chem. Soc.*, 2006, **128**, 11471.
43. M. T. Green, J. H. Dawson and H. B. Gray, *Science*, 2004, **304**, 1653.
44. K. L. Stone, R. K. Behan and M. T. Green, *Proc. Natl. Acad. Sci. USA*, 2006, **103**, 12307.

45. K. L. Stone, L. M. Hoffart, R. K. Behan, C. Krebs and M. T. Green, *J. Am. Chem. Soc.*, 2006, **128**, 6147.

46. V. Starovoitova, T. E. Budarz, G. R. A. Wyllie, W. R. Scheidt, W. Sturhahn, E. E. Alp, E. W. Prohofsky and S. M. Durbin, *J. Phys. Chem. B*, 2006, **110**, 13277.

47. W. Q. Zeng, N. J. Silvernail, D. C. Wharton, G. Y. Georgiev, B. M. Leu, W. R. Scheidt, J. Y. Zhao, W. Sturhahn, E. E. Alp and J. T. Sage, *J. Am. Chem. Soc.*, 2005, **127**, 11200.

48. B. M. Leu, M. Z. Zgierski, G. R. A. Wyllie, M. K. Ellison, W. R. Scheidt, W. Sturhahn, E. E. Alp, S. M. Durbin and J. T. Sage, *J. Phys. Chem. Solids*, 2005, **66**, 2250.

49. B. M. Leu, M. Z. Zgierski, G. R. A. Wyllie, W. R. Scheidt, W. Sturhahn, E. E. Alp, S. M. Durbin and J. T. Sage, *J. Am. Chem. Soc.*, 2004, **126**, 4211.

50. H. Paulsen, V. Schünernann, A. X. Trautwein and H. Winkler, *Coord. Chem. Rev.*, 2005, **249**, 255.

51. H. Paulsen, P. Wegner, H. Winkler, J. A. Wolny, L. H. Bottger, A. X. Trautwein, C. Schmidt, V. Schünemann, G. Barone, A. Silvestri, G. La Manna, A. I. Chumakov, I. Sergueev and R. Ruffer, *Hyperfine Interact.*, 2005, **165**, 17.

52. A. X. Trautwein, P. Wegner, H. Winkler, H. Paulsen, V. Schünemann, C. Schmidt, A. I. Chumakov and R. Ruffer, *Hyperfine Interact.*, 2005, **165**, 295.

53. W. R. Scheidt, S. M. Durbin and J. T. Sage, *J. Inorg. Biochem.*, 2005, **99**, 60.

54. J. T. Sage, C. Paxson, G. R. A. Wyllie, W. Sturhahn, S. M. Durbin, P. M. Champion, E. E. Alp and W. R. Scheidt, *J. Phys. Condens. Matter*, 2001, **13**, 7707.

55. T. Petrenko, W. Sturhahn and F. Neese, *Hyperfine Interact.*, 2007, **175**, 165.

56. K. L. Adams, S. Tsoi, J. S. Yan, S. M. Durbin, A. K. Ramdas, W. A. Cramer, W. Sturhahn, E. E. Alp and C. Schulz, *J. Phys. Chem. B*, 2006, **110**, 530.

57. B. K. Rai, S. M. Durbin, E. W. Prohofsky, J. T. Sage, G. R. A. Wyllie, W. R. Scheidt, W. Sturhahn and E. E. Alp, *Biophys. J.*, 2002, **82**, 2951.

58. B. K. Rai, S. M. Durbin, E. W. Prohofsky, J. T. Sage, M. K. Ellison, A. Roth, W. R. Scheidt, W. Sturhahn and E. E. Alp, *J. Am. Chem. Soc.*, 2003, **125**, 6927.

59. T. E. Budarz, E. W. Prohofsky, S. M. Durbin, T. Sjodin, J. T. Sage, W. Sturhahn and E. E. Alp, *J. Phys. Chem. B*, 2003, **107**, 11170.

60. Y. M. Xiao, H. X. Wang, S. J. George, M. C. Smith, M. W. W. Adams, F. E. Jenney, W. Sturhahn, E. E. Alp, J. O. Zhao, Y. Yoda, A. Dey, E. I. Solomon and S. P. Cramer, *J. Am. Chem. Soc.*, 2005, **127**, 14596.

61. Y. M. Xiao, K. Fisher, M. C. Smith, W. E. Newton, D. A. Case, S. J. George, H. X. Wang, W. Sturhahn, E. E. Alp, J. Y. Zhao, Y. Yoda and S. P. Cramer, *J. Am. Chem. Soc.*, 2006, **128**, 7608.

62. C. B. Bell, S. D. Wong, Y. M. Xiao, E. J. Klinker, A. L. Tenderholt, M. C. Smith, J. U. Rohde, L. Que Jr., S. P. Cramer and E. I. Solomon, *Angew. Chem., Int. Ed.*, 2008, **47**, 9071.
63. T. Petrenko, S. DeBeer George, N. Aliaga-Alcalde, E. Bill, B. Mienert, Y. Xiao, Y. Guo, W. Sturhahn, S. P. Cramer, K. Wieghardt and F. Neese, *J. Am. Chem. Soc.*, 2007, **129**, 11053.
64. R. Davydov, T. M. Makris, V. Kofman, D. E. Werst, S. G. Sligar and B. M. Hoffman, *J. Am. Chem. Soc.*, 2001, **123**, 1403.
65. F. Neese, *Specialist Periodical Reports on EPR Spectroscopy*, Royal Society of Chemistry, Cambridge, 2007.
66. F. Neese, *J. Biol. Inorg. Chem.*, 2006, **11**, 702.
67. F. Neese, *Curr. Opin. Chem. Biol.*, 2003, **7**, 125.
68. F. Neese, in *Biological Magnetic Resonance*, ed. G. Hanson and L. Berliner, Springer, 2009, vol. 28, p. 175.
69. M. Kaupp, *EPR Spectroscopy of Free Radicals in Solids. Trends in Methods and Applications*, Kluwer, Dordrecht, 2002.
70. S. Patchkovskii, *Calculation of NMR and EPR Parameters*, Wiley-VCH Verlag, Weinheim, 2004.
71. B. A. Hess, C. M. Marian, U. Wahlgren and O. Gropen, *Chem. Phys. Lett.*, 1996, **251**, 365.
72. A. Berning, M. Schweizer, H. J. Werner, P. J. Knowles and P. Palmieri, *Mol. Phys.*, 2000, **98**, 1823.
73. F. Neese, *J. Chem. Phys.*, 2005, **122**, 034107.
74. G. Schreckenbach and T. Ziegler, *J. Phys. Chem. A*, 1997, **101**, 3388.
75. O. L. Malkina, J. Vaara, B. Schimmelpfennig, M. Munzarova, V. G. Malkin and M. Kaupp, *J. Am. Chem. Soc.*, 2000, **122**, 9206.
76. M. Kaupp, R. Reviakine, O. L. Malkina, A. Arbuznikov, B. Schimmelpfennig and V. G. Malkin, *J. Comput. Chem.*, 2002, **23**, 794.
77. F. Neese, *J. Chem. Phys.*, 2001, **115**, 11080.
78. E. van Lenthe, P. E. S. Wormer and A. van der Avoird, *J. Chem. Phys.*, 1997, **107**, 2488.
79. K. M. Neyman, D. I. Ganyushin, A. V. Matveev and V. A. Nasluzov, *J. Phys. Chem. A*, 2002, **106**, 5022.
80. O. Vahtras, B. Minaev and H. Agren, *Chem. Phys. Lett.*, 1997, **281**, 186.
81. F. Neese, R. Kappl, J. Huttermann, W. G. Zumft and P. M. H. Kroneck, *J. Biol. Inorg. Chem.*, 1998, **3**, 53.
82. F. Neese, PhD, *Universitat Konstanz*, 1997.
83. J. E. Harriman, *Theoretical Foundations of Electron Spin Resonance*, Academic Press, London, 1978.
84. R. McWeeny, *Methods of Molecular Quantum Mechanics*, Academic Press, London, 1992.
85. A. Schweiger and G. Jeschke, *Principles of Pulse Electron Paramagnetic Resonance*, Oxford University Press, Oxford, UK, 2001.
86. J. S. Griffith, *The Theory of Transition Metal Ions*, Cambridge University Press, Cambridge, 1964.
87. F. Neese, *J. Am. Chem. Soc.*, 2006, **128**, 10213.

88. F. Neese and E. I. Solomon, *Inorg. Chem.*, 1998, **37**, 6568.
89. D. Ganyushin and F. Neese, *J. Chem. Phys.*, 2006, **125**, 024103.
90. F. Neese, *J. Chem. Phys.*, 2007, **127**, 164112.
91. S. Ye, F. Neese, A. Ozarowski, D. Smirnov, J. Krzystek, J. Telser, J. H. Liao, C. H. Hung, W. C. Chu, Y. F. Tsai, R. C. Wang, K. Y. Chen and H. F. Hsu, *Inorg. Chem.*, 2010, **49**, 977.
92. D. G. Liakos, D. Ganyushin and F. Neese, *Inorg. Chem.*, 2009, **48**, 10572.
93. F. Neese and E. I. Solomon, Interpretation and calculation of spin-Hamitonian parameters in transition metal complexes, in *Magnetism: Molecules to Materials IV*, ed. J. S. Miller and M. Drillon, Wiley-VCH Verlag, Weinheim, 2002, pp. 345–446.
94. J. Hüttermann, ENDOR of randomly oriented mononuclear metalloproteins. Toward structural determination of the prosthetic group, in *Biological Magnetic Resonance*, vol. 13, ed. L. J. Berliner and J. Reuben, Plenum Press, New York, 1993, p. 219.
95. H. Kuramochi, L. Noodleman and D. A. Case, *J. Am. Chem. Soc.*, 1997, **119**, 11442.
96. L. Noodleman and E. J. Baerends, *J. Am. Chem. Soc.*, 1984, **106**, 2316.
97. L. Noodleman, C. Y. Peng, D. A. Case and J. M. Mouesca, *Coord. Chem. Rev.*, 1995, **144**, 199.
98. J. M. Mouesca, L. Noodleman, D. A. Case and B. Lamotte, *Inorg. Chem.*, 1995, **34**, 4347.
99. S. Patchkovskii and T. Ziegler, *Inorg. Chem.*, 2000, **39**, 5354.
100. S. Patchkovskii and T. Ziegler, *J. Am. Chem. Soc.*, 2000, **122**, 3506.
101. E. van Lenthe, A. van der Avoird, W. R. Hagen and E. J. Reijerse, *J. Phys. Chem. A*, 2000, **104**, 2070.
102. M. Radoul, M. Sundararajan, A. Potapov, C. Riplinger, F. Neese and D. Goldfarb, *Phys. Chem. Chem. Phys.*, 2010, **12**, 7276.
103. M. Wanner, T. Scheiring, W. Kaim, L. D. Slep, L. M. Baraldo, J. A. Olabe, S. Zalis and E. J. Baerends, *Inorg. Chem.*, 2001, **40**, 5704.
104. M. Li, D. Bonnet, E. Bill, F. Neese, T. Weyhermüller, N. Blum, D. Sellman and K. Wieghardt, *Inorg. Chem.*, 2002, **41**, 3444.
105. R. G. Serres, C. A. Grapperhaus, E. Bothe, E. Bill, T. Weyhermüller, F. Neese and K. Wieghardt, *J. Am. Chem. Soc.*, 2004, **126**, 5138.
106. S. Sinnecker and F. Neese, *J. Comput. Chem.*, 2006, **27**, 1463.
107. M. Stein and W. Lubitz, *Curr. Opin. Chem. Biol.*, 2002, **6**, 243.
108. C. Stadler, A. L. Lacey, Y. Montet, A. Volbeda, J. C. Fontecilla-Camps, J. C. Conesa and V. M. Fernandez, *Inorg. Chem.*, 2002, **41**, 4424.
109. S. Foerster, M. Stein, M. Brecht, H. Ogata, Y. Higuchi and W. Lubitz, *J. Am. Chem. Soc.*, 2003, **125**, 83.
110. B. S. Pierce, J. D. Gardner, L. J. Bailey, T. C. Brunold and B. G. Fox, *Biochemistry*, 2007, **46**, 8569.
111. C. S. Porro, D. Kumar and S. P. de Visser, *Phys. Chem. Chem. Phys.*, 2009, **11**, 10219.
112. J. Krzystek, J. England, K. Ray, A. Ozarowski, D. Smirnov, L. Que Jr. and J. Telser, *Inorg. Chem.*, 2008, **47**, 3483.

113. D. A. Proshlyakov, T. F. Henshaw, G. R. Monterosso, M. J. Ryle and R. P. Hausinger, *J. Am. Chem. Soc.*, 2004, **126**, 1022.
114. J. C. Price, E. W. Barr, B. Tirupati, J. M. Bollinger Jr. and C. Krebs, *Biochemistry*, 2003, **42**, 7497.
115. P. J. Riggs-Gelasco, J. C. Price, R. B. Guyer, J. H. Brehm, E. W. Barr, J. M. Bollinger Jr. and C. Krebs, *J. Am. Chem. Soc.*, 2004, **126**, 8108.
116. J. England, M. Martinho, E. R. Farquhar, J. R. Frisch, E. L. Bominaar, E. Münck and L. Que Jr., *Angew. Chem., Int. Ed.*, 2009, **48**, 3622.
117. J. U. Rohde, J. H. In, M. H. Lim, W. W. Brennessel, M. R. Bukowski, A. Stubna, E. Münck, W. Nam and L. Que Jr., *Science*, 2003, **299**, 1037.
118. T. A. Jackson, J. U. Rohde, M. S. Seo, C. V. Sastri, R. DeHont, A. Stubna, T. Ohta, T. Kitagawa, E. Münck, W. Nam and L. Que Jr., *J. Am. Chem. Soc.*, 2008, **130**, 12394.
119. J. Ridley and M. Zerner, *Theor Chim Acta*, 1973, **32**, 111.
120. J. E. Ridley and M. C. Zerner, *Theor Chim Acta*, 1976, **42**, 223.
121. M. C. Zerner, in *Reviews in Computational Chemistry*, vol. 2, ed. K. B. Lipkowitz and D. B. Boyd, VCH, New York, 1991.
122. S. I. Gorelsky and A. B. P. Lever, *J. Organomet. Chem.*, 2001, **635**, 187.
123. J. C. Slater, in *Advances in Quantum Chemistry*, vol. 6, ed. P.-O. Löwdin, Academic Press, 1972, p. 1.
124. J. F. Janak, *Phys. Rev. B*, 1978, **18**, 7165.
125. J. W. D. Connolly, H. Siegbahn, U. Gelius and C. Nordling, *J. Chem. Phys.*, 1973, **58**, 4265.
126. M. Stener, A. Lisini and P. Decleva, *Chem. Phys.*, 1995, **191**, 141.
127. C. H. Hu and D. P. Chong, *Chem. Phys. Lett.*, 1996, **262**, 729.
128. C. H. Hu and D. P. Chong, *Chem. Phys. Lett.*, 1996, **262**, 733.
129. M. Cavalleri, H. Ogasawara, L. G. M. Pettersson and A. Nilsson, *Chem. Phys. Lett.*, 2002, **364**, 363.
130. W. G. Han, T. Q. Liu, T. Lovell and L. Noodleman, *Inorg. Chem.*, 2006, **45**, 8533.
131. B. O. Roos, K. Andersson, M. P. Fülscher, P. Malmqvist, L. Serrano-Andrés, K. Pierloot and M. Merchán, in *Advances in Chemical Physics*, ed. I. Prigogine and S. A. Rice, John Wiley & Sons, Inc., New York, 1996, vol. XCIII, p 219.
132. M. A. L. Marques and E. K. U. Gross, *Annu. Rev. Phys. Chem.*, 2004, **55**, 427.
133. F. Neese, T. Petrenko, D. Ganyushin and G. Olbrich, *Coord. Chem. Rev.*, 2007, **251**, 288.
134. V. V. Vrajmasu, E. L. Bominaar, J. Meyer and E. Münck, *Inorg. Chem.*, 2002, **41**, 6358.
135. M. Y. M. Pau, M. I. Davis, A. M. Orville, J. D. Lipscomb and E. I. Solomon, *J. Am. Chem. Soc.*, 2007, **129**, 1944.
136. M. Sumimoto, Y. Kawashima, K. Hori and H. Fujimoto, *Dalton. Trans.*, 2009, 5737.
137. A. Decker, M. D. Clay and E. I. Solomon, *J. Inorg. Biochem.*, 2006, **100**, 697.

138. L. E. Grove, J. K. Hallman, J. P. Emerson, J. A. Halfen and T. C. Brunoldt, *Inorg. Chem.*, 2008, **47**, 5762.
139. N. Mitic, M. D. Clay, L. Saleh, J. M. Bollinger Jr. and E. I. Solomon, *J. Am. Chem. Soc.*, 2007, **129**, 9049.
140. A. T. Fiedler and T. C. Brunold, *Inorg. Chem.*, 2005, **44**, 1794.
141. D. Harris, G. Loew and L. Waskell, *J. Inorg. Biochem.*, 2001, **83**, 309.
142. L. E. Grove, J. Xie, E. Yikilmaz, A. Karapetyan, A. F. Miller and T. C. Brunold, *Inorg. Chem.*, 2008, **47**, 3993.
143. A. T. Fiedler, H. L. Halfen, J. A. Halfen and T. C. Brunold, *J. Am. Chem. Soc.*, 2005, **127**, 1675.
144. F. Paulat and N. Lehnert, *Inorg. Chem.*, 2008, **47**, 4963.
145. F. Neese, *J. Chem. Phys.*, 2003, **119**, 9428.
146. A. Decker, J. U. Rohde, L. Que Jr. and E. I. Solomon, *J. Am. Chem. Soc.*, 2004, **126**, 5378.
147. H. H. Zhang, B. Hedman and K. O. Hodgson, in *Inorganic Electronic Structure and Spectroscopy*, ed. E. I .Solomon and A. B. P. Lever, John Wiley & Sons, Inc., New York, 1999, p. 513.
148. S. P. Cramer and K. O. Hodgson, in *Progress in Inorganic Chemistry*, vol. 25, ed. S. J. Lippard, John Wiley & Sons, Inc., New York, 1979.
149. S. DeBeer, D. W. Randall, A. M. Nersissian, J. S. Valentine, B. Hedman, K. O. Hodgson and E. I. Solomon, *J. Phys. Chem. B*, 2000, **104**, 10814.
150. S. DeBeer, P. Wittung-Stafshede, J. Leckner, G. Karlsson, J. R. Winkler, H. B. Gray, B. G. Malmstrom, E. I. Solomon, B. Hedman and K. O. Hodgson, *Inorg. Chim. Acta*, 2000, **297**, 278.
151. S. DeBeer George, M. Metz, R. K. Szilagyi, H. X. Wang, S. P. Cramer, Y. Lu, W. B. Tolman, B. Hedman, K. O. Hodgson and E. I. Solomon, *J. Am. Chem. Soc.*, 2001, **123**, 5757.
152. R. K. Szilagyi, B. S. Lim, T. Glaser, R. H. Holm, B. Hedman, K. O. Hodgson and E. I. Solomon, *J. Am. Chem. Soc.*, 2003, **125**, 9158.
153. A. Dey, T. Glaser, J. J. G. Moura, R. H. Holm, B. Hedman, K. O. Hodgson and E. I. Solomon, *J. Am. Chem. Soc.*, 2004, **126**, 16868.
154. A. Dey, R. K. Hocking, P. Larsen, A. S. Borovik, K. O. Hodgson, B. Hedman and E. I. Solomon, *J. Am. Chem. Soc.*, 2006, **128**, 9825.
155. R. Sarangi, N. Aboelella, K. Fujisawa, W. B. Tolman, B. Hedman, K. O. Hodgson and E. I. Solomon, *J. Am. Chem. Soc.*, 2006, **128**, 8286.
156. K. Ray, S. DeBeer George, E. I. Solomon, K. Wieghardt and F. Neese, *Chem. Eur. J.*, 2007, **13**, 2783.
157. S. DeBeer George, T. Petrenko and F. Neese, *Inorg. Chim. Acta*, 2008, **361**, 965.
158. S. DeBeer George, T. Petrenko and F. Neese, *J. Phys. Chem. A*, 2008, **112**, 12936.
159. S. DeBeer George and F. Neese, *Inorg. Chem.*, 2010, **49**, 1849.
160. J. J. Rehr, E. A. Stern, R. L. Martin and E. R. Davidson, *Phys. Rev. B*, 1978, **17**, 560.
161. T. E. Westre, P. Kennepohl, J. G. DeWitt, B. Hedman, K. O. Hodgson and E. I. Solomon, *J. Am. Chem. Soc.*, 1997, **119**, 6297.

162. M. A. Arrio, S. Rossano, C. Brouder, L. Galoisy and G. Calas, *Europhys. Lett.*, 2000, **51**, 454.
163. A. Chanda, X. Shan, M. Chakrabarti, W. C. Ellis, D. L. Popescu, F. Tiago de Oliveira, D. Wang, L. Que Jr., T. J. Collins, E. Münck and E. L. Bominaar, *Inorg. Chem.*, 2008, **47**, 3669.
164. M. H. Lim, J.-U. Rohde, A. Stubna, M. R. Bukowski, M. Costas, R. Y. N. Ho, E. Münck, W. Nam and L. Que Jr., *Proc. Natl. Acad. Sci. USA*, 2003, **100**, 3665.
165. J. U. Rohde, T. A. Betley, T. A. Jackson, C. T. Saouma, J. C. Peters and L. Que Jr., *Inorg. Chem.*, 2007, **46**, 5720.
166. M. Aliaga-Alcalde, S. DeBeer George, B. Mienert, E. Bill, K. Wieghardt and F. Neese, *Angew. Chem., Int. Ed.*, 2005, **44**, 2908.
167. J. F. Berry, E. Bill, E. Bothe, S. DeBeer George, B. Mienert, F. Neese and K. Wieghardt, *Science*, 2006, **312**, 1937.
168. R. G. Shulman, Y. Yafet, P. Eisenberger and W. E. Blumberg, *Proc. Natl. Acad. Sci. USA*, 1976, **73**, 1384.
169. J. E. Hahn, R. A. Scott, K. O. Hodgson, S. Doniach, S. R. Desjardins and E. I. Solomon, *Chem. Phys. Lett.*, 1982, **88**, 595.
170. G. Dräger, R. Frahm, G. Materlik and O. Brümmer, *Phys. Status Solidi B*, 1988, **146**, 287.
171. P. Glatzel and U. Bergmann, *Coord. Chem. Rev.*, 2005, **249**, 65.
172. G. Smolentsev, A. V. Soldatov, J. Messinger, K. Merz, T. Weyhermüller, U. Bergmann, Y. Pushkar, J. Yano, V. K. Yachandra and P. Glatzel, *J. Am. Chem. Soc.*, 2009, **131**, 13161.
173. Y. Pushkar, X. Long, P. Glatzel, G. W. Brudvig, G. C. Dismukes, T. J. Collins, V. K. Yachandra, J. Yano and U. Bergmann, *Angew. Chem., Int. Ed.*, 2010, **49**, 800.
174. N. Lee, T. Petrenko, U. Bergmann, F. Neese and S. DeBeer, *J. Am. Chem. Soc.*, 2010, **132**, 9715.

Bioinspired Non-heme Iron Catalysts in C–H and C=C Oxidation Reactions

ANNA COMPANY,[a] LAURA GÓMEZ[b] AND MIQUEL COSTAS*[b]

[a] Institute of Chemistry, Metalorganics and Inorganic Materials, Technical University Berlin, D-10623 Berlin, Germany; [b] Grup QBIS, Departament de Química, Parc Científic i Tecnològic de la Universitat de Girona, Pic de Peguera 15, E-17003 Girona, Catalonia, Spain

6.1 Biological Precedents

Selective oxidation of alkanes and alkenes in an environmentally sustainable manner represents an important challenge for chemists. Moreover, the development of catalysts for these reactions can give rise to novel, eventually more efficient synthetic strategies for the preparation of organic molecules. Nature represents an important source of inspiration for catalyst design. There is a myriad of remarkable transition metal-dependent oxidative enzymes, which are capable of activating dioxygen and catalyzing the selective oxidation of C–H (alkane hydroxylation) or C=C bonds (olefin epoxidation and olefin cis-dihydroxylation), see below. The first two transformations are carried out by enzymes with either heme (e.g. cytochrome P450)[1–4] or non-heme iron (but also by other first row transition metals such as copper, manganese, molybdenum, and nickel) centers (e.g. Rieske dioxygenases).[5–9] On the other hand,

Iron-Containing Enzymes: Versatile Catalysts of Hydroxylation Reactions in Nature
Edited by Sam P de Visser and Devesh Kumar
© Royal Society of Chemistry 2011
Published by the Royal Society of Chemistry, www.rsc.org

cis-dihydroxylation is so far a reaction only characteristic of non-heme iron-containing Rieske dioxygenases, which attack arene double bonds in the first step of the biodegradation of arenes by soil bacteria.[10]

6.1.1 Oxidative Iron Proteins

Iron proteins implicated in oxidation processes *via* O_2 activation are present in nature in a wide range of structures. They can be classified in three different groups depending on the chemical structure of their active site: heme oxygenases, mononuclear non-heme, and dinuclear non-heme oxygenases. Table 6.1 shows representative O_2-activating iron proteins that take part in oxidative

Table 6.1 Iron proteins implicated in oxidation reactions by oxygen activation.[6,11–13]

Protein	Catalytic reaction
Heme proteins	
Cytochrome P450	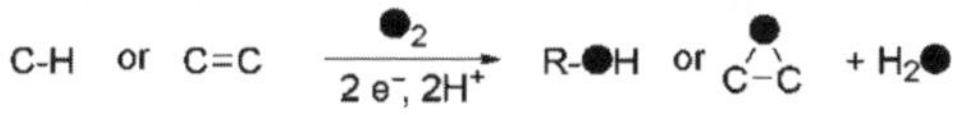 $\text{C-H} \ \text{or} \ \text{C=C} \ \xrightarrow[2\,e^-,\,2H^+]{O_2} \ \text{R-OH} \ \text{or} \ \text{C-C(epoxide)} + H_2O$
Non-heme proteins	
Mononuclear center	
Enzymes with the 2-His-1-carboxylate facial triad motif	
Extradiol-cleaving catechol dioxygenases	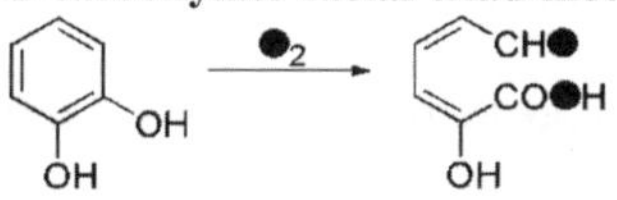 catechol $\xrightarrow{O_2}$ ring-opened product (CHO, COOH)
Rieske dioxygenases	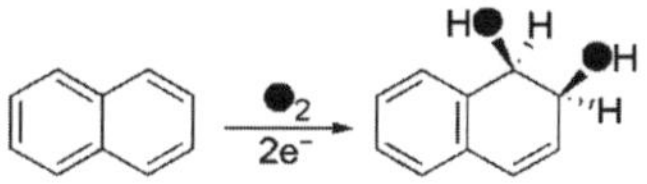 naphthalene $\xrightarrow{O_2,\ 2e^-}$ *cis*-dihydrodiol
α-Ketoglutarate dependent hydroxylases	$\text{R-H} + \text{R'COCOOH} + O_2 \longrightarrow \text{R-OH} + \text{R'COOH} + CO_2$
Pterin-dependent enzymes	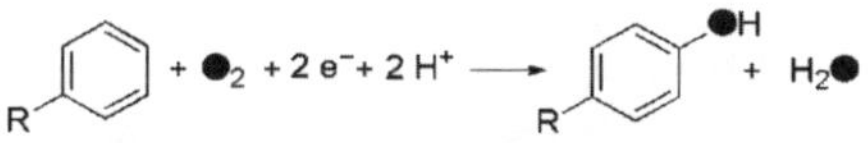 $\text{arene} + O_2 + 2e^- + 2H^+ \longrightarrow \text{arene-OH} + H_2O$
Other Iron(III) and Iron(II) dioxygenases	
Intradiol-cleaving catechol dioxygenases (Fe^{III})	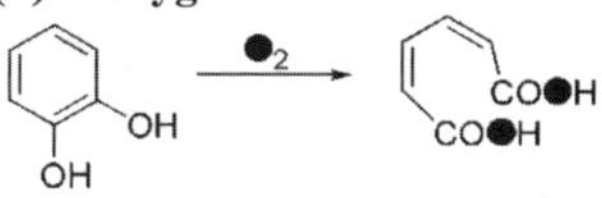 catechol $\xrightarrow{O_2}$ ring-opened product (COOH, COOH)
Lipoxygenases (Fe^{III})	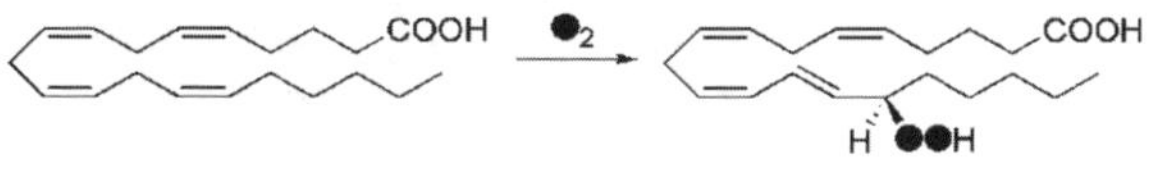 fatty acid $\xrightarrow{O_2}$ hydroperoxide product (H, OOH)
Apocarotene dioxygenase (Fe^{II})	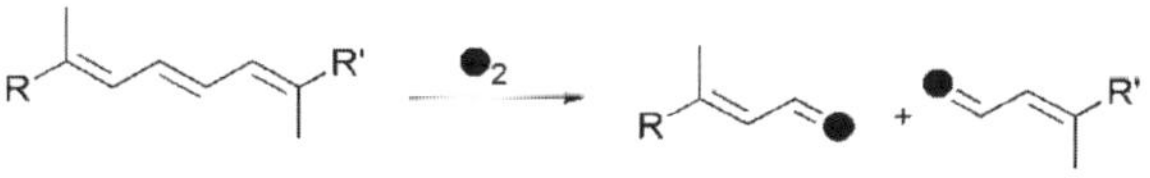 polyene $\xrightarrow{O_2}$ cleavage products
Binuclear center	
Methane monoxygenase (soluble)	$CH_4 + O_2 + 2H^+ + 2e^- \longrightarrow CH_3OH + H_2O$

processes and the reactions they catalyze. Two specific enzymes will be described in some detail as their structure, catalyzed reaction, and mechanistic understanding provide inspiration for the development of catalysts.

6.1.2 Cytochrome P450

The cytochrome P450s (CYP 450s) are an important family of iron enzymes ubiquitous in life forms ranging from bacteria to humans.[1,2,4,14,15] CYP 450 enzymes were first identified and purified approximately 50 years ago and they have been thoroughly studied. They are oxidoreductases that catalyze the oxidation of a wide variety of substrates by means of dioxygen activation. CYP 450 is known to catalyze hydroxylations, epoxidations (Scheme 6.1), N-, S- and O-dealkylations, N-oxidations, sulfoxidations, and dehalogenations. These oxygenation reactions play a central role in biosynthesis, metabolism, and detoxification of harmful substances.

The active site of CYP 450-camphor is known in detail from several X-ray crystal structures (Figure 6.1).[4] It contains a single ferric heme coordinated to a cysteinate sulfur and to a sixth ligand, possibly a water molecule.[4]

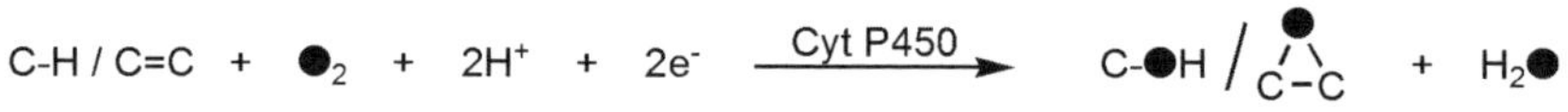

$$\text{C-H / C=C} \; + \; \bullet_2 \; + \; 2\text{H}^+ \; + \; 2\text{e}^- \; \xrightarrow{\text{Cyt P450}} \; \text{C-}\bullet\text{H} \; / \; \overset{\bullet}{\underset{\text{C}-\text{C}}{}} \; + \; \text{H}_2\bullet$$

Scheme 6.1 Hydroxylation and epoxidation reactions catalyzed by cytochrome P450.

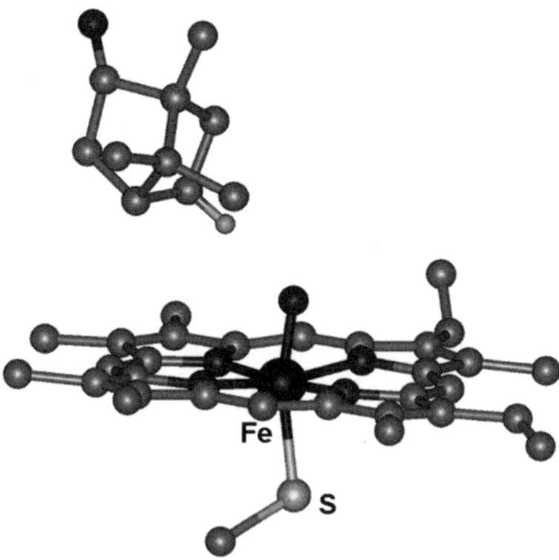

Figure 6.1 Active site of cytochrome P450-camphor from *Pseudomonas putida* (pdb code 2CPP).

The CYP 450 family is considered to be the paradigm for oxygen activation and hydrocarbon oxidation by an iron center due to their versatility and efficiency. For this reason, extensive efforts have been devoted to understand their catalytic mechanism over the past 30 years. Despite the fact that some controversies regarding the possibility of a multiple oxidant scenario remain, the principal features of the catalytic cycle of CYP 450 are nowadays well-established.[1–4] As shown in Scheme 6.2, the first step entails the binding of the alkane substrate to the active site (A → B), which triggers the one-electron reduction of the Fe^{III} (B → C). The subsequent binding of O_2 to the Fe^{II} center generates a ferric CYP 450-superoxide complex. A second electron is transferred to this complex to afford a peroxo-iron(III) complex, which is further protonated to generate a hydroperoxo-iron(III) complex (D). This species undergoes proton assisted heterolytic cleavage of the O–O bond to generate a high-valent oxo-Fe^{IV}-porphyryl radical cation (E) and a water molecule. In the case of alkanes, the oxygen atom is transferred from this oxo complex to the nearby substrate through a two-step process known as "oxygen rebound".[16] Dissociation of the product completes the catalytic cycle. Olefins are epoxidized by the high-valent oxo-Fe^{IV}-porphyryl radical cation (E), and with few exceptions reactions with *cis*-olefins usually occur with retention of stereochemistry, strongly suggesting a concerted mechanism.[2,3,15] Direct cycling between the ferric resting state (B) and the high valent oxidant species (E) can be achieved by using oxidants such as hydro and alkyl peroxides, NaOCl, iodosylbenzene (PhIO) and peracids. This shortcut is known as the "peroxide shunt" (Scheme 6.2) and receives use in catalysis.[15]

6.1.3 Rieske Dioxygenases

In the past 15 years, great strides have been made towards the understanding of mononuclear non-heme iron(II) enzymes. Among them, Rieske dioxygenases are especially efficient and versatile – even more so than the heme-containing CYP 450s. Rieske dioxygenases are the only ones among both heme and non-heme iron enzymes that can carry out enantioselective *cis*-dihydroxylation of arenes (Scheme 6.3).[6,9,10,17,18] This is a novel transformation not observed so far in synthetic organic chemistry. Therefore there is strong interest in these enzymes as biotechnological tools.[19,20] Apart from *cis*-dihydroxylation, Rieske dioxygenases also catalyze several oxidations such as benzylic hydroxylation, desaturation, sulfoxidation, and O- and N-dealkylation.[17] Such versatility has led to Rieske dioxygenases being considered as the non-heme analog of CYP 450.

One of the best studied examples of the Rieske dioxygenases family is naphthalene 1,2-dioxygenase (NDO).[17,21] NDO catalyzes the *cis*-dihydroxylation of naphthalene (Scheme 6.3) and it is known that in the course of catalysis both atoms of O_2 are incorporated into the *cis*-diol product. NDO is representative of other *cis*-dihydroxylating enzymes such as benzoate 1,2-dioxygenase,[22] toluene 2,3-dioxygenase,[23] benzene dioxygenase,[24] and phthalate

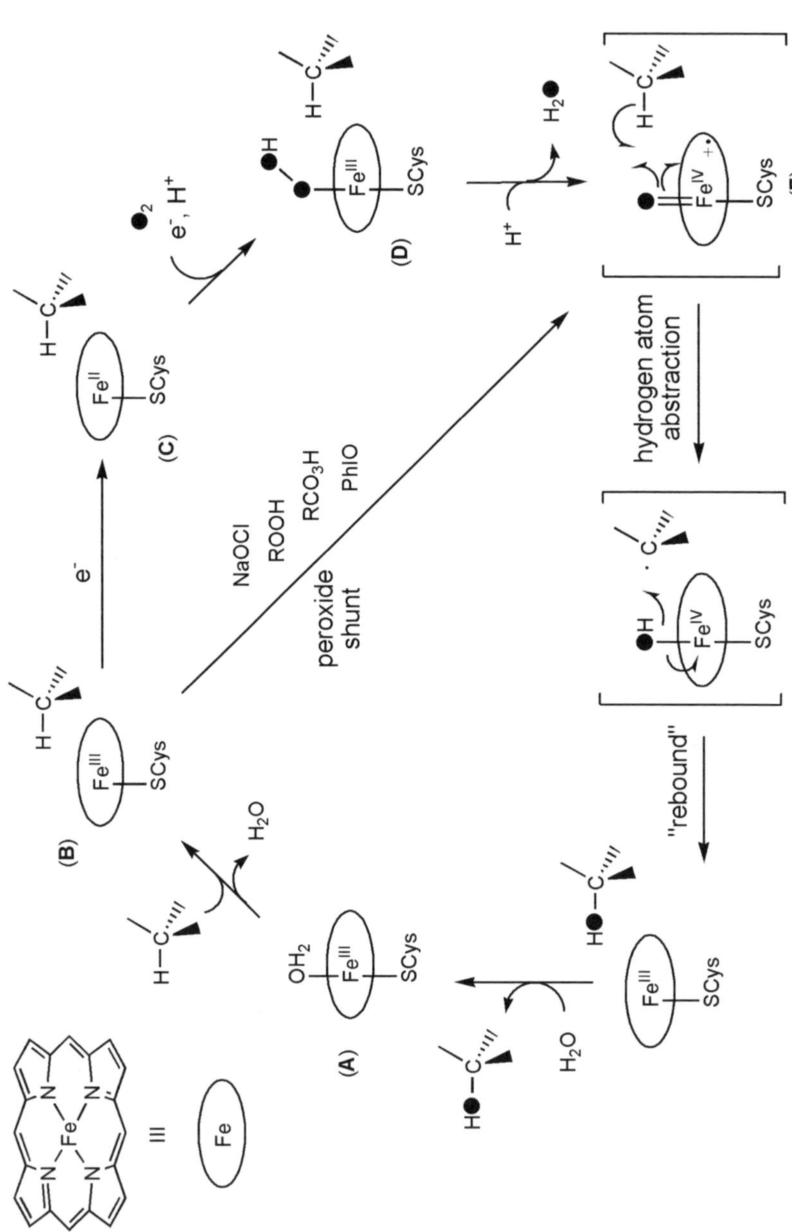

Scheme 6.2 Reaction mechanism of alkane hydroxylation proposed for cytochrome P450.

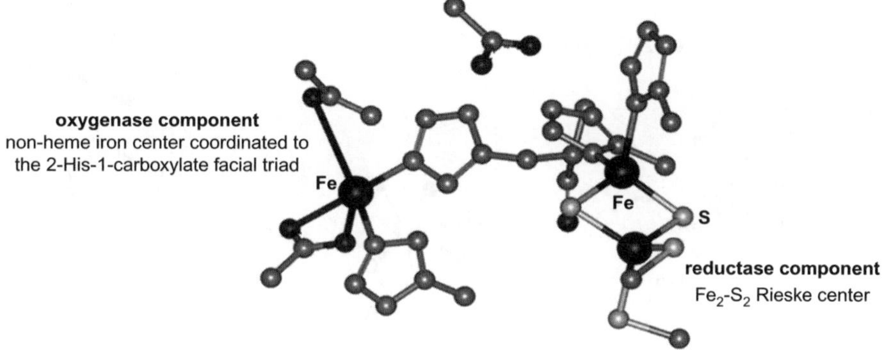

Scheme 6.3   Arene *cis*-dihydroxylation reaction of naphthalene catalyzed by naphthalene 1,2-dioxygenase (NDO).

Figure 6.2   Active center of Rieske dioxygenase showing the oxygenase and reductase components (pdb code 1NDO).

Figure 6.3   Schematic representation of the 2-His-1-carboxylate facial triad found in some mononuclear non-heme iron enzymes.

dioxygenase.[25] The structure of the active site of NDO from *Pseudomonas putida* shows the characteristic structure of the Rieske dioxygenases: an oxygenase component [a mononuclear non-heme iron(II) center] where O_2 activation and substrate dihydroxylation occur and a reductase component (Rieske-type Fe_2S_2 cluster) that mediates electron transfer from NAD(P)H (Figure 6.2).[21]

Moreover, Rieske dioxygenases belong to a superfamily of enzymes that contain an iron(II) center in their active site (oxygenase component) coordinated by the so-called 2-His-1-carboxylate facial triad (Table 6.1).[9,26,27] It consists of three protein residues (two His and an Asp or Glu) that leave three *cis* sites available for exogenous ligands binding (Figure 6.3). In the isolated enzymes, these sites are usually occupied by solvent molecules but they can accommodate both co-substrate (or substrate) and O_2, bringing them into close

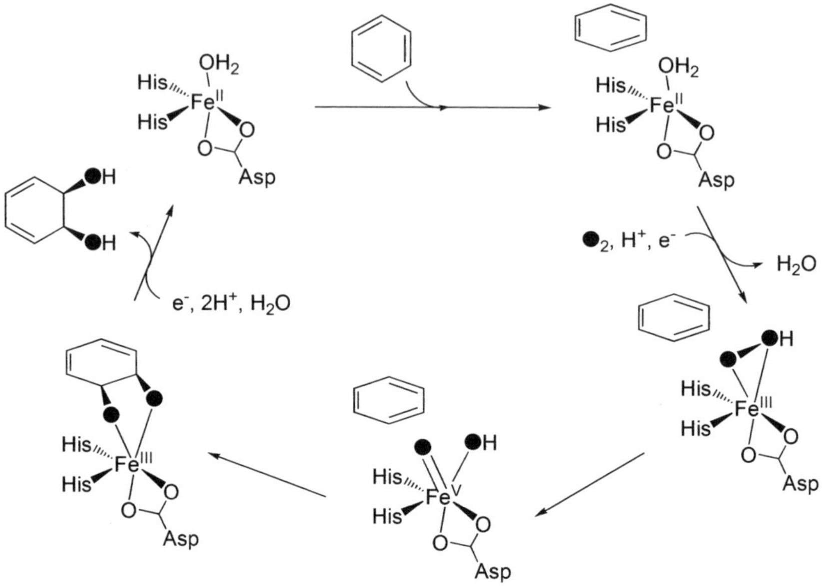

Scheme 6.4   Catalytic cycle proposed for Rieske dioxygenases.[22]

proximity for subsequent reaction, thereby accounting in large part for the chemical versatility of this structural motif.[8]

Time-resolved cryo-crystallography on frozen crystals of NDO exposed to O_2 indicate that a side-on bound peroxo (or hydroperoxo)-iron(III) species is the last detectable intermediate before substrate oxidation occurs.[28] In addition, experiments also reveal that phthalate dioxygenase can elicit catalytic chemistry when H_2O_2 is used as oxidant, thus resembling the "peroxide-shunt" of CYP 450.[29] Overall, there is increasing evidence that points toward a common mechanistic landscape operating in CYP 450 and Rieske dioxygenases and suggests that a high-valent iron species is responsible for the catalysis (Scheme 6.4).

6.2   Non-heme Iron Complexes as Bioinspired Catalysts

Inspired by the active site of iron oxidation enzymes, chemists have designed a wide variety of complexes that could functionally mimic these metalloenzymes, with the final goal of designing oxidation catalysts. Such "bioinspired" synthetic catalysts also constitute a valuable tool to explore the reaction mechanisms by which non-heme enzymes perform their chemistry.[30–34] In the following sections we review recent prominent examples of iron-based alkane and alkene oxidation catalysts. We focus our discussion on iron catalysts developed over the last decade, and especially on those examples where, on the basis of selectivity considerations, a metal-centered oxidation is presumed to take place. For previous work, the reader is referred to early reviews and books.[33,34] Because of their complexity, we have neither included in this work Gif systems[35–38] nor a full discussion on Fenton reactions,[39–42] for which excellent reviews are already available.

6.2.1 Oxidation of Alkanes (C–H Bonds) by Non-heme Iron Complexes

The controlled and selective oxidation of alkanes by cheap oxidants and under mild conditions is nowadays a great challenge that is directly related to industrial applications.[43] Because of economical and environmental considerations, the development of iron-based catalysts for such challenging transformations has received interest.[44,45] Dating back for more than a century, chemists have combined first row metals, especially Fe^{II} but also Cu^{I}, Mn^{II} and Co^{II}, with peroxides to oxidize organic compounds. These reactions have been traditionally known as Fenton reactions.[46,47] With the perspective of a century, Fenton reactions can be understood as a primitive "bioinspired" approach; the combination of O_2 and $2e^-$ commonly occurring at metalloenzymes is replaced by peroxides, which can be understood as a $2e^-$ reduced version of O_2, and transition metal complexes can be seen as the very elemental surrogate of the active site of metalloenzymes. However, after extensive disputes for more than a century over the reaction mechanisms, the current understanding is that Fenton's chemistry is mostly related to non-selective reactions mediated by free-diffusing oxygen atom centered radicals.[41] Such radicals are initially generated by interaction of peroxides with redox active metal ions (Scheme 6.5).

Oxygen-centered radicals are then the active species responsible for alkane C–H oxidation *via* a hydrogen abstraction mechanism (Scheme 6.6).[40,48] The selectivity of these reactions is mainly determined by the energy of the newly formed O–H bond. H-abstraction reactions by hydroxyl radicals form a water molecule, with a very strong H–O bond and a freely diffusing carbon-centered radical. Because of the strength of the O–H bond, little discrimination among different C–H bonds is attained. In contrast, *tert*-butoxyl radicals (commonly formed in reactions with *tert*-butyl hydroperoxide, TBHP) form weaker O–H bonds. Reactions are thus more sensitive to the energy of the C–H bond attacked. This differential reactivity is then translated into C–H bond selectivity, based on relative C–H bond dissociation energy (BDE).[49]

$$M^n + HOOH \longrightarrow M^{n-1} + HOO\cdot + H^+ \quad M^n = Cu^{II}, Mn^{III}, Fe^{III}, Co^{III}$$

$$M^{n-1} + HOOH \longrightarrow M^n + HO\cdot + HO^-$$

Scheme 6.5   Reaction of H_2O_2 with transition metal ions in Fenton reactions.

$$(R)HO\cdot + H\text{-}R \longrightarrow (R)HO\text{-}H + \cdot R$$

Scheme 6.6   Reaction of oxygen atom centered radicals with C–H bonds.

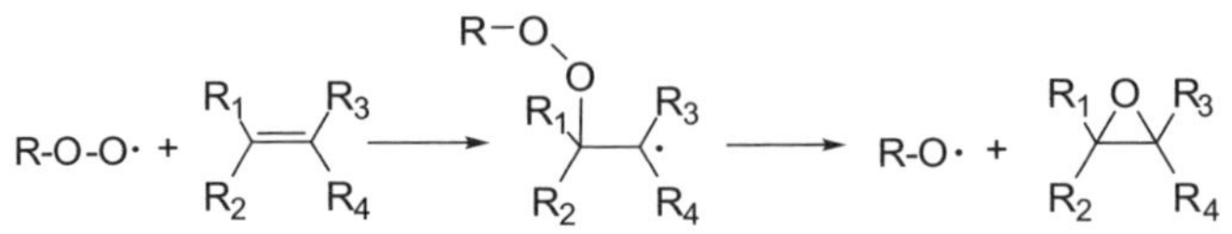

Scheme 6.7 Formation of epoxides *via* a free radical reaction.

Reactions of these oxygen-centered radicals with olefins are more compli-cated and involve allylic H-abstraction, and addition of oxygen-centered radicals to the olefin, among other transformations, which in turn initiate a cascade of different reactions.[48,50] Epoxides can be the result of these free-radical transformations (Scheme 6.7). Consequently, the formation of epoxides cannot be considered as a proof for a metal-centered reaction.

Despite the fact that these sets of reactions are relatively simple, one should keep in mind that high energy radical species are involved. Such species are very reactive and can engage in several diverse reactions with organic and inorganic molecules, which can translate into substrate selectivity patterns that differ from those obtained *via* metal-free radical generation reactions.[40] Such com-plexity means that mechanistic analyses are not straightforward, and these differences can be easily misunderstood as special selectivity characteristic of a metal-based active site.[48] From a more practical viewpoint, these reactions generally suffer from low product yields, and very poor selectivity, which severely limit their use as synthetic tools. However, over the last decade, dif-ferent groups have shown that the introduction of appropriate ligands or additives into the system exert dramatic changes in the chemistry and afford more selective and efficient reactions. In specific cases, the high selectivity exhibited by the reactions is indicative of enzyme-like metal-centered transformations.[44,51]

6.2.1.1 *Alkane Oxidation by Polynuclear Fe^{III} Complexes*

Polynuclear iron compounds, usually bridged by oxo and/or carboxylate ligands, have been traditionally seen as promising candidates for alkane oxi-dation catalysts because of their structural relation to the active site of powerful hydroxylases such as soluble methane monoxygenase,[12,52,53] which mediate the hydroxylation of substrates with strong C–H bonds. The activity of these compounds in oxidation catalysis has been reviewed,[34] but relevant contribu-tions have been published over the last decade.[54–58]

Traditionally, the oxidants of choice for these reactions are generally TBHP and H_2O_2, while other oxidants such as PhIO, peracids, or even O_2 (in com-bination with reducing agents) have seldom been employed.[34] In general, the use of TBHP and peracids affords better efficiencies than H_2O_2 in terms of oxidant equivalents converted into products. This is due to the well-known facile disproportionation of H_2O_2 in the presence of redox active ions. The presence of labile sites appears to be crucial for good catalytic activity, and coordinatively saturated dimers display very poor activity, and most likely reflect the need for oxidant binding to the metal to initiate the generation of the oxidizing species.

In general, reactions with peroxides show product selectivity patterns indicative of the generation of hydroxyl or alkoxyl radicals that are finally responsible for a H-abstraction of the substrate, thus initiating free-diffusion radical paths.[34,54–57] Cyclohexane is oxidized to a roughly 1:1 mixture of cyclohexanol and cyclohexanone, indicative of a Russell-type termination.[59] Yields are very sensitive to the presence of O_2, which is incorporated into products, and finally stereoscrambling in oxidized substrates is observed.

Nevertheless, selected μ-oxo bridged diferric catalysts, especially when reacting with peracids, provide selectivity patterns in their reactions that suggest the involvement of a metal-centered hydroxylation reaction.[58,60] An interesting example is the family of bis-carboxylate-oxo-bridged diferric complexes $[Fe^{III}_2O\text{-}(O_2CR_2)_2(BPR_1Am)_2]^{2+}$ [Figure 6.4, $BPR_1Am = N,N$-bis(2-pyridylmethyl)-benzylamine ($R_1 = Bz$) or N,N-bis(2-pyridylmethyl)-*iso*-butylamine ($R_1 = i$Pr)], which affords large turnover numbers (TN up to 400) in the oxidation of cyclohexane with MCPBA (*meta*-chloroperbenzoic acid). Most remarkably, reactions display high A/K ([alcohol]/[ketone]) ratios (up to 13.7) in the oxidation of cyclohexane, and a normalized 3°/2° C–H selectivity up to 30 in the oxidation of adamantane. Corresponding mononuclear Fe^{III} complexes are also competent for alkane oxidation with similar selectivity patterns. Despite the large TN, in terms of substrate conversion the results are relatively modest ($<10\%$).[58]

In terms of product yields, the most efficient oxo-bridged Fe^{III} complex reported to date is the dinuclear $[Fe^{III}_2(\mu\text{-}O)(NO_3)_2(BBP)_2(CH_3OH)_2](NO_3)_2$, where BBP is a meridional tridentate ligand [Figure 6.4, $BBP = 2,6$-bis(N-methylbenzimidazol-2-yl)pyridine]. The complex efficiently catalyzes the oxidation of cyclohexane with 51% substrate conversion into an equimolar mixture of cyclohexanol and cyclohexanone ($A/K = 1.2$), when using TBHP as oxidant.[54] Competitive oxidation of cyclohexane/cyclohexane-d_{12} exhibits a low kinetic isotope effect (1.8), indicative of a powerful and rather indiscriminate oxidant. Most significantly, oxidation of *cis*-1,2-dimethylcyclohexane affords a mixture of epimeric alcohols, indicating that long-lived alkyl radicals are formed during the reaction.

The complex $[Fe^{III}_2(HPTB)(\mu\text{-}OH)(NO_3)_2](NO_3)_2$ (Figure 6.4, $HPTB = N,N,N',N'$-tetrakis(2-benzimidazolylmethyl)-2-hydroxo-1,3-diaminopropane) is also particularly interesting because in combination with certain amino acids (pyrazine-2-carboxylic acid, pyrazine-2,3-dicarboxylic acid, and picolinic acid), and under mild conditions (room temperature in acetonitrile solution), it catalyzes not only the oxidation of cyclohexane (TN = 140) but also the oxidation of more challenging smaller alkanes such as ethane (TN = 21) and methane (TN = 4) with H_2O_2. Cyclohexane was mainly oxidized to cyclohexyl hydroperoxide (90% selectivity), while ethane was transformed into ethyl hydroperoxide, acetaldehyde, and trace amounts of acetic acid. Methyl hydroperoxide and formaldehyde were obtained from the oxidation of methane. Reactions proceed with stereoscrambling in the oxidation of decalins, and low C–H bond selectivity, which led the authors to propose the implication of hydroxyl radicals.[61]

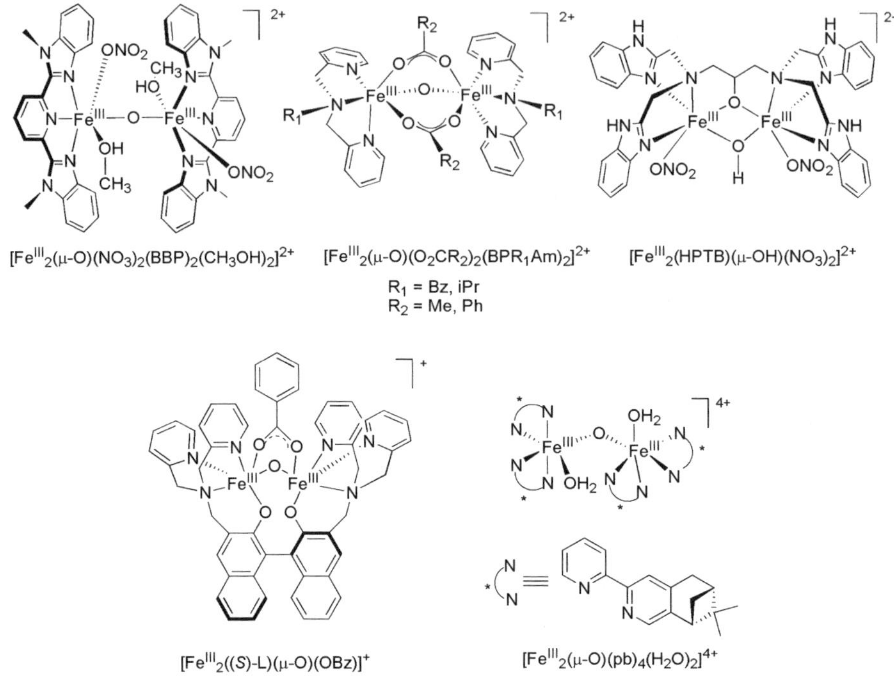

$[Fe^{III}_2(\mu\text{-}O)(NO_3)_2(BBP)_2(CH_3OH)_2]^{2+}$

$[Fe^{III}_2(\mu\text{-}O)(O_2CR_2)_2(BPR_1Am)_2]^{2+}$

$R_1 = Bz, iPr$
$R_2 = Me, Ph$

$[Fe^{III}_2(HPTB)(\mu\text{-}OH)(NO_3)_2]^{2+}$

$[Fe^{III}_2((S)\text{-}L)(\mu\text{-}O)(OBz)]^+$

$[Fe^{III}_2(\mu\text{-}O)(pb)_4(H_2O)_2]^{4+}$

Figure 6.4 Selected oxo-bridged diferric complexes employed in oxidation catalysis.

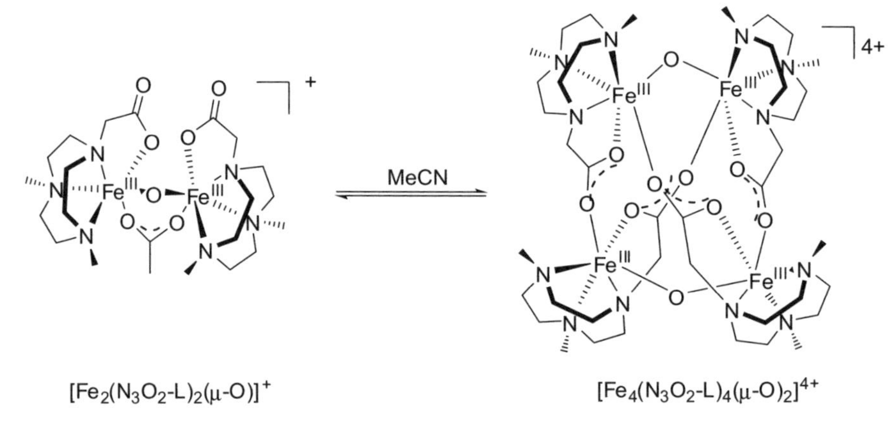

$[Fe_2(N_3O_2\text{-}L)_2(\mu\text{-}O)]^+$

$[Fe_4(N_3O_2\text{-}L)_4(\mu\text{-}O)_2]^{4+}$

Scheme 6.8 Dinuclear and tetranuclear O-bridged iron complexes based on a tacn-carboxylate ligand.[57]

Structurally more complex polymetallic iron complexes have been prepared and studied as catalysts over the last decade. For example, Süss-Fink and coworkers have reported a tetranuclear complex $[Fe^{III}_4(N_3O_2\text{-}L)_4(\mu\text{-}O)_2]^{4+}$ based on a 1,4,7-triazacyclononane (tacn) ring with an appended carboxylate arm (Scheme 6.8).[57] In acetonitrile solution, the complex is in equilibrium with the corresponding dinuclear complex $[Fe^{III}_2(N_3O_2\text{-}L)_2(\mu\text{-}O)]^+$. The tetranuclear complex catalyzes the oxidation of methane with H_2O_2 in water to afford methanol, attaining up to 24 catalytic cycles in 4 h. Larger linear alkanes and cycloalkanes are also oxidized by both the dinuclear and the tetranuclear complexes in acetonitrile solution, with modest TN (up to 43). Cyclohexane is initially oxidized to the corresponding cyclohexyl hydroperoxide, which then evolves into cyclohexanol and cyclohexanone. The observation that the oxidations of *cis*-1,2-dimethylcyclohexane and *trans*-1,2-dimethylcyclohexane produce stereoscrambled tertiary alcohols is indicative that the reactions generate free-diffusing radicals, although the selectivity patterns of the reactions exclude the sole participation of hydroxyl radicals as C–H oxidizing species.

A structurally interesting hexanuclear iron carboxylate $[Fe^{III}_6(\mu_3\text{-}O)_3(\mu_2\text{-}OH)(p\text{-}NO_2C_6H_4CO_2)_{11}(DMF)_4]$ with a very unusual $[Fe^{III}_6(\mu_3\text{-}O)_3(\mu_2\text{-}OH)]^{11+}$ core was studied by Schmid and Arion.[56] The complex catalyzes cyclohexane oxidation using H_2O_2 as an oxidant, affording mixtures of cyclohexanol and cyclohexanone in up to 29.7% yield based on substrate, and turnover numbers reaching 100 when the molar ratio of catalyst/cyclohexane was 1/320.

Diferric oxo-bridged chiral complexes (Figure 6.4 and Scheme 6.9) have been studied with the aim of developing asymmetric C–H oxidation catalysts. Very modest enantiomeric excesses have been obtained so far. The binaphthol based complex $[Fe^{III}_2\{(S)\text{-}L\}(O)(OBz)](ClO_4)$ has been prepared and studied in the catalytic oxidation of alkanes. TBHP and H_2O_2 afforded low TN, but up to 270 TN in cyclohexane oxidation is obtained when MCPBA is used as oxidant. In the latter case, a relatively high A/K ratio (6.5) is obtained. The selectivity for the tertiary C–H bond oxidation of adamantane is also high (17). Both parameters are, therefore, indicative of metal-centered oxidation reactions. Unfortunately, oxidation of tetralin affords the corresponding ketone with a

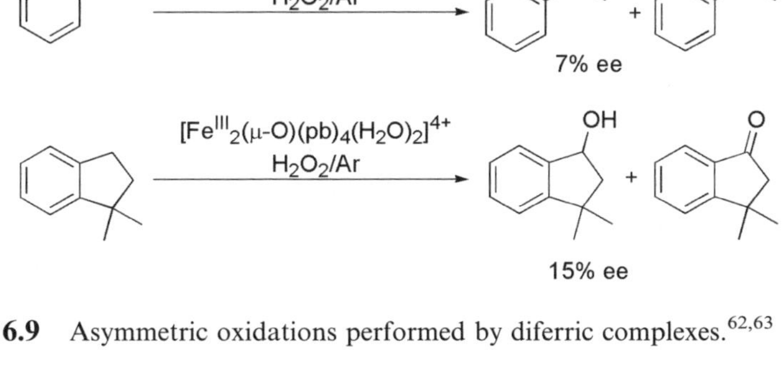

Scheme 6.9 Asymmetric oxidations performed by diferric complexes.[62,63]

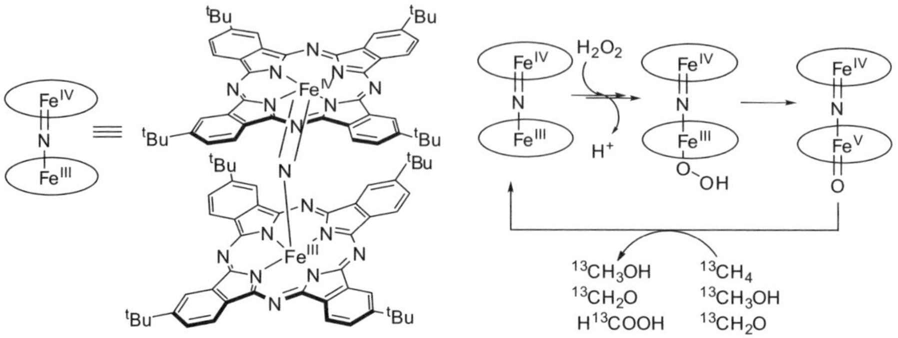

Scheme 6.10 Structural diagram of (FePc^{t}Bu$_4$)$_2$N, and proposed mechanistic scheme.[64]

very modest 9.9% ee.[62] On the other hand, the pineno-bipyridine based diferric complex [Fe$^{III}_2$(O)(pb)$_4$(H$_2$O)$_2$]$^{4+}$ catalyzes the oxidation of ethylbenzene with H$_2$O$_2$ to give a mixture of acetophenone and *sec*-phenethyl alcohol (2.9 TN), the latter with 7% ee. Hydroxylation of dimethylindane affords the corresponding alcohol (1.8 TN, 15% ee) and ketone (TN = 1.5).[63]

Remarkably efficient oxidation of methane with H$_2$O$_2$ at low temperatures (25–60 °C) has been reported by Sorokin *et al.*, using a μ-nitrido diiron phthalocyanine catalyst (Fe^{III}Pc^{t}Bu$_4$)$_2$N (Scheme 6.10). Although this compound is chemically more related to porphyrins rather than non-heme systems, the unique character of its reactivity deserves some comments. Initial methane oxidation experiments with this catalyst were ran in acetonitrile solution, and afforded formic acid as the main product. To discard the possibility that products could arise from oxidation of acetonitrile instead of methane, the catalyst was supported over silica, and reactions were carried out in water. Under these conditions, methane was oxidized to a mixture of methanol, formaldehyde, and formic acid, with total TN over 400. Furthermore, by using isotopically labeled CH$_4$ (47% ^{13}CH$_4$ and 53% ^{12}CH$_4$) as substrate, GC-MS of the formic acid formed indicated it to be 44% H^{13}COOH, in agreement with the isotopic composition of methane. This elegantly demonstrates that formic acid was derived from methane oxidation. The authors propose that a formally described nitride bridged species, Fe^{IV}–N–Fe^V(O), is responsible for the oxidation of methane.[64]

From a mechanistic and biomimetic point of view, a particularly distinct and interesting system has been described by Caradonna and coworkers. The diiron(II) compound shown in Scheme 6.11 catalyzes the hydroxylation of cyclohexane with high turnover numbers and very high A/K ratios.[65] Catalytic efficiency appears to depend critically on the oxidation state of the two iron atoms; high product yields are obtained for the Fe^{II}Fe^{II} complex, but both the Fe^{II}Fe^{III} and the Fe^{III}Fe^{III} complexes proved to be practically inactive in cyclohexane and toluene oxidation reactions. However, the mixed-valence Fe^{II}Fe^{III} compound retained good activity in alkene and sulfide oxidation reactions.

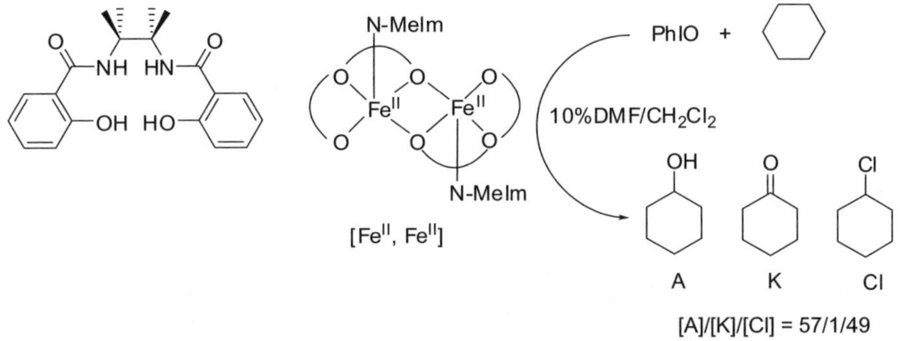

Scheme 6.11   Schematic diagram of the diferrous complex described by Caradonna, and its catalytic oxidation of cyclohexane.[65]

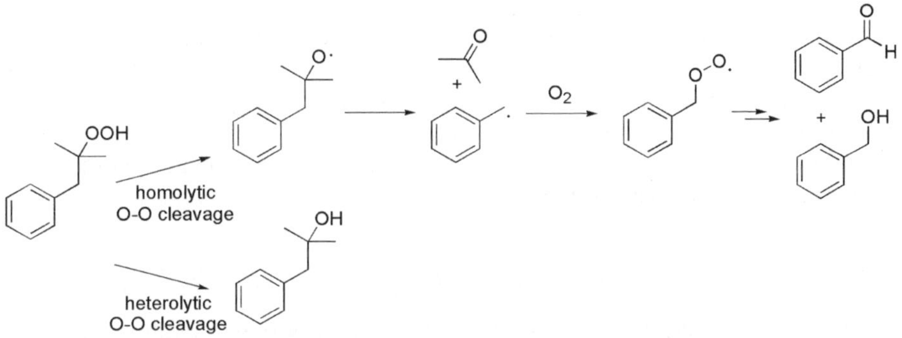

Scheme 6.12   Homolytic and heterolytic cleavage of MPPH (2-methyl-1-phenyl-2-propyl hydroperoxide).

A particularly remarkable aspect of this complex is that it cleaves heterolytically the O–O bond of 2-methyl-1-phenyl-2-propyl hydroperoxide (MPPH), to generate the active species responsible for C–H hydroxylation (Schemes 6.12 and 6.13),[66] which strongly argues against the implication of free-diffusing radical intermediates. MPPH is used as a mechanistic proof to distinguish between homolytic and heterolytic cleavage of the O–O bond. In reactions that involve homolytic cleavage of this peroxide, the resulting radical undergoes very fast β-scission to form acetone and a benzyl radical. Subsequent reaction with O$_2$ ends-up producing benzyl alcohol and benzaldehyde (Scheme 6.12). Such products are not formed in the reaction between MPPH and the [Fe^{II},Fe^{II}] dimer, but instead 2-methyl-1-phenyl-2-propyl alcohol is produced, which is only detected when heterolytic O–O bond cleavage occurs.

On the basis of mechanistic studies, the authors propose the formation of very short lived Fe^{II}Fe^{IV}O mixed-valence intermediates that in the absence of substrates collapse *via* reorganization to a catalytically inactive Fe^{III}–O–Fe^{III} dimer.

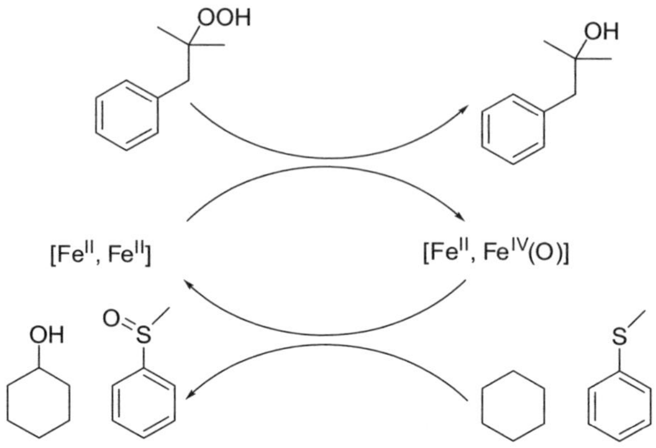

Scheme 6.13 Schematic diagram of catalytic oxidations with MPPH.[66]

Yields obtained with these catalysts in the oxidation of cycloalkanes have been traditionally regarded as very good, especially when compared with the free-radical Co-based process for the oxidation of cyclohexane in the production of adipic acid, which affords values of <10%.[67] However, these yields have become modest compared with recent accomplishments with novel mononuclear iron(II) catalysts (*vide infra*).

6.2.1.2 *Mononuclear Iron Complexes with N-Based Ligands*

The use of mononuclear iron complexes containing N-based ligands, especially those containing Fe^{II} and pyridine heterocycles, has expanded dramatically the efficiency of the oxidation reactions catalyzed by iron species. It has also extended the variety of oxidation reactions that can be mediated by iron. This has also been accompanied by the discovery of a rich diversity of reaction paths. Furthermore, very small changes in the architecture of the iron complex usually result in dramatic changes in the catalytic performance and also in the nature of the oxidizing species. This makes catalyst design a very challenging problem.

A particularly relevant aspect of the use of N-based ligands is that it led to the discovery of iron catalysts that perform the stereospecific hydroxylation of alkanes (Scheme 6.14).

Besides the potential of the transformation in organic synthesis, the reaction has also very important implications from a mechanistic and biomimetic point of view. These reactions require the implication of selective metal-based oxidants analogous to those involved in enzymatic systems, and thus they constitute a clear-cut departure from free-diffusing radical chemistry initiated by $HO^{\bullet}$ or $RO^{\bullet}$ radicals (Fenton-like chemistry). These catalysts are, therefore, valuable functional models for iron-dependent hydroxylases.[6]

The number of iron catalysts that engage in this stereospecific reactivity is still very limited. Accurate design of the ligand is necessary and not obvious.

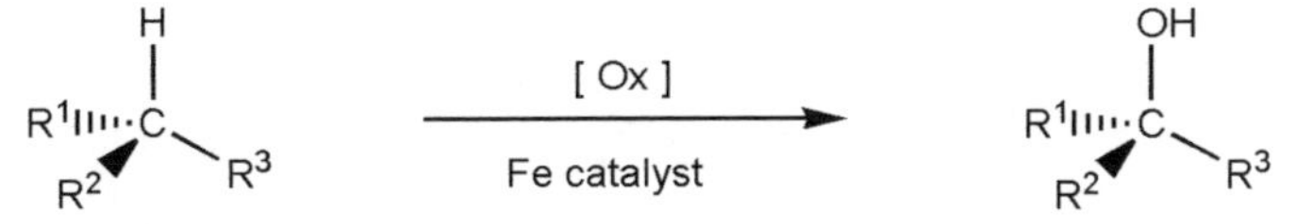

Scheme 6.14 Stereospecific oxidation of alkanes through metal-based oxidants with an iron catalyst.

Fortunately, an attractive feature of non-heme iron catalysts is that ligand modification and catalyst tuning are relatively straightforward, compared with porphyrin-type ligands, and the number of catalysts that can be prepared and studied has virtually no limit.

Iron Complexes with Tetradentate Ligands. The vast majority of iron catalysts that can mediate stereospecific hydroxylation of alkanes contain an Fe^{II} center and a tetradentate N-based ligand that leaves two coordination positions vacant in a *cis* relative disposition. At least one pyridine ring is always present. The two *cis*-positions are labile and are occupied by weakly bound ligands, such as CH_3CN, H_2O, or CF_3SO_3. The lability of these positions is a key aspect to ensure a fast reaction with the oxidant (commonly H_2O_2). Ligands must be oxidatively robust and must bind very tightly to the iron ion to withstand the high oxidative reaction conditions. The use of Fe^{II} instead of Fe^{III} ions is a non-trivial aspect that deserves some comment. Iron(III) ions are very Lewis acidic and tend to form oxo-bridged dimers, which quite often lack available binding sites for the oxidant, and show much limited catalytic activity. On the other hand, in the presence of the oxidant, the iron(II) complexes become rapidly oxidized to very active mononuclear Fe^{III} species. Formation of oxo-bridged Fe^{III} oligomeric species can eventually lead to their deactivation.

Stereospecific hydroxylation of alkanes with H_2O_2 was reported by Que and coworkers in 1997 using $[Fe^{II}(tpa)(CH_3CN)_2]^{2+}$ as catalyst (Figure 6.5 and Scheme 6.15).[68] Reactions were performed under conditions where H_2O_2 (10 equiv. with respect to the catalyst) was the limiting reagent, in the presence of a large excess of substrate (1000 equiv.). This large excess was employed to minimize overoxidation reactions of the alcohol product, which is more easily oxidized than the alkane substrate. In addition, since H_2O_2 can be easily oxidized at faster reaction rates than alkanes, H_2O_2 was added by a syringe pump to minimize its concentration in solution and avoid its disproportionation.

Under these conditions, $[Fe^{II}(tpa)(CH_3CN)_2]^{2+}$ made efficient use of H_2O_2. Using cyclohexane as substrate, more than 32% of the H_2O_2 was converted into alkane oxidation products. Most significant was the observation that cyclohexanol was the main product, and that cyclohexanone was obtained in very small amounts. A high A/K ratio (Table 6.2) constituted a first indication that reactions involved a metal-centered C–H hydroxylation process. Finally, *cis*-1,2-dimethylcyclohexane was hydroxylated stereospecifically at the tertiary C–H position to afford the corresponding tertiary alcohol.

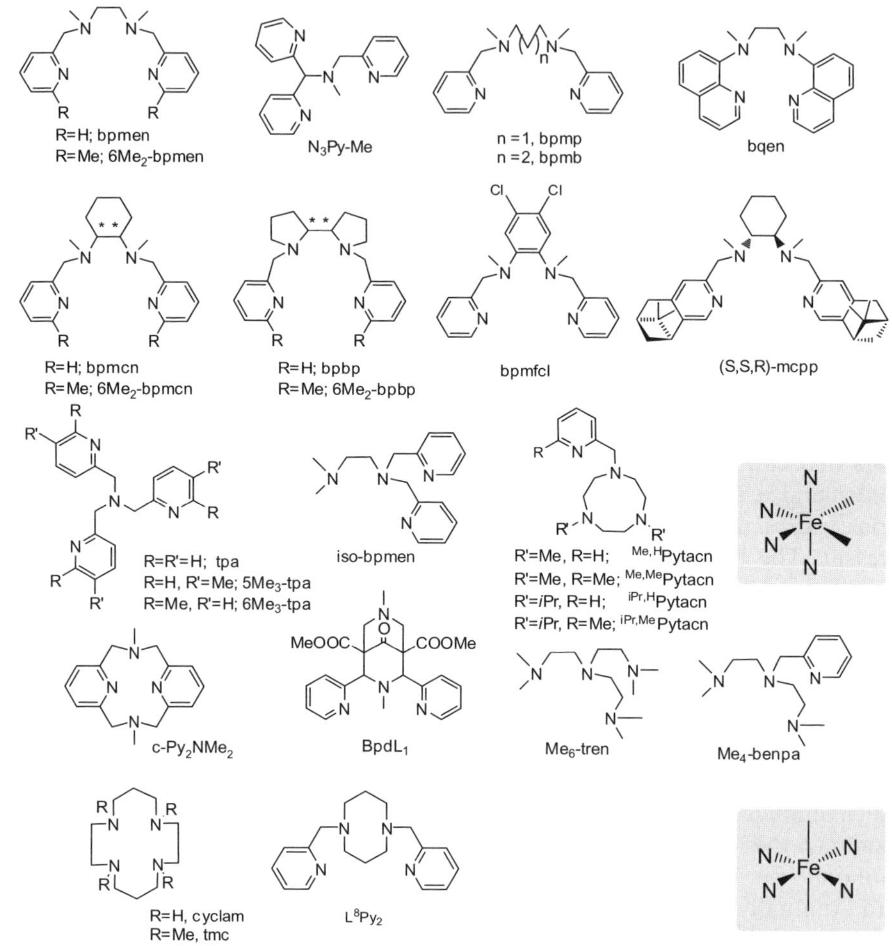

Figure 6.5 Relevant tetradentate ligands used to prepare mononuclear iron(II) complexes to perform alkane and olefin oxidations along with the coordination they adopt around the metal center.

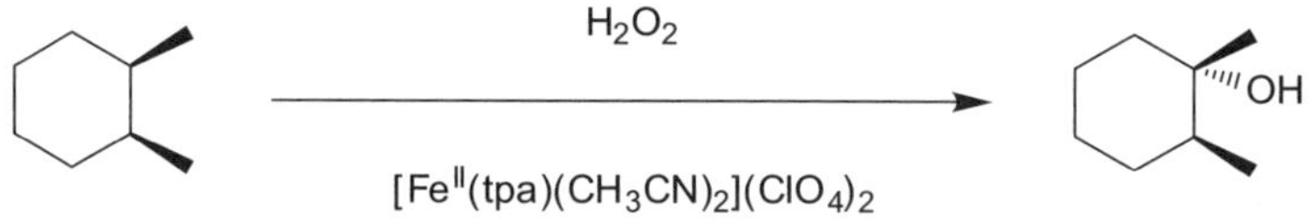

Scheme 6.15 Stereospecific oxidation of *cis*-1,2-dimethylcyclohexane (*cis*-DMCH) by the tpa iron(II) complex reported by Que and coworkers.[68]

Since then, a few non-heme mononuclear iron complexes capable of performing stereospecific alkane oxidation to give alcohols have been described; a schematic representation of the most relevant ones is depicted in Figure 6.5.

Two main families of complexes bearing tetradentate ligands have been most studied in the oxidation of alkanes: tpa (tripodal) and bpmen (linear) families (Figure 6.5; tpa = tris(2-pyridylmethyl)amine and bpmen = N,N'-dimethyl-N,N'-bis(2-pyridylmethyl)ethylene-1,2-diamine).[69] Mononuclear iron(II) complexes derived from these ligands are especially efficient catalysts and the mechanistic probes clearly point towards the mediation of metal-based oxidants (Table 6.2).[32] Oxidation of *cis*-1,2-dimethylcyclohexane occurs at the tertiary C–H position stereospecifically. Adamantane is oxidized with a high $3°/2°$ selectivity, and kinetic isotope effects (KIEs) evaluated in the competitive oxidation of cyclohexane *versus* cyclohexane-d_{12} are relatively high (3–4.3). The KIE values also indicate that the breakage of the C–H bond is involved in the rate-determining step of the reactions. More recently, a new family of tetradentate ligands derived from 1,4,7-triazacyclononane (tacn) was reported, which showed remarkable efficiency and provided evidence of a metal-based oxidant (Table 6.2).[70,71]

The importance of the presence of two *cis* labile sites available for coordination with exogenous ligands (the oxidant and/or the substrate) was studied

Table 6.2 Oxidation of alkanes with H_2O_2 catalyzed by different iron(II) complexes.

| | Cyclohexane | | cis-1,2-DMCH | Adamantane | |
Complex	$A+K$ $(A/K)^a$	KIE (kinetic isotope effect)	RC $(\%)^b$	$3°/2°^c$	Ref.
$[Fe^{II}(tpa)(CH_3CN)_2]^{2+}$	3.2 (5)	3.5	100	17	32
$[Fe^{II}(5Me_3\text{-}tpa)(CH_3CN)_2]^{2+}$	4.0 (9)	3.8	100	21	32
$[Fe^{II}(6Me_3\text{-}tpa)(CH_3CN)_2]^{2+}$	1.4 (3)	3.3	54	15	32
$[Fe^{II}(bpmen)(CH_3CN)_2]^{2+}$	6.3 (8)	3.2	96	15	32
$[Fe^{II}(bqen)(CH_3CN)_2]^{2+}$	5.1 (5)	–	–	–	72
$[Fe^{II}(bpPhOMeen)(CH_3CN)_2]^{2+d}$	3.8 (8)	–	82	–	73
$[Fe^{II}(Cl)_2(bpPhOMeen)]^{d}$	0.6 (1)	–	0	–	73
$\alpha\text{-}[Fe^{II}(CF_3SO_3)_2(bpmcn)]$	5.9 (9)	3.2	>99	15	74
$\beta\text{-}[Fe^{II}(CF_3SO_3)_2(bpmcn)]$	1.9 (0.9)	4.0	68	17	74
$[Fe^{II}(CF_3SO_3)_2(^{Me,H}Pytacn)]$	6.5 (12)	4.3	93	30	71
$[Fe^{II}(CF_3SO_3)_2(^{Me,Me}Pytacn)]$	7.6 (10)	3.4	94	20	71
$[Fe^{II}(BpdL_1)(CH_3CN)_2]^{2+e}$	34 (1.5)	5.2	–	17	75
$[Fe^{II}(CF_3SO_3)_2(iso\text{-}bpmen)]$	3.2 (6.8)	3.4	–	–	79
$[Fe^{II}(CF_3SO_3)_2(Me_4\text{-}benpa)]$	0.34 (3.7)	2.1	–	–	79
$[Fe^{II}(CF_3SO_3)_2(Me_6\text{-}tren)]$	0.32 (1.3)	1.8	–	–	79
$[Fe^{II}(CF_3SO_3)_2(bpmfcl)]$	6.2 (2.4)	–	–	–	76
$[Fe^{II}(CF_3SO_3)_2(bpmp)]$	4.2 (1.4)	–	–	–	76
$[Fe^{II}(CF_3SO_3)_2(bpmb)]$	0	–	–	–	76

Catalyst:H_2O_2:alkane = 1:10:1000; 1:10:10 for adamantane.
^{a}Turnover number (TN, moles of product/moles of catalyst), A = number of moles of cyclohexanol, K = number of moles of cyclohexanone.
^{b}Retention of configuration in the oxidation of *cis*-1,2-DMCH.
$^c3°/2° = 3\times$(1-adamantanol)/(2-adamantanol+2-adamantanone).
^{d}1100 equiv. substrate.
^{e}100 equiv. H_2O_2.

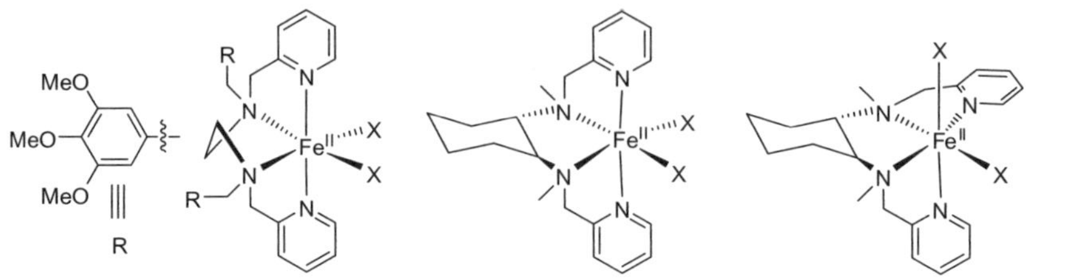

X = Cl, [Fe^{II}(Cl)$_2$(bpPhOMeen)]
X = CH$_3$CN, [Fe^{II}(bpPhOMeen)(CH$_3$CN)$_2$]$^{2+}$ *cis*-α-[Fe^{II}(CF$_3$SO$_3$)$_2$(bpmcn)]   *cis*-β-[Fe^{II}(CF$_3$SO$_3$)$_2$(bpmcn)]

Figure 6.6 Schematic diagram of selected Fe complexes with tetradentate N-based ligands.[73,74]

by Ménage and coworkers for the [Fe^{II}(X)$_2$(bpPhOMeen)] (X = Cl and CH$_3$CN) pair of complexes[73] (Figure 6.6 and Table 6.2). This study concluded that labile acetonitrile ligands in [Fe^{II}(bpPhOMeen)(CH$_3$CN)$_2$]$^{2+}$ led to metal based pathways. Instead, [Fe^{II}(Cl)$_2$(bpPhOMeen)], which contains the more strongly coordinating chloride ligand, led to Fenton-type reactions. The reason is that ligand exchange allowing for peroxide binding to the metal center is a prerequisite for generation of a metal-centered oxidant.

In a series of papers, Britovsek's and Que's groups showed that minor modifications in the ligand architecture can cause important changes in the catalytic performance (Table 6.2). Steric effects were investigated by introducing methyl substituents in the α position of the pyridine ring of tpa ligand.[32] Differently to tpa, 6Me$_3$-tpa generates a catalyst with some non-metal based character as evidenced by the low percentage of retention of configuration (RC) in the oxidation of *cis*-1,2-DMCH and the large percentage of O$_2$ incorporation into products.[32] The latter originates from diffusion controlled O$_2$ trapping of carbon-centered radical intermediates. Electronic effects have been investigated by the introduction of substituents in the pyridyl *para*-positions,[32,77] or by changing the number of pyridines in tripodal ligands.[78] In all cases, similar or lower activity was observed. The effect of the structure and atom donor nature of the backbone was studied for the linear bpmen and bqen ligands.[72,74,79] Rigidity of the ligand backbone appears to be an important factor, since a too flexible backbone gives a mixture of coordination modes, leading to less active catalysts. An increase on the flexibility of the ligand is also described as resulting in weaker ligand fields, which increases the lability of the complexes (in Table 6.2, compare the efficiency of [Fe^{II}(CF$_3$SO$_3$)$_2$(bpmen)], [Fe^{II}(CF$_3$SO$_3$)$_2$(bpmp)] and [Fe^{II}(CF$_3$SO$_3$)$_2$(bpmb)]). Therefore, the activity and selectivity of the catalysts is proposed to be related to the strength of the ligand field and the stability of the catalyst in the oxidizing environment.[76]

A remarkable example of the dramatic differences in the catalytic outcome of an iron complex depending on its coordination structure is the comparison between the *cis*-α and *cis*-β topological isomeric forms of [Fe^{II}(CF$_3$SO$_3$)$_2$(bpmcn)] [Table 6.2 and Figure 6.6, bpmcn = *N*,*N'*-dimethyl-*N*,*N'*-bis(2-pyridylmethyl)*trans*-1,2-diaminocyclohexane]. The former is an efficient catalyst for stereospecific C–H hydroxylation, the latter gives reaction patterns that strongly implicate carbon centered diffusing radicals.[74]

As shown by the aforementioned studies, catalyst stability under the harsh oxidizing conditions required to oxidize alkanes is a major factor in determining the catalytic efficiency of a given catalyst. As proof of concept, very robust ligand scaffolds like bpmen and Pytacn give rise to especially active catalysts [Pytacn = 1-(2-pyridylmethyl)-1,4,7-triazacyclononane].

Iron Complexes with Bidentate, Tridentate and Pentadentate Ligands. Iron complexes containing bidentate, tridentate, and pentadentate ligands have been less explored than their tetradentate counterparts in oxidation catalysis; presumably, because they have proved much less efficient. The combination of $FeCl_3$ with 2,2′-bipyridine used as a catalyst dramatically accelerates oxidation of alkanes with hydrogen peroxide in acetonitrile solution.[80] Turnover numbers for the oxidation of cyclohexane attain 400 after 1 h at 60 °C. Oxidation of cyclohexane yields cyclohexyl hydroperoxide predominantly, which is transformed during the reaction into cyclohexanol and cyclohexanone. Oxidations proceed with low selectivity parameters reminiscent of those determined for hydroxyl-radical generating systems, and stereoscrambling at the tertiary carbon centers is observed in the oxidation of *cis*-1,2-dimethylcyclohexane. The authors proposed that the bipyridine ligand facilitates proton abstraction from a H_2O_2 molecule coordinated to the iron ion.

Tridentate iron(II) complexes shown in Figure 6.7 were prepared and studied in the oxidation of cyclohexane.[78] The complexes afford small amounts of oxidation products ($\leq$5% with respect to H_2O_2). In addition, the *A/K* ratio is low (1–3), and cyclohexyl hydroperoxide is formed as one of the oxidation products. These observations strongly suggest that the oxidation of cyclohexane mediated by these complexes occurs *via* a Fenton-type process, which involves a radical-chain mechanism, initiated by hydroxyl radicals.

Complexes with pentadentate ligands, and incorporating amido binding groups were first explored as structural and functional models of Fe-bleomycin. Fe-bleomycin is a member of a family of antitumor antibiotics, which mediates very fast double strand DNA cleavage.[81] Activated bleomycin, responsible for oxidative DNA cleavage has been spectroscopically characterized as a ferric

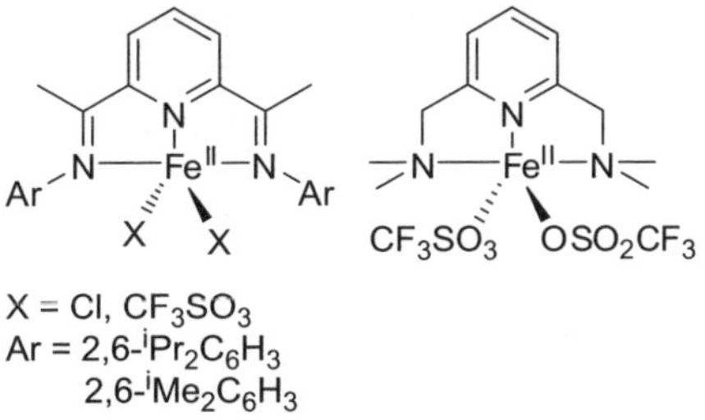

Figure 6.7   Iron catalysts with tridentate N-based ligands.[78]

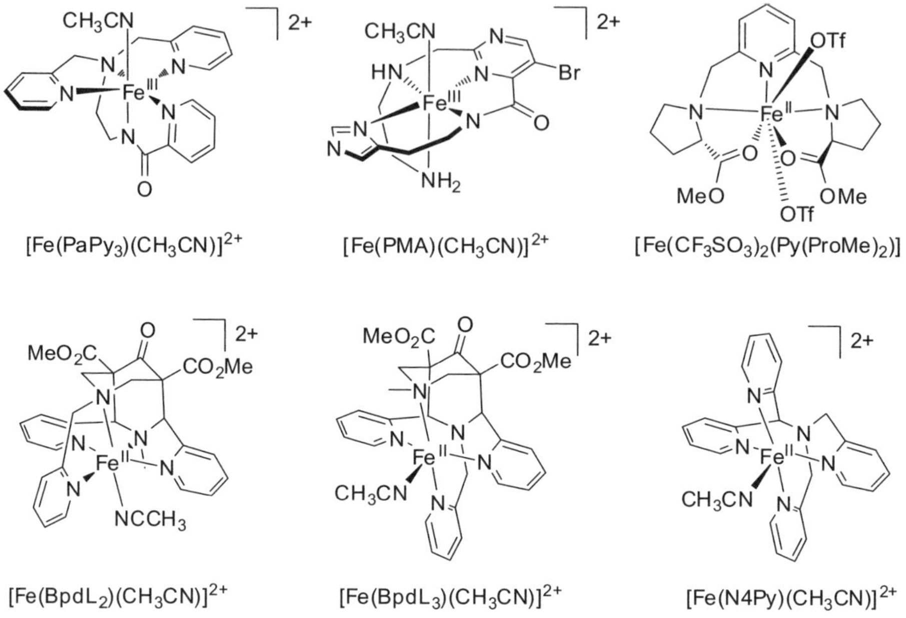

[Fe(PaPy$_3$)(CH$_3$CN)]$^{2+}$

[Fe(PMA)(CH$_3$CN)]$^{2+}$

[Fe(CF$_3$SO$_3$)$_2$(Py(ProMe)$_2$)]

[Fe(BpdL$_2$)(CH$_3$CN)]$^{2+}$

[Fe(BpdL$_3$)(CH$_3$CN)]$^{2+}$

[Fe(N4Py)(CH$_3$CN)]$^{2+}$

Figure 6.8 Iron catalysts with pentadentate ligands.[75,84–88]

hydroperoxide species.[82,83] Synthetic models for this metallodrug have been explored by Mascharak *et al.* as alkane oxidation catalysts using TBHP and H$_2$O$_2$ as oxidants.[84,85] The complexes [Fe^{III}(PaPy$_3$)(CH$_3$CN)]$^{2+}$ and [Fe^{III}(P-MA)(CH$_3$CN)]$^{2+}$ (Figure 6.8, PaPy$_3$H = *N,N*-bis(2-pyridylmethyl)-amine-*N*-ethyl-2-pyridine-2-carboxamide) were studied in the oxidation of cyclohexane. Alcohol and ketone were obtained in close to equimolar ratios, with the TN reaching up to 80, in the case of TBHP oxidations. In addition, *tert*-butyl cyclohexyl peroxide was also obtained as oxidation product. Overall, the product pattern is consistent with oxidations being performed by *tert*-butoxyl radicals.

The bleomycin model complex [Fe^{II}(N4Py)(CH$_3$CN)]$^{2+}$ and also bispidine-based [Fe^{II}(BpdL$_2$)(CH$_3$CN)]$^{2+}$ and [Fe^{II}(BpdL$_3$)(CH$_3$CN)]$^{2+}$, which contain pentadentate ligands (Figure 6.8), have been studied as alkane oxidation catalysts.[75,86,87] H$_2$O$_2$ and peracids were explored as oxidants. Using 1 mol% catalyst loadings of [Fe^{II}(N4Py)(CH$_3$CN)]$^{2+}$ with respect to the oxidant, and large excess of substrate (1000 equiv.), efficiencies in the use of H$_2$O$_2$ between 38 and 43% with respect to the oxidant were obtained. Bispidine based systems afforded slightly lower yields under analogous conditions. Analogous low *A*/*K* ratios (0.8–2.1) in the oxidation of cyclohexane, KIE values of $\sim$1.5–3.8, relatively small 3°/2° ratios in adamantane oxidation (3.1–5), and large incorporation of O$_2$ into oxidation products are also observed when reactions are run under O$_2$. The sum of these observations strongly supports the implication of hydroxyl radicals in these transformations, presumably formed *via* Fenton-type reactions between Fe^{II} complexes and H$_2$O$_2$.

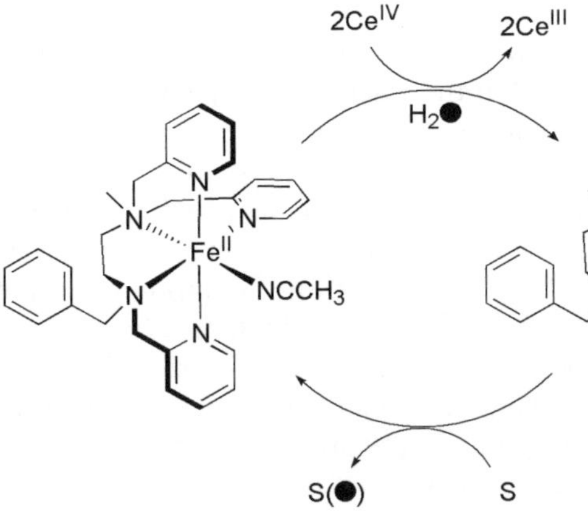

Scheme 6.16 Catalytic oxidation of ethylbenzene with a Fe^{IV}(O) species generated with CAN and water.

In contrast, the use of peracids [MCPBA and peracetic acid (PAA)] gives results that implicate a more selective oxidant.[87] This is evidenced by a larger 3°/2° ratio in the oxidation of adamantane (18 > 3°/2° > 12) a higher KIE (4.5 > KIE > 4.9) and a larger cyclohexanol/cyclohexanone ratio (4.3 > A/K > 7.2). The implication of Fe^{IV}=O species in the C–H oxidation reactions is proposed to account for these observations.

Departing from N-only based systems, Klein Gebbink *et al.* have described a diastereo-pure seven-coordinate iron(II) complex [Fe^{II}(CF$_3$SO$_3$)$_2$\{Py(ProMe)$_2$\}] that contains a pentadentate N$_3$O$_2$-ligand [Py(ProMe)$_2$] [Figure 6.8, Py(ProMe)$_2$ = N,N',N-bis(L-prolinate)pyridine] with proline based arms. The complex adopts an unusual pentagonal bipyramidal coordination geometry.[88] Under conditions of large excess of substrate (500 equiv.), the complex catalyzes ethylbenzene oxidation with TBHP (10 equiv.), to afford acetophenone (TN = 5.4) and 1-phenylethanol (TN = 4) with a modest 6.5% ee.

An original approach has recently been described by Nam, Fukuzumi and coworkers, (Scheme 6.16) who reported the generation of [Fe^{IV}(O)(Bntpen)]$^{2+}$ [Bntpen = N-benzyl-N,N',N'-tris(2-pyridylmethyl)ethane-1,2-diamine] by using cerium(IV)ammonium nitrate (CAN) as oxidant and H$_2$O as oxygen source.[89] The iron(IV)-oxo species can perform the stoichiometric oxidation of an alkyl arene. By using [Fe^{IV}(O)(Bntpen)]$^{2+}$ (0.02 mM) as catalyst and CAN (10 mM) as oxidant, catalytic oxidation of ethylbenzene to acetophenone (TN = 9) and 1-phenylethanol (TN = 1) was accomplished. ^{18}O-labelling studies in the oxidation of thioanisole indicated that water is the oxygen source in the oxidation reactions.

6.2.1.3 Towards Synthetic Applications

The use of iron catalysts in organic transformations in general, and oxidation reactions in particular, is very attractive because of sustainability and economical considerations.[44,45,90,91] However, until very recently, Fe-based oxidation reactions have been performed in conditions of large excess of substrate, which very much hampered their potential as tools in organic synthesis. However, recent work has demonstrated that the reaction of iron complexes with peroxides can provide efficient alkane oxidation methodologies.

The combination of simple iron salts with peroxides can generate efficient oxidation methods. For example, Bolm *et al.* reported that $FeCl_3$ (2 mol%) catalyzes benzylic oxidations with *tert*-butyl hydroperoxide (TBHP, 3 equiv.) as oxidant in pyridine at 82 °C, affording carbonyl compounds in modest (15%) to excellent (>99%) yields.[92] Oxidation of alkylarenes and cycloalkanes to mainly the corresponding ketones was also accomplished by the same group in more modest yields (35–50%) with H_2O_2 (5 equiv.) in acetonitrile at room temperature, using $Fe(ClO_4)_2 \cdot 6H_2O$ (10 mol%) as catalyst and acetic acid (20 mol%) as additive (Scheme 6.17).[93]

In a particularly efficient example, an iron(II) complex with a pentadentate ligand [L-N_3(OMe)$_2$, Scheme 6.18] at a 1% catalyst loading has been reported to convert cyclohexane quantitatively into a mixture of cyclohexanol and cyclohexanone (Scheme 6.18). The system uses H_2O_2 as oxidant and works at 50 °C under argon atmosphere.[94,95] Mechanistic considerations such as the alcohol/ketone ratio, indicative of a Russell-type termination, strongly suggest that reactions proceed *via* formation of cyclohexyl radicals that freely diffuse in solution.

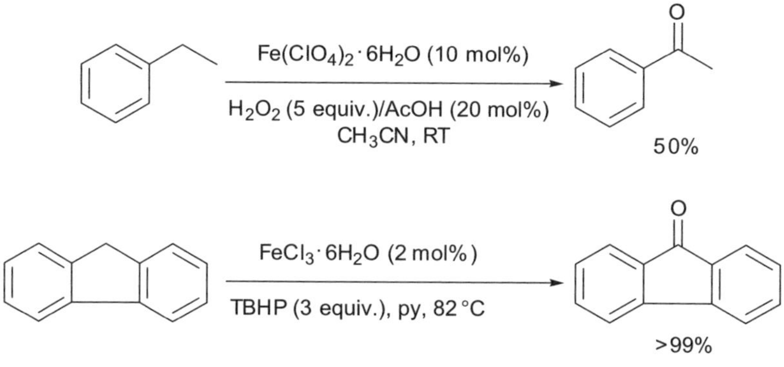

Scheme 6.17　Catalytic oxidation of alkylarenes with iron salts.[92,93]

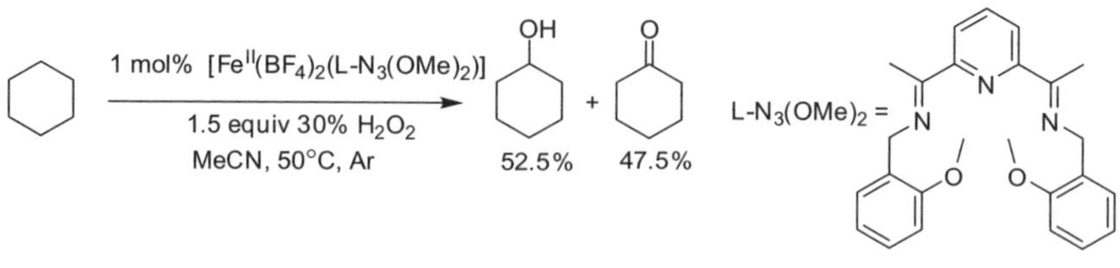

Scheme 6.18　Cyclohexane oxidation catalyzed by [Fe^II(BF$_4$)$_2${L-N_3(OMe$_2$)}].[95]

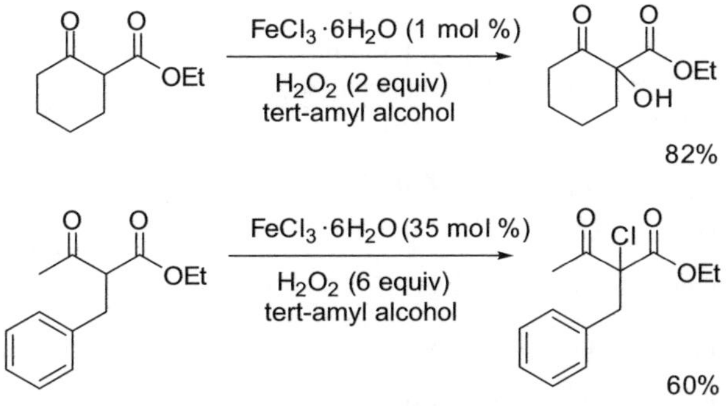

Scheme 6.19 Selective oxidation of β-ketoesters catalyzed by $FeCl_3 \cdot 6H_2O$.[96]

Later studies indicated that the ligand was not strictly required for efficient cyclohexane oxidation, and that the reaction was not substantially sensitive to the nature of the iron source. The use of $Fe(ClO_4)_2$, $Fe(BF_4)_2$, $Fe(ClO_4)_3$, and $FeCl_3$ all led to the efficient oxidation of cyclohexane.[94]

A simple procedure for the selective hydroxylation of β-ketoesters with H_2O_2 catalyzed by $FeCl_3 \cdot 6H_2O$ has been described recently by Beller *et al.*[96] Cyclic β-ketoesters are hydroxylated in excellent yields (75–90%) by using 1 mol% catalyst and 2 equiv. of 30 wt% H_2O_2 in *tert*-amyl alcohol (Scheme 6.19). When a large loading of $FeCl_3 \cdot 6H_2O$ is employed, chlorination instead of hydroxylation takes place.

In a recent landmark study, Chen and White reported the hydroxylation of aliphatic tertiary C–H bonds, achieving high yields and selectivities using a combination of H_2O_2 and acetic acid as the oxidant.[97] In this study the catalyst is a mononuclear iron complex $[Fe^{II}(bpbp)(CH_3CN)_2](SbF_6)_2$ with the tetradentate bis-pyrrolidine based ligand bpbp [see Figure 6.5, bpbp = *N*,*N*′-bis(2-pyridylmethyl)-2,2′-bipyridine]. Impressively, this system showed predictable selectivity based on electronic (Scheme 6.20), steric and directing factors (Scheme 6.21), making it very interesting from a synthetic viewpoint. Most impressively, selective oxidation in complex molecules with multiple tertiary C–H sites was achieved in synthetically useful yields (Scheme 6.22).

The oxidation of 3,7-dimethyl-1X-octane derivates evidences electronic discrimination in the hydroxylation of tertiary C–H bonds (Scheme 6.20). Hydroxylation of 3,7-dimethyloctane affords a 1:1 mixture of the tertiary alcohols. However, when X = F, Cl, or acetate is used the hydroxylation preferentially occurs at the distal site.

Steric effects have also been shown to discriminate reactivity. Owing to the steric hindrance of the tertiary C–H bond in (−)-α-dihydropicrotoxinin (Scheme 6.21, top) it is not effectively oxidized by the catalyst, and the substrate is recovered in 92% yield.

Finally, a carboxylate group can be used to site-direct C–H oxidation, presumably *via* binding to the iron catalyst (Scheme 6.21, bottom). In this case, rapid lactonization follows C–H hydroxylation.

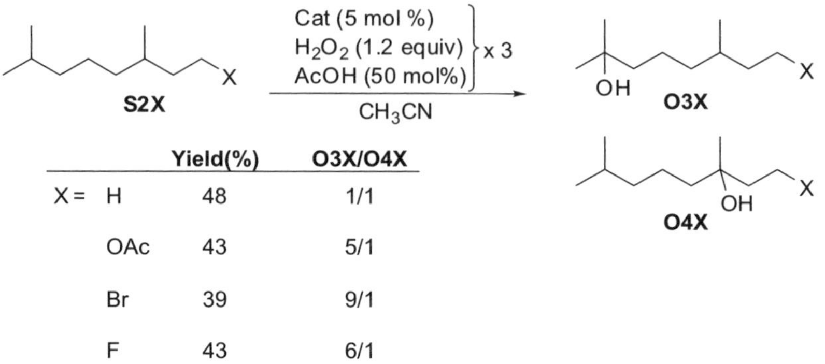

X =	Yield(%)	O3X/O4X
H	48	1/1
OAc	43	5/1
Br	39	9/1
F	43	6/1

Scheme 6.20 Electronic discrimination in the hydroxylation of tertiary C–H bonds catalyzed by [Fe^{II}(bpbp)(CH$_3$CN)$_2$](SbF$_6$)$_2$.[97]

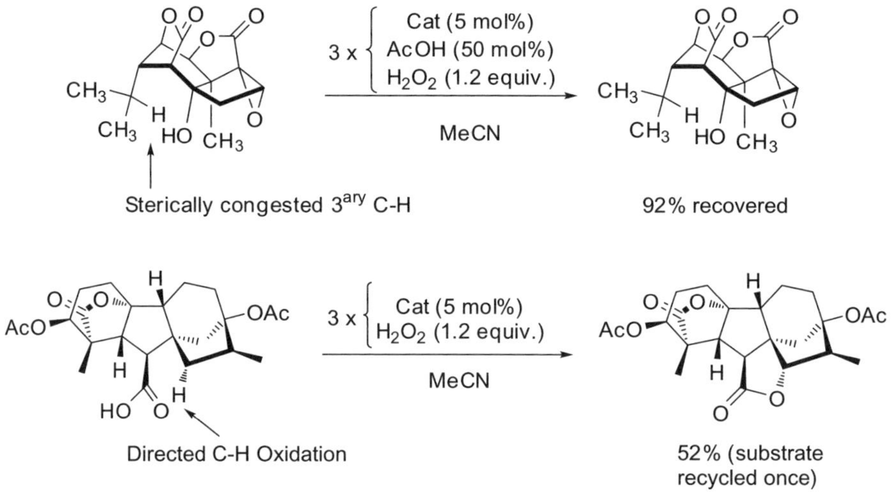

Scheme 6.21 Steric blocking effect (top) and directed hydroxylation of C–H bonds (bottom) catalyzed by [Fe^{II}(bpbp)(CH$_3$CN)$_2$](SbF$_6$)$_2$.[97]

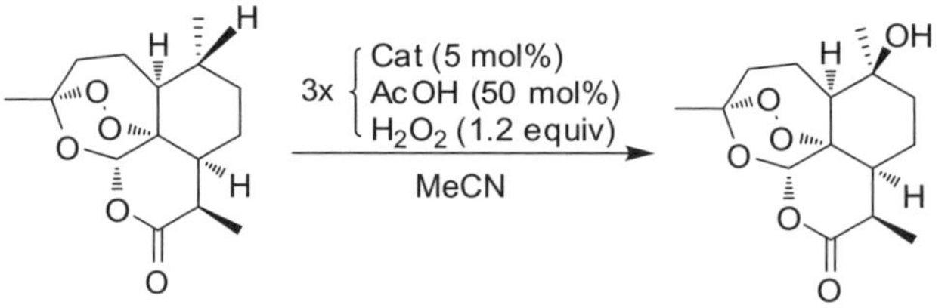

Catalyst	Product Yield
Fe(bpbp)	34%
Recycled	54%
Fe(bpmen)	11%
Recycled	23%

Scheme 6.22 Selective hydroxylation of (+)-artemisin by [Fe^{II}(bpbp)(CH$_3$CN)$_2$]-(SbF$_6$)$_2$ and [Fe^{II}(bpmen)(CH$_3$CN)$_2$](SbF$_6$)$_2$ iron complexes (see Figure 6.5 for ligand structures).[97]

Selective oxidation of complex organic substrates is demonstrated in the hydroxylation of (+)-artemisin. Oxidation occurs preferentially at the most electron-rich and the least sterically hindered 3 °C–H bond, faster and with better isolated yields than when obtained *via* an enzymatic oxidation.

White and Chen's landmark system uses 15% catalyst loadings, and, even though it affords very modest turnover numbers, the efficiencies were substantially better than in any previously reported non-heme catalyst. Indeed, in their work, the authors compare the activity of their catalyst with [Fe^{II}(bpmen)(CH$_3$CN)$_2$](SbF$_6$)$_2$, at that time the most efficient iron catalyst, showing that [Fe^{II}(bpbp)(CH$_3$CN)$_2$](SbF$_6$)$_2$ affords higher yields under the same reaction conditions. In a subsequent study, the authors described experimental procedures that avoid substrate recycling. In the same study, the authors conclude that formation of oxo-bridged oligomeric species appear to be one of the reasons of catalyst deactivation.[99]

The factors that determine the C–H site selectivity in the oxidation reactions catalyzed by [Fe^{II}(bpbp)(CH$_3$CN)$_2$](SbF$_6$)$_2$ have been explored, with the aim of targeting methylene sites.[100] The C–H bond strength is a major factor that directs reactivity in the relative order 3 °C–H > 2 °C–H > 1 °C–H bonds. However, from a steric point of view the order is inverted, and 3 °C–H bonds are the sterically least accessible. Oxidation of tertiary C–H bonds could be restricted by steric crowding due to vicinal groups. A third factor that directs C–H reactivity is the presence of deactivating groups such as carbonyl moieties, which disfavors the oxidation at proximal positions. It was early established for dioxirane mediated oxidations[101] and later recognized in Fe-based oxidations by Eschenmoser and Baran[102] that stereoelectronic factors also need to be considered. For example, axial C–H bonds in cyclohexane skeletons are more readily oxidized than equatorial counterparts because in the former, C–H breakage in the transition state liberates tension strain and 1,3-diaxial interactions. Hydroxylation at adjacent C–H bonds of cyclopropane rings and ether groups is also favored because of hyperconjugation effects.

In simple substrates, when only one of the factors is analyzed, modest selectivity among different C–H bonds is usually attained. However, in complex organic molecules, when several of the effects take place in a concerted manner, exquisite site selective C–H oxidation can take place. Most remarkably, combined effects allow selective oxidation of secondary C–H bonds in the presence of tertiary C–H bonds. As an illustrative example, selective oxidation of methylene groups of several terpenoids was accomplished with impressive selectivity (Scheme 6.23).

The very modest turnover numbers attained with [Fe^{II}(bpbp)(CH$_3$CN)$_2$](SbF$_6$)$_2$ prompted the design of more efficient catalysts. However, the early documented dramatic dependence between catalytic efficiency and small changes in the architecture of related catalysts required the development of innovative strategies for improvement. Costas, Ribas, and coworkers were inspired towards this goal by the well-established principles in oxidation catalysis with heme complexes.[103,104] It is well-known that introduction of aryl groups at the porphyrin *meso*-positions isolates the metal site, which, in turn, limits bimolecular self-decomposition pathways and enhances the catalytic

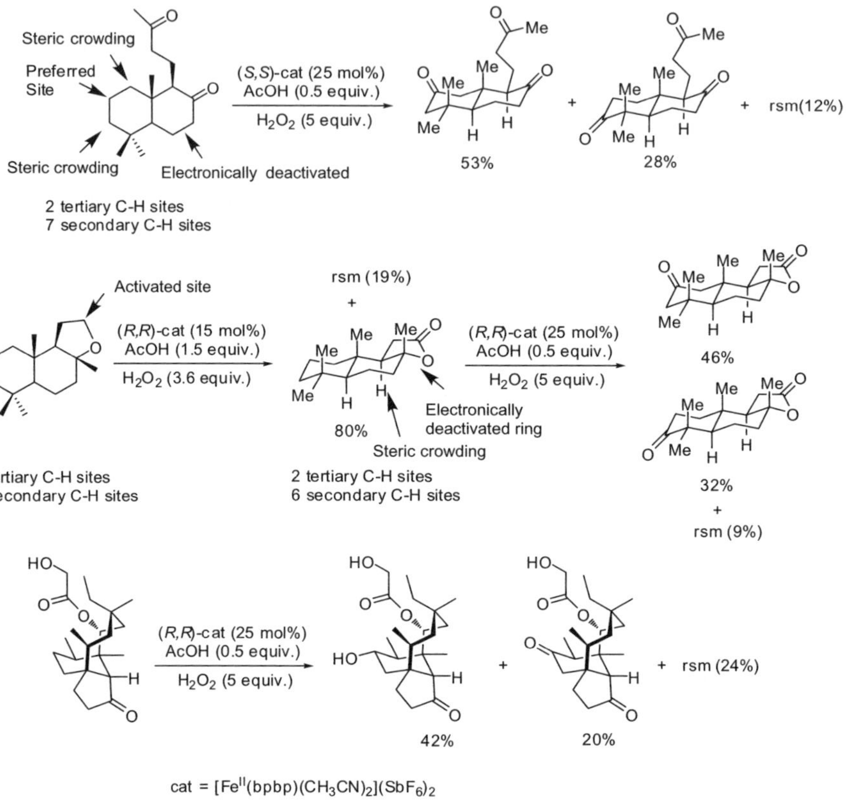

Scheme 6.23 Selective oxidation of methylene groups by [Fe^{II}(bpbp)(CH$_3$CN)$_2$]-(SbF$_6$)$_2$.[100]

activity of the complexes by precluding formation of catalytically non competent oxo-dimers.

In an attempt to apply this principle to non-heme iron complexes, the authors designed a polypyridyl tetradentate iron complex that contained pinene groups at positions 4 and 5 of the pyridine rings.[105] (*S,S,R*)-mcpp (Figures 6.5 and 6.9, mcpp = *N,N'*-dimethyl-*N,N'*-bis[{(*R*)-4,5-pinenepyridin-2-yl}methyl]cyclohexane-1,2-diamine) was targeted because it is structurally related to [Fe^{II}(CF$_3$SO$_3$)$_2$-(bpmen)] and [Fe^{II}(bpbp)(CH$_3$CN)$_2$](SbF$_6$)$_2$, and contains bulky groups that help to isolate the iron site in an oxidatively robust cavity without perturbing its stability. In addition, the chiral *trans*-1,2-cyclohexane diamine backbone imparts rigidity at the ligand, allows predetermination of the topological chirality, and determines the relative orientation of the robust pinene CH$_3$ groups with respect to the metal site. The latter is an important structural aspect aimed at limiting ligand oxidation reactions that could lead to deactivation paths.

The novel catalyst [Fe^{II}(CF$_3$SO$_3$)$_2${(*S,S,R*)-mcpp}] affords comparable yields and C–H site selectivities with [Fe^{II}(bpbp)(CH$_3$CN)$_2$](SbF$_6$)$_2$ in C–H oxidation reactions, albeit the former requires lower catalyst loadings (1–3 mol%).

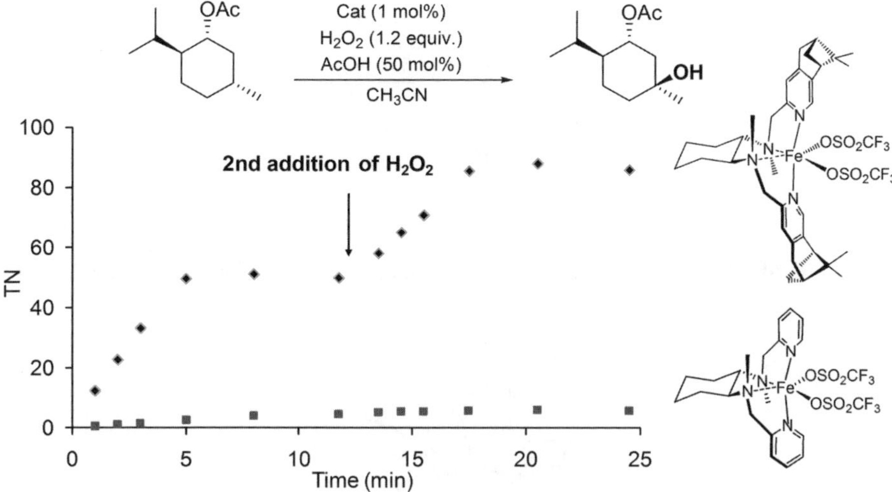

Figure 6.9 Catalytic oxidation of *p*-acetoxymenthane with [Fe^{II}(CF$_3$SO$_3$)$_2${(*S*,*S*,*R*)-mcpp}] (♦) and [Fe^{II}(CF$_3$SO$_3$)$_2$(bpmcn)] (■), showing reactivation of the former catalyst after a second H$_2$O$_2$ addition.

The profound effect that the bulky pinene groups have in the efficiency and stability of the catalysts is evidenced by performing a time profile analysis of the oxidation of *p*-acetoxymenthane catalyzed by [Fe^{II}(CF$_3$SO$_3$)$_2$(bpmcn)] and [Fe^{II}(CF$_3$SO$_3$)$_2${(*S*,*S*,*R*)-mcpp}] (Figure 6.9), which indicates that, unlike [Fe^{II}(CF$_3$SO$_3$)$_2$(bpmcn)], [Fe^{II}(CF$_3$SO$_3$)$_2${(*S*,*S*,*R*)-mcpp}] is not substantially deactivated during the reaction. A second addition of H$_2$O$_2$ and substrate resumes the catalytic activity of the latter, but not of the former. ESI-MS analyses showed that a monomeric [LFe^{III}(OR)(CF$_3$SO$_3$)]$^+$ (OR = OH, OAc) species, presumed to be the precursors of the high valent iron-oxo species responsible for catalytic activity, still remain in the final reaction solutions of [Fe^{II}(CF$_3$SO$_3$)$_2${(*S*,*S*,*R*)-mcpp}], whereas they rapidly disappeared during H$_2$O$_2$ addition when [Fe^{II}(CF$_3$SO$_3$)$_2$(bpmcn)] was used.[105]

Finally, chemically related to alkane oxidation, a very attractive system for the efficient and selective oxidation of arenes with H$_2$O$_2$ by iron catalysts has been reported by the Beller group. By using a combination of FeCl$_3$ · 6H$_2$O:H$_2$Pydic:benzylamine (1:1:2.2, 2.5 mol% Fe), and 3.5 equiv. of H$_2$O$_2$, arenes were oxidized to quinones in modest to good yields and selectivities (H$_2$Pydic = pyridine 2,6-dicarboxylic acid).[106] By using this methodology, vitamin K$_3$ is prepared by oxidation of 2-methylnaphthalene in 55% yield (Scheme 6.24).

6.2.2 Oxidation of Alkenes (C=C Double Bonds) by Non-heme Iron Complexes

Alkenes are substantially more reactive than alkanes, and susceptible to undergo several oxidative transformations. Olefin epoxidation and *cis*-dihydroxylation are important reactions because epoxides and diols are valuable compounds that are amenable for several subsequent transformations. Reliable

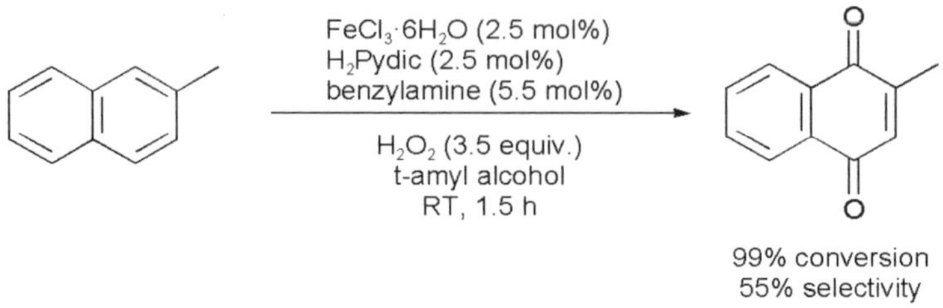

Scheme 6.24 Selective oxidation of 2-methylnaphthalene to form vitamin K3.

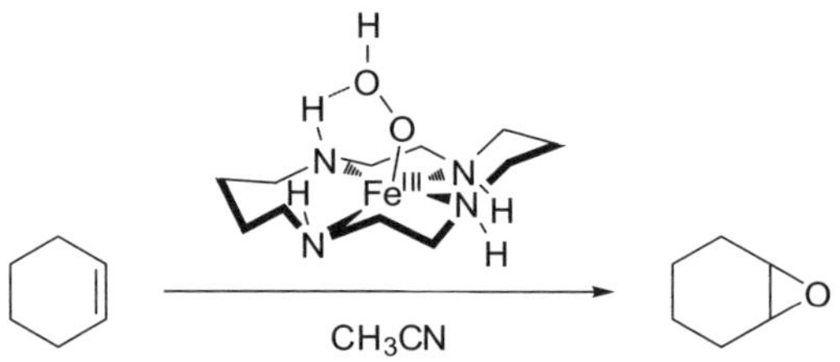

Scheme 6.25 Epoxidation of alkenes catalyzed by [Fe^{II}(CF$_3$SO$_3$)$_2$(cyclam)].

epoxidation and *cis*-dihydroxylation reagents are well established in synthetic chemistry, but they are not sustainable as they are mostly based on stoichiometric methods with low atom economy, and/or expensive and toxic heavy metal oxides. Consequently, environmentally benign methods based on iron catalysts have a huge potential in bulk and fine synthetic chemistry.

6.2.2.1 Epoxidation

One of the first examples of selective epoxidation at a non-heme catalyst was reported by Valentine and coworkers using [Fe^{II}(CF$_3$SO$_3$)$_2$(cyclam)] as catalyst (Figure 6.5, Scheme 6.25) and H$_2$O$_2$ as oxidant.[107] With 2 mol% catalyst, cyclohexene was epoxidized in 40% yield. Poor yields were obtained when TBHP was employed as oxidant. The authors suggested that a Fe^{III}OOH species was formed, where the distal oxygen atom hydrogen bonds to one of the N–H groups of the cyclam.

Following this precedent, several iron complexes have been studied as olefin epoxidation catalysts. Most iron complexes display some degree of epoxidation activity when combined with H$_2$O$_2$, but only few achieve synthetically useful epoxide yields and high substrate conversion. In the following paragraphs selected examples will be discussed.

Francis and Jacobsen used combinatorial chemistry for the screening of a library of 5760 metal-ligand complexes, arising from the combination of 192 ligand libraries, attached to a polystyrene resin, with 29 metal sources with the aim to find highly efficient catalysts for the epoxidation of *trans*-β-methylstyrene as a model substrate (Scheme 6.26).[108] The authors found that specific iron

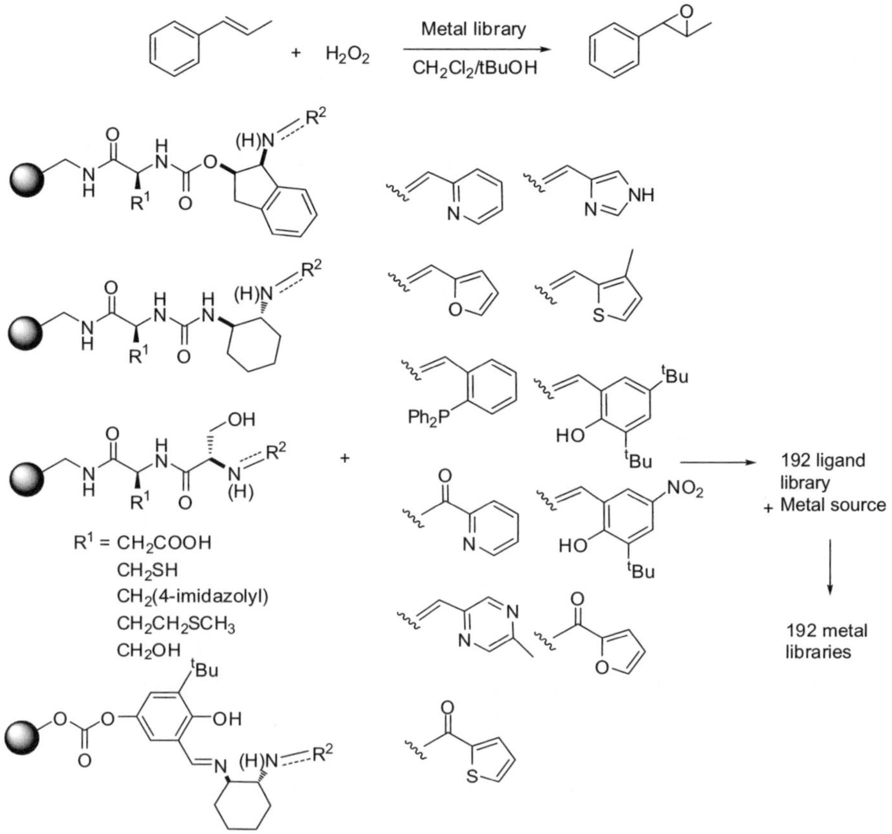

Scheme 6.26 *trans*-β-Methylstyrene epoxidation activity in a parallel library of iron complexes along with the structures of some of the ligands. The terminal balls stand for polystyrene resins.

complexes led to clean epoxide product formation using H_2O_2 as the terminal oxidant. Moreover, parallel libraries were used to determine ligand features important for high catalytic activity and to identify enantioselective catalyst structures.

The study led to the identification of the structures shown in Figure 6.10, in combination with $FeCl_2$, as moderately enantioselective epoxidation catalysts. Up to 78% conversion of the substrate can be obtained (with the epoxide as the sole reaction product) by using the optimized reaction conditions of 1.5 equiv. of 30% H_2O_2 and 5 mol% of the catalyst.

The Jacobsen group also reported a mononuclear iron(II) complex based on the bpmen ligand (Figure 6.11a), which could perform the epoxidation of alkenes using a combination of H_2O_2 and acetic acid as the oxidant (Scheme 6.27).[109] Aliphatic olefins were rapidly epoxidized (< 5 min) in high yields (60–90%). Nowadays this methodology constitutes one of the first examples of a bioinspired catalyst used as a synthetic tool for the synthesis of epoxides.

+ FeCl₂   20% ee

+ FeCl₂   15% ee

+ FeCl₂   20% ee

Figure 6.10   Catalyst structures identified from combinatorial screening.

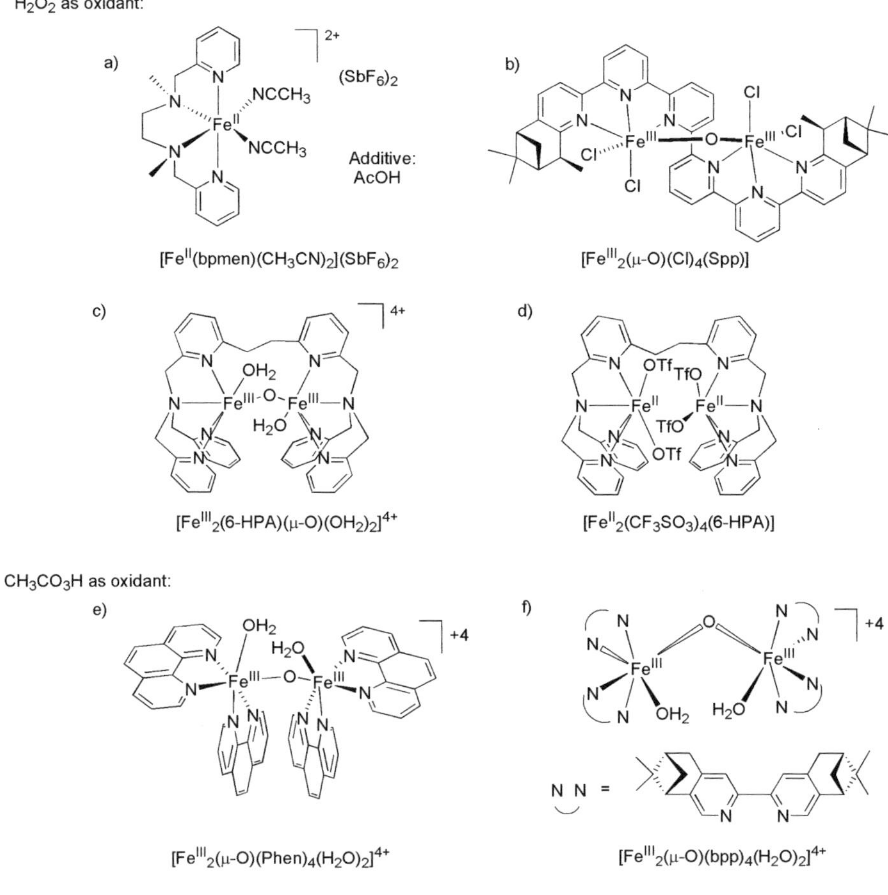

Figure 6.11   Schematic representation of relevant iron catalysts for olefin oxidation: (a) bpmen iron complex reported by Jacobsen;[109] (b) chiral iron sexipyridine reported by Kwong;[111] (c) and (d) dimeric bis-tpa complexes described by Kodera and coworkers;[112] (e) dimeric phenanthroline based complex reported by Stack;[113] and (f) chiral dimeric complex reported by Ménage.[114]

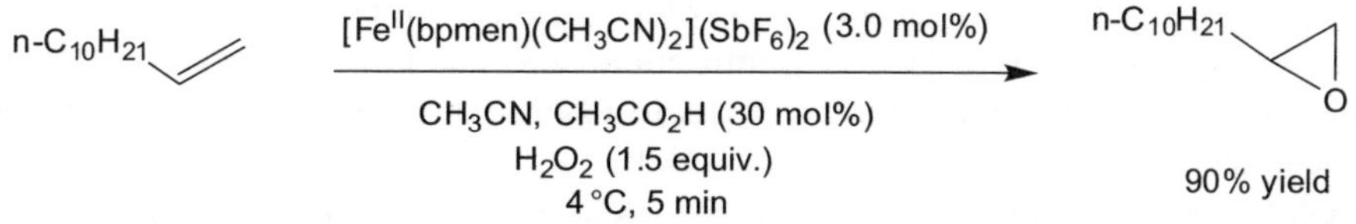

Scheme 6.27 Epoxidation of terminal olefins catalyzed by [Fe^{II}(bpmen)(CH$_3$CN)$_2$]-(SbF$_6$)$_2$.[109]

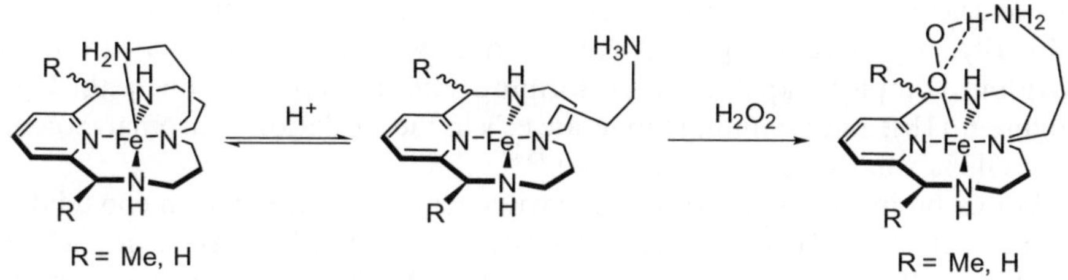

Scheme 6.28 Acid–base and epoxidation activity of [Fe^{II}(CF$_3$SO$_3$)$_2$(Mac-N_5)] complexes.[110]

Jacobsen proposed in the original work that an oxo-acetate bridged diiron(III) species, which could be seen as a structural mimic of methane monoxygenase (MMO), was the active catalyst. However, later work by Que and coworkers on the isolated dimer proved it to be inactive. Instead, Que *et al.* proposed that epoxidation was elicited by mononuclear Fe^{III} species (*vide infra*).

Rybak-Akimova has recently described a family of iron complexes [Fe^{II}(CF$_3$SO$_3$)$_2$(Mac-N_5)] (Scheme 6.28), which contain a pentadentate ligand based on a tetraazamacrocycle ligand with an appended aminopropyl arm.[110] X-Ray analysis of the complex reveals a square-pyramidal geometry, with the aminopropyl arm binding at the axial position. In the presence of poorly coordinating acids like triflic acid, reversible dissociation of the pendant arm occurs. Under these conditions, the complexes efficiently catalyze olefin epoxidation using hydrogen peroxide as the oxidant, under mild conditions. Epoxide yields of 66–89% with 90–98% selectivity were obtained with cyclooctene as substrate in the presence of 5 equiv. of triflic acid *versus* the iron complex, after 5 min at room temperature. However, in the deprotonated form, or in the presence of coordinating acids like HCl, no epoxidation occurs.

On the basis of these observations, the authors propose that epoxidation is triggered by a "pull effect" mechanism that involves a hydrogen bonding interaction between the coordinated peroxide and the pendant arm, which enhances the electrophilicity of the former.

Diiron complexes have also been used for alkene epoxidation. Stack and coworkers reported a μ-oxo-iron(III) dimer also based on nitrogen-donor ligands.[113] Phenanthroline in combination with iron salts produced an efficient epoxidation catalyst for a wide range of olefins, including terminal alkenes, using commercially available 32% peracetic acid as the oxidant (Figure 6.11e).

Low catalyst loadings, *in situ* catalyst preparation from common reagents, fast reaction times (<5 min at 0 °C), and enhanced reaction performance at high substrate concentrations combine to create a synthetically efficient procedure for alkene epoxidation. A structurally related approach has been pursued by Ménage and coworkers,[114] who have synthesized a dinuclear oxo-bridged diiron complex, using chiral bipyridine type of ligands (Figure 6.11f), that enantioselectively catalyzes the conversion of alkenes into epoxides. Unsymmetrical alkenes were oxidized rapidly into their epoxides with high conversions and enantiomeric excesses (ee's) ranging from 9 to 63%. Interestingly, the related mononuclear complex affords lower yields and reduced enantiomeric excesses.

Finally, in a recent report Kwong and coworkers described an iron sexipyridine complex capable of performing efficient epoxidation with H_2O_2 (Figure 6.11b), and obtaining moderate ee's in the oxidation of aromatic olefins (up to 43% was achieved for styrene).[111]

From a bioinorganic viewpoint, a particularly interesting diiron epoxidation catalyst, $[Fe^{III}_2(6\text{-}HPA)(O)(OH_2)_2](ClO_4)_4$, was described as a model for sMMO by Kodera and coworkers (6-HPA = 1,2-bis[2-{bis(2-pyridylmethyl)-aminomethyl}-6-pyridyl]ethane).[112] The complex contains a bis-tetradentate ligand, where the two tetradentate binding sites are connected *via* an ethyl linker, and the two ferric ions are bridged by an oxo group (Figure 6.11c). In the presence of large excess of cyclooctene (1000 equiv.), $[Fe^{III}_2(6\text{-}HPA)(O)(OH_2)_2](ClO_4)_4$ catalyzes the efficient (72% relative to H_2O_2, 150 equiv.) oxidation of olefins with high selectivity towards the epoxide (epoxide: diol > 35:1). Instead, $[Fe^{II}_2(CF_3SO_3)_4(6\text{-}HPA)]$, which lacks the oxo bridge, affords a 3.5:1 epoxide:diol mixture (66% efficiency, 10 equiv. H_2O_2). A metastable peroxo species $[Fe^{III}_2(6\text{-}HPA)(O)(O_2)]^{2+}$ was spectroscopically characterized by UV–vis and ESI-MS when $[Fe^{III}_2(6\text{-}HPA)(O)(OH_2)_2](ClO_4)_4$ and H_2O_2 react at –40 °C. Such species are not kinetically competent for reacting with an olefin.

A particularly interesting aspect of this work is that when the oxo bridge is fully labeled with ^{18}O, and $H_2^{16}O_2$ is employed as oxidant, the corresponding epoxide is 31% ^{18}O-labelled. That observation indicates that O-atom scrambling between the oxo bridge and the peroxide ligand occurs prior to substrate oxidation. The authors propose that scrambling occurs at a putative $Fe^{IV}_2(O)_2(\mu\text{-}O)$ species (Scheme 6.29).

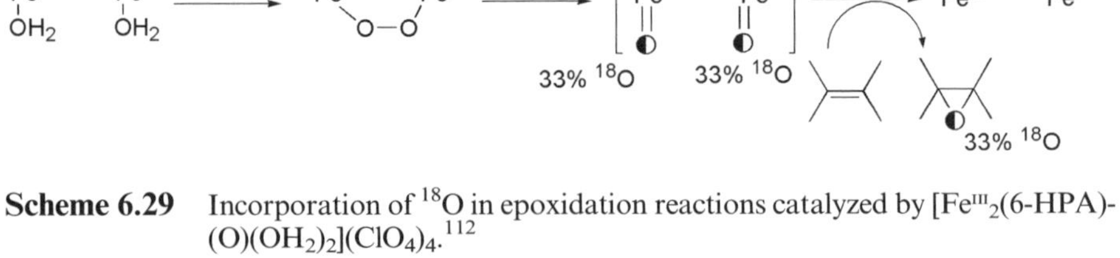

Scheme 6.29 Incorporation of ^{18}O in epoxidation reactions catalyzed by $[Fe^{III}_2(6\text{-}HPA)\text{-}(O)(OH_2)_2](ClO_4)_4$.[112]

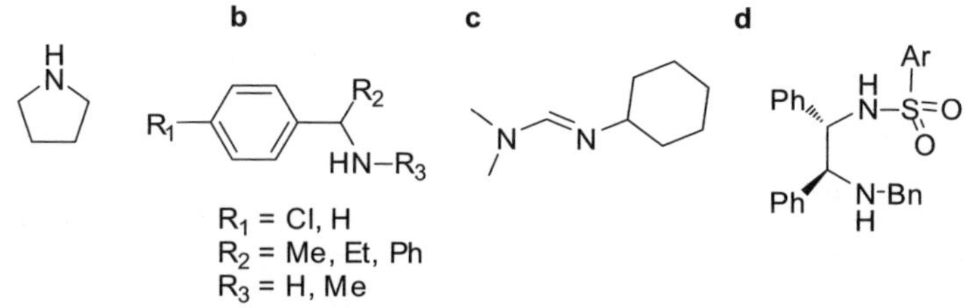

Figure 6.12 Amine additives employed in epoxidation reactions catalyzed by the FeCl$_3$ · 3H$_2$O/H$_2$Pydic/amine system.[115–119]

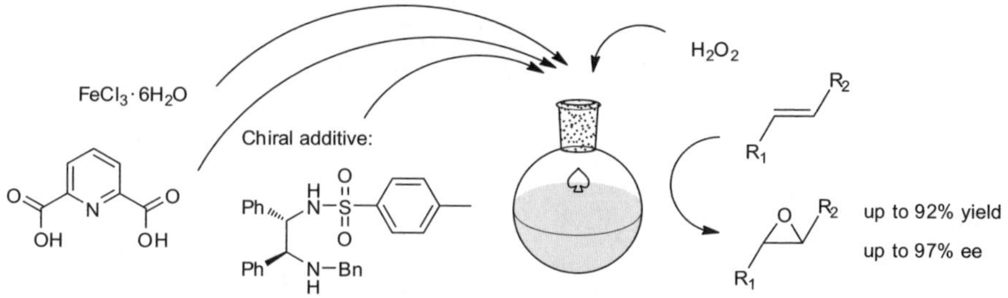

Scheme 6.30 *In situ* generated asymmetric epoxidation catalyst reported by Beller *et al.*[118,119]

Undertaking a very practical approach, Beller and coworkers have developed a simple process (room temperature, aerobic atmosphere) to epoxidize alkenes: commercially available FeCl$_3$ · 3H$_2$O as iron source, pyridine-2,6-dicarboxylic acid (H$_2$Pydic), and simple amines as ligands (Figure 6.12 and Scheme 6.30), are combined *in situ* to generate a catalyst that epoxidizes aromatic alkenes using H$_2$O$_2$ as a terminal oxidant.[115] The original system made use of pyrrolidine (Figure 6.12a) as the amine component. This system was improved by using benzylamine derivatives (Figure 6.12b), which allowed for aliphatic alkene epoxidation.[116,117] By using formamidines (Figure 6.12c), excellent yields in the epoxidation of styrene substrates were obtained. An asymmetric version was achieved by the addition of chiral benzylamines (Figure 6.12d).[118,119] The system achieves good yields and moderate to excellent (up to 97%) enantiomeric excesses. Mechanistic studies on the system containing formamidine ligands suggest the partial implication of hydroxyl radicals, in combination with a metal-centered oxidant.[120]

More recently, the same group reported *in situ* formation of imidazole iron complexes that epoxidize aliphatic and aromatic olefins, and conjugated dienes with high chemoselectivity and moderate to good yields.[120–122] The authors showed that a free 2-position of the imidazole (Im) ligand plays a key role for catalytic activity, as substitution leads to a dramatic lowering of yield and conversion. A mechanistic study led to the identification of

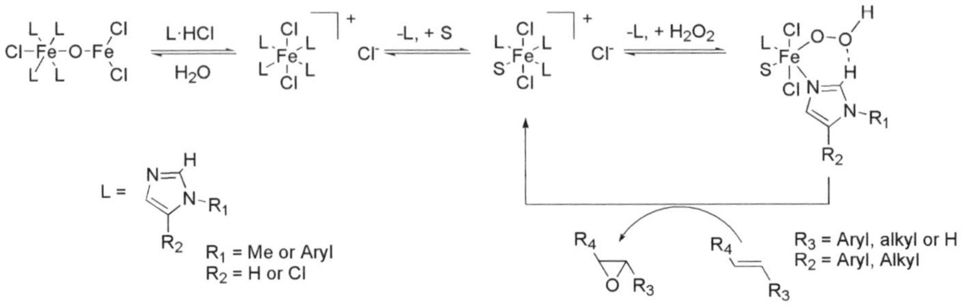

Scheme 6.31 Mechanistic scheme for epoxidation reactions catalyzed by the FeCl$_3$·6H$_2$O/imidazole system.[121]

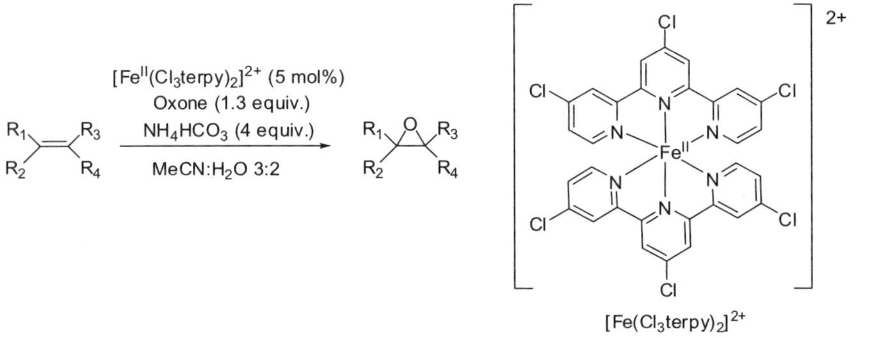

Scheme 6.32 Alkene epoxidation catalyzed by [Fe(Cl$_3$terpy)$_2$]$^{2+}$.[123]

trans-[FeCl$_2$(5-Cl-1-MeIm)$_4$]Cl as a very efficient catalyst, reproducing the results obtained from the *in situ* prepared system.[121] In solution, the complex is in equilibrium with an oxo-dimer complex (Scheme 6.31), formed by reaction with water. The authors propose that a hydroperoxide species, stabilized by H-bonding with the imidazole ligand, is responsible for the epoxidation.

Che and coworkers recently described iron(II) bis-terpyridine complexes (5 mol%) (Scheme 6.32) as highly active epoxidation catalysts using Oxone® (1.3 equiv.).[123] The system shows excellent substrate scope covering aliphatic, aromatic, and terminal, electron-rich and electron-deficient olefins with good to excellent (up to 96%) yields. Epoxidation of *cis*-stilbene occurs without stereoscrambling. The ligand is robust and can be recovered at the end of the reactions, and reused. The Cl-substitution on the ligands is important since [Fe^{II}(terpy)$_2$](ClO$_4$)$_2$ afforded much lower yields (<30%). Finally, reactions can be scaled to multigram scale.

ESI-MS analyses of the reactions show a cluster peak at *m*/*z* 371.9, which could be simulated as [Fe^{IV}(O)(Cl$_3$terpy)$_2$]$^{2+}$. The authors propose that these Fe^{IV}=O species are responsible for the epoxidation activity.

6.2.2.2 cis-*Dihydroxylation*

Iron catalysts have the ability to catalyze not only epoxidation but also olefin *cis*-dihydroxylation, analogously to the reaction performed by Rieske dioxygenases. Concurrently to stereospecific alkane oxidation, the first reported examples of non-heme iron catalysts capable of eliciting both olefin epoxidation and *cis*-dihydroxylation are those containing ligands of the tpa family.[124] Later, other mononuclear complexes based on *N*4 ligands, such as bpmen,[30] bpmcn,[74] bpbp,[125] bispidine,[98] and Pytacn,[71] have also been shown to elicit the same catalytic activity (Figure 6.5).

Table 6.3 collects results in the epoxidation and *cis*-dihydroxylation of olefins catalyzed by different non-heme catalysts. It shows that conversion of the oxidant in *cis*-cyclooctene oxidation can be as high as 93%, and diol:epoxide ratios range from 0.1 to 16.8. Diol product does not derive from the epoxide, as both products are formed at the same time, and the use of epoxide as a potential substrate under the same reaction conditions does not afford *cis*-diol. Clearly, from the results, it can be deduced that the ligand structure exerts a significant control on the diol:epoxide ratio. A general trend for complexes with pyridine containing tetradentate ligands is that the presence of methyl groups in

Table 6.3 Oxidation of *cis*-cyclooctene with H_2O_2 catalyzed by different iron(II) complexes.[a]

Complex	$D + E^b$	$(D/E)^b$	Ref.
$[Fe^{II}(tpa)(CH_3CN)_2](ClO_4)_2$	7.4	1.2	30
$[Fe^{II}(5Me_3\text{-}tpa)(CH_3CN)_2](ClO_4)_2$	6.7	1.4	30
$[Fe^{II}(6Me_3\text{-}tpa)(CH_3CN)_2](ClO_4)_2$	7.1	16.8	30
$[Fe^{II}(N3Py\text{-}Me)(CH_3CN)_2](ClO_4)_2^c$	37	1.6	127
$[Fe^{II}(CF_3SO_3)_2(^{Me,H}Pytacn)]$	8.1	1.0	71
$[Fe^{II}(CF_3SO_3)_2(^{Me,Me}Pytacn)]$	7.1	5.5	71
$[Fe^{II}(CF_3SO_3)_2(^{iPr,H}Pytacn)]$	8.1	3.3	71
$[Fe^{II}(CF_3SO_3)_2(^{iPr,Me}Pytacn)]$	5.0	9.0	71
$[Fe^{II}(bpmen)(CH_3CN)_2](ClO_4)_2$	8.4	0.1	126
$[Fe^{II}(CF_3SO_3)_2(6\text{-}Me_2\text{-}bpmen)]$	7.9	4.3	128
$[Fe^{II}(CF_3SO_3)_2(bpmp)]$	6	0.7	129
$[Fe^{II}(CF_3SO_3)_2(\alpha\text{-}bpmcn)]$	6.5	0.1	74
$[Fe^{II}(CF_3SO_3)_2(\beta\text{-}bpmcn)]$	7.7	1.9	74
$[Fe^{II}(CF_3SO_3)_2(6\text{-}Me_2\text{-}bpmcn)]$	9.3	1.7	128
$[Fe^{II}(CF_3SO_3)_2(bpbp)]^d$	4.3	0.7	125
$[Fe^{II}(CF_3SO_3)_2(6\text{-}Me_2\text{-}bpbp)]^d$	6.5	64.0	125
$[Fe^{II}(cyclam)(CH_3CN)_2](ClO_4)_2$	3.5	0.03	129
$[Fe^{II}(CF_3SO_3)(L^8Py_2)](CF_3SO_3)$	3.9	0.1	129
$[Fe^{II}(CF_3SO_3)_2(Ph\text{-}DPAH)_2]$	7.5	14	130
$[Fe^{II}(PrL1)_2](CF_3SO_3)_2$	4.3	1.7	131
$[Fe^{II}(CF_3SO_3)(Py(ProPh_2OH)_2)](CF_3SO_3)^e$	1.7	1.8	132

[a]Catalyst:H_2O_2:*cis*-cyclooctene = 1:10:1000.
[b]Turnover number (TN, moles of product/moles of catalyst), E = number of moles of epoxide, D = number of moles of diol.
[c]50 equiv. of H_2O_2 instead of 10.
[d]1-Octene was used instead of cyclooctene.
[e]500 Equiv of cyclooctene were used.

the α-position of the pyridine ring results in an increase in the diol : epoxide ratio, although a rational for that behavior is still lacking. In Pytacn systems, *cis*-dihydroxylation *versus* epoxidation selectivity was enhanced by replacing methyl by isopropyl groups in the tacn ring. In most cases, high retention of configuration is observed in the epoxidation and *cis*-dihydroxylation of *cis*-2-heptene. This observation, together with the incorporation of oxygen from water into oxidation products (isotopic labeling studies),[30,71,74,75] clearly indicates that a metal-based oxidant operates in these reactions.

Pentadentate ligands that only leave one labile coordination site at the metal have also been used, but resulted in much less efficient systems.[124,132–134] Stereoscrambling in the diols has been documented for catalytic reactions with iron complexes $[Fe^{II}(BpdL_2)(CH_3CN)]^{2+}$ and $[Fe^{II}(BpdL_3)(CH_3CN)]^{2+}$ (Figure 6.8) containing pentadentate bispidine based ligands. In this case, reaction yields and D/E (moles of diol product/moles of epoxide product) ratios are very sensitive to the presence of O_2. These observations strongly suggest that *cis*-dihydroxylation takes place in a stepwise process, in which carbon-centered radicals are involved.[133]

In addition, tetradentate ligands that adopt an equatorial coordination and leave two labile sites in a relative trans disposition, such as cyclam and L^8Py_2, have also been applied to these catalytic transformations but essentially no *cis*-diol is formed.[124,129,134] It seems clear from these results that the availability of two labile sites in a *cis*-configuration is necessary to model the *cis*-dihydroxylation reaction catalyzed by Rieske dioxygenases in biological systems.

The first ligands employed in these studies provide the metal centre with an all-*N* donor set, which does not accurately reflect the *N,N,O*-ligand environment found at the active site of non-heme enzymes. For this reason, attention has been recently devoted to the development of iron complexes with mixed donor ligands. Oldenburg *et al.* used the ligand Ph-DPAH to obtain a very selective *cis*-dihydroxylation catalyst,[130] and in a related contribution Klein Gebbink and coworkers reported that iron complexes with PrL1 ligands also show epoxidation and enzyme-like *cis*-dihydroxylation activity (see Table 6.3 and Figure 6.13; Ph-DPAH = [di-(2-pyridyl)methyl]benzamide and PrL1 = propyl 3,3-bis(1-methylimidazol-2-yl)propionate).[131]

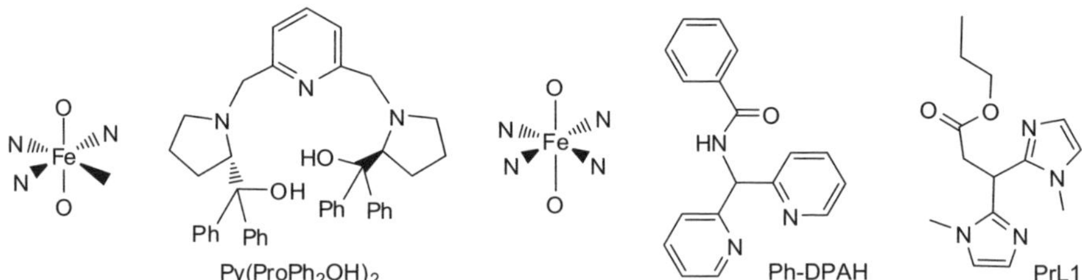

Figure 6.13 Schematic representation of relevant ligands containing mixed *N,O*-binding sites used to prepare mononuclear iron(II) complexes to perform olefin oxidations along with the coordination they adopt around the metal center.

Que and coworkers have described a very accurate structural model for the active site of naphthalene 1,2-dioxygenase (NDO), which also acts as a functional model. By using a Kemps-triacid derivative, a novel ligand set that provides two pyridine N-atoms and one carboxylate oxygen atom as binding units was prepared (Scheme 6.33).[135] Crystallographic characterization of the corresponding Fe^{II} compound $[Fe^{II}(Bpka)(Cl)]$ reveals that the ligand provides a very accurate mimic of the 2-His-1-carboxylate facial triad of the enzyme. The most important difference is the presence of the chloride ligand in the complex, as the same position is occupied by a water molecule in the enzyme. Reaction of the complex with H_2O_2 (10 equiv.) catalyzes the oxidation of 1-octene and produces a TN of 0.26(6) of epoxide and a TN of <0.05 of the corresponding *cis*-diol. However, when the complex was pretreated with $AgCF_3SO_3$, to remove the chloride ligand, the same reaction affords a TN of 0.67 of *cis*-diol and 0.11 of epoxide. The complex appears thus only capable of performing a single turnover reaction. Isotopic labeling experiments indicate that in the *cis*-diol product both oxygen atoms come from the same H_2O_2 molecule, demonstrating that the complex is a true dioxygenase model.

It was not until very recently that Que and coworkers reported the complex $[Fe^{II}(tpa)(CH_3CN)_2]^{2+}$ as the first authentic functional model of naphthalene dioxygenase (Scheme 6.34).[136] The oxidation of naphthalene using H_2O_2 as oxidant afforded four products, *cis*-1,2-dihydro-1,2-naphthalenediol, 1-naphthol, 2-naphthol, and 1,4-naphthoquinone, with the former being the major product as in NDO-catalyzed reaction.

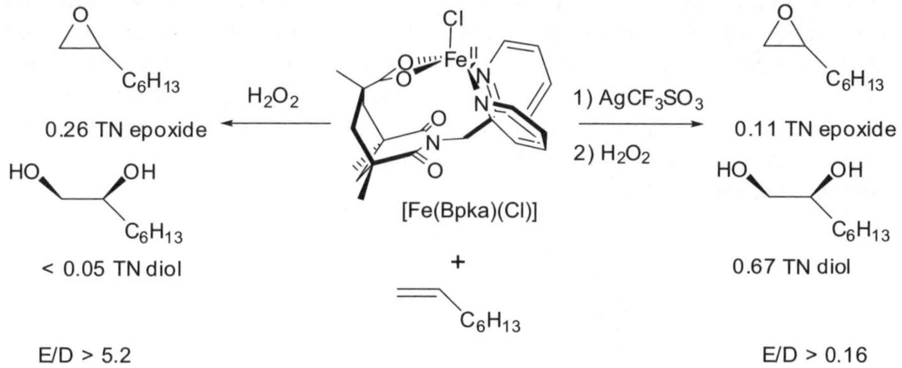

Scheme 6.33 Schematic of alkene oxidation reactions mediated by a 2-His-1-carboxylate model.[135]

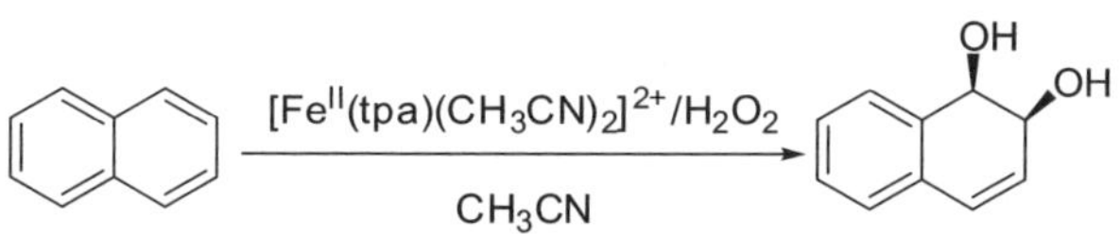

Scheme 6.34 *cis*-Dihydroxylation of an arene by a synthetic catalyst.[136]

The main limitation of the aforementioned systems is the relatively low TN and substrate conversion obtained. An exception is [Fe^{II}(CF$_3$SO$_3$)$_2$(^{Me,Me}Pytacn)] (Figure 6.5), which affords a TN of 141 for diol *versus* 29 for epoxide in the oxidation of cyclooctene using 0.3% of catalyst, and achieved 50% efficiency in the conversion of H$_2$O$_2$ into the oxygenated products.[71] Still, the weakness of this system is the low conversion of substrate. On the other hand, [Fe^{II}(CF$_3$SO$_3$)$_2$(5Me$_3$-tpa)] can catalyze the oxidation of olefins to *cis*-diols (*D/E* ratio around 3 in most cases) under conditions of limiting substrate concentration with high conversion and efficiency, but 3% of catalyst and 4 equiv. of H$_2$O$_2$ are needed.[137]

A very promising alternative has very recently been described by Che and coworkers, who reported [Fe^{II}(Cl)$_2$(*c*-Py$_2$NMe$_2$)] as a fast (5–10 min) and efficient *cis*-dihydroxylation catalyst in combination with Oxone (Scheme 6.35).[138] By using 0.7–3.5 mol% of iron catalyst and 2 equiv. of Oxone, aliphatic and aromatic olefins are *cis*-dihydroxylated with modest to good selectivity. Yields are highly dependent on the nature of the substrate. In general, electron-deficient olefins afford the best yields (56–99%), and dimethyl fumarate constitutes a unique substrate since it is *cis*-dihydroxylated with 99% selectivity and yield to the corresponding *cis*-diol. On the other hand, internal and terminal aliphatic olefins provide more modest yields (16–64%), and reduced selectivity. Epoxides, hydroxycetones and ketones (or aldehydes) resulting from oxidative C–C cleavage are side products of the reaction. Large-scale *cis*-dihydroxylation of methyl cinnamate (9.7 g) in 84% yield (8.4 g) was demonstrated (Scheme 6.35).

ESI-MS analyses show the presence of a cluster ion that could be simulated as [Fe^V(O)$_2$(*c*-Py$_2$NMe$_2$)]. On the basis of DFT analyses, the authors propose that these species are responsible for the *cis*-dihydroxylation reaction, which draws an analogy to OsO$_4$, RuO$_4{}^-$ and MnO$_4{}^-$ mediated reactions. On the other hand, chiral ligands have been used for the development of asymmetric systems capable of giving diols with high enantiomeric excess and with high selectivity (Scheme 6.36).[125,128]

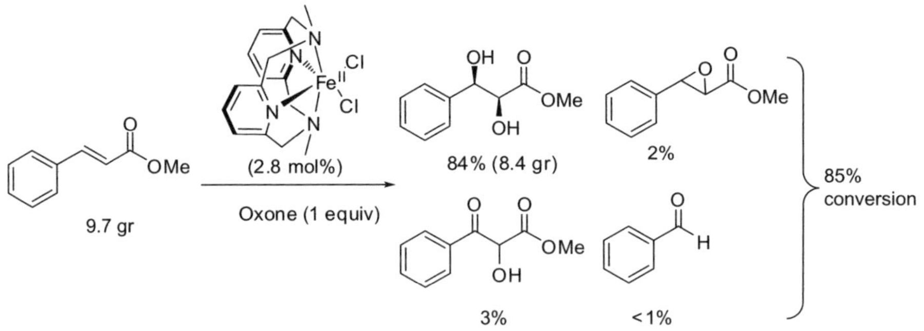

Scheme 6.35 Large-scale oxidation of methyl cinnamate catalyzed by [Fe^{II}(Cl)$_2$-(*c*-Py$_2$NMe$_2$)].[138]

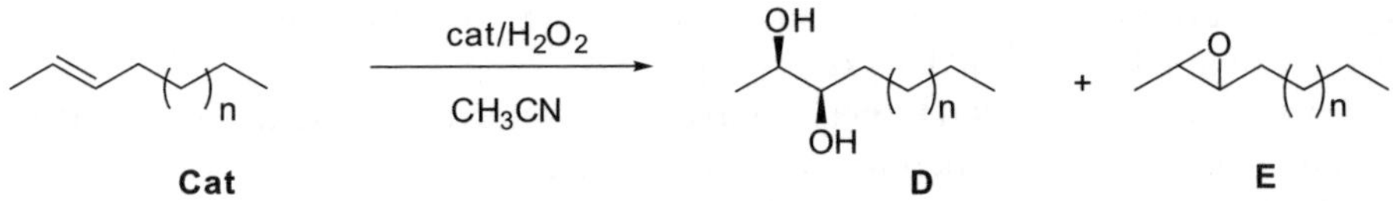

Cat		D	E
$[Fe^{II}(CF_3SO_3)_2(6Me_2\text{-bpmcn})]$	n = 2	82% ee	D/E = 3.3
$[Fe^{II}(CF_3SO_3)_2(6Me_2\text{-bpbp})]$	n = 1	97% ee	D/E = 26

Scheme 6.36   Examples of Fe-catalyzed asymmetric *cis*-dihydroxylation. See Figure 6.5 for catalyst structures.[125,128]

Overall, these mononuclear non-heme iron(II) systems may constitute an attractive environmentally benign alternative to traditional toxic and more expensive osmium reagents used for *cis*-dihydroxylation, although significant improvement in reaction efficiencies is highly required to achieve this goal.

6.3   Reaction Mechanisms in Catalytic C–H and C=C Oxidation Reactions Mediated by Complexes with N-Rich Ligands

Aside from the design of novel synthetically useful oxidation technologies, bioinspired catalysts offer a simple structural platform to study the chemistry associated to biological systems. In this regard, these catalysts can give valuable information about the reaction mechanisms operating in oxygenases.

Mechanistic studies of alkane and alkene oxidation reactions catalyzed by iron complexes containing N-rich ligands have shown a rich and subtle mechanistic scenario. Fenton-type free diffusing radical processes dominate the chemistry in most cases. However, the landmark observation of stereo-specificity in C–H and C=C oxidation reactions for selected catalysts implicates metal-centered oxidants.

The studies of Que and coworkers on the Fe(tpa) family of complexes have built a mechanistic frame that bears resemblance with that of CYP 450, and involves the formation of iron-peroxide species that undergo O–O lysis to generate highly reactive high valent metal-oxo intermediates.[30,32] These landmark studies have been expanded, and in some cases disputed, by studies in other families of iron catalysts.

6.3.1   The Initially Formed Fe^{III}-OOH and its Cleavage Products

The reaction of $[Fe^{II}(tpa)(CH_3CN)_2]^{2+}$ and excess of hydrogen peroxide (or alkyl peroxides) at low temperature in acetonitrile affords metastable purple iron(III)-alkyl(hydro)peroxide (Fe^{III}-OOR(H)) compounds that have been spectroscopically characterized.[68,139–143] Such highly reactive species are interesting *per se* because they are proposed as key intermediates in the catalytic cycle of several iron-based enzymes involved in oxidation reactions such as cytochrome P450,[3,144] or Rieske dioxygesases.[28]

Furthermore, Fe^{III}-OOH species have been detected in the antitumor drug bleomycin,[82,83] a glycopeptide-derived compound that affects the double-stranded oxidative cleavage of DNA in the presence of iron(II) and O_2. Addition of O_2 and one electron to Fe^{II}-bleomycin or, alternatively, direct interaction of Fe^{III}-bleomycin with H_2O_2 furnishes a low spin iron(III)-hydroperoxo species identified by EPR[83] and electrospray mass spectrometry.[82] This compound is the last detectable intermediate prior to DNA cleavage and it can hydroxylate naphthalene and epoxidize olefins.[145] Despite these observations, it is not clear if the observed Fe^{III}-OOH is the real oxidant or if it acts as a precursor of a more powerful oxidant, either a $Fe^V{=}O$ or $Fe^{IV}{=}O + {}^{\bullet}OH$ species, finally responsible for substrate oxidation.

Apart from tpa-based complexes, several other synthetic mononuclear iron complexes have been used as precursors for the preparation of synthetic iron(III)-hydroperoxo species. In contrast to bleomycin, in these systems the formation of Fe^{III}-OOH can only be achieved from iron(II) or iron(III) under conditions of excess H_2O_2.[68,140,146–151] Owing to their thermal instability, there is no solved crystal structure corresponding to a synthetic mononuclear iron(III)-hydroperoxo species and most information about their structure and properties has been gained by spectroscopic methods.[6,152] Particularly relevant are vibrational studies (resonance Raman) on such compounds, which have provided interesting data related to the strength of the O–O and Fe–O bonds, and X-ray absorption analyses, which have provided structural information.[148,151–155]

Studies performed on the mononuclear iron complex with N4Py as a ligand serve as a general model to describe the preparation and spectroscopic characterization of iron(III)-hydroperoxide systems (Scheme 6.37). In this particular case, addition of excess H_2O_2 to $[Fe^{II}(N4Py)(CH_3CN)]^{2+}$ at low temperatures gives rise to a purple low-spin Fe^{III}-OOH intermediate as ascertained by EPR spectroscopy and MS spectrometry.[146] A recent study shows that the compound could also be prepared by the reaction of $[Fe^{II}(N4Py)(CH_3CN)]^{2+}$ with O_2 and a reducing agent.[156]

Two characteristic resonance enhanced features appear in the resonance Raman spectrum of $[Fe^{III}(OOH)(N4Py)]^{2+}$:[140] v(O–O) frequency at 790 cm^{-1}, which downshifts to 746 cm^{-1} when $H_2{}^{18}O_2$ is used rather than $H_2{}^{16}O_2$ ($\Delta = -44$ cm^{-1}), and a v(Fe–O) mode at 632 cm^{-1}, which moves down to 616 cm^{-1} upon

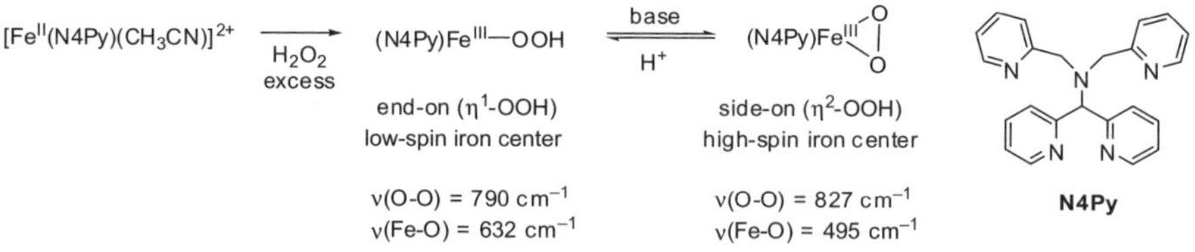

[Fe^{II}(N4Py)(CH$_3$CN)]$^{2+}$ $\xrightarrow{H_2O_2 \text{ excess}}$ (N4Py)Fe^{III}—OOH $\underset{H^+}{\overset{base}{\rightleftharpoons}}$ (N4Py)Fe^{III}(O$_2$)

end-on (η^1-OOH) low-spin iron center — v(O-O) = 790 cm^{-1}, v(Fe-O) = 632 cm^{-1}

side-on (η^2-OOH) high-spin iron center — v(O-O) = 827 cm^{-1}, v(Fe-O) = 495 cm^{-1}

N4Py

Scheme 6.37 Schematic of the preparation and vibrational data for ferric peroxide species.

using ^{18}O-labeled hydrogen peroxide ($\Delta = -16$ cm^{-1}). Mixed isotope labeling experiments unequivocally establish an end-on η^1-OOH structure for this compound. Treatment of this [Fe^{III}(η^1-OOH)(N4Py)]$^{2+}$ with base produces a blue high-spin iron(III)-peroxo species, as ascertained by Mössbauer spectroscopy, which corresponds to the conjugated base of the starting hydroperoxo unit.[148,157] Detailed resonance Raman data involving mixed isotope labeling experiments strongly support a side-on-bound peroxo ligand, thus, formulating the new compound as [Fe^{III}(η^2-O$_2$)(N4Py)]$^+$. Resonance Raman studies indicate that the O–O stretching frequency is higher than in the side-on compound ($v = 827$ cm^{-1}, $\Delta = -46$ cm^{-1}) and the Fe–O vibration is much lower in energy ($v = 495$ cm^{-1}, $\Delta = -17$ cm^{-1}). The value of these vibrational frequencies can be directly correlated to the strength of the corresponding bond: the lower v(O–O) reported for the low-spin [Fe^{III}(η^1-OOH)(N4Py)]$^{2+}$ indicates that the O–O bond is weakened with respect to the high-spin [Fe^{III}(η^2-O$_2$)(N4Py)]$^+$, while the increased v(Fe–O) indicates a stronger iron–oxygen bond. Indeed, detailed inspection of the resonance Raman spectra of several iron-peroxo species reported in the literature leads to the conclusion that the low-spin iron(III) centers present a weaker O–O bond and a stronger Fe–O than their high-spin counterparts (for an exception see ref. 158).[6]

The ability of iron(III) hydroperoxides to carry out olefin and alkane oxidation has been recently discarded on the basis of a study by Nam and coworkers. [Fe^{III}(OOR)(L)] species (R = H, ^{t}Bu; L = tpa, N4Py) were prepared and their reaction with thiosulfide and cyclooctene monitored by UV–vis spectroscopy. These experiments show that reactions are very slow, and incompatible with the fast reactions that take place under standard catalytic oxidation conditions. In addition, a DFT analysis indicates that oxygen atom transfers from Fe^{III}(OOH) species to sulfides and olefins have large activation barriers.[159,160] In conclusion, alternative species must be responsible for alkane and alkene reactions mediated by Fe(tpa) and related catalytic systems.

6.3.2 Olefin Oxidations: Epoxidation and *cis*-Dihydroxylation

Epoxidation and *cis*-dihydroxylation are two prominent reactions by which olefins can be oxidized in biological systems, and product analysis of the catalytic reactions provides important mechanistic information. From this point of view, *cis*-olefins are particularly useful substrates because their reaction pattern gives information about the stereospecificity of the reactions. Labeling studies also provide insight into the reaction mechanisms as the origin of the oxygen atoms that end up into oxidized products can be unambiguously established. Table 6.4 collects relevant mechanistic data in the catalytic oxidation of olefins by different iron catalysts containing tetradentate ligands.

6.3.2.1 *Fe^{III}/Fe^V Mechanism*

Isotopic labeling studies in the *cis*-dihydroxylation process are a valuable tool to address the active species responsible for the observed chemistry (Table 6.4).

Table 6.4 Oxidation of alkenes with H_2O_2 catalyzed by different iron(II) complexes with tetradentate ligands.[a]

	Cyclooctene	cis-Diol	Epoxide	cis-2-Heptene		
Catalyst	$D+E$ (D/E)[b]	$2H_2O_2/H_2O_2 + H_2O/2H_2O$[c]	H_2O_2/H_2O[d]	RC diol[e] (%)	RC epoxide[f] (%)	Ref.
$[Fe^{II}(tpa)(CH_3CN)_2]^{2+}$	7.4 (1.2)	3/97/1	90/9	96	80	30
$[Fe^{II}(5Me_3\text{-}tpa)(CH_3CN)_2]^{2+}$	6.7 (1.4)	2/98/1	93/2	94	83	30
$[Fe^{II}(6Me_3\text{-}tpa)(CH_3CN)_2]^{2+}$	7.1 (16.8)	96/4/0	54/3	93	35	30
$[Fe^{II}(bpmen)(CH_3CN)_2]^{2+}$	8.4 (0.1)	23/73/4	73/30	79	92	126
$[Fe^{II}(CF_3SO_3)_2(6Me_2\text{-}bpmen)]$	7.9 (4.3)	–	–	78	60	128
$\alpha\text{-}[Fe^{II}(CF_3SO_3)_2(bpmcn)]$	6.5 (0.1)	4/88/1	66/15	67	85	74
$[Fe^{II}(CF_3SO_3)_2(bpdL_1)]$	2.5 (1.5)	95/5/0	100/0	–	–	98
$[Fe^{II}(CF_3SO_3)_2(^{Me,H}Pytacn)]$	8.1 (1.0)	3/96/0	25/61	90	93	71, 161
$[Fe^{II}(CF_3SO_3)_2(^{Me,Me}Pytacn)]$	7.1 (5.5)	11/88/0	–	90	91	71
$[Fe^{II}(CF_3SO_3)_2(^{iPr,H}Pytacn)]$	8.1 (3.3)	83/17/0	–	99	88	71

[a]Catalyst:H_2O_2:cyclooctene = 1:10:1000.
[b]Turnover number (TN, moles of product/moles of catalyst), E = number of moles of epoxide, D = number of moles of diol.
[c]Origin of the oxygen atom (%) in the diol: both oxygens from H_2O_2/one oxygen from H_2O_2 and one from water/both oxygens from water.
[d]Origin of the oxygen atom in the epoxide (%): from H_2O_2/from water.
[e]Retention of configuration for the diol.
[f]Retention of configuration for the epoxide.

Que *et al.* divided the iron catalysts into two groups depending on the results obtained for this transformation:[31] class A catalysts are those for which the produced *cis*-diol contains one atom of oxygen that comes from water and one atom of oxygen from H_2O_2. Examples of these catalysts include $[Fe^{II}(X_2)(L)]$ complexes where L = tpa, α-bpmcn, 5Me$_3$-tpa, ^{Me,H}Pytacn, ^{Me,Me}Pytacn and X = CH$_3$CN, OTf. Class B catalysts (L = 6Me$_3$-tpa, BpL1, ^{iPr,H}Pytacn) incorporate both oxygen atoms from the peroxide in the corresponding diol (see Table 6.4 for labeling results). A general trend is that *cis*-dihydroxylation is favored in class B catalysts, while epoxidation is favored in class A. For both classes, high retention of configuration is observed in the *cis*-hydroxylation of *cis*-2-heptene. This observation strongly suggests that a metal-based oxidant is operating in these reactions. Scheme 6.38 summarizes the mechanisms proposed for these two types of catalysts.

In class A catalysts the labeling results suggest the implication of a $Fe^V(O)(OH)$ species formed *via* water assisted O–O heterolysis of the initially generated Fe^{III}-hydroperoxo species (labeled as Fe^{III}/Fe^V *wa* in Scheme 6.38).[30,31] In this scenario one oxygen from the iron(v)-oxo-hydroxo compound comes from hydrogen peroxide and the other from water, and both oxygen atoms end up into the diol. The observation of water incorporation into epoxide products is also an indication of the implication of a $Fe^V(O)$ species. The feasibility of this mechanism is theoretically supported by DFT calculations.[126]

Instead, in class B catalysts there is no water incorporation into oxidized products, and they are proposed to operate *via* a non-water-assisted mechanism

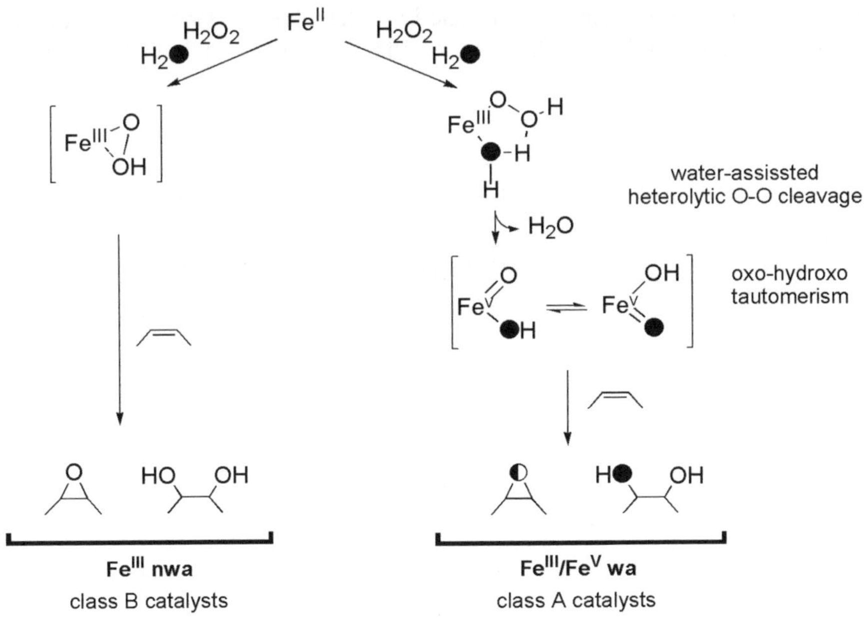

Scheme 6.38   Schematic representation of the Fe^{III}/Fe^V mechanism.

(labeled as Fe^{III} *nwa* in Scheme 6.38).[30] Substrate oxidation is proposed to occur through direct reaction of the alkene with the iron(III)-hydroperoxo intermediate without water mediation. Activation of the hydroperoxide moiety may occur *via* side-on binding to the ferric ion. Alternatively, O–O lysis could occur prior to substrate oxidation, but without the assistance of a water molecule.

The different nature of the active species in class A and class B catalysts is not only evidenced by the origin of the oxygen atoms that end up into the oxidized products but also by their marked different electrophilicity. This phenomenon was made evident in a work reported by Que *et al.*,[162] in which the complexes $[Fe^{II}(tpa)(CH_3CN)_2]^{2+}$ and $[Fe^{II}(6Me_3\text{-}tpa)(CH_3CN)_2]^{2+}$ were selected as prototypes for class A and B, respectively. Competition experiments between electron-rich and electron-deficient olefins using H_2O_2 as the oxidant revealed that the class A catalyst preferentially oxidizes the more electron-rich olefins, while the class B catalyst shows the opposite behavior. For example, in a competition experiment between cyclooctene (electron-rich) and *tert*-butyl acrylate (electron-deficient), $[Fe^{II}(tpa)(CH_3CN)_2]^{2+}$ favors cyclooctene oxidation (epoxidation and *cis*-dihydroxylation) by a factor of more than four, while $[Fe^{II}(6Me_3\text{-}tpa)(CH_3CN)_2]^{2+}$ favors *tert*-butyl acrylate oxidation by a factor of four (Scheme 6.39). These opposite preferences exhibited by both catalysts most likely imply the involvement of distinct oxidants. The reactivity of class A catalysts is consistent with an electrophilic oxidant, such as $Fe^V(O)(OH)$ species, which also can explain the origin of the oxygen atoms incorporated into products and the involvement of the Fe^{III}/Fe^V *wa* mechanism. The nucleophilic behavior of class B could be tentatively associated to a direct oxidation performed by a putative high-spin Fe^{III}-OOH (Fe^{III} *nwa* mechanism) intermediate. However, no *cis*-dihydroxylation reactivity has been reported for the few examples of high spin Fe^{III}-OOH compounds that have been described so far.[147]

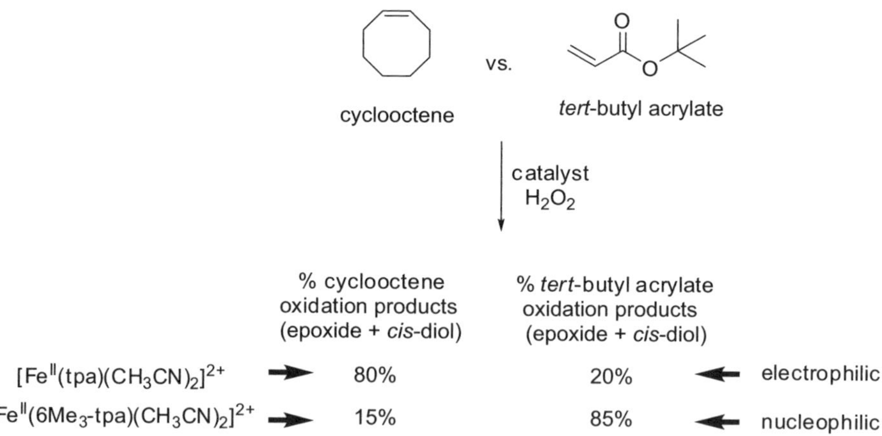

Scheme 6.39 Electrophilic *versus* nucleophilic character of iron-based catalysts in olefin oxidation reactions.

The identity of the *cis*-dihydroxylation reagent in class B catalysts remains then an unresolved question.

Interestingly, recent EPR studies have built on the proposal of Fe^V intermediates as responsible for the epoxidation activity.[163] Under cryoscopic conditions, Talsi, Britovsek and coworkers could trap a highly unstable intermediate formed by reaction of iron complexes $[Fe^{II}(tpa)(CH_3CN)_2]^{2+}$ and $[Fe^{II}(bpmen)(CH_3CN)_2]^{2+}$ with peracids, which is kinetically competent for the epoxidation of olefins. According to its EPR spectrum characteristics, this new species is described as a Fe^V-oxo compound, thus directly supporting the involvement of such species in the catalytic oxidation processes.[163] Notably, though, under the experimental conditions employed in the study no *cis*-dihydroxylation was observed, and thus it is highly possible that the nature of this intermediate differs from that in H_2O_2 reactions.

The exact mechanism by which $[Fe^{II}(tpa)(CH_3CN)]^{2+}$ (class A catalyst) carries out both the epoxidation and *cis*-dihydroxylation of olefins has been studied in detail by DFT calculations.[164] These studies reveal that, in fact, epoxide and diol originate from two different faces of the same oxidant, so that the generated product depends on the initial attack of the iron(v)-oxo-hydroxo over the double bond (Figure 6.14): diol formation occurs when the attack over the olefin starts by the hydroxo group of the $HO–Fe^V=O$ oxidant, while epoxide is generated by direct insertion of the oxo group into the double bond.[164] Both reaction pathways have comparable activation barriers (the *cis*-dihydroxylation is favored by only about 1 kcal mol^{-1}), which is in accordance with the formation of both epoxide and diol products in those reactions catalyzed by $[Fe^{II}(tpa)(CH_3CN)_2]^{2+}$ using H_2O_2 as the primary oxidant.[30]

A rationale for the different chemical reactivity of the two classes of iron catalysts has been suggested on the basis of computational and vibrational

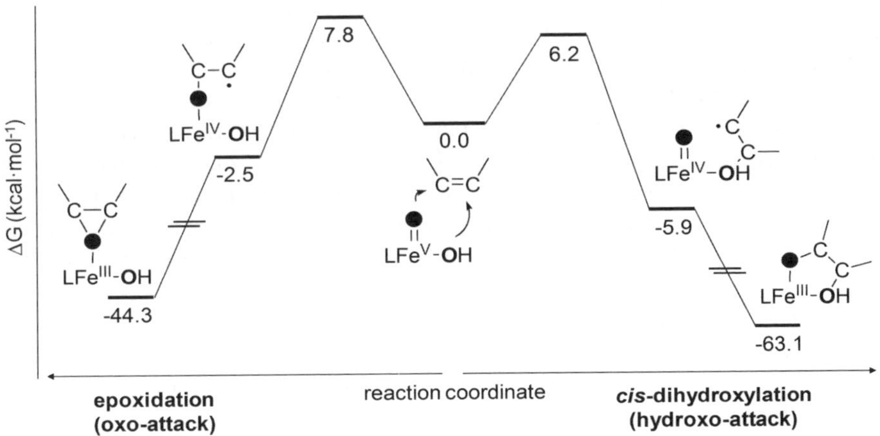

Figure 6.14   Schematic energy diagram of the reaction of olefins with the LFe^V(O)-(OH) species (L = tpa).[164]

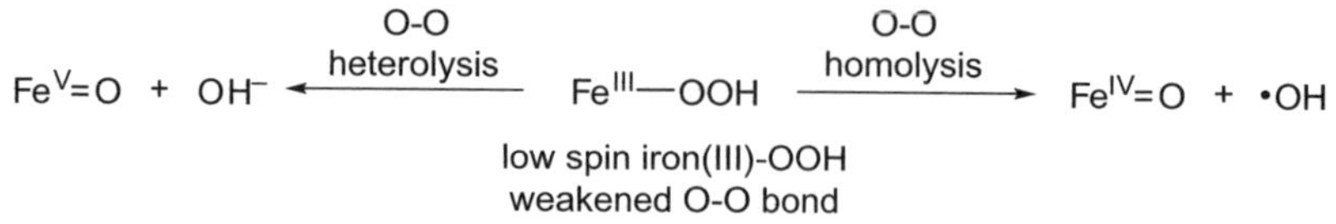

$$\text{Fe}^V{=}O \ + \ OH^- \ \xleftarrow{\overset{\text{O-O}}{\text{heterolysis}}} \ \text{Fe}^{III}{-}OOH \ \xrightarrow{\overset{\text{O-O}}{\text{homolysis}}} \ \text{Fe}^{IV}{=}O \ + \ {}^\bullet OH$$

low spin iron(III)-OOH
weakened O-O bond

Scheme 6.40   Homolytic *versus* heterolytic paths in the O–O breakage of ferric hydroperoxide species.

analyses of mononuclear Fe^{III}-OOR (R = H and alkyl) compounds.[141–143] These studies have shown that the spin state of the iron center has a profound impact on the O–O and Fe–O bond strength, and this is proposed to reflect on their respective reactivity. Low-spin iron(III)-alkyl(hydro)peroxide species exhibit strong Fe–O and weakened O–O bonds.[140,141,143] These features can be understood as an indication that the O–O bond is activated towards lysis, along with the formation of a strong Fe=O bond. In contrast, the non-activated (strong) O–O bond of high-spin analogues disfavors this path.[139,142] Along the same path, computational analyses on the $[Fe^{III}(6Me_x\text{-}tpa)(OO'Bu)]^{2+}$ ($x = 0$–3) families of complexes indicate that the barrier for O–O homolysis is small in the low-spin ($x = 0$) species. However, in high spin analogues ($x = 2, 3$) the barrier is large and the lower barrier corresponds to a Fe^{III}–O bond homolysis.[142]

Homolytic O–O bond cleavage will lead to the formation of a Fe^{IV}=O compound together with a hydroxyl radical (OH$^\bullet$), while the heterolytic breakage would generate a formally Fe^{V}=O compound and a hydroxide (OH$^-$) (Scheme 6.40). The selective and efficient reactivity observed for selected non-heme iron catalysts (Tables 6.3 and 6.4) is not consistent with the rather indiscriminate chemistry of hydroxyl radicals,[40] and alternatively selectivity properties are better described by assuming O–O bond heterolysis to generate Fe^{V}=O species.[30] In line with this, theoretical studies on the hydroperoxo complex $[Fe^{III}(OOH)(tpa)(H_2O)]^{2+}$ indicate that both the heterolytic and the homolytic pathways are feasible and have barriers of comparable size.[165]

Further subtleties have been evidenced in mechanistic studies on epoxidation reactions catalyzed by $[Fe^{II}(CF_3SO_3)_2(^{Me,H}Pytacn)]$. Isotopic labeling experiments showed that remarkably large percentages of water are incorporated into epoxide products (33–71%). The level of water incorporation proved to be independent of substrate concentration, but instead it depends on the particular structure of the substrate. The observation that oxygen atoms from water end up into the epoxide products should be understood as indirect evidence that iron-oxo species are responsible for the oxygen atom transfer event. However, unlike in synthetic heme systems, these observations also discard a competition between the reactions of water exchange and oxygen atom transfer. The latter is the classical behavior observed in heme systems, where water exchange is explained *via* the so-called oxo-hydroxo tautomerism.[166] To account for these results, Costas *et al.* have proposed that two tautomeric $Fe^{V}(O)(OH)$ species, which are in rapid equilibrium, mediate the epoxidation reaction (Scheme 6.41). One of the two tautomers contains an oxo ligand that originates from a water

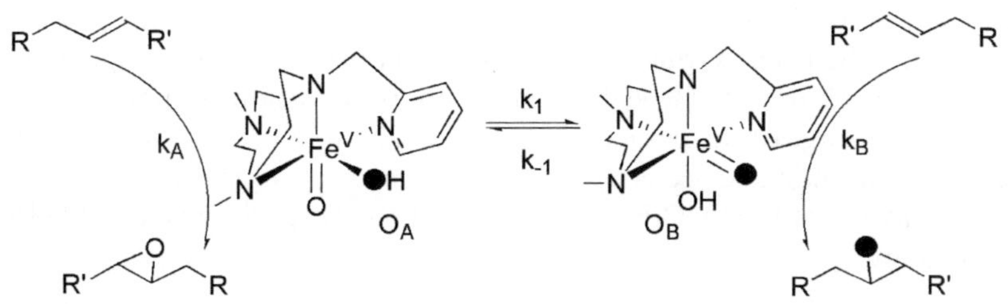

Scheme 6.41 Proposed tautomeric $Fe^V(O)$ species involved in catalytic epoxidations.[161]

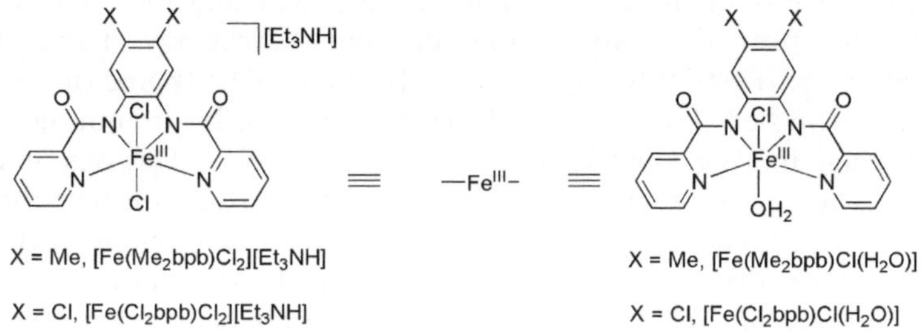

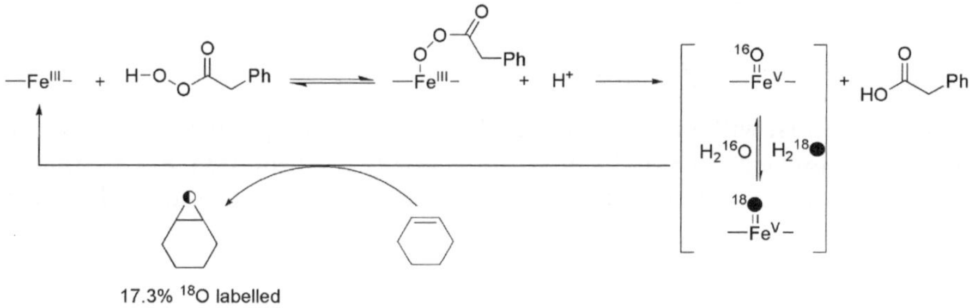

X = Me, [Fe(Me$_2$bpb)Cl$_2$][Et$_3$NH]

X = Cl, [Fe(Cl$_2$bpb)Cl$_2$][Et$_3$NH]

X = Me, [Fe(Me$_2$bpb)Cl(H$_2$O)]

X = Cl, [Fe(Cl$_2$bpb)Cl(H$_2$O)]

17.3% ^{18}O labelled

Scheme 6.42 Mechanistic scheme for the formation of $Fe^V(O)$ species in bis-imido based complexes.[167]

molecule, but in the other isomer the oxo ligand comes from the peroxide oxidant. In this scenario, the level of water incorporation reflects the relative reactivity of a given substrate with each of the two tautomeric species.[161]

Involvement of non-heme $Fe^V(O)$ species in catalytic epoxidations with peracids has also been proposed by Kim and coworkers. The authors prepared a series of iron complexes (Scheme 6.42) that contain a tetradentate dianionic ligand that occupies four positions in the same plane of the iron coordination sphere, leaving two *trans* sites occupied by water and/or chloride anions.[167]

The complexes catalyze the epoxidation of a series of olefins, and a high degree of stereoretention was observed in the epoxidation of *cis*-stilbene. When reactions were run in the presence of H$_2$^{18}O, epoxides show ^{18}O incorporation (up to 17.3% in the epoxidation of cyclohexene), which is an indication of the implication of iron-oxo species that can exchange the oxygen atom with $H_2^{18}O$. In addition, when peroxyphenylacetic acid was used as oxidant, phenylacetic acid was formed in $\sim 84\%$ yield. The latter observation is indicative of the heterolytic cleavage of the peracid O–O bond. The sum of all these observations support the mechanistic scenario shown in Scheme 6.42 (bottom), where heterolytic cleavage of the O–O bond leads to a $Fe^V=O$ species responsible for the catalytic activity.

Role of Acetic Acid in Catalytic Epoxidation Reactions. The beneficial role of the acetic acid in alkene epoxidation reactions has been recently studied by Que and coworkers. Acetic acid increases the epoxidation efficiency and inhibits olefin *cis*-dihydroxylation (*vide infra*), switching the olefin oxidation selectivity to afford the epoxide product almost selectively in the cases studied (Table 6.5, $[Fe^{II}(CF_3SO_3)_2(tpa)]$ and $[Fe^{II}(CF_3SO_3)_2(bpmen)]$ complexes). Que *et al.* proposed that acetic acid plays a key role in promoting fast O–O bond cleavage in a $Fe^{III}(OOH)$ species (Scheme 6.43).[168] This reaction affords a $Fe^V(O)(OAc)$ species finally responsible for the epoxidation reaction. Along a parallel path, homolytic cleavage of the peroxide O–O bond in the hydroperoxide intermediate, or $1e^-$ reduction of the $Fe^V(O)$ species, produces a $Fe^{IV}(O)$ intermediate that was characterized spectroscopically, and demonstrated to be a very sluggish reactant against olefins.

6.3.2.2 Fe^{II}/Fe^{IV} Mechanism

In sharp contrast with the "Fe^{III} *wa*" and "Fe^{III}/Fe^V *nwa*" pathways proposed by Que *et al.* (Scheme 6.38), Comba and coworkers proposed an alternative mechanism to explain the experimental results obtained in the oxidation of

Table 6.5 Catalytic oxidation of alkenes with H_2O_2 catalyzed by $[Fe^{II}(CF_3SO_3)_2(tpa)]$ and $[Fe^{II}(CF_3SO_3)_2(bpmen)]$ in the presence and absence of acetic acid.[a]

		No acid		*100 equiv. AcOH*	
Catalyst	*Olefin*	*Epoxide (TN)[b]*	*Diol (TN)[b]*	*Epoxide (TN)[b]*	*Diol (TN)[b]*
$[Fe^{II}(CF_3SO_3)_2$-	Cyclooctene	7.5	0.5	8.5	0.2
(bpmen)]	*cis*-2-Heptene	6.5	0.4	7.4	0.1
	1-Octene	3.6	0.8	5.8	0.2
	^{t}Butyl acrylate	0.6	0.9	3.8	0.7
$[Fe^{II}(CF_3SO_3)_2$-	Cyclooctene	3.1	4.4	7.6	0.7
(tpa)]	*cis*-2-Heptene	1.9	3.5	7.6	1.2
	1-Octene	1.3	6.1	4.2	0.6
	^{t}Butyl acrylate	<0.1	2.6	<0.1	1.3

[a]Catalyst:H_2O_2:olefin = 1:10:1000.
[b]Turnover number (TN, moles of product/moles of catalyst).

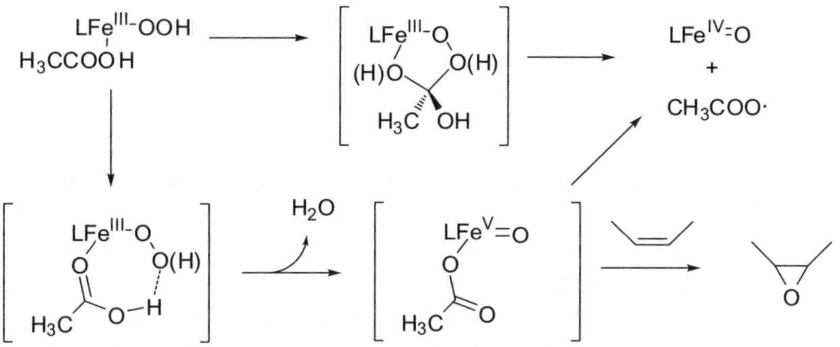

Scheme 6.43 Mechanistic scheme for AcOH promoted O–O heterolysis in iron-catalyzed epoxidations.[168]

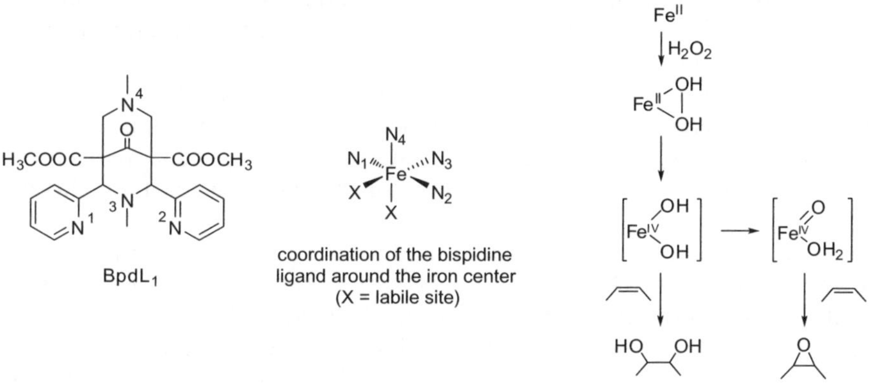

Scheme 6.44 Fe^{II}/Fe^{IV} mechanism proposed for alkene oxidation reactions catalyzed by iron bispidine complexes.[98]

olefins catalyzed by a mononuclear iron complex based on a tetradentate bispidine ligand using H_2O_2 as the oxidant (Scheme 6.44). This alternative pathway suggests that the high valent species responsible for the oxidation is a Fe^{IV} compound formed by O–O homolysis of a $Fe^{II}(H_2O_2)$ intermediate.[98] In particular, it was found by DFT calculations that two high-valent isomers are energetically favored: $Fe^{IV}(OH)_2$ and $H_2O–Fe^{IV}=O$.[98] This study concludes that the former is responsible for the *cis*-dihydroxylation and the latter affords the epoxide product (Fe^{II}/Fe^{IV} mechanism, Scheme 6.44). The authors do not put in doubt the viability of the "Fe^{III}/Fe^{V} *wa*" mechanisms, but they find that in their particular bispidine system $[Fe^{II}(CF_3SO_3)_2(BpdL1)]$ it is less probable due to the unfavorable energetics calculated for the putative $Fe^{V}(O)(OH)$, the key oxidant in Que's mechanism. Interestingly, isotopic labeling experiments observed for the *cis*-dihydroxylation reaction catalyzed by $[Fe^{II}(CF_3SO_3)_2(BpdL_1)]$ show that both oxygen atoms originate from the same H_2O_2 molecule. Therefore, this catalyst would belong to the class B classification proposed by Que.

6.3.3 Alkane Oxidations

Selective and efficient C–H hydroxylation reactivity has been observed for a limited number of H_2O_2-oxidation reactions catalyzed by bioinspired non-heme iron catalysts. Most generally, such complexes belong to the class A family (*vide supra*). Oxidation reactions catalyzed by this class of complexes are characterized by relatively high kinetic isotope effects measured in the competitive catalytic oxidation of cyclohexane/cyclohexane-d_{12}, and large normalized 3°/2° C–H selectivity ratios in the oxidation of adamantane (Table 6.6). These reactivity patterns are incompatible with the chemistry of the highly unselective hydroxyl radical,[40] but instead are indicative of the involvement of a selective oxidant. More interesting is the observation that the oxidation of *cis*-1,2-dimethylcyclohexane is stereospecific. Isotopic labeling experiments indicate minimum incorporation of O_2 in the oxidation products, demonstrating that carbon-centered radicals are not significantly formed. Instead, the origin of the oxygen atoms incorporated into products is the oxidant (H_2O_2) and water. Since peroxide type of species cannot exchange their oxygen atoms with water, the incorporation of oxygen from water into the alcohol product derived from alkane oxidation is taken as an indication for the involvement of a high-valent iron-oxo species.

As can be seen in Table 6.6, slight modifications in the ligand architecture can cause important changes in the catalytic results. Class B catalysts, for which $[Fe^{II}(6Me_3\text{-tpa})(CH_3CN)_2]^{2+}$ and $[Fe^{II}(CF_3SO_3)_2(^{iPr,H}Pytacn)]$ catalysts are prototypical examples, mediate reactions with some radical character as evidenced by the low percentage of retention of configuration (RC) in the oxidation of *cis*-1,2-dimethylcyclohexane (*cis*-1,2-DMCH) and by the high level of O_2 incorporation, a clear indication that free radical processes take place.

On the bases of these observations, the proposed mechanism for the oxidation of alkanes by class A type of complexes is analogous to that depicted in Scheme 6.38 for the oxidation of olefins. In this scenario, a high-valent $Fe^V(O)(OH)$ species, formed *via* water assisted heterolytic cleavage of the O–O bond in a $Fe^{III}OOH$ species (Scheme 6.45) is directly responsible for the selective C–H oxidation reaction. Water incorporation into the oxo group is proposed to occur *via* a oxo-hydroxo tautomerism, analogous to that occurring in heme systems.[166] C–H oxidation by the $Fe^V(O)(OH)$ species occurs *via* a rebound type mechanism. In the first step, the ferryl species abstracts the hydrogen atom from the substrate to form an alkyl radical, which rapidly rebounds with the newly formed hydroxo group to give the corresponding alcohol (Scheme 6.45).

Through this mechanism the maximum amount of oxygen from water that can be incorporated into the alcohol would be 50%, depending on the $Fe^V(O)(OH)$ tautomer involved. In addition, because of the competition between oxo-hydroxo tautomerism and substrate oxidation, the extent of water

Table 6.6 Oxidation of alkanes with H_2O_2 catalyzed by Fe^{II} complexes.[a]

| *Catalyst* | *Cyclohexane* | | *cis-1,2-DMCH* | *Adamantane* | *Oxygen source* | *Ref.* |
	$A+K\ (A/K)$[b]	*KIE*	$RC\ (\%)$[c]	$3°/2°$[d]	$H_2O/H_2O_2/O_2$[e]	
$[Fe^{II}(tpa)(CH_3CN)_2]^{2+}$	3.2 (5)	3.5	100	17	27/70/3	32, 68
$[Fe^{II}(CF_3SO_3)_2(bpmen)]$	6.3 (8)	3.2	96	15	18/84/0	32, 69
$[Fe^{II}(5Me_3\text{-}tpa)(CH_3CN)_2]^{2+}$	4.0 (9)	3.8	100	21	38/69/0	32
$[Fe^{II}(6Me_3\text{-}tpa)(CH_3CN)_2]^{2+}$	1.4 (3)	3.3	54	15	1/22/77	32
$[Fe^{II}(CF_3SO_3)_2(^{Me,H}Pytacn)]$	6.5 (12)	4.3	93	30	45/47/8	70, 71
$[Fe^{II}(CF_3SO_3)_2(^{Me,Me}Pytacn)]$	7.6 (10)	3.4	94	20	11/85/4	71
$[Fe^{II}(CF_3SO_3)_2(^{iPr,H}Pytacn)]$	2.4 (4)	4.8	86	14	8/59/33	71

[a]Catalyst:H_2O_2:alkane = 1:10:1000; 1:10:10 for adamantane.
[b]Turnover number (TN, moles of product/moles of catalyst), A = number of moles of cyclohexanol, K = number of moles of cyclohexanone.
[c]Retention of configuration in the oxidation of *cis*-1,2-DMCH.
[d]$3°/2° = 3\times$(1-adamantanol)/(2-adamantanol+2-adamantanone).
[e]Origin of the oxygen atom (%) in cyclohexane oxidation: from water/hydrogen peroxide/O_2.

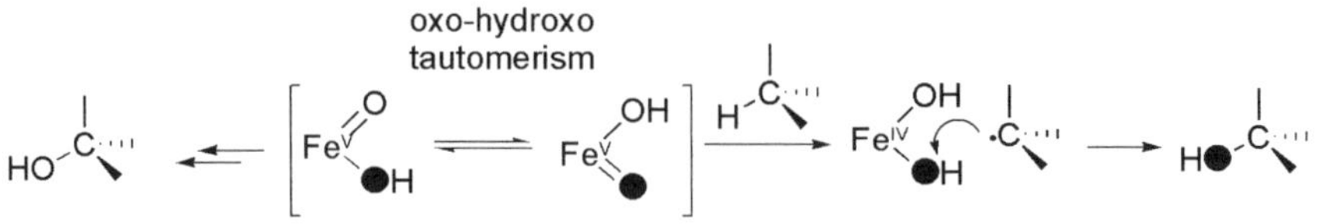

Scheme 6.45 Mechanism of alkane hydroxylation by HO–Fe^V=O. One tautomer affords incorporation of oxygen from H_2O_2 into the alcohol product while the other involves incorporation of one oxygen from water.

incorporation into products is directly related to the strength of the hydroxylated C–H bond, as observed in oxidation reactions catalyzed by selected non-heme iron complexes.[32] The involvement of a non-water assisted pathway ("Fe^{III} *nwa*" pathway in Scheme 6.38) has also been postulated in some cases to explain some labeling results for alkane oxidation.

Key aspects of the rebound-mechanism involved in the "Fe^{III}/Fe^V *wa*" mechanism have been reconsidered in the work performed in the group of Costas using the [Fe^{II}(CF$_3$SO$_3$)$_2$(^{Me,H}Pytacn)] catalyst.[70] In this work it was detected that the level of water incorporation into products was highly dependent on the specific alkane: while secondary C–H bonds afforded 42(2)% of alcohol containing the oxygen from water this value increased up to 76(3)% for tertiary C–H bonds. Although these high levels of water incorporation into alcohols are unprecedented among the reported bioinspired models, such a situation is reminiscent to the results obtained for selected enzymes within the Rieske dioxygenase family. Indeed the hydroxylation of indane to 1-indanol by toluene dioxygenase[169] occurs with a remarkably large amount of oxygen from water incorporated into the hydroxylated product (68%).

To explain the unprecedented high levels of water incorporation when [Fe^{II}(CF$_3$SO$_3$)$_2$(^{Me,H}Pytacn)] is used, an alternative rebound-like mechanism was proposed. The observation that weaker C–H bonds afford higher levels of water incorporation and that this value is independent of substrate concentration led to the conclusion that the newly formed alkyl radical (formed by initial hydrogen-atom abstraction) can "rebound" with any of the non-equivalent OH groups in the Fe^{IV}(OH)$_2$ disposed in a relative *cis* configuration (Scheme 6.46). This suggestion differs from the commonly accepted mechanism involving hydrogen-atom abstraction and rebound by the same oxygen atom. According to the results reported, it seems that 2°-alkyl radicals do not discriminate between the two OH groups, but 3°-alkyl radicals favor the rebound with the OH group derived from water. Steric effects may provide a rational for this preference since the sphere surrounding the two *cis*-positions is different.

Interestingly, the presence of at least two *cis*-labile sites in non-heme iron catalysts is also strictly required for stereospecific C–H oxidation.[32] Remarkably, the presence of *cis*-labile sites is a very common structural feature in several non-heme iron oxygenases. In this regard, the mechanistic scenario arising from this study may be related to aspects of the catalytic

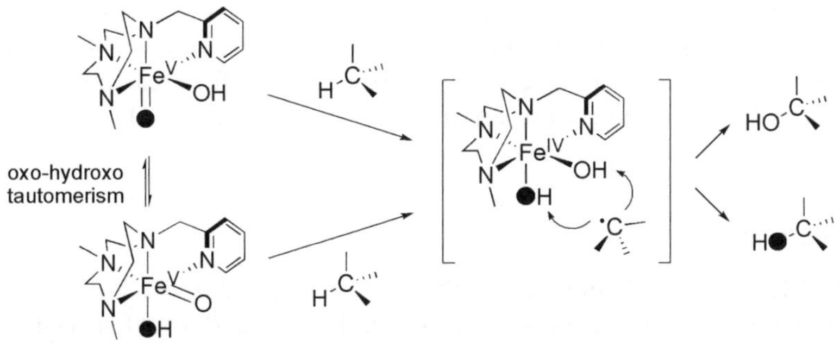

Scheme 6.46   Alternative rebound mechanism for alkane hydroxylation proposed for the Fe(Pytacn) system.[70]

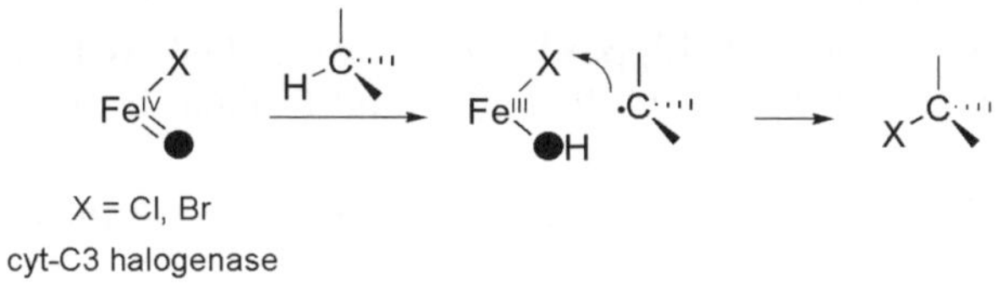

Scheme 6.47   Reaction mechanism for halogenations in α-ketoglutarate dependent halogenases.

cycle of some non-heme iron enzymes. For example, reminiscences can be found with the α-ketoglutarate-dependent CytC3 halogenase,[170] in which two high-spin $Fe^{IV}O(X)$ (X = Cl or Br) intermediates have been characterized by rapid-freeze-quench Mössbauer experiments and found to be directly responsible for the C–H activation event. Following hydrogen-atom abstraction by the oxo group, rebound does not take place with the hydroxide ligand but instead with the adjacent X group (Scheme 6.47). It is therefore likely that the *cis*-labile site characteristic of several non-heme iron dependent oxygenases can be the basis of an oxidative reactivity that is even more versatile than that of hemes.

6.4   Conclusions

The last decade has viewed spectacular advances in the understanding of C–H and C=C oxidation at non-heme iron sites. The discovery that iron complexes can mediate enzyme-like reactivity such as stereospecific C–H hydroxylation, olefin epoxidation, and *cis*-dihydroxylation of olefins constitutes a clear-cut departure from Fenton chemistry. Our understanding of the mechanisms underlying these reactions has advanced very much but it is still in its infancy. Reaction landscapes appear remarkably rich, and very sensitive to the nature of the iron centers. On the other hand, the exquisite selectivity offered by some of these reactions is approaching, and in some case is already surpassing, that attained by traditional oxidation reagents. Consequently, and because of the very attractive use of iron as a catalyst, these reactions bear a huge potential in chemical synthesis, which has just been started to be uncovered. Successful development of iron-based catalytic technologies for selective oxidation reactions is thus changing from an attractive hypothesis to a flourishing field.

References

1. B. Meunier and J. Bernadou, *Struct. Bonding*, 2000, **97**, 1.
2. B. Meunier, S. P. de Visser and S. Shaik, *Chem. Rev.*, 2004, **104**, 3947.
3. P. R. Ortiz de Montellano, *Cytochrome P450: Structure, Mechanism and Biochemistry*, Kluwer Academic/Plenum Publishers, New York, 2005.
4. I. Schlichting, J. Berendzen, K. Chu, A. M. Stock, S. A. Maves, D. E. Benson, R. M. Sweet, D. Ringe, G. A. Petsko and S. G. Sligar, *Science*, 2000, **287**, 1615.
5. E. Y. Tshuva and S. J. Lippard, *Chem. Rev.*, 2004, **104**, 987.
6. M. Costas, M. P. Mehn, M. P. Jensen and L. Que, Jr., *Chem. Rev.*, 2004, **104**, 939.
7. M. M. Abu-Omar, A. Loaiza and N. Hontzeas, *Chem. Rev.*, 2005, **105**, 2227.
8. E. I. Solomon, T. C. Brunold, M. I. Davis, J. N. Kemsley, S.-K. Lee, N. Lehnert, F. Neese, A. J. Skulan, Y.-S. Yang and J. Zhou, *Chem. Rev.*, 2000, **100**, 235.
9. P. C. A. Bruijnincx, G. v. Koten and R. J. M. Klein Gebbink, *Chem. Soc. Rev.*, 2008, **12**, 2716.
10. D. T. Gibson and V. Subramanian, in *Microbial Degradation of Aromatic Hydrocarbons*, ed. D. T. Gibson, Marcel Dekker, New York, 1984, p. 181.
11. T. A. Jackson and L. Que Jr., in *Concepts and Models in Bioinorganic Chemistry*, ed. H.-B. Kraatz and N. Metzler-Nolte, Wiley-VCH Verlag, Weinheim, 2006, p. 259.
12. M. Merkx, D. A. Kopp, M. H. Sazinsky, J. L. Blazyk, J. Müller and S. J. Lippard, *Angew. Chem., Int. Ed.*, 2001, **40**, 2782.
13. B. J. Wallar and J. D. Lipscomb, *Chem. Rev.*, 1996, **96**, 2625.
14. F. P. Guengerich, *Chem. Res. Toxicol.*, 2001, **14**, 611.
15. M. Sono, M. P. Roach, E. D. Coulter and J. H. Dawson, *Chem. Rev.*, 1996, **96**, 2841.
16. J. T. Groves and G. A. McCluskey, *J. Am. Chem. Soc.*, 1976, **98**, 859.
17. D. T. Gibson, S. M. Resnick, K. Lee, J. M. Brand, D. S. Torok, L. P. Wackett, M. J. Schocken and B. E. Haigler, *J. Bacteriol.*, 1995, **177**, 2615.
18. M. D. Wolfe, J. V. Parales, D. T. Gibson and J. D. Lipscomb, *J. Biol. Chem.*, 2001, **276**, 1945.
19. D. T. Gibson and R. E. Parales, *Curr. Opin. Biotechnol.*, 2000, **11**, 236.
20. T. Hudlicky, D. Gonzalez and D. T. Gibson, *Aldrichim. Acta*, 1999, **32**, 35.
21. B. Kauppi, K. Lee, E. Carredano, R. E. Parales, D. T. Gibson, H. Eklund and S. Ramaswamy, *Structure*, 1998, **6**, 571.

22. M. D. Wolfe, D. J. Altier, A. Stubna, C. V. Popescu, E. Münck and J. D. Lipscomb, *Biochemistry*, 2002, **41**, 9611.
23. D. T. Gibson, W.-K. Yeh, T.-N. Liu and V. Subramanian, in *Oxygenases and Oxygen Metabolism*, ed. M. Nozaki, S. Yamamoto, Y. Ishimura, M. J. Coon, L. Ernster and R. W. Estabrook, Academic, New York, 1982, p. 51.
24. S. E. Crutcher and P. J. Geary, *Biochem. J.*, 1979, **177**, 393.
25. E. G. Pavel, L. J. Martins, W. R. Ellis Jr. and E. I. Solomon, *Chem. Biol.*, 1994, **1**, 173.
26. E. L. Hegg and L. Que Jr., *Eur. J. Biochem.*, 1997, **250**, 625.
27. K. Koehntop, D. J. P. Emerson and L. Que Jr., *J. Biol. Inorg. Chem.*, 2005, **10**, 87.
28. A. Karlsson, J. V. Parales, R. E. Parales, D. T. Gibson, H. Eklund and S. Ramaswamy, *Science*, 2003, **299**, 1039.
29. M. D. Wolfe and J. D. Lipscomb, *J. Biol. Chem.*, 2003, **278**, 829.
30. K. Chen, M. Costas, J. Kim, A. K. Tipton and L. Que Jr., *J. Am. Chem. Soc.*, 2002, **124**, 3026.
31. K. Chen, M. Costas and L. Que Jr., *J. Chem. Soc., Dalton Trans.*, 2002, 672.
32. K. Chen and L. Que Jr., *J. Am. Chem. Soc.*, 2001, **123**, 6327.
33. *Biomimetic Oxidations Catalyzed by Transition Metal Complexes*, ed. B. Meunier, Imperial College Press, London, 2000.
34. M. Costas, K. Chen and L. Que Jr., *Coord. Chem. Rev.*, 2000, **200/202**, 517.
35. M. J. Perkins, *Chem. Soc. Rev.*, 1996, **25**, 229.
36. P. Stavropoulos, R. Celenligil-Cetin and A. E. Tapper, *Acc. Chem. Res.*, 2001, **34**, 745.
37. D. H. R. Barton, *Chem. Soc. Rev.*, 1996, **25**, 229.
38. D. H. R. Barton, *Tetrahedron*, 1998, **54**, 5805.
39. D. T. Sawyer, *Coord. Chem. Rev.*, 1997, **165**, 297.
40. C. Walling, *Acc. Chem. Res.*, 1975, **8**, 125.
41. C. Walling, *Acc. Chem. Res.*, 1998, **31**, 155.
42. F. Gozzo, *J. Mol. Cat. A: Chem.*, 2001, **171**, 1.
43. G. Dyker, *Handbook of C–H Transformations*, Wiley-VCH Verlag, Weinheim, 2005.
44. A. Correa, O. G. Mancheño and C. Bolm, *Chem. Soc. Rev.*, 2008, **8**, 1108.
45. S. Enthaler, K. Junge and M. Beller, *Angew Chem., Int. Ed.*, 2008, **47**, 3317.
46. D. T. Sawyer, A. Sobkowiak and T. Matsushita, *Acc. Chem. Res.*, 1996, **29**, 409.
47. P. A. MacFaul, D. D. M. Wayner and K. U. Ingold, *Acc. Chem. Res.*, 1998, **31**, 159.
48. K. U. Ingold and P. A. MacFaul, in *Biomimetic Oxidations Catalyzed by Transition Metal Complexes*, ed. B. Meunier, Imperial College Press, London, 2000, p. 45.
49. G. A. Russell, in *The Chemistry of Alkanes and Cycloalkanes*, ed. S. Patai and Z. Rappoport, John Wiley & Sons, Inc., New York, 1992, pp. 963.

50. F. Minisci, F. Fontana, S. Araneo, F. Recupero, S. Banfi and S. Quici, *J. Am. Chem. Soc.*, 1995, **117**, 226.
51. L. Que Jr. and W. B. Tolman, *Nature*, 2008, **455**, 333.
52. S. Friedle, E. Reisner and S. J. Lippard, *Chem. Soc. Rev.*, 2010, **39**, 2768.
53. I. Siewert and C. Limberg, *Chem. Eur. J.*, 2009, **15**, 10316.
54. X. Wang, S. Wang, L. Li, E. B. Sundberg and G. P. Gacho, *Inorg. Chem.*, 2003, **42**, 7799.
55. M. C. Esmelindro, E. G. Oestreicher, H. Márquez-Alvarez, C. Dariva, S. M. Egues, C. Fernandes, A. J. Bortoluzzi, V. Drago and O. A. Antunes, *J. Inorg. Biochem.*, 2005, **99**, 2054.
56. G. Trettenhahn, M. Nagl, N. Neuwirth, V. B. Arion, W. Jary, P. Pöchlauer and W. Schmid, *Angew. Chem., Int. Ed.*, 2006, **45**, 2794.
57. V. B. Romakh, B. Therrien, G. Süss-Fink and G. B. Shul'pin, *Inorg. Chem.*, 2007, **46**, 3166.
58. K. Visvaganesan, E. Suresh and M. Palaniandavar, *Dalton Trans.*, 2009, 3814.
59. G. A. Russell, *J. Am. Chem. Soc.*, 1957, **79**, 3871.
60. R. Mayilmurugan, H. Stoeckli-Evans, E. Suresh and M. Palaniandavar, *Dalton Trans.*, 2009, 5101.
61. G. V. Nizova, B. Krebs, G. Süss-Fink, S. Schindler, L. Westerheide, L. G. Cuervo and G. B. Shul'pin, *Tetrahedron*, 2002, **58**, 9231.
62. T. Nagataki, Y. Tachi and S. Itoh, *J. Mol. Cat. A*, 2005, **225**, 103.
63. Y. Mekmouche, C. Duboc-Toia, S. Ménage, C. Lambeaux and M. Fontecave, *J. Mol. Catal. A: Chem.*, 2000, **156**, 85.
64. A. B. Sorokin, E. V. Kudrika and D. Bouchub, *Chem. Commun.*, 2008, 2562.
65. S. Mukerjee, A. Stassinopoulos and J. P. Caradonna, *J. Am. Chem. Soc.*, 1997, **119**, 8097.
66. T. L. Foster and J. P. Caradonna, *J. Am. Chem. Soc.*, 2003, **125**, 3678.
67. T. Punniyamurthy, S. Velusamy and J. Iqbal, *Chem. Rev.*, 2005, **105**, 2329.
68. C. Kim, K. Chen, J. Kim and L. Que Jr., *J. Am. Chem. Soc.*, 1997, **119**, 5964.
69. K. Chen and L. Que Jr., *Chem. Commun.*, 1999, 1375.
70. A. Company, L. Gómez, M. Güell, X. Ribas, J. M. Luis, L. Que Jr. and M. Costas, *J. Am. Chem. Soc.*, 2007, **129**, 15766.
71. A. Company, L. Gómez, X. Fontrodona, X. Ribas and M. Costas, *Chem. Eur. J.*, 2008, **14**, 5727.
72. J. England, G. J. P. Britovsek, N. Rabadia and A. J. P. White, *Inorg. Chem.*, 2007, **46**, 3752.
73. Y. Mekmouche, S. Ménage, C. Toia-Duboc, M. Fontecave, J.-B. Galey, C. Lebrun and J. Pecaut, *Angew. Chem., Int. Ed.*, 2001, **40**, 949.
74. M. Costas and L. Que Jr., *Angew. Chem., Int. Ed.*, 2002, **12**, 2179.
75. P. Comba, M. Maurer and P. Vadivelu, *Inorg. Chem.*, 2009, **48**, 10389.
76. J. England, C. R. Davies, M. Banaru, A. J. P. White and G. J. P. Britovsek, *Adv. Synth. Catal.*, 2008, **350**, 883.

77. J. England, M. Martinho, E. R. Farquhar, J. R. Frisch, E. L. Bominaar, E. Münck and L. Que Jr., *Angew. Chem., Int. Ed.*, 2009, **48**, 3622.
78. G. J. P. Britovsek, J. England, S. K. Spitzmesser, A. J. P. White and D. J. Williams, *Dalton Trans.*, 2005, 945.
79. G. J. P. Britovsek, J. England and A. J. P. White, *Inorg. Chem.*, 2005, **44**, 8125.
80. G. B. Shul'pin, C. C. Golfeto, G. Süss-Fink, L. S. Shul'pina and D. Mandelli, *Tetrahedron Lett.*, 2005, **46**, 4563.
81. R. M. Burger, *Chem. Rev.*, 1998, **98**, 1153.
82. J. W. Sam, X.-J. Tang and J. Peisach, *J. Am. Chem. Soc.*, 1994, **116**, 5250.
83. A. Veselov, H. Sun, A. Sienkiewicz, H. Taylor, R. M. Burger and C. P. Scholes, *J. Am. Chem. Soc.*, 1995, **117**, 7508.
84. J. M. Rowland, M. M. Olmstead and P. K. Mascharak, *Inorg. Chem.*, 2001, **40**, 2810.
85. C. Nguyen, R. J. Guajardo and P. K. Mascharak, *Inorg. Chem.*, 1996, **35**, 6273.
86. G. Roelfes, M. Lubben, R. Hage, L. Que Jr. and B. L. Feringa, *Chem. Eur. J.*, 2000, **6**, 2152.
87. T. A. van den Berg, J. W. de Boer, W. R. Browne, G. Roelfes and B. L. Feringa, *Chem. Commun.*, 2004, 2550.
88. S. Gosiewska, H. P. Permentier, A. P. Bruins, G. v. Koten and R. J. M. Klein Gebbink, *Dalton Trans.*, 2007, 3365.
89. Y.-M. Lee, S. N. Dhuri, S. C. Sawant, J. Cho, M. Kubo, T. Ogura, S. Fukuzumi and W. Nam, *Angew. Chem. Int. Ed.*, 2009, **48**, 1803.
90. L. Liang-Xian, *Curr. Org. Chem.*, 2010, **14**, 1099.
91. *Iron Catalysis in Organic Chemistry: Reactions and Applications*, ed. B. Plietker, Wiley-VCH Verlag, Weinheim, 2008.
92. M. Nakanishi and C. Bolm, *Adv. Synth. Catal.*, 2007, 861.
93. C. Pavan, J. Legros and C. Bolm, *Adv. Synth. Catal.*, 2005, **347**, 703.
94. B. Retcher, J. S. Costa, J. K. Tang, R. Hage, P. Gamez and J. Reedijk, *J. Mol. Catal. A Chem.*, 2008, **286**, 1.
95. J. Tang, P. Gamez and J. Reedijk, *Dalton Trans.*, 2007, 4644.
96. D. Li, K. Schröder, B. Bitterlich, M. K. Tse and M. Beller, *Tetrahedron Lett.*, 2008, **49**, 5976.
97. M. S. Chen and M. C. White, *Science*, 2007, **318**, 783.
98. J. Bautz, P. Comba, C. L. d. Laorden, M. Menzel and G. Rajaraman, *Angew. Chem. Int. Ed.*, 2007, **46**, 8067.
99. N. A. Vermeulen, M. S. Chen and M. C. White, *Tetrahedron*, 2009, **65**, 3078.
100. M. S. Chen and M. C. White, *Science*, 2010, **327**, 566.
101. R. Curci, L. D'Accolti and C. Fusco, *Acc. Chem. Res.*, 2006, **39**, 1.
102. K. Chen, A. Eschenmoser and P. S. Baran, *Angew. Chem. Int. Ed.*, 2009, **48**, 9705.
103. B. Meunier, *Chem. Rev.*, 1992, **92**, 1411.
104. J. L. McLain, J. Lee and J. T. Groves, in *Biomimetic Oxidations Catalyzed by Transition Metal Complexes*, ed. B. Meunier, Imperial College Press, London, 2000, p. 91.

105. L. Gomez, I. Garcia-Bosch, A. Company, J. Benet-Buchholz, A. Polo, X. Sala, X. Ribas and M. Costas, *Angew. Chem., Int. Ed.*, 2009, **48**, 5720.
106. K. Möller, G. Wienhöfer, K. Schröder, B. Join, K. Junge and M. Beller, *Chem. Eur. J.*, 2010, **16**, 10–300.
107. W. Nam, R. Y. N. Ho and J. S. Valentine, *J. Am. Chem. Soc.*, 1991, **113**, 7052.
108. M. B. Francis and E. N. Jacobsen, *Angew. Chem., Int. Ed.*, 1999, **38**, 937.
109. M. C. White, A. G. Doyle and E. N. Jacobsen, *J. Am. Chem. Soc.*, 2001, **123**, 7194.
110. S. Y. Taktak, W. Herrera and E. V. Rybak-Akimova, *Inorg. Chem.*, 2007, **46**, 2929.
111. H.-L. Yeung, K.-C. Sham, C.-S. Tsang, T.-C. Lau and H.-L. Kwong, *Chem. Commun.*, 2008, 3801.
112. M. Kodera, M. Itoh, K. Kano, T. Funabiki and M. Reglier, *Angew. Chem. Int. Ed.*, 2005, **44**, 7104.
113. G. Dubois, A. Murphy and T. D. P. Stack, *Org. Lett.*, 2003, **5**, 2469.
114. C. Marchi-Delapierre, A. Jorge-Robin, A. Thibon and S. Ménage, *Chem. Commun.*, 2007, 1166.
115. G. Anilkumar, B. Bitterlich, F. G. Gelalcha, M. K. Tse and M. Beller, *Chem. Commun.*, 2007, 289.
116. B. Bitterlich, G. Anilkumar, F. G. Gelalcha, B. Spilker, A. Grotevendt, R. Jackstell, M. K. Tse and M. Beller, *Chem. Asian J.*, 2007, **2**, 521.
117. B. Bitterlich, K. Schroeder, M. K. Tse and M. Beller, *Eur. J. Org. Chem.*, 2008, 4867.
118. F. G. Gelalcha, G. Anilkumar, M. K. Tse, A. Brückner and M. Beller, *Chem. Eur. J.*, 2008, **14**, 7687.
119. F. G. Gelalcha, B. Bitterlich, G. Anilkumar, M. K. Tse and M. Beller, *Angew. Chem. Int. Ed.*, 2007, **46**, 7293.
120. K. Schröder, S. Enthaler, B. Join, K. Junge and M. Beller, *Adv. Synth. Catal.*, 2010, **352**, 1771.
121. K. Schröder, S. Enthaler, B. Bitterlich, T. Schulz, A. Spannenberg, M. K. Tse, K. Junge and M. Beller, *Chem. Eur. J.*, 2009, **15**, 5471.
122. K. Schröder, X. Tong, B. Bitterlich, M. K. Tse, F. G. Gelalcha, A. Brückner and M. Beller, *Tetrahedron Lett.*, 2007, **48**, 6339.
123. P. Liu, E. L.-M. Wong, A. W.-H. Yuen and C.-M. Che, *Org. Lett.*, 2008, **10**, 3275.
124. K. Chen and L. Que Jr., *Angew. Chem. Int. Ed.*, 1999, **38**, 2227.
125. K. Suzuki, P. D. Oldenburg and L. Que Jr., *Angew. Chem. Int. Ed.*, 2008, **47**, 1887.
126. D. Quinonero, K. Morokuma, D. G. Musaev, R. Mas-Balleste and L. Que Jr., *J. Am. Chem. Soc.*, 2005, **127**, 6548.
127. M. Klopstra, G. Roelfes, R. Hage, R. M. Kellogg and B. L. Feringa, *Eur. J. Inorg. Chem.*, 2004, 846.
128. M. Costas, A. K. Tipton, K. Chen, D.-H. Jo and L. Que Jr., *J. Am. Chem. Soc.*, 2001, **123**, 6722.

129. R. Mas-Ballesté, M. Costas, T. van den Berg and L. Que Jr., *Chem. Eur. J.*, 2006, **12**, 7489.
130. P. D. Oldenburg, A. A. Shteinman and L. Que Jr., *J. Am. Chem. Soc.*, 2005, **127**, 15672.
131. P. C. A. Bruijnincx, I. L. C. Buurmans, S. Gosiewska, M. A. H. Moelands, M. Lutz, A. L. Spek, G. van Koten and R. J. M. Klein Gebbink, *Chem. Eur. J.*, 2008, **14**, 1228.
132. S. Gosiewska, M. Lutz, A. L. Spek and R. J. M. Klein Gebbink, *Inorg. Chim. Acta*, 2007, **360**, 405.
133. M. R. Bukowski, P. Comba, A. Lienke, C. Limberg, C. L. de Laorden, R. Mas-Balleste, M. Merz and L. Que Jr., *Angew. Chem. Int. Ed.*, 2006, **45**, 3446.
134. P. D. Oldenburg, R. Mas-Balleste and L. Que Jr., in *Mechanisms in Homogeneous and Heterogeneous Epoxidation Catalysis*, ed. S. T. Oyama, Elsevier, Amsterdam, 2008, p. 217.
135. P. D. Oldenburg, C.-Y. Ke, A. A. Tipton, A. A. Shteinman and L. Que Jr., *Angew. Chem., Int. Ed.*, 2006, **45**, 7975.
136. Y. Feng, C.-y. Ke, G. Xue and L. Que Jr., *Chem. Commun.*, 2009, 50.
137. J. Y. Ryu, J. Kim, M. Costas, K. Chen, W. Nam and L. Que Jr., *Chem. Commun.*, 2002, 1288.
138. T. W.-S. Chow, E. L.-M. Wong, Z. Guo, Y. Liu, J.-S. Huang and C.-M. Che, *J. Am. Chem. Soc.*, 2010, **132**, 13229.
139. Y. Zang, T. E. Elgren, Y. Dong and L. Que Jr., *J. Am. Chem. Soc.*, 1993, **115**, 811.
140. R. Y. N. Ho, G. Roelfes, B. L. Feringa and L. Que Jr., *J. Am. Chem. Soc.*, 1999, **121**, 264.
141. N. Lehnert, R. Y. N. Ho, L. Que Jr. and E. I. Solomon, *J. Am. Chem. Soc.*, 2001, **123**, 8271.
142. N. Lehnert, R. Y. N. Ho, L. Que Jr. and E. I. Solomon, *J. Am. Chem. Soc.*, 2001, **123**, 12802.
143. N. Lehnert, F. Neese, R. Y. N. Ho, L. Que Jr. and E. I. Solomon, *J. Am. Chem. Soc.*, 2002, **124**, 10810.
144. P. R. Ortiz de Montellano and J. J. De Voss, *Nat. Prod. Rep.*, 2002, **19**, 477.
145. J. Stubbe and J. W. Kozarich, *Chem. Rev.*, 1987, **87**, 1107.
146. M. Lubben, A. Meetsma, E. C. Wilkinson, B. Feringa and L. Que Jr., *Angew. Chem., Int. Ed. Engl.*, 1995, **34**, 1512.
147. A. Wada, S. Ogo, S. Nagatomo, T. Kitagawa, Y. Watanabe, K. Jitsukawa and H. Masuda, *Inorg. Chem.*, 2002, **41**, 616.
148. G. Roelfes, V. Vrajmasu, K. Chen, R. Y. N. Ho, J.-U. Rohde, C. Zondervan, R. M. la Crois, E. P. Schudde, M. Lutz, A. L. Spek, R. Hage, B. L. Feringa, E. Münck and L. Que Jr., *Inorg. Chem.*, 2003, **42**, 2639.
149. M. R. Bukowski, P. Comba, C. Limberg, M. Merz, L. Que Jr. and T. Wistuba, *Angew. Chem., Int. Ed.*, 2004, **43**, 1283.
150. M. Martinho, F. Banse, J. Sainton, C. Philouze, R. Guillot, G. Blain, P. Dorlet, S. Lecomte and J.-J. Girerd, *Inorg. Chem.*, 2007, **46**, 1709.

151. M. Martinho, P. Dorlet, E. Rivière, A. Thibon, C. Ribal, F. Banse and J.-J. Girerd, *Chem. Eur. J.*, 2008, **14**, 3182.
152. J.-J. Girerd, F. Banse and A. J. Simaan, *Struct. Bonding*, 2000, **97**, 143.
153. X. Shan, J.-U. Rohde, K. D. Koehntop, Y. Zhou, M. R. Bukowski, M. Costas, K. Fujisawa and L. Que Jr., *Inorg. Chem.*, 2007, **46**, 8410.
154. J.-U. Rohde, S. Torelli, X. Shan, M. H. Lim, E. J. Klinker, J. Kaizer, K. Chen, W. Nam and L. Que Jr., *J. Am. Chem. Soc.*, 2004, **126**, 16750.
155. K. D. Koehntop, J.-U. Rohde, M. Costas and L. Que Jr., *Dalton Trans.*, 2004, 3191.
156. S. Hong, Y.-M. Lee, W. Shin, S. Fukuzumi and W. Nam, *J. Am. Chem. Soc.*, 2009, **131**, 13910.
157. R. Y. N. Ho, G. Roelfes, R. Hermant, R. Hage, B. L. Feringa and L. Que Jr., *Chem. Commun.*, 1999, 2161.
158. D. Krishnamurthy, G. D. Kasper, F. Namuswe, W. D. Kerber, A. A. N. Sarjeant, P. Moënne-Loccoz and D. P. Goldberg, *J. Am. Chem. Soc.*, 2006, **128**, 14222.
159. M. S. Seo, T. Kamachi, T. Kouno, K. Murata, M. J. Park, K. Yoshizawa and W. Nam, *Angew. Chem., Int. Ed.*, 2007, **46**, 2291.
160. M. J. Park, J. Lee, Y. Suh, J. Kim and W. Nam, *J. Am. Chem. Soc.*, 2006, **128**, 2630.
161. A. Company, Y. Feng, M. Güell, X. Ribas, J. M. Luis, L. Que Jr. and M. Costas, *Chem. Eur. J.*, 2009, **15**, 3359.
162. M. Fujita, M. Costas and L. Que Jr., *J. Am. Chem . Soc.*, 2003, **125**, 9912.
163. O. Y. Lyakin, K. P. Bryliakov, G. J. P. Britovsek and E. P. Talsi, *J. Am. Chem. Soc.*, 2009, **131**, 10798.
164. A. Bassan, M. R. A. Blomberg, P. E. M. Siegbahn and L. Que Jr., *Angew. Chem. Int. Ed.*, 2005, **44**, 2939.
165. A. Bassan, M. R. A. Blomberg, P. E. M. Siegbahn and L. Que Jr., *J. Am. Chem. Soc.*, 2002, **124**, 11056.
166. J. Bernadou and B. Meunier, *Chem. Commun.*, 1998, 2167.
167. S. H. Lee, J. H. Han, H. Kwak, S. J. Lee, E. Y. Lee, H. J. Kim, J. H. Lee, C. Bae, S. N. Lee, Y. Kim and C. Kim, *Chem. Eur. J.*, 2007, **133**, 9393.
168. R. Mas-Balleste and L. Que Jr., *J. Am. Chem. Soc.*, 2007, **129**, 15964.
169. L. P. Wackett, L. D. Kwart and D. T. Gibson, *Biochemistry*, 1988, **27**, 1360.
170. D. P. Galonić, E. W. Barr, C. T. Walsh, J. M. Bollinger and C. Krebs, *Nat. Chem. Biol.*, 2007, **3**, 113.

Application of Magnetic Circular Dichroism, X-Ray Absorption Spectroscopy and Extended X-Ray Absorption Fine Structure in Determining Geometric and Electronic Structure of Non-heme Iron(IV)-oxo Enzymatic Intermediates and Related Synthetic Models

SOMDATTA GHOSH DEY* AND ABHISHEK DEY*

Department of Inorganic Chemistry, Indian Association for the Cultivation of Science, Jadavpur, Kolkata, 700032, India

7.1 Introduction

Non-heme iron systems are versatile oxygen activating agents of nature. They catalyze very important biological transformations by either activating the substrate (Scheme 7.1a) or molecular oxygen (Scheme 7.1b) to produce reactive intermediates that can affect various processes. Their malfunction has been

Iron-Containing Enzymes: Versatile Catalysts of Hydroxylation Reactions in Nature
Edited by Sam P de Visser and Devesh Kumar
© Royal Society of Chemistry 2011
Published by the Royal Society of Chemistry, www.rsc.org

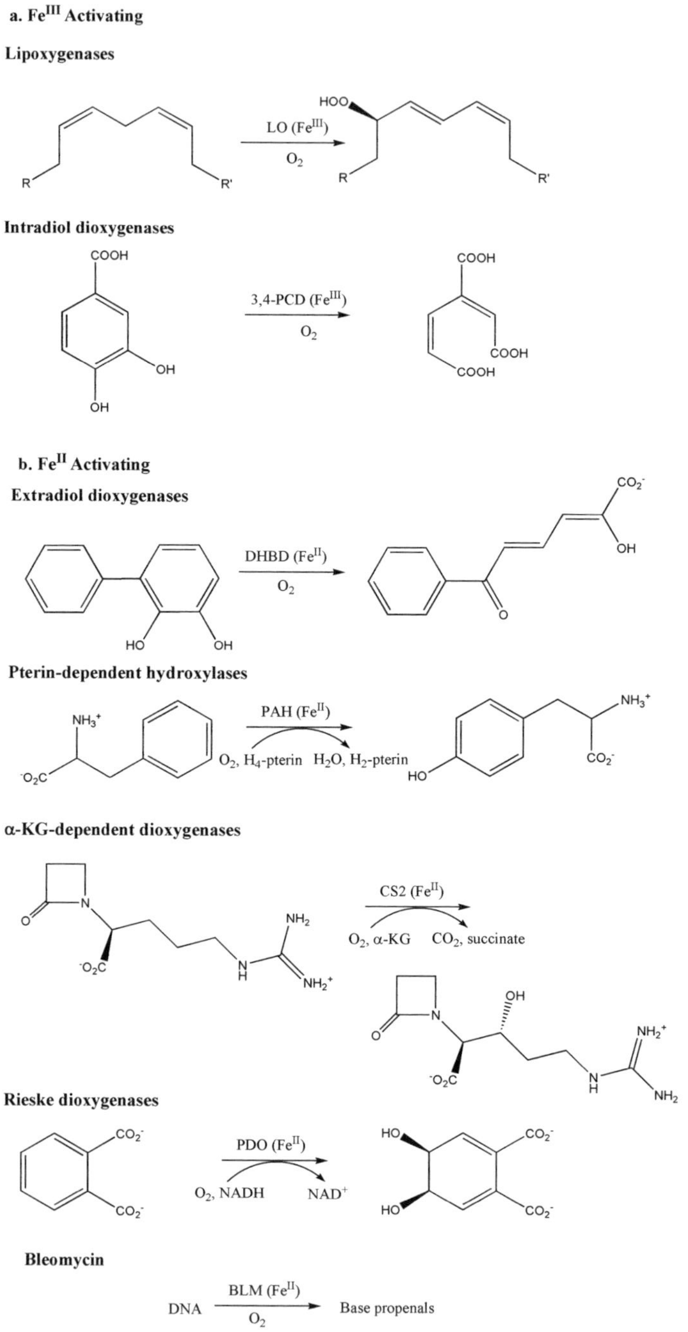

Scheme 7.1   Common classes of reactions catalyzed by non-heme iron enzymes and an anticancer drug, bleomycin (BLM). The enzyme abbreviations are defined as follows: lipoxygenase (LO), protocatechuate 3,4-dioxygenase (3,4-PCD), 2,3-dihydroxybiphenyl 1,2-dioxygenase (DHBD), phenylalanine hydroxylase (PAH), clavaminate synthase (CS2) and phthalate dioxygenase (PDO).

associated with several medical conditions that can be life-threatening or fatal. Non-heme iron active sites have always interested chemists of different fields due to the challenges involved in studying them. Their structural modeling is difficult as the biological ligand set, two histidines and a carboxylate with three exchangeable water ligands, is difficult to synthesize and manipulate. The absence of charge transfer (CT) transitions in the visible region and characteristic vibrational features make spectroscopic characterization of active sites and catalytic cycle intermediates a formidable task. The fact that these active sites are involved in some of the key transformations in biology, as illustrated in Scheme 7.1, has inspired scientific creativity over several decades, and techniques, both spectroscopic and synthetic, have been invented or adapted specifically to investigate these systems. The last decade has observed a spike in research activity in this field as these methodologies reached a state where they could be used to obtain unprecedented insight into these systems.

Fundamentally, all non-heme iron active sites that are α-ketoglutarate (α-KG) dependent are believed to follow the same mechanistic scheme (Scheme 7.2).[1,2] The resting ferrous site has an Fe^{II} bound to two histidine amino acids and one carboxylate (also called the facial triad) as well as three water molecules. Two of the three water ligands bound to Fe^{II} are replaced by α-KG (the co-substrate) followed by substrate binding. Initially the substrate binding was

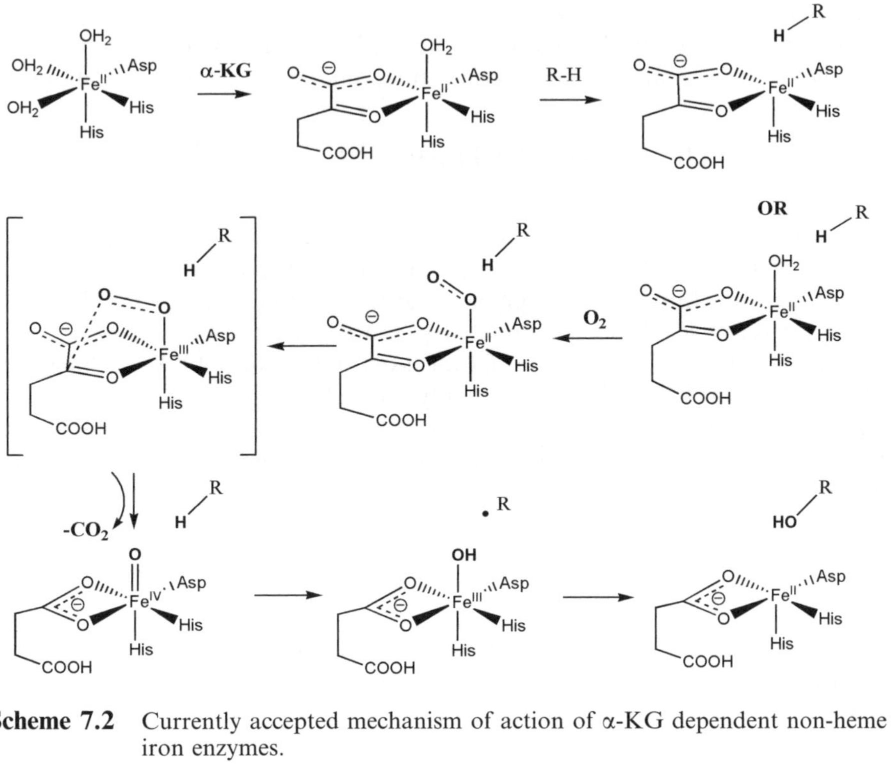

Scheme 7.2 Currently accepted mechanism of action of α-KG dependent non-heme iron enzymes.

proposed to create a five-coordinate active site ready for O_2 binding.[1] Although in the original proposed mechanism this was found to be the case for several non-heme iron enzymes, the presence of a six-coordinate ferrous site after substrate binding was also reported in several cases.[1–3] The O–O bond cleavage leads to the formation of a highly reactive iron(IV)-oxo species that can abstract hydrogen atoms from C–H bonds and hydroxylate them *via* a rebound mechanism. Thus, quite logically, understanding the nature of this chemically "hot" iron(IV)-oxo species and modeling it to produce viable artificial catalysts took center stage in bioinorganic chemistry.

A general problem with the non-heme iron active site is its spectroscopic characterization. While crystallography provides in-depth knowledge of the overall protein structure and iron binding residues, it does not provide very accurate geometries of the active site (*e.g.* the error bar in determining bond lengths is often ± 0.2 Å). In the absence of intense absorption features in the visible region (except for the metal-ligand CT bands of the α-KG bound ferrous state), absorption spectroscopy or resonance Raman spectroscopy does not yield satisfactory results. Thus, many of the mechanistic proposals had to await validation till appropriate spectroscopic tools could be developed. Magnetic circular dichroism (MCD), X-ray absorption spectroscopy (XAS) and extended X-ray fine structure (EXAFS) analysis were some vital tools that have enhanced development of geometric and electronic structure–function correlations of non-heme iron enzymes in the last two decades. The present chapter focuses on the use of these techniques in elucidating geometric and electronic structure of enzymatic iron(IV)-oxo intermediates and their synthetic model complexes in conjunction with density functional theory (DFT) calculations.

7.1.1 Magnetic Circular Dichroism (MCD)

Non-heme iron enzyme active sites are less well characterized as compared to their heme counter parts due to the lack of intense ligand $\pi \rightarrow \pi^*$ transitions. The oxidized ferric (d^5, $S = \frac{5}{2}$) active sites can be probed by absorption *via* CT transitions and electron paramagnetic resonance (EPR) spectroscopy. The high-spin ferrous (d^6, $S = 2$) non-heme active sites are less accessible by spectroscopic techniques. They do not exhibit CT transitions in the normal UV–visible region of the spectrum (since those are usually located above $33\,000$ cm^{-1}) and are EPR silent (in the conventional perpendicular mode). The d→d transitions in the near-IR region are Laportè or parity forbidden and are often unobserved in the absorption spectrum. However, these active sites can be probed with the circular dichroism (CD) technique, for optically active metal sites, and MCD. Both these spectroscopic tools are governed by different selection rules, which allow for intense d→d transitions, unlike absorption spectroscopy. MCD provides information about the ligand field (LF) excited states and variable temperature variable field (VTVH) MCD determines the ground state sublevel splitting. This methodology has been applied to obtain geometric and electronic structures of various mononuclear and binuclear non-heme ferrous enzymes, which have provided insight into their catalytic function.

7.1.1.1 MCD Theory

MCD and CD spectroscopic techniques measure the difference in the extinction coefficients between the left and right circularly polarized light, which is given by Equation (7.1):[4]

$$\Delta\varepsilon = \varepsilon_L - \varepsilon_R = \Delta A/cl \qquad (7.1)$$

In Equation (7.1), ΔA represents the difference between the absorbance of the left and right circularly polarized light (*via* Beer's law), c is the concentration in mol l^{-1} and l is the path length in cm. The units of $\Delta\varepsilon$ are M^{-1} cm^{-1}.

There are many advantages of using CD and MCD spectroscopies for the study of non-heme ferrous enzymes, such as:

1. As $\Delta\varepsilon$ has either positive or negative sign, the data allows the resolution of overlapping bands in the absorption spectra.
2. CD spectroscopy is a differential technique using phase sensitive detection, allowing detection of signals from systems with $\varepsilon \approx 1$–10 M^{-1} cm^{-1} and $\Delta\varepsilon \approx 10^{-6}$ M^{-1} cm^{-1}.
3. MCD spectra of high spin ferrous systems show intense signals at low temperatures (related to the C-term, *vide infra*) even if their CD signal is not that much enhanced. The MCD intensity of the ferrous active site is usually amplified by two or three orders of magnitude relative to the protein background and aqueous/solvent medium.
4. The temperature and field dependence of the MCD signals can provide information about the energy splittings of the ground state sublevels. This technique essentially provides information usually achieved from EPR spectroscopy, but for EPR inactive sites.

According to the selection rule for CD activity, an optical transition from the ground state to the excited state must be both electric and magnetic dipole allowed along the same molecular direction.[5] This derives from the fact that circularly polarized light excites electrons with helical motion and hence undergoes both translation (electric dipole) and rotation (magnetic dipole) operations. Only optically active molecules described by point groups C_n, D_n, O, T and I can have non-zero projection of the electric and magnetic dipole moments of a given transition and are CD active. Thus, low symmetry protein molecules having C_1 site symmetry can be detected by CD spectroscopy.

MCD spectroscopic technique is essentially a CD experiment in the presence of a magnetic field applied parallel to the propagation direction of circularly polarized light. It probes the Zeeman splittings of the ground and excited states and magnetic field induced mixing between states. MCD intensity for a transition can be expressed by Equation (7.2),[5–8] where ΔA represents the difference in the absorbance of the left and right circularly polarized light, $E = h\nu$ is the energy of the incident light, β is the Bohr magneton, H is the strength of the applied magnetic field and $f(E)$ is the absorption band shape. $\mathbf{A}_1$, $\mathbf{B}_0$ and $\mathbf{C}_0/kT$ are called the **A**, **B** and **C** terms, respectively, and describe the three different mechanisms that give rise to the MCD intensity. Since, $\partial f(E)/\partial E$ is the derivative of the absorption band shape, the shape of the **A** term is a derivative. On a similar basis, the **B** and **C** terms have an absorption band shape. The equation is valid only when the thermal energy is greater than the energy splittings of the sublevels of the ground state:

$$\frac{\Delta A}{E} = \text{const} \bullet \beta H \left[\mathbf{A}_1 \left(-\frac{\partial f(E)}{\partial E} \right) + \left(\mathbf{B}_0 + \frac{C_0}{kT} \right) f(E) \right] \qquad (7.2)$$

The **A**-term is expressed in detail by Equation (7.3),[4–8] with the electronic degeneracy of the ground state $|A\rangle$ given as d_A, $|J\rangle$ is an excited state, $\mu_Z = L_z + 2S_z$ is the Zeeman operator, and M_+ and M_- are the electric dipole transition moments for left and right circularly polarized light, respectively. For a system to show an **A**-term intensity, there must exist a degenerate ground or excited state. Figure 7.1 illustrates possible cases that can give rise to **A**-term signals in MCD spectroscopy:

$$\mathbf{A}_1 = \frac{1}{d_A} \sum \left(\langle J | \mu_Z | J \rangle - \langle A | \mu_Z | A \rangle \right) \left(\left[M_-^{AJ} \right]^2 - \left[M_+^{AJ} \right]^2 \right) \qquad (7.3)$$

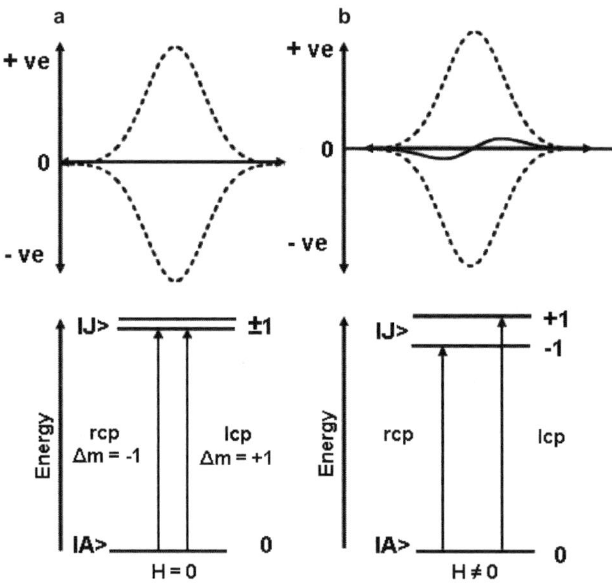

Figure 7.1 (a) In the absence of a magnetic field, the transitions from a nondegenerate ground state **A** to a doubly degenerate excited state **J** caused by right and left circularly polarized light (rcp and lcp, respectively) cancel each other, resulting in no **A** term intensity. (b) In the presence of a magnetic field the doubly degenerate excited state **J** becomes a nondegenerate state, resulting in two transitions that effectively produce a derivative shaped **A**-term in the magnetic circular dichroism (MCD) signal.

The **B**-term intensity is given by Equation (7.4).[4–9] Note that both the **A** and **B** terms are temperature independent. For a transition to have **B**-term intensity, it requires the presence of a non-degenerate state that can mix with either the non-degenerate ground state or an excited state upon the application of magnetic field. The possible mechanisms for **B**-term intensity are explained in Figure 7.2:

$$\mathbf{B}_0 = \frac{2}{d_A} \mathrm{Re} \sum \left[\sum_{K;K\neq J} \frac{\langle J|\mu_z|K\rangle}{\Delta E_{KJ}} \left([M_-^{AJ}] \bullet [M_+^{KA}] - [M_+^{AJ}] \bullet [M_-^{KA}] \right) \right.$$

$$\left. + \sum_{K;K\neq J} \frac{\langle K|\mu_z|A\rangle}{\Delta E_{KA}} \left([M_-^{AJ}] \bullet [M_+^{JK}] - [M_+^{AJ}] \bullet [M_-^{JK}] \right) \right] \qquad (7.4)$$

The **C**-term intensity is expressed by Equation (7.5),[4–9] whereby the intensity requires the presence of a degenerate ground state. Figure 7.3 demonstrates the different possibilities that can result in **C**-term intensity in an MCD spectrum. The **C**-term intensity is inversely proportional to temperature. Therefore, at low temperatures the MCD spectra of paramagnetic protein active sites or

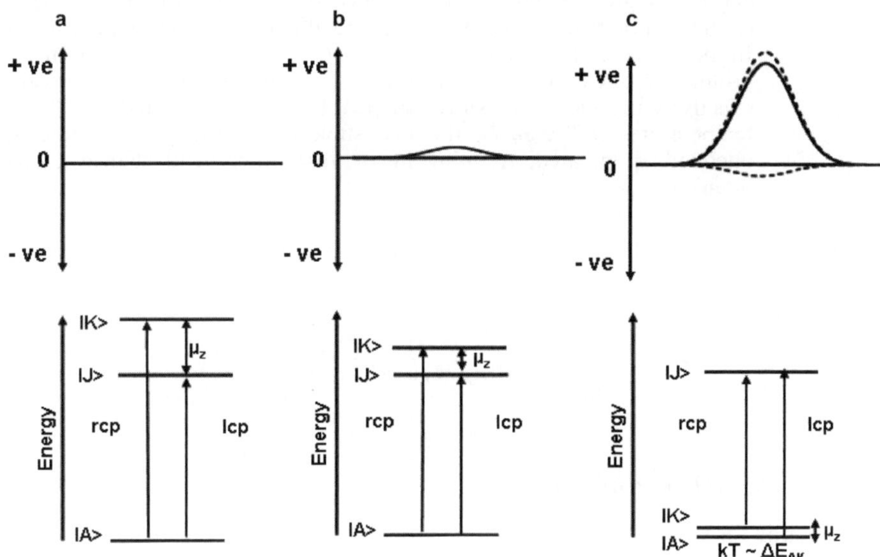

Figure 7.2   The amount of mixing of the **K** state with other states for several possibilities: (a) where the state **K** is much higher in energy than both the ground state **A** and excited states **J**, which results in no mixing and practically no **B**-term intensity; (b) where the state **K** is close to state **J**, which results in more mixing and a (temperature independent) **B**-term MCD signal; (c) the state **K** is close in energy to **A** ($\Delta E \approx kT$), which results in mixing between the **K** and **A** states and a substantial population of **K** with temperature, which gives a large temperature dependent **B**-term intensity.

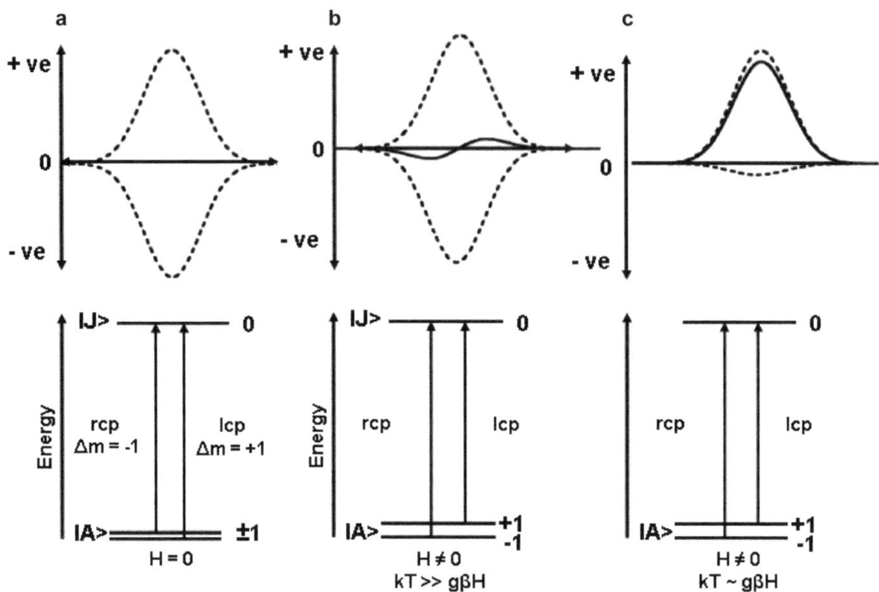

Figure 7.3   (a) In the absence of a magnetic field, the transitions from a doubly degenerate ground state **A** to an excited state **J** caused by right and left circularly polarized light are canceled, resulting in no **C**-term intensity. (b) In the presence of a magnetic field and $kT >> g\beta H$, the doubly degenerate ground state **A** becomes nondegenerate, resulting in two transitions that effectively produce a derivative shaped **A**-term MCD signal. (c) At low temperatures $(kT \approx g\beta H)$, the two sublevels of the ground state are unequally populated, resulting in an absorption band shaped **C**-term intensity.

model complexes are dominated by the **C**-term intensity, unlike the **A**- and **B**-terms:

$$\mathbf{C}_0 = -\frac{1}{d_A} \sum \langle A | \mu_Z | A \rangle \left(\left[M_{-}^{AJ} \right]^2 - \left[M_{+}^{AJ} \right]^2 \right) \tag{7.5}$$

7.1.1.2   MCD Methodology

A high-spin ferrous (Fe^{II}, d^6) free ion with four unpaired electrons has a ^{5}D ground state with degenerate d orbitals. The 3d orbitals split in energy depending on the ligand environment around the Fe^{II} center. An octahedral ligand field splits the ^{5}D into a triply degenerate ^{5}T$_{2g}$ ground state and a doubly degenerate ^{5}E$_g$ excited state. The ^{5}T$_{2g}$ state consists of the atomic 3d$_{xy}$, 3d$_{yz}$ and 3d$_{xz}$ orbitals and is 10Dq(O_h) (Dq is the cubic field splitting parameter) lower in energy than the ^{5}E$_g$ excited state that contains the 3d$_{x2-y2}$ and 3d$_{z2}$ orbitals. For biologically relevant oxygen and nitrogen ligands, this 10Dq value is approximately 10 000 cm^{-1}. There is only one spin allowed transition from the ^{5}T$_{2g}$

ground state to the ^5E_g excited state. Since this transition corresponds to the $(t_{2g})^4(e_g)^2 \rightarrow (t_{2g})^3(e_g)^3$ one-electron excitation, the splitting of the ^5E_g excited state reflects the separation of the e_g orbitals, which is sensitive to the geometry of the site. The orbital degeneracy of the ground and excited states are lost in case of low symmetry environment of protein active sites. The $^5T_{2g}$ and ^5E_g states further split in energy, depending on the coordination number and ligand geometry around the Fe^{II} center.[10–13] For a six-coordinate distorted octahedral ligand field, the doubly degenerate ^5E_g state splits up by <2000 cm^{-1} into non-degenerate d orbitals (Figure 7.4). This results in two LF transitions from the $^5T_{2g}$ ground state to the two excited state E_g orbitals centered around 10 000 cm^{-1}, and split by <2000 cm^{-1}.[9] The removal of an axial ligand produces a five-coordinate square pyramidal geometry. This causes a higher splitting of the ^5E_g state by ~5000 cm^{-1}, resulting in two LF transitions from the $^5T_{2g}$ state at $\sim10\,000$ and 5000 cm^{-1}. Distortion of the five-coordinate square pyramidal structure to a five-coordinate trigonal bipyramidal geometry causes a change in the LF environment, resulting in two lower energy transitions at $<10\,000$ and <5000 cm^{-1}.[9] Removal of another ligand gives rise to a four-coordinate distorted tetrahedral geometry. Since $10Dq(T_d) \approx -4/9\ 10Dq(O_h)$, the LF transitions are lower in energy and are observed at ~5000–7000 cm^{-1}.[9]

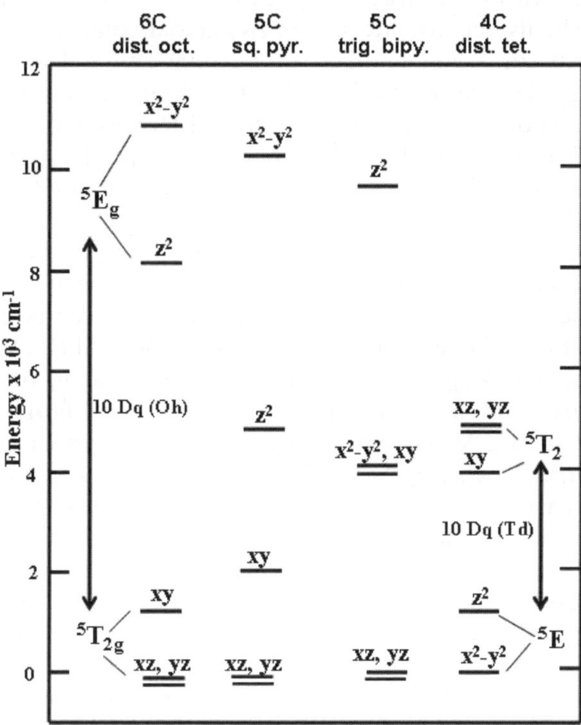

Figure 7.4 Ligand field theory splitting of the d orbitals of a Fe^{II} center with different coordination environments.

As mentioned above, these Laportè forbidden ligand field transitions are weak in intensity and often unobserved in the absorption spectrum. Note that metal sites in proteins usually have low symmetry, and hence lack the inversion symmetry. Thus, the d→d transitions become weakly allowed through mixing with higher energy electric dipole-allowed charge transfer excited states. Moreover, they usually appear in the near-IR region (5000–12 000 cm^{-1}) and are obscured by vibrations and overtones from the protein and buffered solutions. The $^5T_{2g}$ ground state splittings are smaller than the LF transitions and cannot be probed by the more conventional spectroscopic techniques such as EPR. This is because high-spin Fe^{II} is a non-Kramers ion (integer spin, $S = 2$), where the lowest spin doublet is split by δ due to zero field splitting (ZFS, *vide infra*).[9] This splitting (δ) is usually larger than the microwave frequency used in EPR, making these high-spin ferrous centers EPR inactive despite being paramagnetic. MCD spectroscopy has been applied on paramagnetic non-heme ferrous systems, which due to the **C**-term behavior (*vide infra*) show enhanced intensities at low temperatures, making this technique unique to study non-heme ferrous active sites.

Low temperature MCD data obtained for a series of 25 structurally defined non-heme ferrous model complexes show transitions that are consistent with the ligand field predictions (Figure 7.5).[12] The distorted six-coordinate octahedral sites show two LF transitions split by <2000 cm^{-1}, centered at around 10 000 cm^{-1}. The five-coordinate square pyramidal complexes exhibit a transition at $\sim$5000 cm^{-1} and another at >10 000 cm^{-1}. The five-coordinate trigonal bipyramidal complexes show two transitions at <10 000 and <5000 cm^{-1}. The four-coordinate distorted tetrahedral ferrous complexes show only low energy LF transitions at $\sim$6000 cm^{-1} in the MCD spectra. Thus, the energies of the ^5E excited state stores information about the coordination environment of the ferrous active site.

The $^5T_{2g}$ splitting of the ground state involves the d$_\pi$ orbitals and provides information about the π-bonding description of the ligand with the metal orbitals.[13] The energy splittings of the $^5T_{2g}$ ground state can be experimentally obtained from the variable temperature and variable field (VTVH) dependent behavior of the MCD signal.[4,10,14,15] For an $S = \frac{1}{2}$ system, the MCD signal intensity increases with increasing magnetic field, Figure 7.6(a), and decreasing temperature (**C**-term). When the intensity of such a system is plotted as a function of $\beta H/2kT$ (where β is the Bohr magneton and k is the Boltzmann constant) at a particular wavelength, it increases rapidly with increasing field and decreasing temperature and then saturates, giving rise to a saturation magnetization curve.[16] This saturation behavior can be explained by the schematic representation shown in the inset of Figure 7.6(b). In the presence of a magnetic field, the $S = \frac{1}{2}$ ground and excited states give Zeeman splitting proportional to $g\beta H$. According to the MCD selection rule, for the left circularly polarized light, $\Delta M = +1$, which corresponds for the transition from the $-\frac{1}{2}$ ground state to the $+\frac{1}{2}$ excited state. For the right circularly polarized light, the MCD selection rule gives $\Delta M = -1$, corresponding to the transition from the $+\frac{1}{2}$ sublevel of the ground state to the $-\frac{1}{2}$ excited state. At high

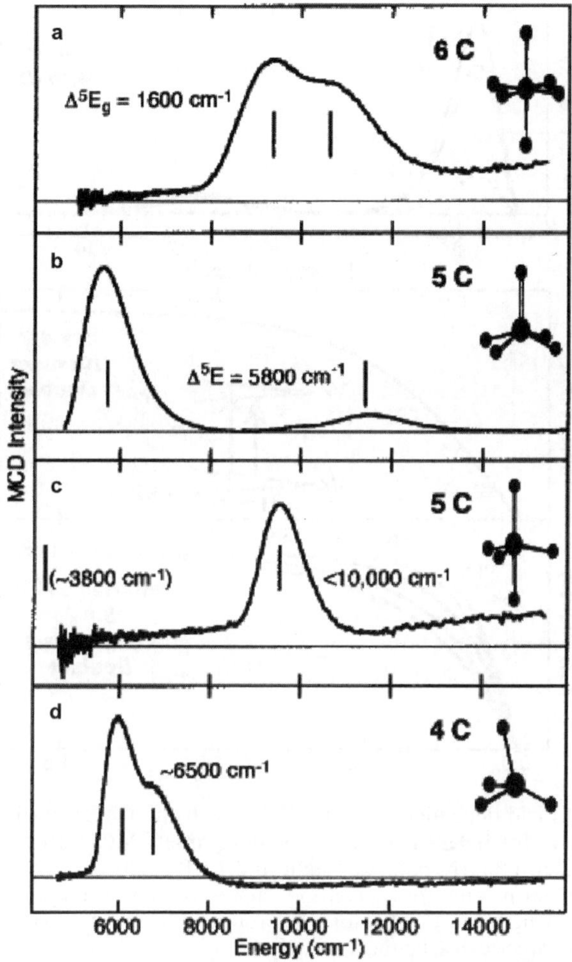

Figure 7.5 Low-temperature MCD spectra of structurally characterized model complexes of different geometries and coordination numbers: (a) [Fe(H$_2$O)$_6$](SiF$_6$), (b) [Fe{HB(3,5-iPr$_2$py)$_3$}(OAc)], (c) [Fe{tris(2-dimethyl-aminoethyl)amine}Br]$^+$ and (d) [Fe(HB(3,5-iPr$_2$pz)$_3$)(Cl)]. Note, HB(3,5-iPr$_2$pz)$_3$ is hydro-tris(3,5-di-iso-propyl-1-pyrazolyl)borate. (Reproduced from ref. 9 with permission from the American Chemical Society.)

temperature and low magnetic field, the two sublevels of the ground state will be equally populated leading to cancellation of these transitions, resulting in a weak MCD signal. However, at lower temperatures and higher magnetic fields, the lower sublevel of the ground state has a Boltzmann population, which prevents the signal from canceling out, and generates a C-term MCD signal intensity. Eventually, at very low temperature and high magnetic field (the $M_s = -\frac{1}{2}$ sublevel is fully populated), the signal intensity maximizes and no longer increases upon further lowering of the temperature and leads to

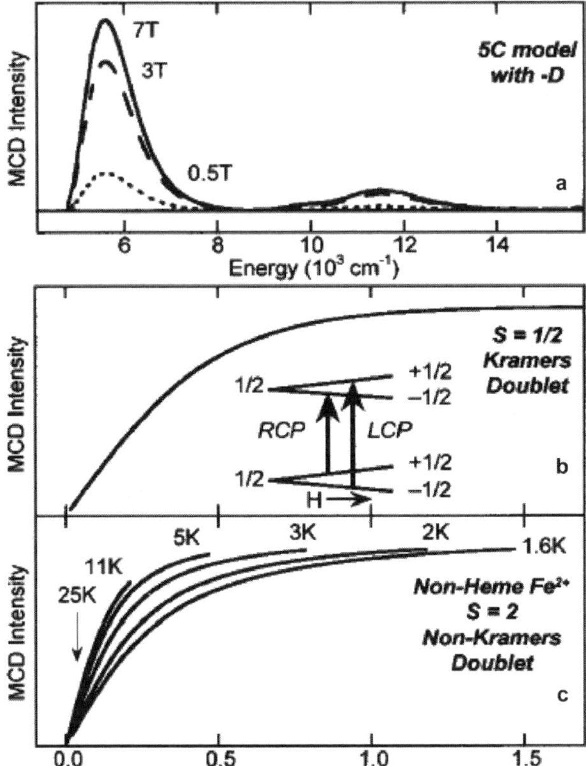

Figure 7.6 (a) Field dependence of an MCD spectrum. (b) Saturation magnetization curve for a Kramers ($S=\frac{1}{2}$) system; inset: MCD selection rule (RCP is right circularly polarized light and LCP is left circularly polarized light). (c) Saturation magnetization curve for a non-Kramers ($S=2$) system showing nesting behavior. (Reproduced from ref. 9 with permission from the American Chemical Society.)

saturation of the signal. This generates MCD saturation magnetization curves and can be expressed by Equations (7.6–7.8), with E being the energy of the incident light and θ and Φ the angles between the incident light and molecular z-axis and the xy-plane, respectively. The orientations of the magnetic field relative to the molecular x-, y- and z-axes are given as l_x, l_y and l_z and $g(x/y/z)$ are the molecular g-values. The products of the relative polarizations of two transitions are given as M_{ij}^{eff} and γ is expressed in Equation (7.7). MCD intensities obtained at different temperatures fall on the same saturation curve for the Kramers (half-integer spin) $S=\frac{1}{2}$ system (Figure 7.6b).[16] However, VTVH MCD on non-heme ferrous active sites yields nesting of the saturation magnetization curves, *i.e.* non-superimposable isotherms (Figure 7.6c).[17] This nesting behavior is reflective of non-Kramers (integer spin) systems.[4,10,18] The shapes of the VTVH isotherms are related to the g-values of the ground state and products of relative polarizations (M_{ij}) for an $S=\frac{1}{2}$ system. In the case

$S > \frac{1}{2}$, fitting of the VTVH data generates g-values, spin Hamiltonian parameters D and E (*vide infra*) and the products of relative polarizations:[9]

$$\frac{\Delta\varepsilon_{av}}{E} = \frac{const}{4\pi} \int\limits_{\theta}\int\limits_{\varphi} \tanh\left(\frac{\gamma\beta H}{2kT}\right)$$
$$\times \frac{\sin\theta}{\gamma}\left(l_x^2 g_x M_{yz}^{eff} + l_y^2 g_y M_{xz}^{eff} + l_z^2 g_z M_{xy}^{eff}\right)d\varphi d\theta \qquad (7.6)$$

$$\gamma = \sqrt{G_x^2 + G_y^2 + G_z^2} \qquad (7.7)$$

$$G_i = l_i g_i \qquad (7.8)$$

with $i = x, y, z$.

The $S = 2$ ground state of a non-heme ferrous site is five-fold degenerate with $M_s = 0, \pm 1, \pm 2$. The ground state sublevels will split even in the absence of a magnetic field in a low symmetry protein environment due to ZFS. ZFS is the cause of lack of EPR activity and nesting of saturation magnetization curves of non-Kramers ions. The zero field energy splitting is given by Equation (7.9):[19]

$$H = D\left(S_z^2 - \tfrac{1}{3}S^2\right) + E\left(S_x^2 - S_y^2\right) \qquad (7.9)$$

In Equation (7.9), the axial and rhombic ZFS parameters are given as D and E, respectively, while S are the spin operators. When the axial ligand field along the z-axis is strong, the value of D is positive (left-hand-side of Figure 7.7) and the $M_s = 0$ sublevel is lowest in energy. In case of a weak axial ligand field, however, D is negative, and the $M_s = \pm 2$ are the lowest in energy (right-hand-side of Figure 7.7). For a non-heme protein environment, the rhombic ZFS parameter $E \neq 0$, resulting in the splitting of the degenerate $M_s = \pm 2$ by a magnitude δ, even in the absence of a magnetic field. This splitting is equal to $g_{\parallel}\beta H$ and increases with the size of the magnetic field and eventually results in a pure $M_s = -2$ state for the lowest energy sublevel, while the $M_s = +2$ becomes the higher energy sublevel. Thus, application of a magnetic field results in

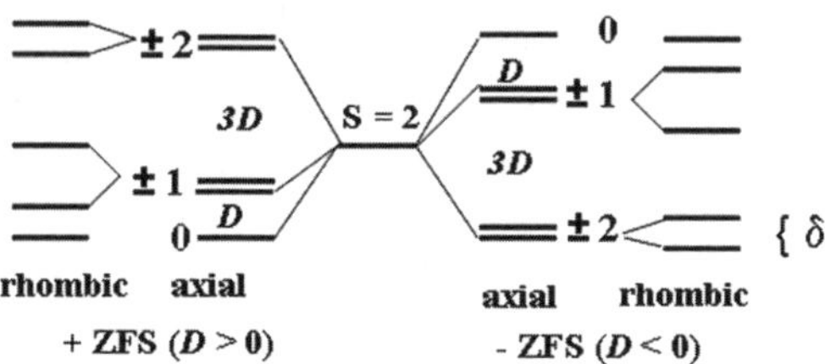

Figure 7.7   Energy splittings of a non-Kramers $S = 2$ state with sublevels for positive (left) and negative (right) ZFS. –ZFS is further elaborated in Figure 7.8.

splitting the doublet and ultimately provides $g_{||}$ values. This splitting occurs only for non-Kramers systems. VTVH MCD, therefore, helps to evaluate the spin Hamiltonian EPR parameters ($g_{||}$, δ) of the ground state of EPR inactive systems.[4]

The spin Hamiltonian parameters can be used to obtain orbital splittings of the $^5T_{2g}$ ground state of a high-spin d^6 ferrous system. The T_{2g} state has three-fold orbital degeneracy, with an extra electron in the d_{xy}, d_{yz} or d_{xz} orbital. This leads to an orbital angular momentum contribution to the ground state, which spin orbit couples (SOC) to the quintet ($S = 2$) spin state. This splits the $^5T_{2g}$ state into 7, 5 and 3 degenerate spin orbit sublevels in an octahedral symmetry, (Figure 7.8). The many-electron SOC parameter is designated as λ, which has a value of approximately 80 cm^{-1} for Fe^{II} complexes. In systems with much lower symmetry, as observed in protein active sites, the axial splitting of the d_{xy} orbital relative to d_{xz} and d_{yz} is denoted by Δ. The rhombic splitting between the d_{xz} and d_{yz} orbitals is described by V. The combined effects of SOC and axial and rhombic d-orbital energy splittings is illustrated on the right-hand-side in Figure 7.8. Thus, the spin Hamiltonian parameters $g_{||}$ and δ obtained experimentally can be used to obtain these energy splittings Δ and V, and hence the energy splittings of the T_{2g} d_π orbitals in a non-heme ferrous active site.[13]

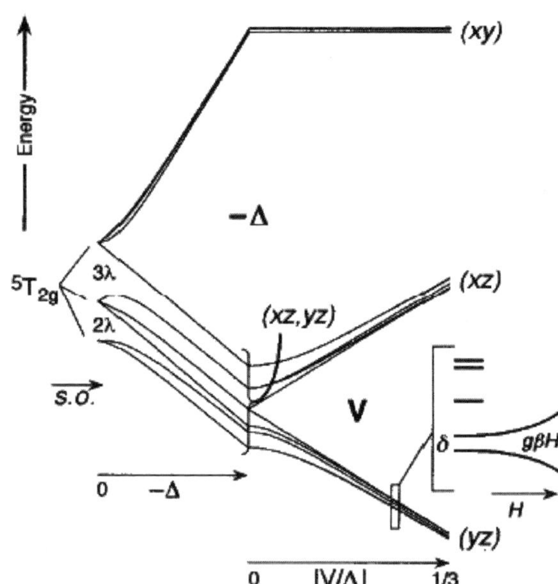

Figure 7.8 Ligand field theory splitting of the $^5T_{2g}$ ground state for a negative ZFS Fe^{II} system. (Reproduced from ref. 9 with permission from the American Chemical Society.)

7.1.2 X-Ray Absorption Spectroscopy and Extended X-Ray Absorption Fine Structure

X-Ray absorption spectroscopy (XAS) and extended X-ray absorption fine structure (EXAFS) are two different techniques resulting from differences in treatments of the same dataset.[20–22] A metal 1s ionization edge can be divided into three regions. The first region is the pre-edge, which represents metal $1s \rightarrow 3d$ transition, while the second region is the rising edge, which represents for the $1s \rightarrow 4p$ transitions. The third region gives the EXAFS oscillations (Figure 7.9).

7.1.2.1 XAS Pre-edge

The $1s \rightarrow 3d$ transition is electric dipole forbidden and hence is weak. However, lack of spherical symmetry in complexes allows mixing of 4p contributions into 3d orbitals. Since the $1s \rightarrow 4p$ transition is electric dipole allowed, even weak mixing of the 4p orbitals with the 3d orbitals causes significant increase in the pre-edge intensity.[23] This was demonstrated in a series of iron complexes in both ferrous and ferric oxidation states where the pre-edge intensity increased

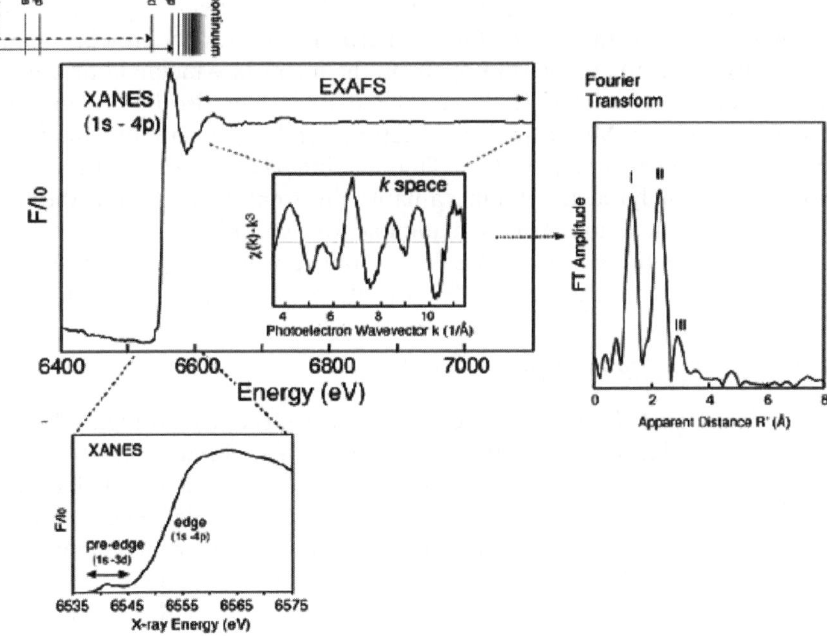

Figure 7.9 Schematic representations of the events in a *K*-edge XAS. Top: the energy level diagram shows the $1s \rightarrow 3d$ pre-edge and $1s \rightarrow 4p$ rising-edge transitions. The fluorescence/incident current (F/I_0) plot shows typical edge data. The inset at the bottom shows the weak pre-edge transitions. The EXAFS oscillations obtained from the post-edge region and its Fourier transform (FT) are also shown. (Reproduced from ref. 22 with permission from Springer Publishers.)

as the symmetry of the metal center was reduced from O_h to C_{4v} and T_d.[23] Large pre-edge intensities were recorded when the symmetry of some ferric complexes was reduced to C_{3v} symmetry.[24] The pre-edge transitions can be analyzed to yield information about the ligand field around the metal. Since the excited state is $1s^1 3d^{n+1}$ it will give rise to several many-electron excited states. These states need to be coupled to the core 1s hole (which has A_g symmetry) to yield the final states. This analysis has been performed for O_h, C_{4v} and T_d symmetry complexes.[23] For Fe^{II} complexes with O_h symmetry three symmetry allowed transitions ($^4T_{1g}$, $^4T_{2g}$ and $^4T_{1g}$) contribute to the pre-edge (Figure 7.10a). The relative energies of these transitions can be estimated from the Tanabe–Sugano matrices using an excited state 10Dq value which is 0.8 times the ground state 10Dq value.[23] For complexes with T_d symmetry, four many-electron states were predicted ($^4A_2$, $^4T_2$, $^4T_1$ and $^4T_1$). However, only the higher energy $^4T_1$ state could be resolved with XAS methods while the others were all under a low energy envelope (Figure 7.10b). Likewise, the allowed transitions for a system with C_{4v} symmetry give the ^4E, $^4B_2$ and $^4A_2$ states (Figure 7.10c). Similar methodology is also established for Fe^{III} complexes with varying ligand fields. Normally the ligand field excited states can be observed in absorption spectroscopy in the UV–visible region, but the presence of intense charge transfer bands in the same region of the d–d bands often complicates assignment of the ligand field excited states. Alternatively, in XAS pre-edge analysis the intensity is due to charge transfer from the metal 1s → metal 3d orbitals and is not complicated by LMCT. These excited states can also be obtained from the C-terms of the MCD spectra (Section 7.1.1.2), where the intensity is dominated by metal centered SOC. Thus the intensity pattern of the pre-edge region can be analyzed to obtain valuable information regarding the ligand field around the metal ion. This is invaluable for Fe^{II} and Fe^{III} complexes, where ligand field transitions are too weak, *i.e.* outside the visible spectral range, suppressed by intense charge transfer transitions or symmetry forbidden.

7.1.2.2 *XAS Rising Edge*

The dominant contribution in the edge region, *i.e.* the second region from the XAS/EXAFS data, is the transition from the 1s → 4p orbitals (Figure 7.9, lower inset), which is called the rising edge and is electric dipole allowed. The energy position of this directly reflects the Z_{eff} on the metal that is being investigated and is frequently used to assign the oxidation state of metal centers in inorganic complexes and metalloenzyme active sites. In general, an increase in oxidation state increases the value of Z_{eff} but lowers the energy of the 1s orbital and thereby shifts this transition to higher energy.[25]

7.1.2.3 *EXAFS*

Beyond the edge where the electron is ionized (Figure 7.9), the behavior is that of a photoelectron with kinetic energy $(E - E_0) = \frac{1}{2} m_e v^2$, where E is the absorbed energy, E_0 the ionization energy and m_e the mass of a free electron. These

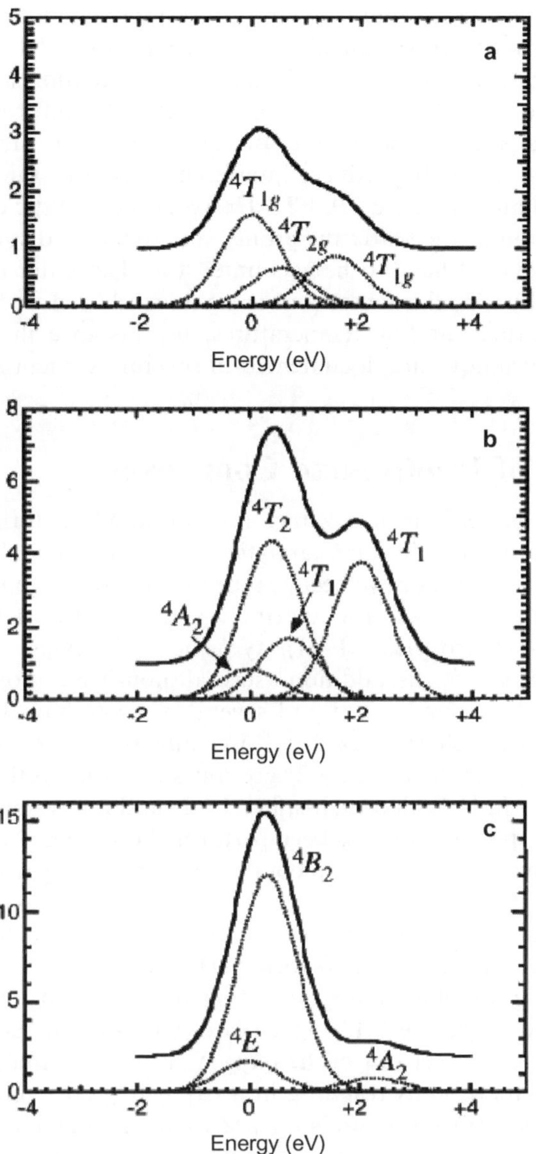

Figure 7.10 Pre-edge region and the excited state contributions to it for ferric complexes having (a) octahedral (O_h), (b) tetrahedral (T_d) and (c) square pyramidal (C_{4v}) ligand fields. (Reproduced from ref. 23 with permission from the American Chemical Society.)

photoelectrons are scattered by neighboring atoms. The EXAFS oscillations occur due to interference between the ejected photoelectron and the back-scattering waves. EXAFS represents the difference between the oscillation component of the absorption coefficient observed beyond the edge between the sample being interrogated and a free atom. In most common representations these oscillations are transformed to k-space (Figure 7.9, inset), where k represents the wave vector given by the De Broglie equation with units of Å^{-1}. When the data of EXAFS spectra are fitted this yields information about the scattering atoms, such as the number of scattering atoms and their distances from the emitted atom. It produces interference patterns typically up to 5 Å from the central atom (Figure 7.9, FT). However, both static disorders (due to spatial distribution of the scattering atoms) and dynamic disorders (as a result of thermal vibrations) have to be accounted for. The latter is also expressed through the Debye–Waller factor and is significantly reduced when measurements are performed at low temperatures, for instance in liquid helium.[25] Details of this technique are documented in several excellent reviews.[22,26–30]

7.2　MCD of Iron(IV)-oxo Complexes

Mononuclear non-heme iron enzymes are proposed to form a high valent iron(IV)-oxo intermediate in their catalytic cycle.[31] So far, only four high-spin iron(IV)-oxo intermediates have been trapped in the α-ketoglutarate-dependent enzymes, namely, the one for taurine/α-ketoglutarate (KG) dioxygenase (TauD), prolyl-4-hydroxylase (P4H), tyrosine hydroxylase (TyrH) and the halogenase CytC3.[32–37] In addition, several non-heme iron(IV)-oxo model complexes have been synthesized and several of those have been structurally and spectroscopically characterized.[38–49] In contrast to the enzyme iron(IV)-oxo species that have a high-spin ($S = 2$) ground state, the synthetic iron(IV)-oxo model complexes have both a high-spin ($S = 2$) and a low-spin ($S = 1$) ground state.[50,51] MCD spectroscopy has been performed on some of these structurally characterized low-spin iron(IV)-oxo complexes. The first synthetic iron(IV)-oxo complex, $[\text{Fe}^{IV}(\text{O})(\text{TMC})(\text{NCCH}_3)]^{2+}$ with TMC = 1,4,8,11-tetramethyl-1,4,8,11-tetraazacyclotetradecane, was synthesized and structurally characterized by Münck, Nam, Que and coworkers (Figure 7.11).[52]

The qualitative bonding description of such a non-heme iron(IV)-oxo complex is presented in Figure 7.12, which is consistent with DFT calculations. There are six valence electrons in the oxygen 2p orbitals and four electrons in the iron 3d atomic orbitals resulting in a $(d_{xy})^2 (d_{xz/yz})^2$ configuration in the ground state. The $\sigma(\text{Fe–O})$ bond is formed by the interaction of the p_z orbital on oxygen with the unoccupied d_{z^2} orbital on iron. The interaction of the oxygen p_x and p_y atomic orbitals with the half-filled Fe $d_{xz/yz}$ pair results in the formation of two half-filled π-bonds. These half-filled $d_{xz/yz}$ π* orbitals are the lowest energy orbitals that are not fully occupied and therefore should principally be involved in electrophilic attack.[53,54]

7.2.1　$[\text{Fe}^{IV}{=}\text{O(TMC)(NCCH}_3)]^{2+}$

The absorption spectrum of the $[\text{Fe}^{IV}{=}\text{O(TMC)(NCCH}_3)]^{2+}$ complex has a weak, broad band at $\sim 12\,000$ cm^{-1}, a weaker band at $\sim 18\,000$ cm^{-1} and an

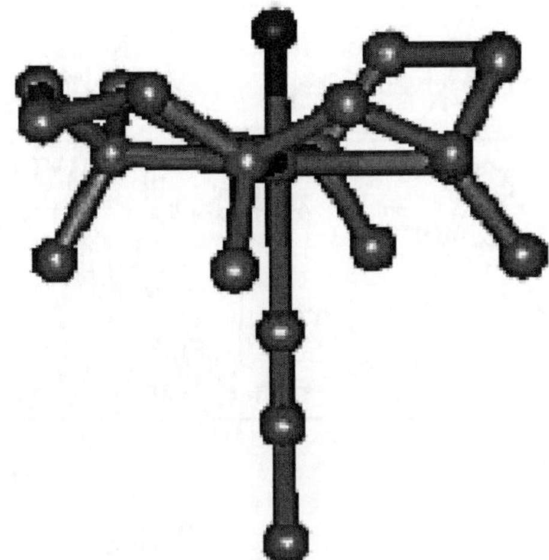

Figure 7.11 Structure of $[Fe^{IV}(O)(TMC)(NCCH_3)](OTf)_2$ with key bond lengths (Å): Fe–O, 1.646(3); Fe–NTMC, 2.067(3), 2.069(3), 2.109(3), 2.117(3); Fe–$N_{acetonitrile}$, 2.058(3). OTf $= CF_3SO_3^-$.

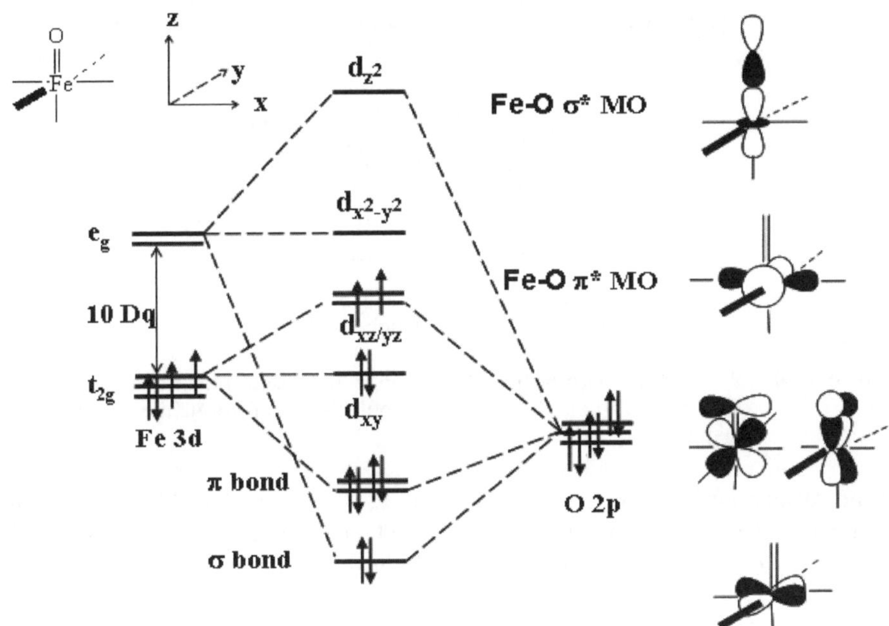

Figure 7.12 Fe–O bonding in $Fe^{IV}=O$, $S=1$ system, where the Fe–O bond is along the z-axis (left), and Fe d-based frontier molecular orbitals (right).

intense band $\sim 24\,000$ cm^{-1} (Figure 7.13a).[55] The MCD spectra of this system are shown in Figure 7.13(b) with vibronic fine structure. The temperature dependent MCD spectra show that there are three bands (I, II, III) under the

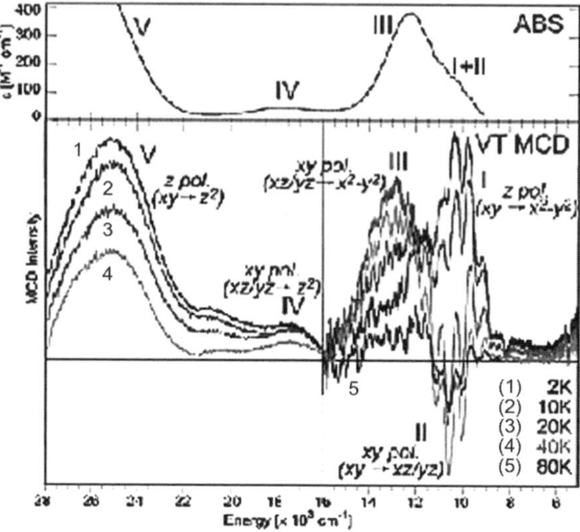

Figure 7.13 $[Fe^{IV}(O)(TMC)(NCCH_3)](OTf)_2$ spectra: (a) absorption spectrum at –40 °C in acetonitrile; (b) mull MCD spectra at 7 T and variable temperatures and the temperature dependence MCD spectra of band V. (Reproduced from ref. 54 with permission from Elsevier Ltd.)

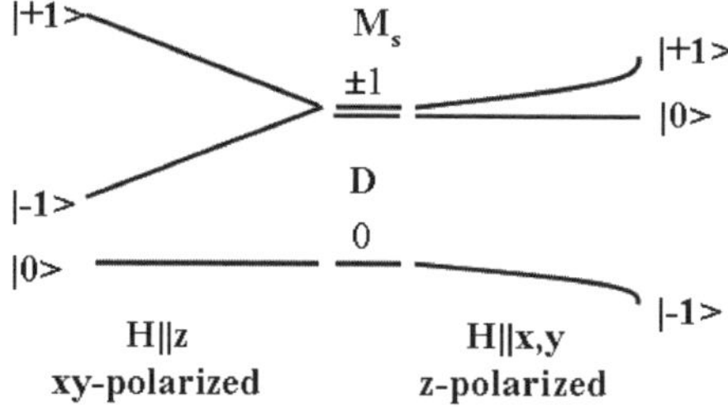

Figure 7.14 $S = 1$ system with positive ZFS: for xy-polarized transitions the field (H) is along the z-axis (left) and for z-polarized transitions the field is in the xy-plane (right).

broad absorption band centered at around 12 000 cm⁻¹. The MCD spectrum obtained at 2 K resolves band I ($\sim$ 9000–12 000 cm⁻¹), band IV ($\sim$ 17 000 cm⁻¹) and band V ($\sim$ 25 000 cm⁻¹). Bands I and V decrease in intensity with increasing temperature, showing normal MCD C-term behavior (Section 7.1.1). However, two new bands (II at $\sim$ 10 500 cm⁻¹ and III at $\sim$ 13 000 cm⁻¹) grow in intensity with increasing temperatures, but decrease again after $\sim$ 20 K.[55,56]

Mössbauer spectroscopy indicates that the $[Fe^{IV}=O(TMC)(NCCH_3)]^{2+}$ complex is an $S = 1$ system with a positive ZFS and $D = 29$ cm⁻¹ between $M_s = 0$ and $M_s = \pm 1$.[52] From temperature dependent MCD spectra one can determine the polarization of the electronic transitions. The xy-polarized transitions require the magnetic field to be along the z-direction (Figure 7.14, left-hand side) for MCD intensity. The lowest energy sublevel split by the Zeeman effect in this case has $M_s = 0$, which does not mix with $M_s = +1$ or $M_s = -1$, hence no MCD intensity can be observed here. However, an increase in temperature leads to the population of the $M_s = \pm 1$ states and consequently gives rise to MCD intensity. With further increase in temperature the MCD signal intensity starts to decrease due to population of the $M_s = +1$ sublevel. For z-polarized transitions, the magnetic field must be aligned along the xy-plane (Figure 7.14, right) to obtain MCD intensity. In this particular case, the $M_s = 0$ state mixes with the $M_s = \pm 1$ states and induces a $M_s = -1$ state as the lowest energy sublevel – hence MCD intensity is detected at low temperatures. At higher temperatures, in contrast, the $M_s = 0$ and $+1$ states are populated and result in a lowering of the MCD intensity. Thus, z-polarized transitions decrease the MCD signal intensity with increasing temperature, while xy-polarized transitions increase the MCD intensity with increasing temperature, followed by decrease in intensity with further increase in temperature.[55,56]

Using this information the polarizations to the bands were assigned. Bands I and V are z-polarized, whereas bands II and III are xy-polarized. Band IV, however, is weak and overlaps with bands III and V, hence the assignment of its polarization is not unambiguous. Group theory predicts the iron(IV)-oxo complex (d^4, $S = 1$) with C_{4v} symmetry to have a $^3A_2$ ground state. There are five spin and electric dipole allowed d $\rightarrow$ d ligand field transitions (Figure 7.15). Therefore, band I can be assigned as the lowest energy z-polarized $d_{xy} \rightarrow d_{x2-y2}$ transition. This transition occurs at a much lower energy than that expected from Figure 7.12, because the electron originates from a fully occupied metal d-orbital and experiences electron–electron repulsion. The electron–electron repulsion is significantly reduced when the excitation is into the empty d_{x2-y2} orbital, which is partially delocalized into equatorial ligands. This is a drawback of the one-electron MO diagram, which does not take the electron–electron repulsion into account. Band II is negative in the MCD spectrum with vibronic progression and is xy-polarized. The lowest energy xy-polarized transition is from $d_{xy} \rightarrow d_{xz/yz}$. Excitation of an electron from the nonbonding d_{xy} orbital into the π^* $d_{xz/yz}$ orbital weakens the Fe–O bond and can be associated with the reactivity of iron(IV)-oxo complexes. The transition energy also reflects the covalency and Fe–O π-bond strength. The fine structure in the MCD spectrum is associated with a change in this bond in the excited state. Band III is the next xy-polarized $d_{xz/yz} \rightarrow d_{x2-y2}$ transition. Although this transition is a ligand field transition, it gives an intense signal in the absorption spectrum, as a result of mixing of a low lying ligand to metal charge transfer (LMCT) excited state from the equatorial nitrogen atoms to the d_{x2-y2} orbital. Band IV can be assigned as the highest energy xy-polarized $d_{xz/yz} \rightarrow d_{z2}$ transition. The intensity of band V is too low to be assigned as the lowest energy LMCT transition ($O_\pi \rightarrow d_{xz/yz}$) and, therefore, is assigned as the highest energy z-polarized $d_{xy} \rightarrow d_{z2}$ transition and is associated with the Fe–O σ-bond strength. The splitting of the e_g set by ~ 4500 cm^{-1}, which is the energy difference between bands III and IV, reflects the strength of the σ-bond, which is stronger than that found for typical Fe–N σ-bonds. The MCD data analysis reveals that the π-bonds are also strong covalent bonds.

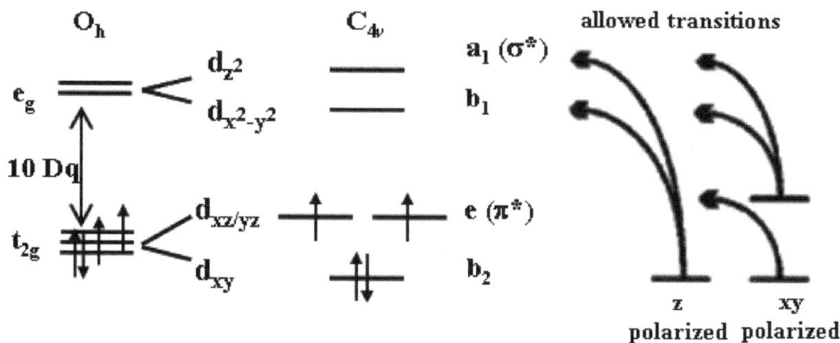

Figure 7.15 Ligand field splitting diagram for Fe^{IV}=O complex (d^4, $S=1$) with C_{4v} symmetry, showing the allowed d→d ligand field transitions.

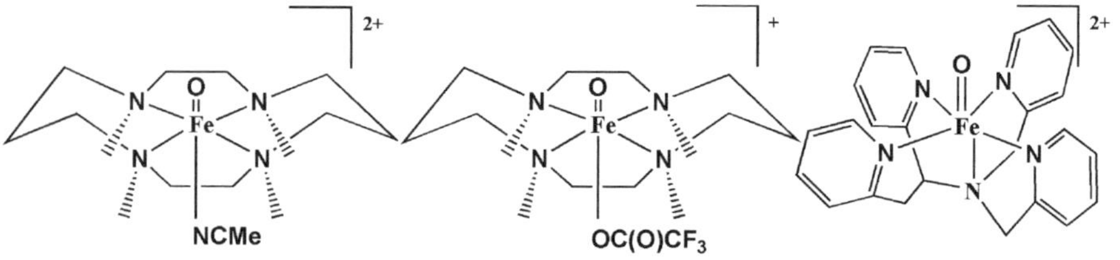

Figure 7.16 Structure of three Fe^{IV}=O complexes: left, [Fe^{IV}(O)(TMC)(NCCH$_3$)]$^{2+}$ (**1**); middle, [Fe^{IV}(O)(TMC)(OC(O)CF$_3$)]$^+$ (**2**); and right, [Fe^{IV}(O)-(N4Py)]$^{2+}$ (**3**).

7.2.2 Iron(IV)-oxo MCD: Varying Axial and Equatorial Ligands

The ligand *trans* to the oxo group is called the axial ligand and experimental studies have been performed that replaced the axial acetonitrile ligand in [Fe^{IV}=O(TMC)(NCCH$_3$)]$^{2+}$ (**1**) with a trifluoroacetate group in [Fe^{IV}=O(TMC)(OC(O)CF$_3$)]$^+$ (**2**) (Figure 7.16).[57] Further studies included replacement of the TMC ligand in **1** with a ligand with pyridine-based groups: [Fe^{IV}=O(N4Py)]$^{2+}$ (**3**) with N4Py = *bis*-(2-pyridylmethyl)-*bis*-(2-pyridyl)-methylamine.[47] Complexes **1**–**3** show very different reactivity patterns. Whereas the TMC complexes (**1** and **2**) are less reactive and can only abstract hydrogen atoms from weak C–H bonds such as dihydroanthracene, the N4Py complex is much more reactive and activates, for instance, the C–H bond of cyclohexane.

Figure 7.17 presents the absorption and MCD spectra of compounds **1**–**3**. Similar to the absorption spectrum of **1** (Figures 7.13 and 7.17), the absorption spectrum of **2** gives a broad band at $\sim$12 000 cm^{-1} followed by a sharp increase in intensity at $\sim$25 000 cm^{-1}. In addition, the spectrum displays a shoulder at $\sim$11 000 cm^{-1} with fine structure. The intensity of the band at $\sim$12 000 cm^{-1} is lower in **2** as compared to **1**. The absorption spectrum of **3**, on the other hand, is similar to that found for **1** and **2**, although the band at $\sim$11 000 cm^{-1} is blue-shifted to $\sim$14 400 cm^{-1}.[56]

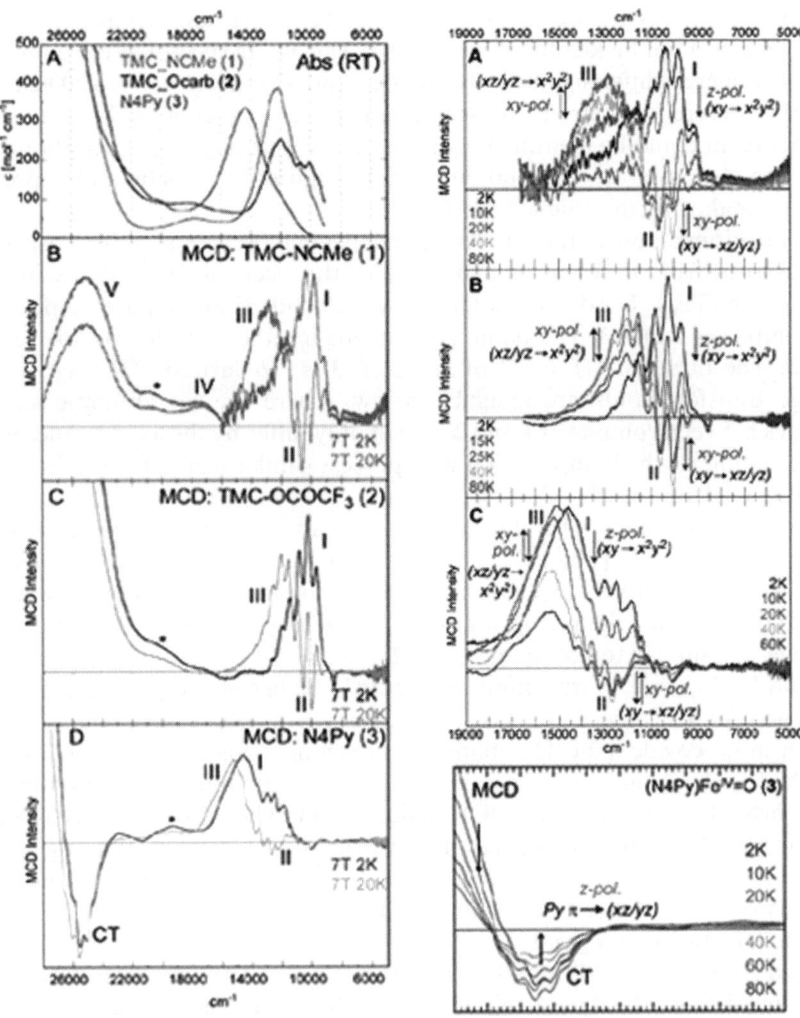

Figure 7.17 Left-hand panel: overlay of absorption spectra of **1–3** (a), 7 T MCD spectra at 2 K and 20 K for **1** (b), **2** (c) and **3** (d). Right-hand panel: variable temperature MCD spectra of **1** (a), **2** (b) **3** (c) and **3** at higher energy (d). (Reproduced from ref. 56 with permission from the American Chemical Society.)

The MCD spectra of the complexes are temperature dependent (Figure 7.17). Complex **2**, much like complex **1**, has three bands (I, II and III) under the broad absorption feature in the near-IR region. Complexes **1** and **2** have absorption bands at similar energies, whereby bands II and III are shifted by ∼ 500 and 900 cm^{-1}, respectively. In contrast, complex **3** has absorption bands blue-shifted relative to the TMC complexes, and band I appears at a higher energy than band II and results in different shapes of the MCD spectrum. Unlike complex **1**, band IV could not be identified for **2** and **3**. The MCD spectra of the complexes at higher energy differ significantly. While band V appears at $\sim 25\,000$ cm^{-1} in complex **1** there is a sharp increase in intensity in this region for **2**, although no distinct band could be identified. Complex **3** shows a negative absorption band with fine structure at $\sim 25\,600$ cm^{-1} (CT, Figure 7.17d, left) and a positive MCD signal at further higher energies.

As mentioned above, the temperature dependent behavior of the absorption bands arises from different polarizations of the electronic transitions. In analogy to complex **1**, band I is z-polarized while bands II and III are xy-polarized for complexes **2** and **3**. The band assignments are the same for complexes **1**, **2** and **3**. The negative CT band of complex **3** is z-polarized. The high energy charge transfer bands are assigned below. There are no dramatic spectral changes between complexes **1** and **2**, which only differ by the axial ligand bound to the complex, which implies that they possess similar ligand field splitting and bonding descriptions.[56]

Band I is associated with the $d_{xy} \rightarrow d_{x2-y2}$ transition and is similar in energy for complex **1** and **2** (Table 7.1). This transition involves the excitation of a non-bonding electron from a d_{xy} orbital to the antibonding σ^* orbital between the iron and the equatorial nitrogen atoms, *i.e.* the d_{x2-y2} orbital. These transition energies are similar for complex **1** and **2**.[56]

Band II ($d_{xy} \rightarrow d_{xz/yz}$ transition) is 500 cm^{-1} higher in energy in complex **2** as compared to **1** (Table 7.1). This higher energy transition might be associated with a more covalent Fe–O π bond, *i.e.* a stronger Fe–O π^* interaction, in **2** (Table 7.1). It could also be because of enhanced π^* interactions between the $d_{xz/yz}$ orbitals with the $^-$OC(O)CF$_3$ axial ligand. The vibronic analysis shows similar Fe–O π-bonding interaction between the complexes, thus the higher energetic transitions in **2** reflects greater π^* interactions between the $d_{xz/yz}$ orbitals with the $^-$OC(O)CF$_3$ axial ligand. Note that complex **1** has NCCH$_3$, a σ-donor, as the axial ligand.

Band III is red-shifted by ~ 900 cm^{-1} for complex **2** and involves the $d_{xz/yz} \rightarrow d_{x2-y2}$ transition. The $d_{xz/yz}$ orbital is destabilized in **2**, while the d_{x2-y2} orbital is

Table 7.1   Transition energies determined from absorption and variable temperature MCD of complexes **1–3**; z-polarized transitions are in bold.

		1	*2*	*3*	*Assignment*
Absorption		$\sim 12\,100$	$\sim 12\,000$ (shoulder $\sim 11\,000$)	$\sim 14\,400$	
MCD	Band I	$\sim 10\,400$	$\sim 10\,300$	$\sim 14\,000$	$\mathbf{d_{xy} \rightarrow d_{x2-y2}}$
	Band II	$\sim 10\,600$	$\sim 11\,100$	$\sim 12\,600$	$d_{xy} \rightarrow d_{xz/yz}$
	Band III	$\sim 12\,900$	$\sim 12\,000$	$\sim 15\,100$	$d_{xz/yz} \rightarrow d_{x2-y2}$
	Band IV	$\sim 17\,600$	–	–	$d_{xz/yz} \rightarrow d_{z2}$
	Band V	$\sim 24\,900$	–	–	$\mathbf{d_{xy} \rightarrow d_{z2}}$
	CT	–	–	$\sim 25\,600$	

stabilized relative to **1**. This is also consistent with a red-shift of band I and a blue-shift of band II in complex **2** relative to **1**.

Bands IV and V are observed in complex **1** because of ferric impurities. An intense carboxylate to iron LMCT (assigned from DFT calculations)[56] is observed in complex **2**.

The spectral changes between complexes **3** and **1** are more dramatic, because **3** has a pentadentate N4Py ligand. The ligand field transitions are blue-shifted in **3** relative to **1**.[56] Band I represents the $d_{xy} \rightarrow d_{x2-y2}$ transition and is ~ 3600 cm^{-1} higher in energy in complex **3** than in **1**, indicating stronger equatorial ligands. The pyridine groups of the N4Py system, in fact, are stronger σ-donors than the tertiary amine groups of the TMC ligand set; therefore, the $d_{xy} \rightarrow d_{x2-y2}$ transition is significantly blue-shifted in **3** relative to **1**, so that it no longer is the lowest energy transition.

Band II is blue-shifted by ~ 2000 cm^{-1} in **3** relative to **1**, probably as a result of destabilization of the $d_{xz/yz}$ pair of orbitals due to stronger Fe–O π-bonding that leads to stronger Fe–O π^* interactions. The destabilization of the acceptor orbitals could also have arisen from enhanced π^* interactions of the Fe–O orbitals with the pyridine ligands in the equatorial plane. Note that the tertiary amine groups of the N4Py ligand system are σ-donors and, therefore, cannot undergo π interactions with the $d_{xz/yz}$ orbitals. From the vibronic structure analysis, it follows that complex **3** has stronger Fe–O π bonding than **1**. Thus, the former reason accounts for the high energy band II.

The $d_{xz/yz} \rightarrow d_{x2-y2}$ transition (band III) is blue-shifted by ~ 2200 cm^{-1} in **3**, which indicates that the d_{x2-y2} orbital is destabilized more than the $d_{xz/yz}$ orbitals in complex **3** as compared to **1**.

The high-energy region of the MCD spectra of complexes **1** and **3** show significant differences. Band V appears as a positive peak at $\sim 24\,900$ cm^{-1} and is assigned as the z-polarized $d_{xy} \rightarrow d_{x2-y2}$ transition for complex **1**. The band, however, is not observed in complex **3**, but instead a negative z-polarized CT transition is observed at $\sim 25\,600$ cm^{-1}. This band shows fine structure similar to band II and has been assigned as a pyridine $\pi \rightarrow d_{xz/yz}$ CT transition. It has a iron-oxo stretch vibration ν_{es}(Fe–O) at about 500 cm^{-1}, which implies weakening of the Fe–O π bond in the excited state. The X-ray structure of complex **3** shows that the pyridine rings of the N4Py ligand are tilted: two with angles of around 24° and the other two with angles of approximately 12°. The pyridine group tilting makes it possible for the pyridine π orbitals to mix with the iron $d_{xz/yz}$ orbitals. Band V shows similarity to band II, due to excitation of an electron from the non-bonding pyridine π orbital to the $d_{xz/yz}$ Fe–O π^* orbitals that results in vibronic structure in this LMCT transition.

7.2.3 Vibronic Progression in MCD

A very interesting feature in most known $S = 1$ iron(IV)-oxo complexes is the presence of vibronic fine structure in the $d_{xy} \rightarrow d_{xz/yz}$ transition (band II), which requires promotion of an electron from the nonbonding d_{xy} orbital into one of the antibonding $d_{xz/yz}$ Fe–O π^* orbitals. An additional electron in an Fe–O π^* orbital

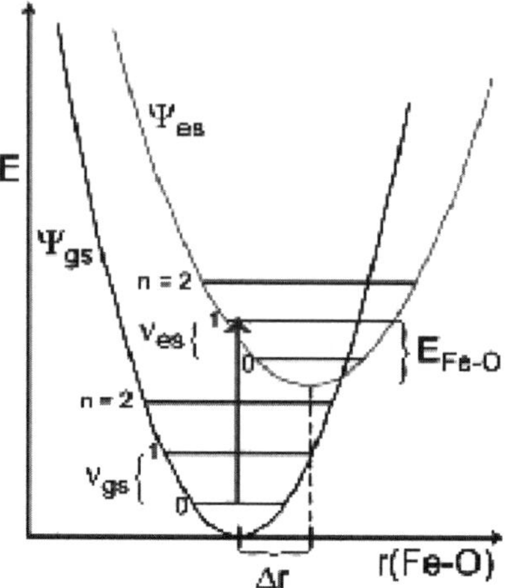

Figure 7.18   Vibronic transitions between ground state (ψ_{gs}) and distorted excited state (ψ_{es}). (Reproduced from ref. 54 with permission from Elsevier Ltd.)

weakens the Fe–O π bond and leads to a distortion of the Fe–O bond length in the excited state, as described by Equation (7.10) and shown in Figure 7.18. This also leads to a decrease in the iron-oxo force constant (k) in the excited state (v_{es}) relative to that of the ground state (v_{gs}). The distortion of the excited state geometry relative to that of the ground state (Δr) lowers the energy of the excited state (ψ_{es}) relative to that of the ground state (ψ_{gs}) by a value $\Delta E(\text{Fe–O}) = \frac{1}{2}k\Delta r^2$, Equation (7.11). The vibronic structure of band II has been deconvoluted and plotted as a positive progression in Figure 7.19. The Gaussian fits of bands I and III were subtracted from the MCD spectra to isolate band II:

$$\Delta r(\text{Fe–O}) = r_{es}(\text{Fe–O}) - r_{gs}(\text{Fe–O}) \tag{7.10}$$

$$E_{\text{Fe–O}} = \tfrac{1}{2}k\Delta r^2 \tag{7.11}$$

The average spacing of the progression reflects the Fe–O stretching frequency in the excited state, $v_{es}(\text{Fe–O})$. The intensities of the Gaussian contributions to the vibronic progression were determined from the spectra to obtain the Huang–Rhys ($S_{\text{Fe–O}}$) factor, which gives the band-shape from Equation (7.12). The Franck–Condon energy $E_{\text{Fe–O}}$ is obtained from Equation (7.13), whereas the distortion in the excited state $\Delta r(\text{Fe–O})$ is determined from Equation (7.10). Table 7.2 presents the ground and excited state parameters for the Franck–Condon analysis of complexes **1** and **3**:

$$\frac{I_{0\rightarrow n}}{I_{0\rightarrow 0}} = \frac{(S_{\text{Fe–O}})_n}{n!} \tag{7.12}$$

$$S_{\text{Fe–O}} = \frac{E_{\text{Fe–O}}}{\nu_{es}(\text{Fe–O})} \tag{7.13}$$

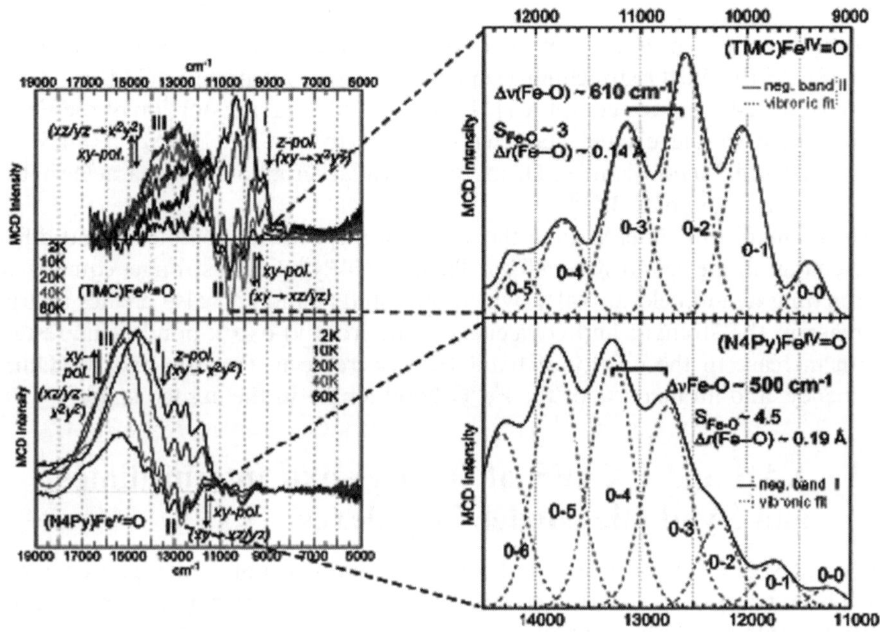

Figure 7.19 Gaussian contributions to the vibronic progression of band II of complexes **1** and **3**. The sign of band II has been reversed. (Reproduced from ref. 54 with permission from Elsevier Ltd.)

Table 7.2 Ground state and excited state parameters for complexes **1** and **3**.

Ground state	*1*	*3*	*Excited state (Fe–O π*)*	*1*	*3*
v_{gs}(Fe–O) (cm^{-1})	834	814	v_{es}(Fe–O) (cm^{-1})	610	·500
v_{gs}(Fe–O) (Å)	1.65	1.64	S_{Fe-O}	3.0	4.5
			E_{Fe-O} (cm^{-1})	1810	2259
			Δr(Fe–O) (Å)	0.14	0.19

The energy spacings of the vibronic progression are 610 and 500 cm^{-1} for complexes **1** and **3**, respectively, and reflect the Fe–O stretch vibration in the excited state. These Fe–O stretch vibrations decrease relative to those of the ground state; for example, they are 834 cm^{-1} for complex **1** and 814 cm^{-1} for **3** (Table 7.2). This is consistent with the observed elongation of the Fe–O bond lengths by 0.14 and 0.19 Å in complexes **1** and **3** respectively. The observed decrease in the Fe–O stretching frequency and increase of the corresponding bond length is a consequence of an excitation of an electron from the $(d_{xy})^2(d_{xz/yz})^2$ ground state to the $(d_{xy})^1(d_{xz/yz})^3$ excited state, which results in the weakening of the Fe–O π bond in the excited state. The bond order for the excited state is reduced by one half due to the promotion of an electron from a nonbonding orbital to a Fe–O π* orbital. As can be seen from Table 7.2, the ground state parameters of complexes **1** and **3** are similar; however, there are quantitative differences in the excited state parameters. Complex **3** shows a larger decrease of the Fe–O stretching frequency in the excited state, (314 cm^{-1} for complex **3**, 224 cm^{-1} for complex **1**). This leads to a larger distortion of the Fe–O bond length in the excited state with respect to complex **1** ($S_{\text{Fe–O}} = 4.5$ for complex **3** and 3.0 for complex **1**), and implies a larger π bonding character in the Fe–O bond in complex **3**.[56,58]

In summary, the differences in the absorption spectra of complexes that differ by a change in the ligand system from TMC to N4Py reflect a stronger equatorial, σ-donating ligand field in N4Py, which blue-shifts band I. It also creates a large increase in the strength and covalency of the iron(IV)-oxo π-bond. This results in an increase in the energy of band II, a decrease in the iron-oxo stretching frequency, and an increase of the Fe–O bond length in the excited state.

7.3 XAS and EXAFS of Iron(IV)-oxo Intermediates and Synthetic Model Complexes

In recent years, several iron(IV)-oxo intermediates of non-heme iron enzymes have been trapped and characterized – namely, those of the enzymes taurine α-ketoglutarate dioxygenase (TauD) and prolyl-4-hydroxylase (P4H), which both belong to the α-ketoglutarate dependent dioxygenases as well as for the α-ketoglutarate dependent halogenase CytC3. Of these the ferryl intermediates of TauD, CytC3 and SyrB2 have been characterized using EXAFS, as will be described in the following section.

7.3.1 Enzymatic Catalytic Cycle Intermediates

7.3.1.1 *Taurine α-Ketoglutarate Dioxygenase*

Taurine α-ketoglutarate dependent dioxygenase (TauD) belongs to the broad class of non-heme iron enzymes that use α-ketoglutarate as a cofactor (see Chapters 1 and 3 for further details). Taurine is produced *via* the cysteine sulfinic acid pathway and is a major component of the bile. It is converted into sulfite and aminoacetaldehyde by TauD (Scheme 7.3). It is expressed from *E. coli* and exists in solution as a homodimer with a molecular weight of 81 kDa.[59] The native protein has a K_{M} of 55 μM for taurine and a V_{max} of 4.1 units per mg of protein. The enzyme has also been shown to be able to oxidize other sulfonic acid containing substrates, such as butanesulfonic acid and pentanesulfonic

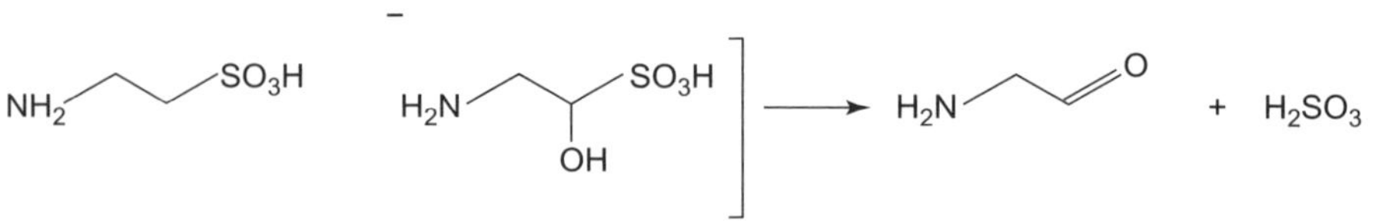

Scheme 7.3   Reaction catalyzed by TauD enzymes.

acid although with much higher K_M and lower V_{max}.[59] The crystal structure of the protein indicated the presence of two histidine amino acids and one carboxylate based ligand in the active site iron forming a facial triad.[60] MCD data on TauD are consistent with the presence of a distorted six-coordinate active site with three water ligands in addition to this facial triad.[61]

Initial stopped-flow studies of the Hausinger group in Michigan indicated formation of an intermediate that is maximized after around 20–25 ms but has a decay rate constant of 42 s^{-1}.[62] The Penn State group of Bollinger reported the presence of a reactive intermediate that abstracts a hydrogen atom from the C^1 position of taurine with a very large primary kinetic isotope effect of $28 < k_H/k_D < 50$.[63,64] This intermediate was characterized with an absorption feature at 318 nm and Mössbauer parameters of $\delta = 0.31$ mm s^{-1} and $\Delta E_Q = -0.88$ mm s^{-1}, which implicated the formation of a high-spin iron(IV) species.[63] Further studies of the Michigan group using resonance Raman spectroscopy indicated oxygen sensitive bands with vibrational shifts in the $^{18}O_2$ spectrum at 821 and 583 cm^{-1}. These spectroscopic studies alluded to the formation of an iron(IV)-oxo species, but structural confirmation was not available.

The Penn State group was successful in generating large quantities (71%) of this fleeting intermediate required for XAS/EXAFS experiments.[33] The XAS data showed that the rising edge was shifted to 7123 eV (Figure 7.20a), which is ~ 2 eV higher in energy than the resting state form of this enzyme and consistent with an increase of the oxidation state of the iron center. The pre-edge transition (Fe 1s $\rightarrow$ Fe 3d) is centered around 7113 eV and shows a remarkable increase in intensity (Figure 7.20a, inset) compared to the iron(II) resting state. This dramatic increase in pre-edge intensity is a hallmark of the ferryl species and is currently thought to originate from the loss of symmetry that allows mixing of the 4p orbitals into the 3d orbitals of iron.[24] Unlike the iron 1s $\rightarrow$ 3d transition, which is electric dipole forbidden, the iron 1s $\rightarrow$ 4p transition is electric dipole allowed and consequently gives two orders of magnitude larger intensity than the former. Thus, a small extent of 4p mixing into a wave function produces a dramatic increase in pre-edge intensity.

The scientific challenge at that stage for the research groups was focused on the detection of the Fe–O vector in the EXAFS data. Although the FT of the EXAFS data did not show obvious indications of the presence of a short Fe–O vector (Figure 7.20b), it was significantly different from the iron(II) resting state (Figure 7.20b, inset), and the fitting could only be satisfactorily achieved by including a very short Fe–O distance of 1.62 Å. The EXAFS analysis indicated the presence of three shells: one very short Fe–O bond of 1.62 Å, four N/O ligand atoms at around 2.06 Å and one N/O atom at 2.42 Å.[33] The short Fe–O bond length is consistent with the data obtained on structurally defined model complexes containing an Fe^{IV}=O unit.

Many computational studies of the active site and catalytic mechanism of TauD have been reported (see Chapters 1, 3 and 4 for further details).[65] The EXAFS characterization of TauD was followed by detailed theoretical calculations[66] that compared several experimental parameters, including molecular geometry and Mössbauer spectra. Several possible models of the TauD active site

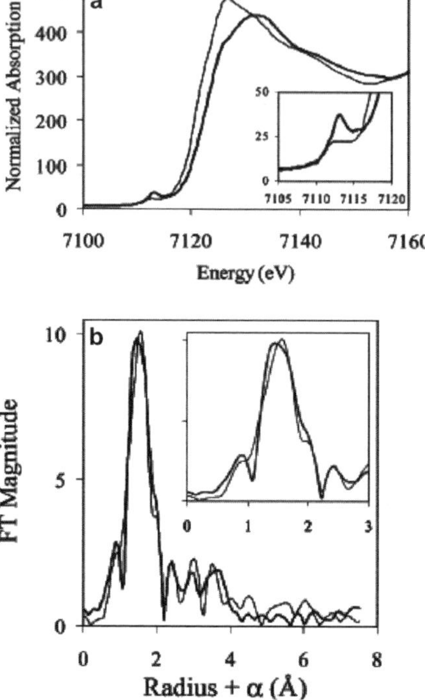

Figure 7.20 (a) Iron *K*-edge XAS of the α-KG and substrate bound iron(II) (thin line) and intermediate complex (bold line). The pre-edge region is magnified in the inset. (b) Fourier transform (FT) of the iron *K*-edge XAS of the α-KG and substrate bound iron(II) (thin line) and intermediate (bold line) complexes. FT of the short-range scattering is magnified in the inset. (Reproduced from ref. 33 with permission from the American Chemical Society.)

were calculated and it was concluded that it occupies a pseudo-octahedral geometry where the aspartate ligand binds as a monodentate ligand while the succinate ligand is a bidentate ligand. Subsequent QM/MM calculations elaborated this model to include hydrogen bonding interactions toward the oxo and carboxylato ligands present in the active site (Figure 7.21).[67] These results clearly indicated a strong hydrogen bonding network mediated by the Arg_{270} residue, which interacts with the sulfonate group of the substrate and anchors it into the active site. The arginine side chain also forms hydrogen bonds with one of the oxygen atoms of the carboxylate group of the succinate co-product and prevents its coordination to the iron(IV)-oxo unit. Thus these calculations further refined the structure of the intermediate to that of a trigonal bipyramid with both the carboxylates binding in a monodentate fashion.[67]

7.3.1.2 *CytC3 Halogenase*

The ansa-bridged cyclotrienin (Figure 7.22) produced by the soil bacteria *Streptomyces* contains a 1-aminocyclopropane-1carboxylate moiety. The

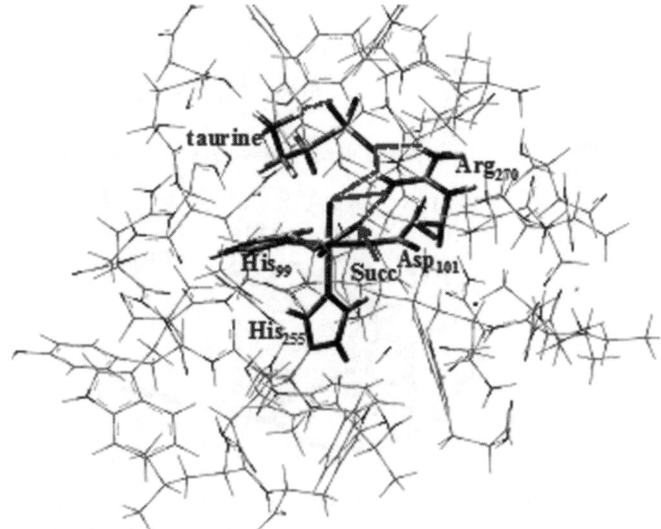

Figure 7.21 QM/MM energy minimized structure of the $Fe^{IV}=O$ intermediate formed in the active site of TauD.

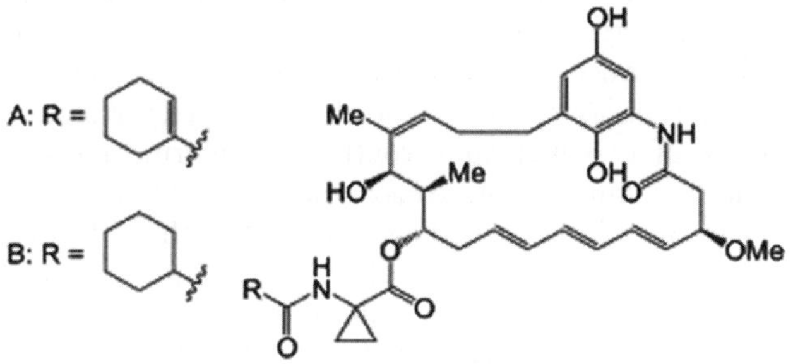

Figure 7.22 Structures of cyclotrienin A and cyclotrienin B.

synthesis of this cyclopropane ring proceeds through dichlorination of an unactivated C–H bond that is catalyzed by the enzyme CytC3.[68] The enzyme catalyzes the chlorination of a wide range of amino acids and contains a non-heme iron center.[68] Experiments using labeled $Na^{36}Cl$ established free Cl^- as the source of chloride.[68] The crystal structure of the CytC3 enzyme established that the non-heme iron center is coordinated by two histidine amino acids but in contrast to the α-ketoglutarate dependent dioxygenases lacks the carboxylate ligand (Figure 7.23).[69] Importantly, no Cl^- ion was found bound to the Fe in this α-KG bound active site.

CytC3 binds Fe^{II}, α-KG and substrate and reacts with O_2 rapidly to produce an intermediate with an absorbance maximum at 318 nm with a formation rate constant of $18\,s^{-1}$ at $5\,°C$.[36] The chlorination reaction proceeds through

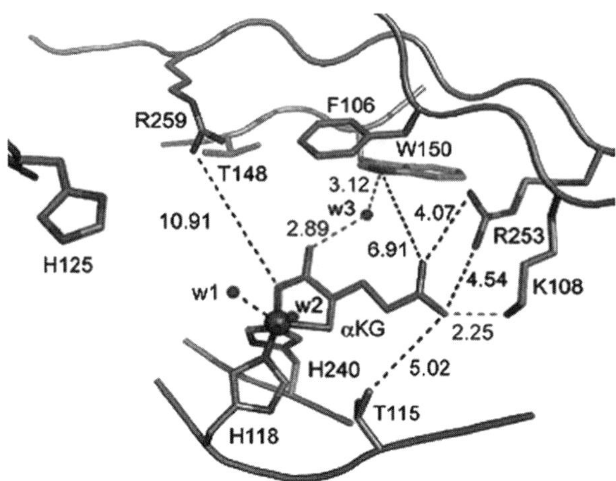

Figure 7.23 Crystal structure of CytC3. The sphere at the center represents the iron. The coordinating His_{118} and His_{240} ligands are indicated. (Reproduced from ref. 67 with permission from the American Chemical Society.)

hydrogen atom abstraction, as indicated by the very large kinetic isotope effect (KIE) associated with this step. In particular, the deuterated substrate decays at least two orders of magnitude slower than the regular substrate. The concentration of this intermediate maximizes after ~ 300 ms under the reaction conditions and has a decay rate constant of 0.2 s^{-1}. The Mössbauer data indicated the presence of two high-spin iron(IV)-oxo intermediates in equilibrium with $\delta = 0.30$ and 0.22 mm s^{-1} and $\Delta E_Q = 1.09$ and 0.70 mm s^{-1}, respectively.

Despite the fact that the intermediate involved in the chlorination reaction catalyzed by the CytC3 could not be characterized using XAS/EXAFS, the intermediate formed during an analogous bromination reaction was trapped and characterized. Mössbauer data on this intermediate, which took 2 s to form, indicated $\sim 84\%$ conversion into an iron(IV)-oxo species. Furthermore, there was evidence of the presence of two iron(IV)-oxo species at slightly different proportions during the chlorination reaction.[36] XAS data on these intermediates gives a 3 eV shift in the rising edge inflection point (Figure 7.24) indicating oxidation of the Fe^{II} center (7120 eV) in the resting state of the enzyme to Fe^{IV} (7123 eV), which is consistent with the Mössbauer data.[37] A similar shift in the rising edge was also observed in the TauD intermediate. The pre-edge region of the data shows an increase in intensity (Figure 7.24, inset) that is characteristic for an iron(IV)-oxo species.

The EXAFS fits to the experimental data were complicated by the presence of at least three species. Within experimental error, a short Fe–O distance of 1.62 Å, two Fe–O distances of 1.86 Å, two Fe–N bonds of 2.07 Å and a Fe–Br bond of 2.43 Å were resolved. The short Fe–O bond of 1.62 Å is consistent with the formation of an iron(IV)-oxo species. The Fe–O distance of 1.86 Å is typical for the interactions of the metal with a carboxylate oxygen atom (such as those

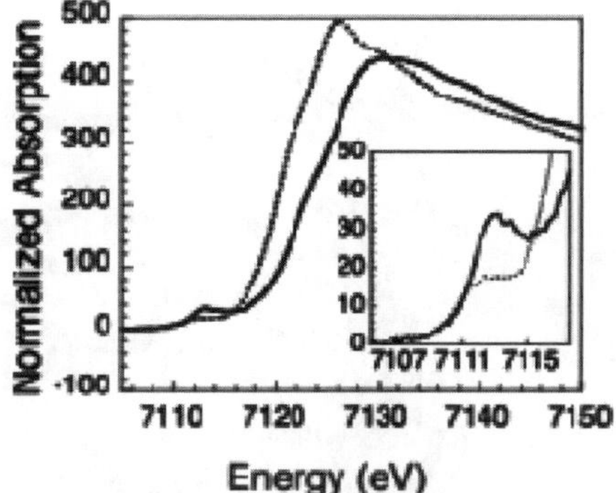

Figure 7.24 XAS data on the α-KG bound resting (dotted) enzyme and the ferryl intermediate (bold). Inset: the pre-edge region. (Reproduced from ref. 37 with permission from the American Chemical Society.)

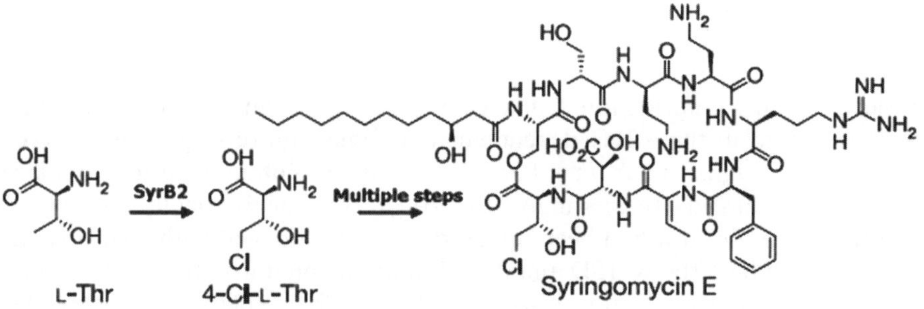

Scheme 7.4 Halogenation of threonine catalyzed by SyrB2.

of succinate), whereas the Fe–N bonds of 2.07 Å reflect two histidine ligated groups to iron. The Fe–Br distance of 2.43 Å provided clear evidence that the Br⁻ is bound to the catalytic intermediate, which is an important observation since halide ions were not observed in the crystal structure of the resting state of the enzyme.[69]

7.3.1.3 *SyrB2 Halogenase*

The phytotoxin syringomycin E (Scheme 7.4) includes a nine-residue containing lipopeptide that is biosynthesized in *Pseudomonas syringae*. The ninth residue is a 4-chloro-L-threonine residue and its chlorination is vital for its activity. The chlorination reaction is catalyzed by the non-heme iron enzyme SyrB2,[70] which is a protein 66 kDa in size with a $K_M = 3.1$ mM for L-threonine and a k_{cat} of 29 min^{-1}.[70] The crystal structure of SyrB2 in the α-KG bound form (Figure 7.25) shows that the iron is coordinated by two histidine residues from the protein backbone, a bidentate bound α-KG group, two water

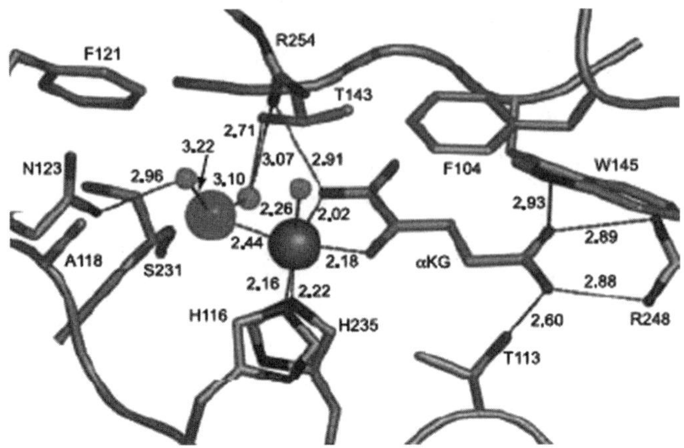

Figure 7.25 Active site structure of the α-KG bound SyrB2. The grey sphere at the center is the Fe^{II}. The density indicated by light grey near the iron is Cl$^-$. (Reproduced from ref. 71 with permission from the Nature Publishing Company.)

molecules and a chloride ion.[71] The geometry is quite different from the facial triad observed in the α-KG dependent non-heme iron dioxygenases, where a carboxylate, originating from either an aspartate or glutamate residue, replaces the chloride ligand. Interestingly, the Ala$_{118}$ residue is ideally located for binding to the Fe^{II} if it had been replaced by an amino acid with a carboxylate side chain. In fact, the A118D and A118E mutants bind iron but do not exhibit halogenation activity.[71] Thus, it seems that nature has replaced the carboxylate side chain containing amino acids with an alanine group to create space for the chloride to bind the iron.[71]

Stopped-flow experiments on the SyrB2 enzyme indicated the formation of an intermediate species with an absorbance at 318 nm and a formation rate constant of 48 s^{-1}.[72] This species was found to have a lifetime of $\sim$100 s at 5 °C, which is the longest lifetime obtained for an enzymatic iron(IV)-oxo species so far. The lifetime can be extended to 1000 s through the substitution of the methyl hydrogen atoms of the substrate by deuterium atoms. The hydrogen abstraction step has a KIE of 94.[72] Substitution of the –CH$_2$CH(OH)– group of serine by cyclopropane actually extends the lifetime of the intermediate to 10 000 s.[72] Thanks to its unprecedented stability the intermediate can be generated at >90% purity and has allowed its characterization using several spectroscopic techniques. Mössbauer spectra on this intermediate shows the presence of two oxidized species with $\delta = 0.30$ and 0.23 mm s^{-1} and $\Delta E_Q = 1.09$ and 0.76 mm s^{-1}, respectively. The results obtained for SyrB2 are qualitatively similar to those obtained for the other non-heme iron halogenase CytC3. However, the ratio of the two oxidized species is 1:1 in SyrB2 and 4:1 in CytC3.

XAS data on the intermediate gives a 1.9 eV shift in the rising edge (Figure 7.26a), which is similar to that observed in TauD and CytC3.[72] There is a

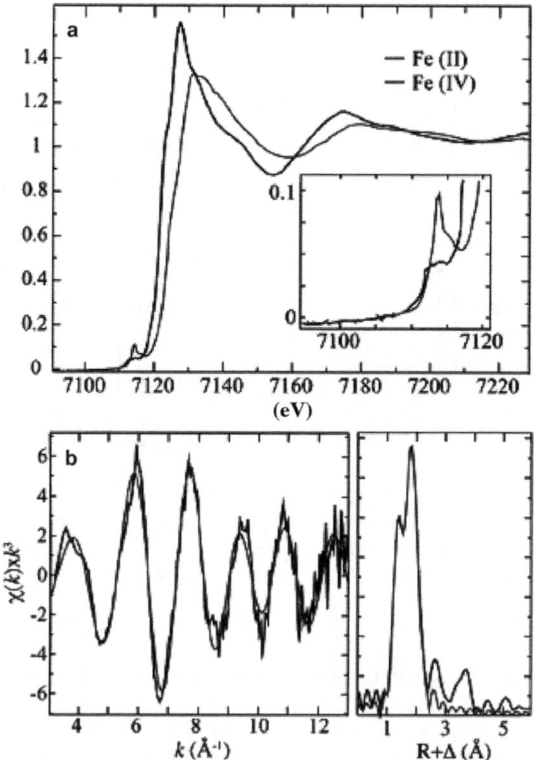

Figure 7.26 (a) XAS data of the SyrB2-αKG and substrate bound Fe^{II} form (darker line) and the Fe^{IV}=O intermediate (lighter line). The pre-edge region is expanded in the inset. (b) EXAFS data and the FT (darker line) and their corresponding fits (lighter). (Reproduced from ref. 72 with permission from the American Chemical Society.)

remarkable increase in pre-edge intensity in SyrB2 (by 2.5 times) as had been observed in other cases. It must be emphasized that the very high purity of the high-valent intermediate (>90%) and large concentrations (>1.8 mM) of the protein allowed unambiguous characterization of these species with XAS. The EXAFS data were fit (Figure 7.26b) with a three shell model. The first shell has a short Fe–O distance at 1.66–1.67 Å, the second shell has four Fe–N/O bonds of 2.12 Å and the third shell has a Fe–Cl distance at 2.31 Å. Notably, the error of fits achieved was larger than those obtained with the dioxygenases.[72]

DFT calculations on the SyrB2 enzyme were reported by Siegbahn and coworkers,[76] and provided an explanation for the two halogenase intermediates detected in the spectroscopic studies. The theoretical model included several second sphere residues responsible for α-KG binding in addition to the active site ligands and the substrate (Figure 7.27). It was proposed that the two different ferryl species resulted from two possible conformations. In one conformation the oxo ligand is *trans* to His$_{235}$ and the chloride ligand is *trans* to the

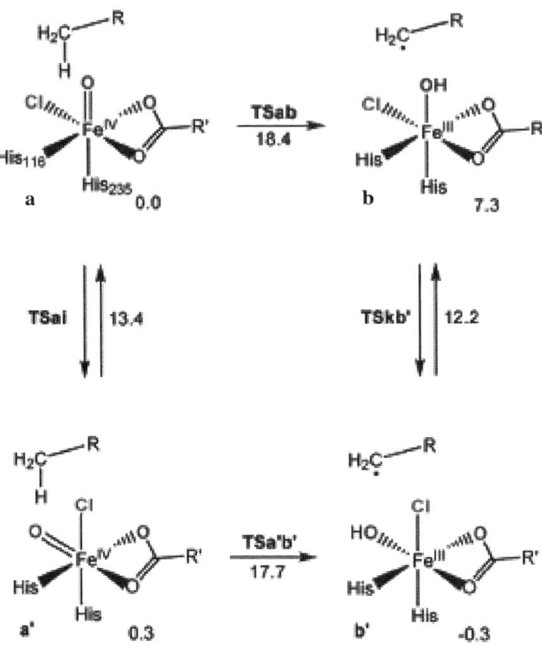

Figure 7.27   Proposed reactivity of the SyrB2 active site. The two conformations of the $Fe^{IV}=O$ intermediate are **a** and **a′**. Energies are in kcal^{-1}. Transition state energies are symbolized as **TS**. (Reproduced from ref. 76 with permission from the American Chemical Society.)

succinate group (**1**), while in the other conformation the oxo ligand is *trans* to the succinate and the chloride ligand is *trans* to His$_{235}$ (**2**). The calculated Mössbauer δ and ΔE_Q values for these two structures were $\delta = 0.27$ and 1.31 mm s^{-1} and $\Delta E_Q = 0.21$ and 0.94 mm s^{-1}, respectively.[73] The obtained values are in reasonable agreement with those reported by the Penn State group.[72] Experimentally, the δ and ΔE_Q values of one ferryl species are lower by 0.07 and 0.33 mm s^{-1}. The calculations indicated that the δ and ΔE_Q values of (**2**) were lower by 0.06 and 0.37 mm s^{-1}, which is in very good agreement with the experimental values.[73]

Most non-heme iron α-KG dependent enzymes utilize the high-valent ferryl-oxo intermediate to hydroxylate C–H bonds. Halogenases utilize the same reactive intermediate to halogenate those bonds. Thus, the relevant question is what determines the halogenation *versus* hydroxylation reactivity of these enzymes. Early computational studies suggested that the iron(III)-hydroxo intermediates are neutralized prior to the radical rebound step to generate alcohol products. One suggestion for this is CO_2 trapping with a product from an earlier step in the catalytic cycle.[74] An alternative suggestion is protonation of the hydroxyl radical as a route to avoid the hydroxylation reaction.[75] However, the transition state energies calculated were of the order of 6–8 kcal mol^{-1}, which are values below the diffusion limit and, therefore, do not agree with the long lifetime of these intermediates observed experimentally. The computational study that reproduced most of the experimental results indicates that the selectivity arises from small differences in the transition state energies of C–H bond abstraction by the two isomers (**a** and **a′**) of the ferryl-oxo intermediate (Figure 7.27).[73,76,77] In principle since these isomers are calculated to interconvert with low energy barriers, the small energy difference between the transition states (0.7 kcal mol^{-1}) may favor chlorination over hydroxylation.[76] However, given the accuracy of DFT calculations, this small difference in transition state energies is negligible. Recent biochemical studies with several mutants of SyrB2 and alternative substrates suggest that the selectivity arises from differences in substrate orientations.[19] The study stresses the role of second sphere interactions between the substrate and the active site and the orientation of the substrate, thereby, dictates the reactivity.[77]

7.3.2 Model Complexes

Prior to the characterization of the ferryl-oxo intermediates in enzymatic systems, significant progress has been made in synthetic modeling of high-valent iron coordination complexes.[78] Collins and coworkers reported structural and spectroscopic characterization of several high-valent iron(IV) species that varied in coordination number and spin state.[79–81] These complexes had chloride as the axial ligand. Que and Nam and coworkers reported the first structurally characterized iron(IV)-oxo complex in 2003.[39] The strategy developed was through oxidation of a Fe^{II} complex with an oxygen atom donor such as iodosylbenzene (PhIO). This procedure later enabled them to synthesize a series of these high-valent iron(IV)-oxo complexes.[82]

7.3.2.1 S = 1 Ferryl Complexes

Low-spin six-coordinate iron(IV)-oxo complexes have been stabilized using various polydentate ligand frameworks (Figure 7.28). Table 7.3 lists relevant geometric, spectroscopic and reactivity parameters of the reported $S = 1$ iron(IV)-oxo complexes. The pre-edge intensities on these complexes vary from 20 to 40 units, which is significantly larger than that found for the starting iron(II) and decayed iron(III) complexes. Thus, similar to metalloenzymes, the increased pre-edge intensity of biomimetic $S = 1$ ferryl-oxo species provides a diagnostic marker of the presence of iron(IV)-oxo species. There is a systematic variation in the pre-edge intensities, which was demonstrated by Jackson and Que.[82] Using TMC as a supporting chelate ligand they systematically varied the axial ligand from weak donors, such as CF_3COO^-, to strong donors like CN^- and OH^-. The work clearly demonstrated a decrease in pre-edge intensity from 38 in the CF_3COO^- ligated complex to 22 in the OH^- bound complex. While this effect is not quantitatively understood, the disruption of spherical symmetry due to a short Fe–O bond is the generally accepted theory.

The Fe–O distances in these complexes vary from 1.64 to 1.71 Å depending on the axial ligands of the complexes.[82] Dependence of the Fe–O distance on

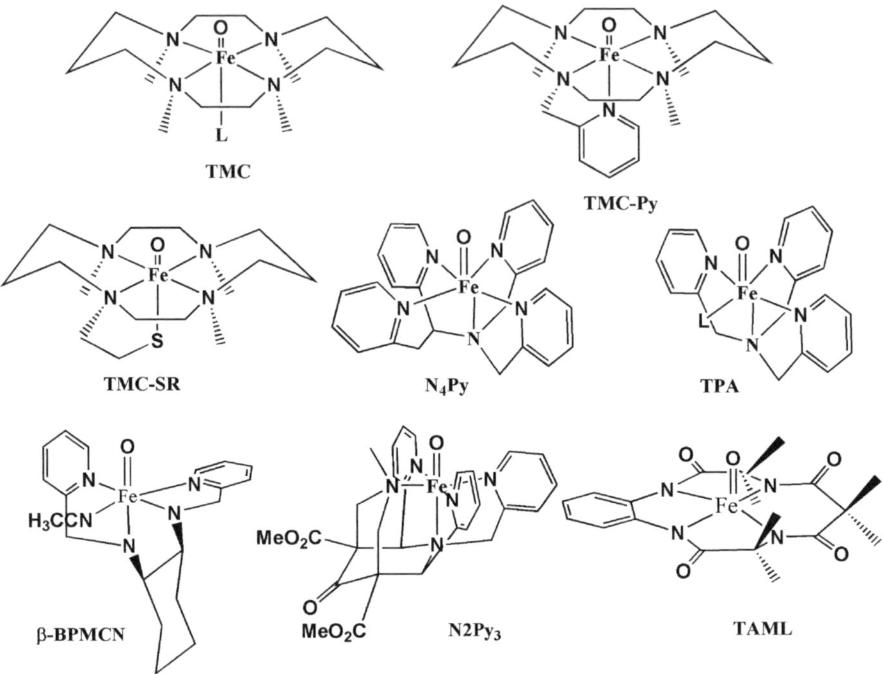

Figure 7.28 Ligands used to stabilize synthetic $S = 1$ $Fe^{IV}=O$ complexes.

Table 7.3 Spectroscopic parameters of reported $S = 1$ $Fe^{IV}=O$ complexes determined by EXAFS/XAS and stability details.

	Fe–O (Å)	Fe–N (Å)	Fe–X (Å)	Pre-edge area (units)	T (°C)	$t_{1/2}$ (RT)
$[Fe^{IV}(O)(TMC)(NCMe)]^{2+}$	1.64	2.08	2.08	26.2	−20	10 h
$[Fe^{IV}(O)(TMC)(N_3)]^+$	1.66	2.08	2.08	29.6	−20	15 min
$[Fe^{IV}(O)(TMC)(NCS)]^+$	1.65	2.07	2.07	29.1	−20	30 min
$[Fe^{IV}(O)(TMC)(NCO)]^+$	1.67	2.07	2.07	30.5	−20	30 min
$[Fe^{IV}(O)(TMC)(O_2CCF_3)]^+$	1.64	2.08	N.a.	38.6	−20	1 h
$[Fe^{IV}(O)(TMC)(CN)]^+$	1.66	2.08	N.a.	26.2	−20	1.5 h
$[Fe^{IV}(O)(TMC)(OH)]^+$	1.68	2.10	1.94	21.7	−20	Seconds
$[Fe^{IV}(O)(TMC-SR)]^+$	1.70	2.09	2.33	20.0	−60	5 min
$[Fe^{IV}(O)(TMC-Py)]^{2+}$	1.64	2.08	2.12	N.a.	RT	7 h
$[Fe^{IV}(O)(TPA)(NCMe)]^{2+}$	1.65	1.98	1.98	25.4	−40	1 h
$[Fe^{IV}(O)(TPA)(O_2CCF_3)]^+$	1.66	1.98	1.98	24–27	−40	20 min
$[Fe^{IV}(O)(TPA)(Cl)]^+$	1.65	1.98	2.29	24–27	−40	2 min
$[Fe^{IV}(O)(TPA)(Br)]^+$	1.66	1.98	2.43	24–27	−40	2 min
$[Fe^{IV}(O)(BPMCN)(NCMe)]^{2+}$	1.66	2.01	N.a.	27	−45	N.a.
$[Fe^{IV}(O)(N4Py)]^{2+}$	**1.64**	**1.96**	**2.03**	25.2	RT	60 h
$[Fe^{IV}(O)(BnTPEN)]^{2+}$	1.67	1.90	N.a.	29.3	RT	6 h
$[Fe^{IV}(O)(BQEN)(NCMe)]^{2+}$	1.67	1.97			0	30 min[a]
$[Fe^{IV}(O)(TAML)]$	1.69	1.90		41	RT	13 h

[a] $t_{1/2}$ at 0 °C.

the ligand structure is complicated by the differences in coordination number and steric requirements of the different ligands used. However, the TMC series with different *trans* axial ligands provides an excellent opportunity to evaluate the dependence of the Fe–O distance on the axial ligand.[83] In this series the Fe–O distance varies dramatically from 1.64 Å for the complex with CH_3CN axial ligand to 1.71 Å for the complex with RS^- axial ligand.[83] Organizing the complexes in increased order of Fe–O bond strength gives the series $CH_3CN = CF_3COO^- = Py < NCS^- < N_3^- = CN^- < NCO^- < HO^- < RS^-$. This series does not conform to the predictions of ligand field theory where a series of $RS^- < NCS^- < N_3^- < OH^- < NCS^- < CH_3CN < Py < CN^-$ would have been obtained. A small but noticeable effect is also observed in the TPA series whereupon varying a *cis* ligand results in a 0.1 Å change in the Fe–O bond lengths (Table 7.3). While the reasons for the latter are not known yet, it is clear that varying the axial ligand is a convenient way of tuning the Fe–O bond strength and consequently its reactivity.

The increased pre-edge intensity in the XAS data is a hallmark feature suggesting the presence of an iron(IV)-oxo species and its intensity varies from 21 units for the TMC-SR complex[38] to 41 units in the TAML complex.[84] There is a remarkable correlation between the experimentally observed Fe–O bond length from EXAFS and the pre-edge area (Figure 7.29). This is best demonstrated by the TMC series, where the replacement of the *trans* ligand produces dramatic change in the Fe–O bond length.[83] Analysis of published data reveals that the pre-edge intensity reduces with increasing bond length with a slope of 2.5 units per 0.01 Å. This correlation is also observed for the DFT optimized Fe–O bond lengths for the same series. A correlation between the observed pre-edge intensity and the Mössbauer quadrupole splitting parameter (ΔE_Q) was

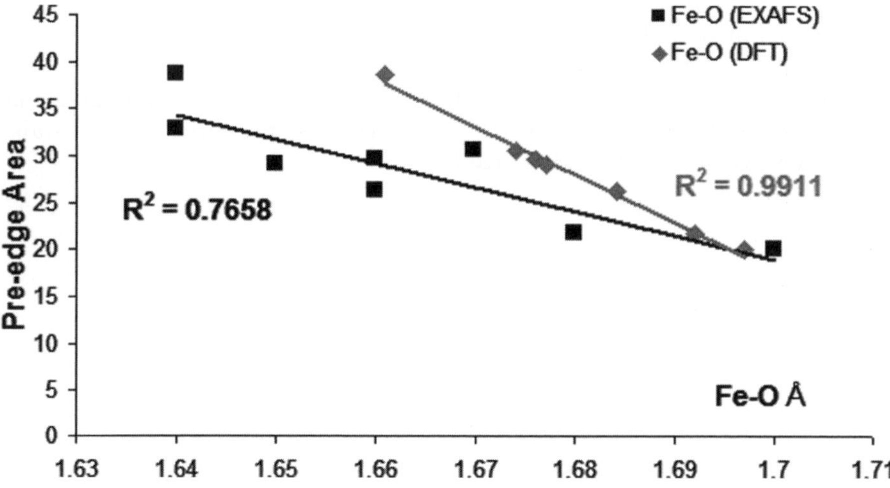

Figure 7.29 Correlation between EXAFS and DFT optimized Fe–O bond lengths and experimental pre-edge area obtained from XAS.

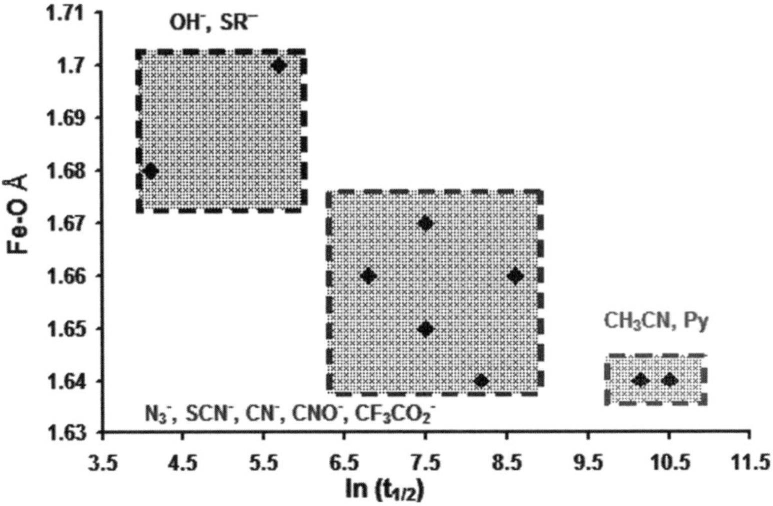

Figure 7.30   Plot of Fe–O bond length against $\ln(t_{\frac{1}{2}})$; $t_{\frac{1}{2}}$ values are in seconds.

reported.[83] Qualitatively, these correlations reflect the distortions from spherical symmetry due to short Fe–O bonds.

An intriguing question is: What controls the stability of these complexes? Since the preparation conditions are identical for the TMC series with varying axial ligands they are ideal for investigating any systematic changes. The stability of complexes varies dramatically and ranges from seconds for the complex with OH^- as an axial ligand to several hours for the one with CH_3CN axial ligand.[83] A plot of the experimental determined Fe–O bond length *versus* the natural logarithm of the half-life (ln $t_{\frac{1}{2}}$) shows that the relative stabilities are clustered in three regions (Figure 7.30). Neutral ligands like Py and CH_3CN produce the most stable iron(IV)-oxo species with a half-life of $t_{\frac{1}{2}}$ of 7–10 h. Anionic ligands (N_3^-, SCN^-, CN^-, *etc.*) are weak π-donors and give intermediate $t_{\frac{1}{2}}$ values. The third region contains complexes with anionic ligands that are very strong π-donors, such as RS^- and OH^-, and produces the least stable or, in other words, most reactive iron(IV)-oxo species.

7.3.2.2   *S = 2 Ferryl Complexes*

The iron(IV)-oxo intermediates trapped and characterized from non-heme iron enzyme active sites all possess a $S = 2$ ground state, whereas most synthetic model complexes have an $S = 1$ ground state. The electronic structure of these complexes indicated that, owing to the very strong bonding in the iron(IV)-oxo group, the spin state is determined by the equatorial ligand field (Figure 7.31). Using experimentally calibrated DFT calculations it was predicted that a trigonal bipyramidal (TBP) ligand field will stabilize a $S = 2$ iron(IV)-oxo complex.[24] The presence of an aqueous iron(IV)-oxo species was reported by Bakac

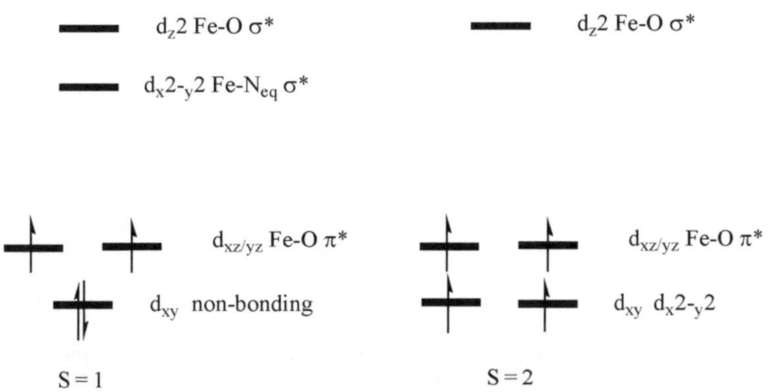

d_{z^2} Fe-O σ^*

d_{z^2} Fe-O σ^*

$d_{x^2-y^2}$ Fe-N$_{eq}$ σ^*

$d_{xz/yz}$ Fe-O π^*

d_{xy} non-bonding

$d_{xz/yz}$ Fe-O π^*

d_{xy} $d_{x^2-y^2}$

$S = 1$

$S = 2$

Figure 7.31 Comparative MO diagram for octahedral or square pyramidal $S = 1$ and trigonal bipyramidal $S = 2$ Fe^{IV}=O complexes (energy not on scale).

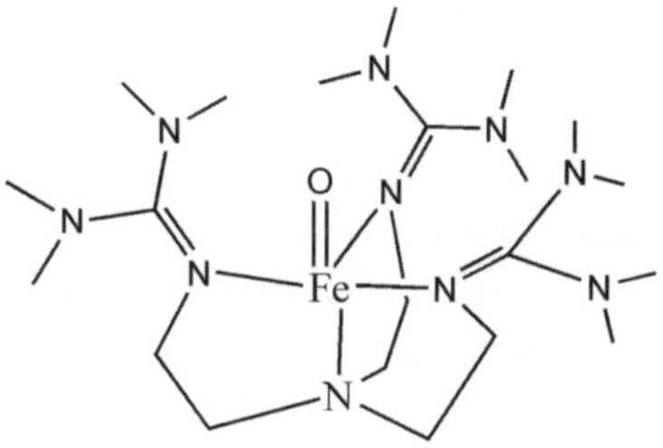

Figure 7.32 The $S = 2$ Fe^{IV}=O complex stabilized by using a TBP ligand framework.

and coworkers, who obtained it by oxidizing [Fe^{II}(H$_2$O)$_6$] with ozone.[50] The Mössbauer spectra indicated the presence of an $S = 2$ center.[50] The XAS data showed a pre-edge with 65 $\pm$ 5 units indicating the loss of spherical symmetry of the starting aqua complex. The rising-edge (*i.e.* 7127 eV) showed a reverse trend as it is at higher energy than those reported for the known $S = 2$ enzymatic intermediates as well as the $S = 1$ biomimetic iron(IV)-oxo complexes.[50]

The first crystallographically characterized $S = 2$ iron(IV)-oxo species was reported by the Que group who utilized a trianionic TBP ligand TMG3tren (Figure 7.32). The TBP ligand field enforced by this ligand allowed isolation and characterization of a $S = 2$ iron(IV)-oxo species.[85,86] The Mössbauer spectra indicated the presence of a $S = 2$ species with $\delta = 0.09$ mm s^{-1} and $\Delta E_Q = -0.29$ mm s^{-1}. These values are very similar to those reported for the $S = 2$ iron(IV)-oxo intermediate of TauD.[63,86] The rising-edge is at 7123.2 eV, *i.e.* it is 2 eV shifted relative to that found for the Fe^{II} complex with the same ligand (Figure 7.33a).[86] The XAS data show an intense pre-edge band at 7114.7 eV with 27 units of intensity (Figure 7.33a, inset). The Fe–O bond length was determined from the EXAFS spectra to be 1.65 Å (Figure 7.33b), which is consistent with earlier theoretical predictions.[24] This complex was crystallized ingenuously by

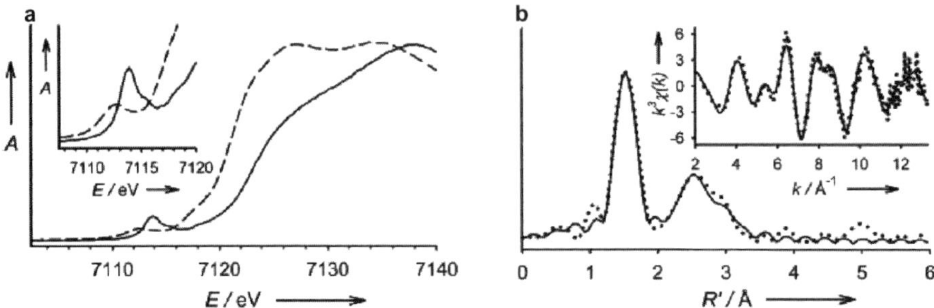

Figure 7.33   (a) XAS data of the Fe^{II} complex (dashed line) and the iron(IV)-oxo complex (solid line). Inset: the pre-edge area. (b) FT of the EXAFS data (solid line) and its fit (dotted line). Inset: the raw EXAFS data (bold) and its fit (dotted). (Reproduced from ref. 86 with permission from Wiley-VCH Verlag GmbH & Co. KGaA.)

using a deuterated ligand that slows down the self-decay process of this very reactive intermediate.[85]

Notably, the pre-edge intensity of the complex presents an interesting case as the pre-edge intensities of the Fe^{III}–X complex with similar ligands are reported to be as high as 21–28 units.[24] Thus, the characteristic and dramatic increase in pre-edge intensity upon formation of an iron(IV)-oxo complex with a distorted octahedral geometry is absent in the TBP geometry – a matter that requires further investigation.

7.4   Parting Thoughts

The pursuit of the unknown, however arduous, often yields benefits that surpass expectations. Such is the case of research in the area of iron(IV)-oxo species. After decades of concerted effort in honing the skills required for cryogenically trapping these short-lived intermediates, developing and perfecting spectroscopic techniques specifically to identify and characterize these species and designing ligands that can either stabilize these highly reactive iron(IV)-oxo species without being themselves digested or tune the reactivity of these species to allow structural and spectroscopic characterization, the bioinorganic community has been successful in providing a port hole to explore the iron(IV)-oxo landscape. The rich chemistry of these species has dwarfed predictions and anticipations. Unprecedented functions, reactions and selectivities are being reported by experimentalists on a regular basis, which leads to improved structure–function correlations and inspires detailed theoretical studies. More intriguing results are now anticipated as the bridge between heme and non-heme iron(IV)-oxo species is fast closing and both communities are attempting to generate iron(IV)-oxo from molecular oxygen, catalytically. It is a mutual feeling of the co-authors of this chapter that as far as research on iron(IV)-oxo chemistry is concerned, the game has just begun.

References

1. H. M. Hanauske-Abel and V. Günzler, *J. Theor. Biol.*, 1982, **94**, 421.
2. M. L. Neidig, A. Decker, O. W. Choroba, F. Huang, M. Kavana, G. R. Moran, J. B. Spencer and E. I. Solomon, *Proc. Natl. Acad. Sci. USA*, 2006, **103**, 12966.
3. M. L. Neidig, M. Kavana, G. R. Moran and E. I. Solomon, *J. Am. Chem. Soc.*, 2004, **126**, 4486.
4. E. I. Solomon, E. G. Pavel, K. E. Loeb and C. Campochiaro, *Coord. Chem. Rev.*, 1995, **144**, 369.
5. S. B. Piepho and P. N. Schatz, *Group Theory in Spectroscopy: With Applications to Magnetic Circular Dichroism*, John Wiley & Sons, New York, 1983.
6. P. J. Stephens, *J. Chem. Phys.*, 1970, **52**, 3489.
7. P. J. Stephens, *Annu. Rev. Phys. Chem.*, 1974, **25**, 201.
8. P. J. Stephens, *Adv. Chem. Phys.*, 1976, **35**, 197.
9. E. I. Solomon, T. C. Brunold, M. I. Davis, J. N. Kemsley, S.-K. Lee, N. Lehnert, F. Neese, A. J. Skulan, Y.-S. Yang and J. Zhou, *Chem. Rev.*, 2000, **100**, 235.
10. J. W. Whittaker and E. I. Solomon, *J. Am. Chem. Soc.*, 1988, **110**, 5329.
11. Y. Zhang, M. S. Gebhard and E. I. Solomon, *J. Am. Chem. Soc.*, 1991, **113**, 5162.
12. E. G. Pavel, N. Kitajima and E. I. Solomon, *J. Am. Chem. Soc.*, 1998, **120**, 3949.
13. M. L. Neidig and E. I. Solomon, *Chem. Commun.*, 2005, 5843.
14. C. Campochiaro, E. G. Pavel and E. I. Solomon, *Inorg. Chem.*, 1995, **34**, 4669.
15. E. I. Solomon, A. Decker and N. Lehnert, *Proc. Natl. Acad. Sci. USA*, 2003, **100**, 3589.
16. E. I. Solomon, *Inorg. Chem.*, 2001, **40**, 3656.
17. A. J. Thomson and M. K. Johnson, *Biochem. J.*, 1980, **191**, 411.
18. J. W. Whittaker and E. I. Solomon, *J. Am. Chem. Soc.*, 1986, **108**, 835.
19. A. Abragam and B. Bleaney, *Electron Paramagnetic Resonance of Transition Ions*, Dover, New York, 1986.
20. D. E. Sayers, E. A. Stern and F. Lytle, *Phys. Rev. Lett.*, 1971, **27**, 1204.
21. P. Eisenberger and B. M. Kincaid, *Science*, 1978, **200**, 1441.
22. J. Yano and V. K. Yachandra, *Photosynth. Res.*, 2009, **102**, 241.
23. T. E. Westre, P. Kennepohl, J. G. DeWitt, B. Hedman, K. O. Hodgson and E. I. Solomon, *J. Am. Chem. Soc.*, 1997, **119**, 6297.
24. A. Dey, R. K. Hocking, P. Larsen, A. S. Borovik, K. O. Hodgson, B. Hedman and E. I. Solomon, *J. Am. Chem. Soc.*, 2006, **128**, 9825.
25. R. G. Shulman, Y. Yafet, P. Eisenberger and W. E. Blumberg, *Proc. Natl. Acad. Sci. USA*, 1976, **73**, 1384.
26. L. R. Sharpe, W. R. Heineman and R. C. Elder, *Chem. Rev.*, 1990, **90**, 705.
27. J. E. Penner-Hahn, *Struct. Bonding*, 1998, **90**, 1–36.
28. I. Ascone and R. Strange, *J. Synchrotron Radiat.*, 2009, **16**, 413.

29. A. Gerson, D. Cookson and K. Prince, in *Handbook of Surface and Interface Analysis*, CRC Press, 2nd edn, 2009, pp. 193–221.
30. R. W. Strange and M. C. Feiters, *Curr. Opin. Struct. Biol.*, 2008, **18**, 609.
31. M. Costas, M. P. Mehn, M. P. Jensen and L. Que Jr., *Chem. Rev.*, 2004, **104**, 939.
32. J. C. Price, E. W. Barr, B. Tirupati, J. M. Bollinger Jr. and C. Krebs, *Biochemistry*, 2003, **42**, 7497.
33. P. J. Riggs-Gelasco, J. C. Price, R. B. Guyer, J. H. Brehm, E. W. Barr, J. M. Bollinger Jr. and C. Krebs, *J. Am. Chem. Soc.*, 2004, **126**, 8108.
34. L. M. Hoffart, E. W. Barr, R. B. Guyer, J. M. Bollinger Jr. and C. Krebs, *Proc. Natl. Acad. Sci. USA*, 2006, **103**, 14738.
35. B. E. Eser, E. W. Barr, P. A. Frantom, L. Saleh, J. M. Bollinger Jr., C. Krebs and P. F. Fitzpatrick, *J. Am. Chem. Soc.*, 2007, **129**, 11334.
36. D. P. Galonic, E. W. Barr, C. T. Walsh, J. M. Bollinger Jr. and C. Krebs, *Nat. Chem. Biol.*, 2007, **3**, 113.
37. D. G. Fujimori, E. W. Barr, M. L. Matthews, G. M. Koch, J. R. Yonce, C. T. Walsh, J. M. Bollinger Jr., C. Krebs and P. J. Riggs-Gelasco, *J. Am. Chem. Soc.*, 2007, **129**, 13408.
38. M. R. Bukowski, K. D. Koehntop, A. Stubna, E. L. Bominaar, J. A. Halfen, E. Münck, W. Nam and L. Que Jr., *Science*, 2005, **310**, 1000.
39. J.-U. Rohde, J.-H. In, M. H. Lim, W. W. Brennessel, M. R. Bukowski, A. Stubna, E. Münck, W. Nam and L. Que Jr., *Science*, 2003, **299**, 1037.
40. J. Kaizer, E. J. Klinker, N. Y. Oh, J.-U. Rohde, W. J. Song, A. Stubna, J. Kim, E. Münck, W. Nam and L. Que Jr., *J. Am. Chem. Soc.*, 2004, **126**, 472.
41. M. H. Lim, J.-U. Rohde, A. Stubna, M. R. Bukowski, M. Costas, R. Y. N. Ho, E. Münck, W. Nam and L. Que Jr., *Proc. Natl. Acad. Sci. USA*, 2003, **100**, 3665.
42. C. A. Grapperhaus, B. Mienert, E. Bill, T. Weyhermuller and K. Wieghardt, *Inorg. Chem.*, 2000, **39**, 5306.
43. V. Balland, M. F. Charlot, F. Banse, J. J. Girerd, T. A. Mattioli, E. Bill, J. F. Bartoli, P. Battioni and D. Mansuy, *Eur. J. Inorg. Chem.*, 2004, 301.
44. M. Martinho, F. Banse, J. F. Bartoli, T. A. Mattioli, P. Battioni, O. Horner, S. Bourcier and J. J. Girerd, *Inorg. Chem.*, 2005, **44**, 9592.
45. E. J. Klinker, J. Kaizer, W. W. Brennessel, N. L. Woodrum, C. J. Cramer and L. Que Jr., *Angew. Chem., Int. Ed.*, 2005, **44**, 3690.
46. M. P. Jensen, M. Costas, R. Y. N. Ho, J. Kaizer, A. M. I. Payeras, E. Münck, L. Que Jr., J.-U. Rohde and A. Stubna, *J. Am. Chem. Soc.*, 2005, **127**, 10512.
47. J.-U. Rohde and L. Que Jr., *Angew. Chem., Int. Ed.*, 2005, **44**, 2255.
48. J. Bautz, M. R. Bukowski, M. Kerscher, A. Stubna, P. Comba, A. Lienke, E. Münck and L. Que Jr., *Angew. Chem., Int. Ed.*, 2006, **45**, 5681.
49. C. V. Sastri, M. J. Park, T. Ohta, T. A. Jackson, A. Stubna, M. S. Seo, J. Lee, J. Kim, T. Kitagawa, E. Münck, L. Que Jr. and W. Nam, *J. Am. Chem. Soc.*, 2005, **127**, 12494.

50. O. Pestovsky, S. Stoian, E. L. Bominaar, X. P. Shan, E. Münck, L. Que Jr. and A. Bakac, *Angew. Chem., Int. Ed.*, 2005, **44**, 6871.

51. O. Pestovsky, S. Stoian, E. L. Bominaar, X. Shan, E. Münck, L. Que Jr. and A. Bakac, *Angew. Chem., Int. Ed.*, 2006, **46**, 340.

52. J.-U. Rohde, J.-H. In, M. H. Lim, W. W. Brennessel, M. R. Bukowski, A. Stubna, E. Münck, W. Nam and L. Que Jr., *Science*, 2003, **299**, 1037.

53. A. Decker and E. I. Solomon, *Angew. Chem., Int. Ed.*, 2005, **44**, 2252.

54. E. I. Solomon, S. D. Wong, L. V. Liu, A. Decker and M. S. Chow, *Curr. Opin. Chem. Biol.*, 2009, **13**, 99.

55. A. Decker, J.-U. Rohde, L. Que Jr. and E. I. Solomon, *J. Am. Chem. Soc.*, 2004, **126**, 5378.

56. A. Decker, J.-U. Rohde, E. J. Klinker, S. D. Wong, L. Que Jr. and E. I. Solomon, *J. Am. Chem. Soc.*, 2007, **129**, 15983.

57. E. J. Klinker, J. Kaizer, W. W. Brennessel, N. L. Woodrum, C. J. Cramer and L. Que Jr., *Angew. Chem., Int. Ed.*, 2005, **44**, 3690.

58. E. I. Solomon, S. D. Wong, L. V. Liu, A. Decker and M. S. Chow, *Curr. Opin. Chem. Biol.*, 2009, **13**, 99.

59. E. Eichhorn, J. R. van der Ploeg , M. A. Kertesz and T. Leisinger, *J. Biol. Chem.*, 1997, **272**, 23031.

60. J. M. Elkins, M. J. Ryle, I. J. Clifton, J. C. Dunning Hotopp, J. S. Lloyd, N. I. Burzlaff, J. E. Baldwin, R. P. Hausinger and P. L. Roach, *Biochemistry*, 2002, **41**, 5185.

61. M. L. Neidig, C. D. Brown, K. M. Light, D. G. Fujimori, E. M. Nolan, J. C. Price, E. W. Barr, J. M. Bollinger Jr., C. Krebs, C. T. Walsh and E. I. Solomon, *J. Am. Chem. Soc.*, 2007, **129**, 14224.

62. M. J. Ryle, R. Padmakumar and R. P. Hausinger, *Biochemistry*, 1999, **38**, 15278.

63. J. C. Price, E. W. Barr, B. Tirupati, J. M. Bollinger Jr. and C. Krebs, *Biochemistry*, 2003, **42**, 7497.

64. J. C. Price, E. W. Barr, T. E. Glass, C. Krebs and J. M. Bollinger Jr., *J. Am. Chem. Soc.*, 2003, **125**, 13008.

65. T. Borowski, A. Bassan and P. E. M. Siegbahn, *Chem. Eur. J.*, 2004, **10**, 1031.

66. S. Sinnecker, N. Svensen, E. W. Barr, S. Ye, J. M. Bollinger Jr., F. Neese and C. Krebs, *J. Am. Chem. Soc.*, 2007, **129**, 6168.

67. E. Godfrey, C. S. Porro and S. P. de Visser, *J. Phys. Chem. A*, 2008, **112**, 2464.

68. M. Ueki, D. P. Galonic, F. H. Vaillancourt, S. Garneau-Tsodikova, E. Yeh, D. A. Vosburg, F. C. Schroeder, H. Osada and C. T. Walsh, *Chem. Biol.*, 2006, **13**, 1183.

69. C. Wong, D. G. Fujimori, C. T. Walsh and C. L. Drennan, *J. Am. Chem. Soc.*, 2009, **131**, 4872.

70. F. H. Vaillancourt, J. Yin and C. T. Walsh, *Proc. Natl. Acad. Sci. USA*, 2005, **102**, 10111.

71. L. C. Blasiak, F. H. Vaillancourt, C. T. Walsh and C. L. Drennan, *Nature*, 2006, **440**, 368.

72. M. L. Matthews, C. M. Krest, E. W. Barr, F. H. Vaillancourt, C. T. Walsh, M. T. Green, C. Krebs and J. M. Bollinger Jr., *Biochemistry*, 2009, **48**, 4331.
73. P. K. Grzyska, E. H. Appelman, R. P. Hausinger and D. A. Proshlyakov, *Proc. Natl. Acad. Sci. USA*, 2010, **107**, 3982.
74. S. P. de Visser and R. Latifi, *J. Phys. Chem. B*, 2009, **113**, 12.
75. S. Pandian, M. A. Vincent, I. H. Hillier and N. A. Burton, *Dalton Trans.*, 2009, **6201**.
76. T. Borowski, H. Noack, M. Radoń, K. Zych and P. E. M. Siegbahn, *J. Am. Chem. Soc.*, 2010, **132**, 12887.
77. M. L. Matthews, C. S. Neumann, L. A. Miles, T. L. Grove, S. J. Booker, C. Krebs, C. T. Walsh and J. M. Bollinger Jr., *Proc. Natl. Acad. Sci. USA*, 2009, **106**, 17723.
78. D.-L. Popescu, A. Chanda, M. Stadler, O. F. de Tiago, A. D. Ryabov, E. Münck, E. L. Bominaar and T. J. Collins, *Coord. Chem. Rev.*, 2008, **252**, 2050.
79. T. J. Collins, K. L. Kostka, E. Münck and E. S. Uffelman, *J. Am. Chem. Soc.*, 1990, **112**, 5637.
80. T. J. Collins, B. G. Fox, Z. G. Hu, K. L. Kostka, E. Münck, C. E. F. Rickard and L. J. Wright, *J. Am. Chem. Soc.*, 1992, **114**, 8724.
81. K. L. Kostka, B. G. Fox, M. P. Hendrich, T. J. Collins, C. E. F. Rickard, L. J. Wright and E. Münck, *J. Am. Chem. Soc.*, 1993, **115**, 6746.
82. L. Que Jr., *Acc. Chem. Res.*, 2007, **40**, 493.
83. T. A. Jackson, J.-U. Rohde, M. S. Seo, C. V. Sastri, R. DeHont, A. Stubna, T. Ohta, T. Kitagawa, E. Münck, W. Nam and L. Que Jr., *J. Am. Chem. Soc.*, 2008, **130**, 12394.
84. A. Chanda, X. Shan, M. Chakrabarti, W. C. Ellis, D. L. Popescu, O. F. de Tiago, D. Wang, L. Que Jr., T. J. Collins, E. Münck and E. L. Bominaar, *Inorg. Chem.*, 2008, **47**, 3669.
85. J. England, Y. Guo, E. R. Farquhar, V. G. Young Jr., E. Münck and L. Que Jr., *J. Am. Chem. Soc.*, 2010, **132**, 8635.
86. J. England, M. Martinho, E. R. Farquhar, J. R. Frisch, E. L. Bominaar, E. Münck and L. Que Jr., *Angew. Chem., Int. Ed.*, 2009, **48**, 3622.

Structure, Mechanism and Function of Cytochrome P450 Enzymes

KIRSTY J. MCLEAN, HAZEL M. GIRVAN, AMY E. MASON, ADRIAN J. DUNFORD AND ANDREW W. MUNRO*

Manchester Interdisciplinary Centre, Faculty of Life Sciences, University of Manchester, 131 Princess Street, Manchester M1 7DN, UK

8.1 Introduction

The cytochromes P450 (commonly termed P450s or CYPs) are a superfamily of heme-binding enzymes that are widely studied as a consequence of their multitude of physiological functions in organisms from prokaryotes through to man and, increasingly, due to their biotechnological potential.[1] The cytochromes P450 catalyse the reduction of dioxygen bound to their heme iron and the insertion of an atom of oxygen into organic substrates bound in their active site and proximal to the heme.[2] The ability to perform this function in a regio- and (usually) stereoselective manner is highly prized chemically, since such regioselective oxidation of non-activated C–H bonds is extremely difficult to achieve using synthetic chemistry methods. The cytochromes P450 are best known for their functions in the metabolism of drugs and other xenobiotics in humans, where the hepatic isoforms (in particular) play pivotal roles in chemical detoxification and health. However, the last 50 years of research in this field has unravelled details of structure and mechanism of the cytochromes P450, and has

---

Iron-Containing Enzymes: Versatile Catalysts of Hydroxylation Reactions in Nature
Edited by Sam P de Visser and Devesh Kumar

Published by the Royal Society of Chemistry, www.rsc.org

shown them to be adapted to a huge range of biochemical functions. This chapter will review the discovery of the cytochromes P450, leading into studies that have revealed their structural properties and mechanisms by which they achieve their many physiological functions. Turning to the future, the biotechnological potential of these enzymes (*e.g.* for exploitation in the synthesis of oxidized molecules for fine chemical and pharmaceutical applications) will be discussed.

8.2 Cytochromes P450 – A Brief History

With more than 65 000 papers returned by a search on PubMed with the term "cytochrome P450" at the end of 2010, the volume of research activity in this field becomes obvious. As will be covered in detail in the remainder of this chapter, the cytochromes P450 are heme *b*-containing oxygenase enzymes, for which the current number of recognized genes from various organisms is close to 20 000.[3] The enzymes that we now know as cytochromes P450 have their origins in studies dating to the 1940s and 1950s, and a detailed historical perspective in this area is provided by Ron Estabrook in his elegant review of the early days of cytochrome P450 research and the discovery of its role in oxygen chemistry.[4] Key events leading to the discovery of the cytochromes P450 and their roles as hemoprotein monoxygenases include pioneering studies by Mueller and Miller in demonstrating requirements for both NADPH and oxygen in the metabolism of certain azo dyes in particulate fractions from rat liver.[5,6] The Nobel Laureate Julius Axelrod also performed seminal studies demonstrating the requirements for NADPH and oxygen in drug metabolism in rabbit liver.[7] Studies at this time helped to define key roles for oxidative enzymes involved in the metabolism of xenobiotics and located in the membranous fractions of liver cells. The name cytochrome P450 likely originates from a pigment (P) that has an absorption maximum at 450 nm in presence of reductant and the noxious gas carbon monoxide (CO). The earliest report of this phenomenon came from work by Martin Klingenberg on liver microsomal fractions.[8] At this stage, a hemoprotein was an obvious candidate for the CO-binding pigment. However, previous studies had indicated that CO-binding to hemoproteins resulted in blue-shifts of their absorption maxima to shorter wavelengths, whereas Klingenberg's study revealed an unusual red-shift in absorption.[3] Later studies indicated the presence of the same type of 450 nm pigment (and the hemoprotein cytochrome b_5) in adrenal gland microsomes, raising the possibility of a role for this pigment as a catalyst for steroid oxygenations.[3,9] The confirmation that cytochrome P450 was a hemoprotein came in 1962 with the paper of Tsuneo Omura and Ryo Sato, in which the name "cytochrome" also became associated with the pigment.[10] There followed further intense research in the area and the establishment of key roles for cytochrome P450 in drug metabolism, including studies by Remmer, Cooper and others that showed the increased quantity of cytochrome P450 in liver microsomes that accompanied induction of drug metabolizing enzymes, and thus proved the relationship between cytochrome P450 and the oxidative metabolism of various drugs and steroids.[9,11,12] Early research on cytochromes P450 was performed with the premise that there were limited numbers of types of cytochrome P450. Key studies were done by the laboratory of Irwin Gunsalus (*e.g.* ref. 13) on the camphor hydroxylase system $P450_{cam}$ from the bacterium *Pseudomonas putida*, by Judd Coon and coworkers on liver microsomal cytochrome P450 systems (*e.g.* ref. 14), and by Kimura and Suzuki on adrenal mitochondrial cytochromes P450.[15,16] These, and other, studies helped to reveal important features of the different cytochrome P450 systems, including the fact that the eukaryotic cytochromes P450 are membrane associated proteins whereas prokaryotic cytochromes P450 are almost invariably soluble, cytoplasmic enzymes.[17] In addition, cytochromes P450 require electron transport protein systems to deliver the NAD(P)H-derived electrons that they require for reductive activation of oxygen, and distinctive systems are present in eukaryotes and prokaryotes (see Section 8.6). The last 40 years have seen massive efforts in the analysis of cytochrome P450 structure, mechanism, physiology/pharmacology and biodiversity. Important properties revealed are discussed in detail in the following sections.

8.3 Optical and Spectroscopic Features

As indicated above, an important aspect of the cytochromes P450 that first drew attention to their unusual nature is their UV–visible absorption spectrum, and particularly the red-shifted absorbance maximum in the ferrous/CO-bound form. In their resting state, the cytochromes P450 bind heme *b* in its ferric state (ferriprotoporphyrin IX). The ferric cytochromes P450 typically occupy either a low-spin (LS) or a high-spin (HS) state, related to the organization of electrons in the 3d orbitals of the iron. In the LS state, there is an overall $S = \frac{1}{2}$ state with the five electrons maximally paired and all residing in the three lower energy t_{2g} orbitals (*i.e.* with a single unpaired electron). In the HS state, $S = \frac{5}{2}$ and the five electrons are maximally unpaired and singly occupy each of the five t_{2g} and e_g orbitals. The LS form is typical for substrate-free cytochrome P450 enzymes and is usually associated with the heme iron being hexacoordinate. Four equatorial ligands to the heme iron originate from nitrogen atoms in the heme pyrrole rings. The fifth ligand comes from a cysteine thiolate side chain, where this proximal cysteine is completely conserved in the cytochrome P450 enzymes.[18] The electron-donating nature of the thiolate ligand is considered essential for the cytochromes P450 to retain activity, and is associated with the Fe(II)CO complex having its spectral maximum at ~ 450 nm.[19] The sixth ligand originates from the oxygen of a distal water molecule.

The LS cytochrome P450 typically has its major (Soret) absorption maximum at ~ 418 nm, with two other smaller heme bands (the α- and β-bands, or Q-bands) in the 500–600 nm region (Figure 8.1). The major factors responsible for UV–visible spectra of cytochromes P450 (and other hemoproteins) are π–π^* transitions in the heme macrocycle, with contributions from the mixing of sulfur p-electrons and heme π molecular orbitals also influencing cytochrome P450 spectral properties. The binding of substrate (typically a lipophilic

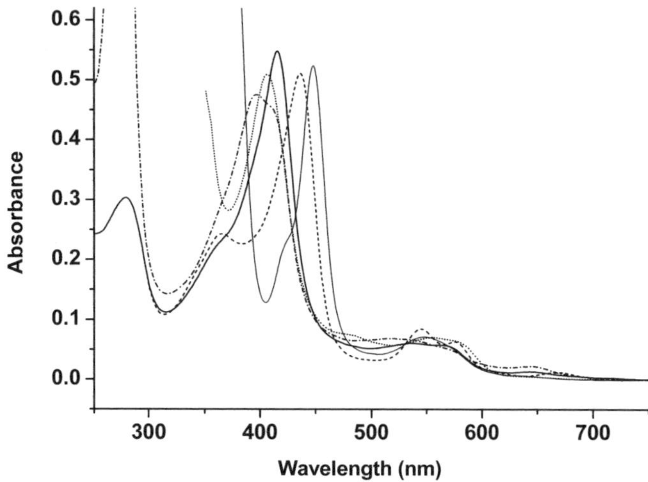

Figure 8.1 Optical properties of cytochrome P450. UV–visible absorption spectral features are shown for a typical cytochrome P450 – the *Mycobacterium tuberculosis* CYP121, which catalyses the oxidative coupling of tyrosine side chain carbon atoms in the generation of a novel secondary metabolite from the cyclic dipeptide cyclo-L-Tyr-L-Tyr (cYY).[99,100] The ferric resting form of CYP121 ($\sim$ 5.5 µM) is shown as a thick solid line ($A_{max} = 416$ nm), with the dithionite reduced (ferrous) form as a dotted line ($A_{max} = 406$ nm), the ferrous-CO adduct as a thin solid line ($A_{max} = 448$ nm), the ferric-NO adduct as a dashed line ($A_{max} = 436$ nm), and the ferric, cYY-bound form as a dot-dashed line ($A_{max} = 395$ nm). Spectral differences among these species are also obvious in the Q-band region between 500 and 600 nm.

molecule such as a prescribed drug, steroid or fatty acid) often causes the displacement of the weakly bound water and an electronic rearrangement in the d-orbitals, leading to the formation of HS ferric iron. The optical spectrum for the HS form is quite distinctive, with a Soret shift to $\sim$ 390 nm (Figure 8.1). The HS shift (also referred to as a type I cytochrome P450 shift) and large associated absorbance changes provide the basis for determining affinity of substrates by optical titrations. For instance, binding titrations of the polyunsaturated fatty acid arachidonic acid with the high activity *Bacillus megaterium* fatty acid hydroxylase cytochrome P450 BM3 (CYP102A1) provided a K_d value of 3.6 µM for this lipid.[20] The HS shift and associated electronic reorganization of the heme iron electrons also leads to a more positive ferric heme iron reduction potential, and for "well evolved" cytochromes P450 this can be an excellent means of thermodynamically regulating electron transfer to the heme iron, such that electron flow occurs only when there is substrate available for oxidation. A positive shift in potential of $\sim$ 130–150 mV is typical when an extensively LS cytochrome P450 is converted into a mainly HS species on substrate binding, as seen for CYP102A1 on binding arachidonic acid and for cytochrome P450$_{cam}$ (CYP101A1) in complex with camphor.[21,22] However, many (particularly eukaryotic) cytochromes P450 are isolated in a mixed spin-state (*i.e.* proportions of both LS and HS forms) and may not undergo further HS shift on substrate binding. In such cases, phenomena such as cytochrome P450 substrate-dependent structural reorganization may contribute towards controlling interactions with redox partner proteins and regulation of electron transfer. Some molecules (typically inhibitors) also induce shifts from LS to HS (reverse type I shifts), which should be related to their ability to restore an aqua distal ligand to the heme iron.[23] However, the major types of cytochrome P450 inhibitors are those which coordinate to the heme iron (in place of the distal water), cause red-shifts in the Soret band, and are referred to as type II inhibitors. These include molecules such as imidazole, cyanide and nitric oxide (which also binds to ferrous iron), and also imidazole- and triazole-derivatives (azoles), such as clotrimazole, fluconazole and econazole, which coordinate to the heme iron *via* a nitrogen atom from their imidazole/triazole ring (Figure 8.1).[24] The azole inhibitors have found clinical use as antifungals by inhibiting the action of the fungal sterol demethylase cytochrome P450 (CYP51) that is essential for the oxidative demethylation of lanosterol to form ergosterol, leading to loss of integrity of the cell membrane.[25] They have also raised interest recently as potential antitubercular therapeutics.[26] Of course, the best known cytochrome P450 inhibitor is still carbon monoxide, and CO binds exclusively to the reduced (ferrous) heme iron. The ~450 nm peak in the Fe(II)CO spectrum is associated with the distal coordination of CO to the iron and the retention of cysteine thiolate as the proximal ligand (Figure 8.1). However, a different species with Soret maximum at ~420 nm (P420) is also often observed. This is often interpreted as being due to the presence of inactivated or structurally disrupted cytochrome P450, and is likely due to protonation of the heme thiolate ligand to the thiol form.[18,27] However, P420 formation can simply be associated with instability of the ferrous or ferrous/CO-bound forms of the cytochrome P450 to cysteinate protonation, and there are excellent examples, for instance, of reconversion of a P420 into a P450 enzyme by addition of substrate to the Fe(II)CO complex.[28,29]

The heme cofactor in the cytochromes P450 can also be analysed by several other spectroscopic methods. Prominent among these is electron paramagnetic resonance (EPR), which provides important information on the nature of the heme ligands and any heterogeneity therein, and can also report on the spin-state of the iron [although at appropriate temperatures for heme EPR, approximately 10 K, HS signals are often considerably decreased in favour of LS form(s)] (Figure 8.2).[30] Resonance Raman spectroscopy is also a powerful method for exploring heme iron oxidation- and spin-state, as well as for analysis of the conformation of the bound cofactor and its peripheral substituent groups.[31,32] Magnetic circular dichroism (MCD) is another useful tool for probing heme iron ligation, and was used to good effect to disprove an early hypothesis that histidine was the cytochrome P450 heme iron proximal ligand, and (together with other methods) that an oxygen ligand (*i.e.* H_2O) occupied the distal ligation position.[33,34] Other spectroscopic methods often applied in cytochrome P450 studies are fluorimetry (particularly useful for analysis of fluorescent ligand binding), NMR and Mössbauer spectroscopy, the latter

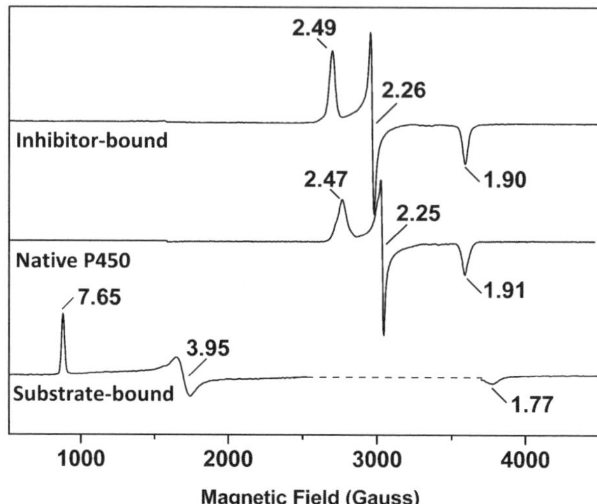

Figure 8.2 EPR spectroscopy of cytochrome P450 enzymes. The central spectrum is a typical X-band EPR spectrum for a low-spin (LS), ferric cytochrome P450 enzyme. The enzyme shown is CYP121 from *Mycobacterium tuberculosis*. CYP121 has typical LS g-values of $g_z = 2.47$, $g_y = 2.25$ and $g_x = 1.91$. The upper EPR spectrum is for the royal demolition explosive (hexahydro-1,3,5-trinitro-1,3,5-triazine or RDX) degrading XplA from *Rhodococcus rhodochrous* strain 11Y, in complex with the heme-coordinating inhibitor imidazole. The binding of the nitrogenous ligand induces shifts in the LS g-values to 2.46, 2.26 and 1.90. The lower spectrum shows XplA in complex with its RDX substrate. RDX binding induces the displacement of the (distal) axial water ligand and the generation of a high-spin (HS) ferric heme iron. The distinctive HS g-values are at $g_z = 7.65$, $g_y = 3.95$ and $g_x = 1.77$. For clarity, remnant signals from LS XplA heme iron have been removed.

being valuable for analysis of the electronic state and chemical environment of the heme iron.[35] Collectively, UV–visible and other spectroscopic tools have proven critical in the development of our understanding of the properties of the heme centre in cytochrome P450 enzymes, and continue to be important tools for, for example, investigating and quantifying ligand binding and for analysis of the properties of reactive and transient species in the cytochrome P450 catalytic cycle.

8.4 Cytochrome P450 Catalytic Cycle

This chapter comes at an exciting time for the cytochrome P450 field, with the recent publication of the first truly compelling characterization of a reactive intermediate (named compound I) that is considered to be the major species responsible for oxygen insertion into cytochrome P450 substrates.[36] As shown in Figure 8.3, the classic cytochrome P450 cycle starts with a ferric heme species with cysteine thiolate (indicated by a sulfur atom, S) as the proximal ligand and

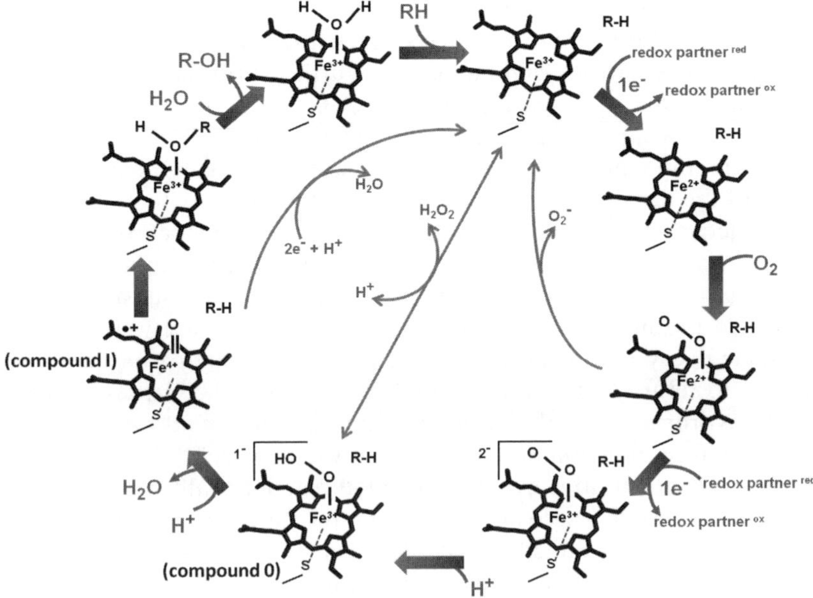

Figure 8.3 Cytochrome P450 catalytic cycle. The catalytic cycle of a typical cytochrome P450 enzyme is shown, with intermediates as described in Section 8.4. Successive events of substrate (RH) binding, distal water ligand dissociation, single-electron reduction by a redox partner, dioxygen binding, second single-electron reduction and protonation generate the transient compound 0 (ferric-hydroperoxo) species. Further protonation of this intermediate leads to O–O bond scission and the production of a molecule of water, leaving the ferryl-oxo compound I species as the catalyst for oxygen insertion to generate ROH product. Departure of the product and return of a water molecule as a sixth heme ligand restores the hexacoordinated resting form of the enzyme. Different non-productive pathways for oxy-iron species collapse are also shown, all resulting in the return to the pentacoordinate ferric enzyme, and (according to the species that decays) production of superoxide, peroxide or water molecules.

a water molecule as a distal ligand to the heme iron. The binding of a substrate (RH) then induces water ligand displacement and formation of HS ferric heme iron with a more positive potential for the $Fe(III)/Fe(II)$ redox couple.[37] The next step is the delivery of a single electron from a redox partner protein (Section 8.6), reducing the heme iron to the ferrous state and enabling binding of dioxygen. This ferrous-oxy complex (usually considered to be ferric-superoxo in character) is, for most cytochromes P450, the last step in the catalytic cycle that is readily spectroscopically observed at ambient temperature – typically having a lifetime of several seconds. The ferric-superoxo complex is diamagnetic and EPR silent, but evidence for its ferric nature comes from both Mössbauer spectroscopy (a Mössbauer quadrupole splitting consistent with ferric iron in the complex) and from the low frequency (1140 cm^{-1}) of the O–O stretch observed in the $P450_{cam}$ resonance Raman spectrum.[2,38,39] The ferric-superoxo complex then receives a second electron from the redox partner, leading to formation of a ferric-peroxo complex, which then becomes protonated to a ferric hydroperoxo form (compound 0). Further protonation at the distal oxygen atom leads to heterolysis of the O–O bond, with production of a molecule of water, and the generation of a reactive and highly transient ferryl-oxo species (compound I). Protons may originate from water molecules, and are delivered to the distal oxygen in a mechanism that involves the side chain of a conserved active site threonine (or serine) residue in the cytochromes P450.[40,41] Compound I is often described as an oxo-ferryl porphyrin π-cation species, and is the species considered to effect oxygen transfer to the substrate, forming hydroxylated product (ROH). Also shown in Figure 8.3 are well known oxy-intermediate decay pathways, by which (i) the ferric-superoxo species releases superoxide to return to the ferric state, (ii) compound 0 collapses to the ferric state with hydrogen peroxide formation, and (iii) compound I collapses to the ferric state with production of a molecule of water. In case (ii), this mechanism has been exploited to drive cytochrome P450 reactions (albeit generally inefficiently) using hydrogen peroxide (or other substitutes, *e.g.* cumene hydroperoxide) in the productive direction – the so-called "peroxide shunt" mechanism.[42] The decay pathways occur in absence of substrate or, for example, in presence of poor or inappropriately positioned substrates in the active site, or through failure of timely delivery of electrons or protons. The catalytic cycle, its intermediates and their spectroscopic and catalytic properties are covered in detail in the excellent review of Denisov *et al.*[2]

The extremely reactive and transient nature of, particularly, compound 0 and compound I in cytochrome P450 enzymes has resulted not only in problems generating these species for spectroscopic analysis, but also in vigorous debate as to their relative roles and importance in substrate oxidation. In particular, experimental studies (including analysis of cytochromes P450 in which the conserved threonine implicated in oxy complex protonation is mutated) have raised the possibility that compound 0 could have oxidative functions, at least in less "challenging" reactions such as epoxidation of olefins.[43,44] However, computational studies have indicated that the strongly basic nature of the ferric-hydroperoxo species is conducive to a barrier-free protonation on the distal oxygen to generate compound I, and this could occur (albeit less efficiently) even in absence of the catalytic threonine.[45] Moreover, energetic barriers to ethylene oxidation were determined to be much higher for compound 0 than for compound I, thus favouring the latter as the catalytically relevant species for such epoxidation reactions.[44,46] Further computational studies led to the development of a two-state reactivity model for cytochrome P450 compound I activity, whereby at least two virtually degenerate ground state species can form, each generating its own set of products with associated differences in rate constants, and regio- and stereoselectivity of substrate oxidation. The reactivity pattern of the species can further be "tuned" by the substrate and, for example, the protein environment and hydrogen bonding patterns in the active site.[47] Thus, for instance, density functional theory (DFT) calculations predicted that competing sulfoxidation and monoxygenation (leading to N-dealkylation) reactions observed in studies of CYP102A1 and its T268A catalytic threonine mutant (where the latter showed enhanced sulfoxidation) likely result from competing HS and LS compound I pathways, as opposed to contributions from both compound 0 and compound I species.[48,49] Despite the high reactivity of compounds 0 and I, recent research has led to their spectroscopic capture. For instance, studies by Hoffman and coworkers demonstrated compellingly the formation of P450$_{cam}$ compound 0 by γ-irradiating cryoreduced ferric-superoxo enzyme at 77 K, and then raising the sample temperature (annealing) to 170 K to identify compound 0 using EPR and electron-nuclear double resonance (ENDOR) spectroscopy. The presence of substrate was found to considerably stabilize the compound 0 state.[50] The "holy grail", however, was truly revealed in late 2010 when Rittle and Green presented solid spectroscopic (UV–visible and Mössbauer) and catalytic data showing the formation of compound I. This key landmark in cytochrome P450 science was achieved by mixing the thermostable CYP119 enzyme from *Sulfolobus acidocaldarius* with *m*-chloroperbenzoic acid, and with validation of its reactivity by substrate turnover measurements (*e.g.* determining an apparent rate constant of CYP119 compound I-mediated lauric acid oxidation of $1.1 \times 10^7 \, M^{-1} \, s^{-1}$).[36]

8.5 Biological Diversity

Early studies on the cytochromes P450 focused on mammalian liver microsomal and adrenal mitochondrial enzymes, and on the *Pseudomonas putida* P450$_{cam}$ enzyme. However, over the last 40 years it has become clear that these systems are really only the tip of the iceberg, and that cytochrome P450 enzymes (and often large numbers thereof) are found in unicellular organisms such as archaea and bacteria (*e.g. Sulfolobus solfataricus* and various *Streptomyces* species) through to higher plants and mammals.[51] The escalating numbers of *CYP* genes identified, particularly in recent years with the advent of whole genome sequencing, necessitated the creation of a classification system. To this end, the cytochromes P450 became the first gene superfamily, with enzymes having ≥ 40% amino acid sequence identity placed in the same family, and generally showing related substrate selectivity consistent with a relatively close evolutionary relationship.[52] A recent (2011) study indicated that 12 456 cytochromes P450 had received names and cytochrome P450 superfamily classifications, with a further ∼ 6000 recognized but not classified.[3] The cytochrome P450 homepage of David Nelson (http://drnelson.uthsc.edu/ cytochromeP450.html) provides a regularly updated overview of the cytochrome P450 superfamily and distribution in various organisms. The numbers of cytochromes P450 encoded varies widely between organisms, with (for instance) 57 CYP genes in humans, 273 in *Arabidopsis thaliana* (thale cress) and 33 in the bacterium *Streptomyces avermitilis* (which produces the anti-helmintic drug avermectin).[53] Higher eukaryotes generally have large numbers of cytochromes P450, and well known roles for human cytochromes P450 include xenobiotic detoxification by several hepatic isoforms, cholesterol side chain cleavage by CYP11A1 to produce pregnenolone (leading to synthesis of steroid hormones) and fatty acid ω-hydroxylation by CYP4A11 in human liver and kidney.[54] However, although certain prokaryotes (such as *Escherichia coli*) are devoid of cytochromes P450, others (*e.g.* Actinobacteria) typically have several. A notable example is the human pathogen *Mycobacterium tuberculosis*, which has 20 cytochromes P450.[55] Recent studies on the *M. tuberculosis* cytochromes P450 have revealed (or predicted) unexpected functions in host cholesterol oxidation, branched chain fatty acid and menaquinone oxidation, and secondary metabolite synthesis.[55,56]

A key difference between eukaryotic and prokaryotic/archaeal cytochromes P450 is that the former enzymes are integral membrane proteins, anchored by an N-terminal transmembrane helical segment, whereas the latter are generally soluble, cytoplasmic enzymes. Interestingly, although *E. coli* does not possess its own cytochromes P450, it has been shown that eukaryotic cytochromes P450 heterologously expressed in the bacterium are catalytically active due to electron transfer from a bacterial redox system. This was shown to be the *E. coli* NADPH-flavodoxin reductase/flavodoxin system.[57] These two flavoproteins are homologues of the two different component domains of the eukaryotic diflavin enzyme NADPH-cytochrome P450 reductase (CPR), which naturally supports the catalytic function of most eukaryotic cytochromes P450.[58] In the next section, the biological diversity of redox partner systems that support cytochrome P450 function is discussed.

8.6 Cytochrome P450 Redox Partner Systems

As illustrated in Figure 8.3 and described above, the typical monooxygenation function of the cytochromes P450 requires the consecutive reduction of the cytochrome P450 ferric heme iron by two single electrons, and two protonation events to enable the formation of the catalytic compound I, according to the equation:

$$RH + O_2 + 2e^- + 2H^+ \rightarrow ROH + H_2O \tag{8.1}$$

Early studies on cytochrome P450 systems identified two distinct classes of redox partner systems, both of which used NAD(P)H coenzyme as the reductant. The class 1 system is typified by the P450$_{cam}$ redox apparatus, which uses a NADH-dependent FAD-containing reductase (putidaredoxin reductase, PdR) and an iron sulfur protein binding a 2Fe-2S cluster (putidaredoxin, Pd) to transport electrons to the camphor hydroxylase cytochrome P450.[59] The PdR abstracts two electrons (as a hydride ion) from NADH to reduce its FAD cofactor to the fully reduced hydroquinone form. The electrons are then ferried one at a time to the cytochrome P450 with consecutive PdR-dependent reductions of the Pd 2Fe-2S cluster and Pd's reoxidation by P450$_{cam}$. This three-component cytochrome P450 redox system (PdR, Pd, P450$_{cam}$) is typical of that found for many other prokaryotic cytochromes P450, although the nature of the iron-sulfur cluster is variable. For example, 3Fe-4S and even 4Fe-4S ferredoxins are known to support bacterial function, and at least one example of a FMN-binding flavodoxin as the electron conduit between reductase and cytochrome P450 is known.[59–62] Although the adrenal mitochondrial steroid hydroxylase systems (*e.g.* CYP11A1) also use a class 1 redox system (NADH-dependent FAD-binding adrenodoxin reductase [AdR] and 2Fe-2S cluster-binding adrenodoxin [Ad]), the cytochrome P450 and AdR are membrane proteins.[63] The second major redox partner system (class 2) is the diflavin CPR protein, which sources electrons from NADPH to reduce its FAD cofactor, and then shuttles electrons one at a time through its flavodoxin-like FMN-binding domain to the cognate cytochrome P450.[64] The class 2 system is used for the mammalian hepatic cytochromes P450, and is generally also found for other membrane associated non-mitochondrial eukaryotic cytochromes P450.[1]

The realization that nature did not exclusively use class 1 and 2 systems for all cytochromes P450 came with the characterization of the cytochrome P450 BM3 (CYP102A1, BM3) system from *B. megaterium*. BM3 is a soluble fatty acid hydroxylase cytochrome P450 fused to a soluble CPR.[65] The fusion arrangement enables highly efficient electron transfer between the CPR/cytochrome P450 partners, and BM3 has the highest monoxygenase activity yet reported for a cytochrome P450 enzyme ($\sim 285\,\mathrm{s}^{-1}$ with arachidonic acid).[20] Recent studies showed that BM3 is a dimer, with electron transfer between the reductase of one monomer and the heme (cytochrome P450) domain of the other likely representing the major electron transport pathway.[66,67] Several other homologues of the BM3-type arrangement have subsequently been identified in other prokaryotes and lower eukaryotes.[68] Further novelty in cytochrome P450 redox partner systems was identified in other bacteria (including a *Rhodococcus* strain) for the CYP116B family, in which a soluble cytochrome P450 (N-terminal) is fused to a FMN- and 2Fe-2S cluster-binding module resembling the enzyme phthalate dioxygenase reductase and reduced by NAD(P)H.[1,69,70] Following these findings, the impact of genome sequencing has been such that several different types of cytochrome P450-redox partner (and other cytochrome P450-partner protein fusions) have been identified. Figure 8.4 illustrates various examples of cytochrome P450 redox systems and cytochrome P450-redox partner fusion enzymes. These include characterized cytochrome P450-flavodoxin and cytochrome P450-ferredoxin fusion proteins,[71,72] and as yet uncharacterized cytochrome P450 fusions to modules resembling acyl CoA dehydrogenase (CYP221 family). Other "non-standard" cytochrome P450 enzymes include cytochrome P450-heme peroxidase fusions,[73] a fungal nitric oxide reductase cytochrome P450 (P450nor, CYP55A1, originally found in *Fusarium oxysporum*) that catalyses NADH-dependent reduction of two molecules of NO to N_2O in a (likely) NO detoxification role and without redox partner involvement,[74] and cytochromes P450 that catalyse molecular rearrangements of bound substrates without requirement for electrons or oxygen chemistry. An example of the latter type of cytochrome P450 is platelet thromboxane synthase (CYP5A1) that converts prostaglandin H_2 into thromboxane A_2.[75] The recent identification of several new types of cytochrome P450-partner fusion systems offers exciting prospects of uncovering new cytochrome

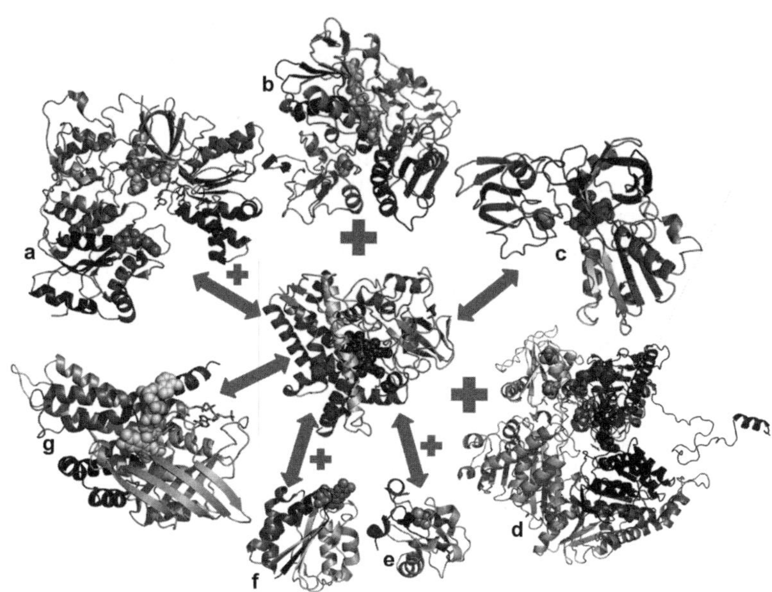

Figure 8.4 Cytochrome P450 redox partner systems. The central image is that of a representative cytochrome P450 (*M. tuberculosis* CYP125, PDB code 3IVY). Heme is shown in dark grey spheres, with the central iron atom in lighter grey. The I helix (running along the face of the heme) is in light grey. Surrounding the central cytochrome P450 are redox partners/cytochrome P450 fusion proteins. Double headed arrows represent cytochrome P450-fusion enzymes expressed in single polypeptide chains. Double headed arrows with small plus (+) signs indicate cytochrome P450 redox systems that are found either fused to the indicated protein or as independently interacting partners in other cytochrome P450 systems. Large plus signs indicate independent interacting cytochrome P450 redox partners. (a) Class 2 CPR (rat structure, PDB: 3ESP) with FAD and FMN shown as dark and light grey spheres, respectively, and bound $NADP^+$ also in light grey sticks adjacent to the FAD. The cytochrome P450 BM3-type (CYP102 family) enzymes are natural fusions of cytochrome P450 and CPR modules. (b) Class 1 type partners as found in the *Pseudomonas putida* cytochrome $P450_{cam}$ system. Redox partners are putidaredoxin reductase (PdR) and putidaredoxin (Pd), shown as the structure of a crosslinked PdR-Pd complex (PDB: 3LB8). FAD in PdR is in light grey spheres and the 2Fe-2S cluster in Pd is in light grey/dark grey spheres. (c) Phthalate dioxygenase reductase (PDOR) from *Burkholderia cepacia* (PDB: 2PIA) with FMN in dark grey and the 2Fe-2S cluster in light grey/dark grey spheres. A PDOR-like module is fused to the cytochrome P450 in the CYP116B enzyme family). (d) *Sulfolobus solfataricus* CYP119 pyruvate-dependent system, as modelled by structure of pyruvate-ferredoxin oxidoreductase (*Desulfovibrio africanus*, PDB: 2PDA), and shown with the TPP cofactor in sticks and three different iron sulfur clusters in spheres. Pyruvate is also shown in sticks, stacking with TPP. (e) Ferredoxin (Fdx), shown as the *Azotobacter vinelandii* 3Fe-4S protein (PDB: 1FDA). A cytochrome P450 is also fused to a 3Fe-4S ferredoxin in the cytochrome P450-Fdx fusion protein from *Methylococcus capsulatus*.

P450-mediated chemistry and novel physiological roles for cytochromes P450, as well as providing the basis for structural studies that could provide insights into modes of interaction between cytochromes P450 and redox partners. While such protein–protein complex structural data are currently lacking in the field, the past decade has been an extremely successful one in terms of providing a tranche of novel crystal structures for both eukaryotic and prokaryotic cytochrome P450 enzymes.

8.7 Cytochrome P450 Structure

The first cytochrome P450 to yield its crystal structure was the intensively studied cytochrome P450$_{cam}$ (CYP101A1), with the initial structure reported in 1985.[76,77] For several years this enzyme provided the template for modelling other cytochromes P450, with the second structure to be crystallized being that of the heme (cytochrome P450) domain of the bacterial cytochrome P450-CPR fusion enzyme BM3 (CYP102A1) in 1993.[78] The structures of cytochrome P450$_{cam}$ and P450$_{BM3}$ (in substrate-free and various substrate- and ligand-bound forms) helped to define what is now clearly known to be a typical cytochrome P450 fold. As seen in Figure 8.5 for the *M. tuberculosis* CYP121 structure, the cytochromes P450 have two major domains.[79] The first of these is the larger α-domain, which consists mainly of α-helical segments (at bottom of Figure 8.5). The smaller β-domain has substantial beta sheet content (top of Figure 8.5). These domains sandwich the heme cofactor in a relatively hydrophobic cavity. The overall architecture of a cytochrome P450 resembles a triagonal prism, and helical and sheet segments are largely conserved throughout the cytochromes P450, albeit with alterations in their positioning and size in many cases, and with structural conservation generally greater in proximity to the heme.[80] The I and L helices make contact with the heme, and the I helix typically contains residues important for substrate interactions and catalysis, including the threonine involved in iron-oxo complex protonation. A β-bulge section that precedes the L helix on the proximal face of the heme is highly conserved in the cytochromes P450. This section contains the cysteine proximal ligand, as well as other highly conserved residues in a relatively rigid arrangement required to hold the cysteine in place and to enable it to accept hydrogen bonds from surrounding peptide NH groups.[80] One of the strongly conserved residues in this segment is a phenylalanine that is typically seven residues before the cysteine. Studies of mutations at the relevant residue in

The 3Fe-4S centre is shown in spheres. (f) Flavodoxin (FLD), shown as the *Desulfovibrio desulfuricans* FLD (PDB: 3KAP). The flavodoxin cindoxin partners the cineole oxidase P450cin. A cytochrome P450 is also fused to a FLD in the *R. rhodochrous* XplA enzyme (CYP170 family). FMN is shown in dark grey spheres. (g) Acyl CoA dehydrogenase (ACAD, shown as *Sus scrofa* ACAD, PDB: 3MDE) representing cytochrome P450-ACAD fusions (CYP221 family). FAD is shown in light grey spheres and the octanoyl CoA substrate is shown as sticks in the ACAD protein.

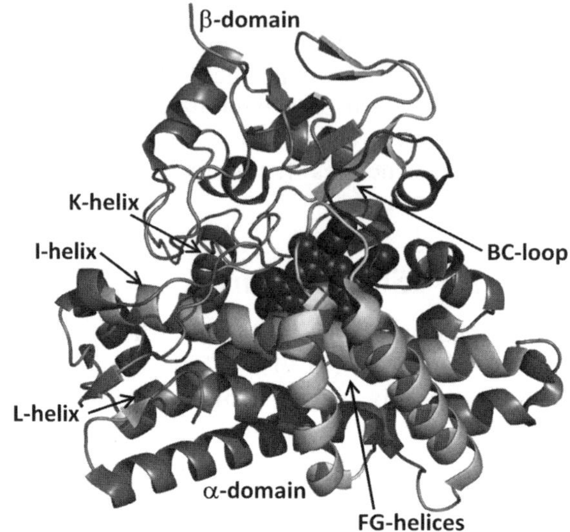

Figure 8.5 Structural properties of the cytochrome P450 CYP121. The *Mycobacterium tuberculosis* cYY oxidizing CYP121 was solved at 1.06 Å resolution (PDB: 1N40).[79] Heme is shown in dark spheres in the centre of the molecule. The beta sheet-rich β-domain is shown at the top of the figure, and the (larger) helix-rich α-domain is at the bottom. The heme is sandwiched between these domains. Structural elements described in the text (Section 8.7) are indicated by arrows, including the long I helix containing several substrate-contacting and catalytically important residues, and the FG-helices (light grey) and BC-loop (in dark grey, linking the B'- and C-helices) regions that likely help to define substrate/product entry/exit channels to the CYP121 active site.

CYP102A1 (F393) showed that F393A and F393H mutants had major effects on the heme iron Fe(III)/Fe(II) redox couple, stabilized the ferrous-oxy form of the enzymes and caused blue-shifts in the Fe(II)-CO absorption maximum to 444 and 445 nm, respectively.[81] This residue is thus likely to play an important role in regulating the electronic properties of the heme system in cytochrome P450 oxygenases. An EXXR motif is also near-completely conserved in the cytochrome P450 K helix, and considered to be involved in salt-bridging interactions for structural stabilization. However, most members of the *Streptomyces* CYP157C cytochrome P450 subfamily lack this motif, and *S. coelicolor* A3(2) CYP157C1 was shown to be a "normal" cytochrome P450 with Fe(II)-CO complex spectral maximum at 448 nm, suggesting the cysteinate heme proximal ligand may be the only amino acid conserved in every cytochrome P450 enzyme.[82]

The eukaryotic cytochromes P450 are usually larger enzymes than those in prokaryotes and archaea, and possess a 30–50 amino acid N-terminal transmembrane segment that tethers them to the endoplasmic reticulum or to the mitochondrial inner membrane.[80] This necessitates the purification of such cytochromes P450 using non-denaturing detergents, and provided obstacles to overcome in the crystallization and structural elucidation of eukaryotic cytochromes P450. However, the removal of the amino terminus membrane anchor region and (frequently) further engineering at the N-terminal end of the truncated cytochromes P450 can produce more soluble entities, and in pioneering work Williams *et al.* produced the first mammalian cytochrome P450 structure when the crystal structure of the rabbit CYP2C5 progesterone hydroxylase was reported in 2000.[83] This breakthrough has been followed by several other eukaryotic cytochrome P450 structures, including that of the important human hepatic drug metabolizing cytochrome P450 CYP3A4. Crystal structures of CYP3A4 in complex with the drugs ketoconazole and erythromycin revealed large structural adaptations to accommodate the molecules, and multiple binding sites for ketoconazole.[84] Although the scale of these particular changes were large for a cytochrome P450, the conformational flexibility of the cytochromes P450 is now well recognized, and distinct conformational changes are frequently observed on binding substrates/ligands – *e.g.* in the case of CYP102A1 on binding to palmitoleic acid.[85] While mechanisms for substrate entry/product exit may vary between cytochromes P450, crystal structure and associated studies have consistently pointed to involvements of the regions including the BC loop (between B′ and C helices) and the FG helices in conformational changes to facilitate substrate entry for catalysis.[80] With novel structures of both prokaryotic and eukaryotic cytochromes P450 now being routinely solved, challenges for structural biologists in the area must include the determination of structures for the physiologically relevant membrane-bound forms of the eukaryotic cytochromes P450, as well as obtaining a better understanding of the conformational dynamics in cytochrome P450 enzymes that drive substrate association, active site binding, catalysis and product release.

8.8   Physiological Roles of Cytochromes P450

Detailed descriptions of the multiple roles of cytochrome P450 enzymes in the physiology of eukaryotic and prokaryotic organisms are beyond the scope of this chapter, and the reader is directed to several useful reviews in the area relating to physiological roles for cytochromes P450 in several organisms from bacteria through to man.[26,54,86–88] As an example, this section will focus on the activities of selected human cytochromes P450.

As a superfamily, the range of chemical reactions achievable by cytochrome P450 enzymes is vast, and includes (but is not limited to) alkene epoxidation, epoxide reduction, oxidative C–C bond cleavage, dehydration, dehydrogenation, hydrocarbon hydroxylation, aromatic hydroxylation, N-, O- and S-dealkylation, N-hydroxylation, N- and S-oxidation, reductive dehalogenation, isomerization and oxidative deamination.[89,90] Several of these chemical activities are typical of members of the 57-strong human

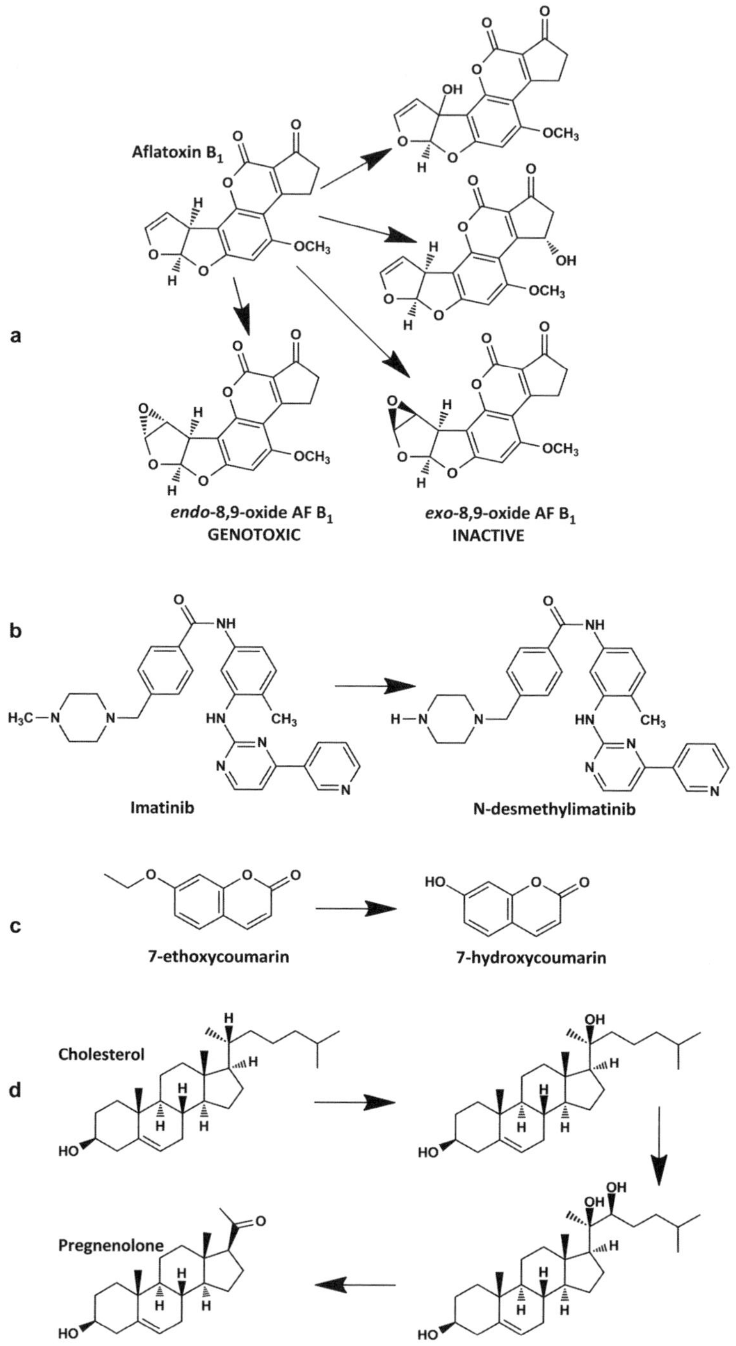

Figure 8.6 Examples of important human cytochrome P450-catalysed reactions. (a) Oxidation of the *Aspergillus* spp. aflatoxin B1 can be performed by CYP3A4, CYP1A2 and other cytochromes P450, resulting in

cytochrome P450 cohort, and exemplary reactions for human (and other mammalian) cytochromes P450 are shown in Figure 8.6. However, there are several other crucial roles for cytochromes P450 in humans, including: thromboxane A_2 synthesis from prostaglandin H_2 by platelet CYP5A1 (causing vasoconstriction and platelet aggregation), cholesterol 7α-hydroxylation by liver CYP7A1 (the rate-limiting step in bile acid synthesis), retinoic acid 4-hydroxylation by CYP26B1 in brain, vitamin D activation through 1α-hydroxylation by kidney mitochondrial CYP27B1, and cholesterol 24-hydroxylation by CYP46A1 in brain (producing an activator of the Liver X receptor [LXR] family of nuclear hormone receptors).[54] A huge array of oxidative reactions on drugs and other xenobiotics (aimed at detoxification and to enable drug excretion) are also catalysed by the hepatic cytochromes P450 (including CYP1A2, CYP2C9, CYP2C19, CYP2D6 and CYP3A4), with well-known reactions including 4-hydroxylation of the antihypertensive debrisoquine (CYP2D6) and N^3-demethylation of the stimulant caffeine (CYP1A2).[54] With structural biology now becoming a tool more readily applicable to human and other mammalian cytochromes P450, the binding modes of substrates/inhibitors can be probed to (for instance) rationalize mechanism or enable enhanced drug design. As discussed above, cholesterol is an important substrate for human cytochromes P450, with cholesterol oxidation reactions leading to synthesis of steroid hormones and bile acid, and to nuclear receptor activation. Figure 8.7 shows active site structures for three human cytochrome P450 enzymes that metabolize cholesterol or derivatives thereof, and the structure of a *M. tuberculosis* cytochrome P450 cholesterol oxidase (CYP125) in complex with an oxidized cholesterol derivative (cholest-4-en-3-one). In each case the images show the different binding modes of the substrates in the respective active sites and the distinctive amino acids side chains of proximal amino acids (including the prospective proton donor threonine residue) reflecting the evolution of the different cytochromes P450 to achieve distinctive oxidative transformations of cholesterol and cholesterol-derived substrates (see Figure 8.7 legend). In the case of *M. tuberculosis* CYP125 (Figure 8.7b), the pathogen has evolved at least two distinct cholesterol 27-hydroxylases (CYP125 and CYP142) that enable the bacterium to initiate host cholesterol breakdown for energy utilization and to sustain infection when it is engulfed in the human macrophages.[29,91]

---

hydroxylations at different positions and epoxidations. This can result in either inactive or genotoxic/carcinogenic products. (b) N-demethylation of the anticancer drug imatinib (Glivec®) by CYP3A4 and CYP2C8. (c) O-dealkylation of 7-ethoxycoumarin to 7-hydroxycoumarin (providing the basis for a simple fluorimetric assay) performed by several human cytochromes P450 (*e.g.* CYP1, 2 and 3 families).[101] (d) The three-step oxidation of cholesterol by CYP11A1 (P450scc or side chain cleavage enzyme) involving two successive cholesterol side chain hydroxylation reactions followed by a C–C bond cleavage reaction to yield pregnenolone in a committed pathway to steroid synthesis.

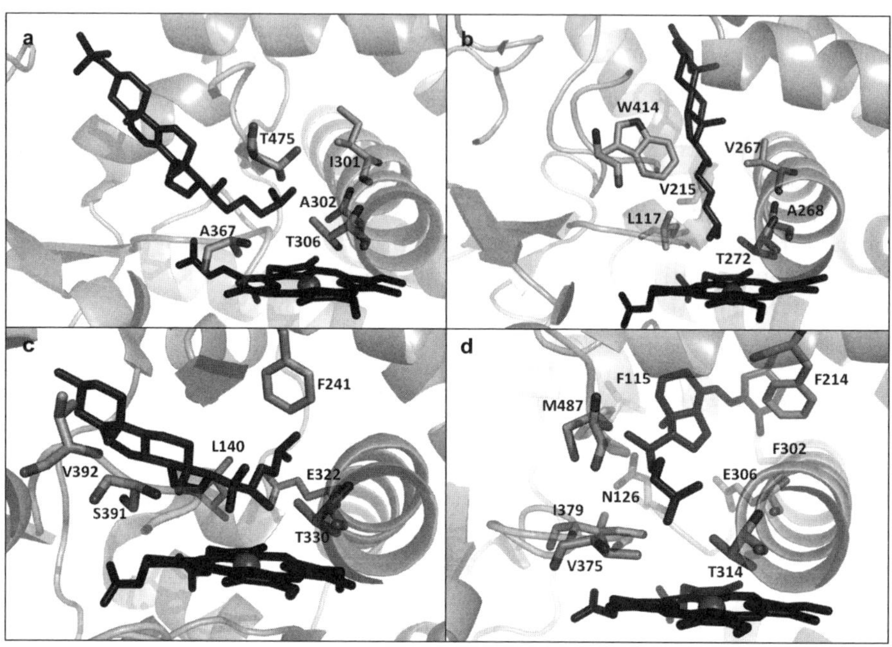

Figure 8.7 Active sites of cytochrome P450 enzymes involved in metabolism of cholesterol and derived compounds. Images are shown of different cytochrome P450 enzymes in complex with their cholesterol or cholesterol derivative substrates. All structures are orientated such that the heme is relatively planar at the bottom of each panel, and with the view as seen along the direction of the I helix. Secondary structure is shown as a light grey cartoon in each case, with substrates coloured dark and the heme macrocycles in black sticks with iron atoms as light grey spheres. Proximal active site residues are shown in each case, including the conserved proton donating threonines. Different sites of hydroxylation of the substrates can be inferred from the distinct binding modes and orientations of the substrates with respect to the heme. (a) Human CYP46A1 in complex with cholesterol sulfate (cholesterol 24-hydroxylase, PDB: 2Q9F).[102] (b) *M. tuberculosis* CYP125 in complex with 4-cholesten-3-one (cholesterol 27-hydroxylase, PDB: 2X5W).[103] (c) Human CYP11A1 in complex with 20,21-dihydroxycholesterol (cholesterol side chain cleavage, PDB: 3NA0). (d) Human CYP2R1 in complex with vitamin D3 (vitamin D3 25-hydroxylase, PDB: 3C6G).[104]

8.9 Cytochrome P450 Medicine and Biotechnology

Previous sections of this chapter have focused on aspects of chemistry, structure, spectroscopy, electron transfer and physiology of the cytochromes P450. However, an overview of the cytochromes P450 would not be complete without a short examination of the prospects for exploitation of the cytochromes P450 for biotechnological uses, and of their importance in medicine both as drug targets and as a consequence of CYP isoform mutations being associated with disease.

From a biotechnological perspective, much excitement still surrounds the prospects of using engineered cytochromes P450 systems to generate regio- (and possibly stereo-) specifically oxygenated molecules as high value intermediates in synthetic pathways, or as reagents in their own right. As an example, engineering of the yeast *Saccharomyces cerevisiae*, including introduction of genes encoding a mammalian cholesterol side chain cleavage cytochrome P450 (CYP11A1) and its associated redox partners, facilitated the production of pregnenolone and progesterone from the yeast cells.[92] In addition, microbial systems expressing an appropriate compactin hydroxylase cytochrome P450 have been used industrially as a more efficient and cost-effective route to production of the cholesterol-lowering drug pravastatin, also overcoming (at least partially) problems encountered with different stereoisomers of products that are formed during chemical synthesis.[93] More recent studies have used directed evolution (random or semi-random mutagenesis coupled to an activity screen) to engineer cytochromes P450 for desired activities. A popular target enzyme for such treatment is CYP102A1, which has been evolved, for instance, into specific variants capable of producing selected drug metabolites typical of those generated by human cytochromes P450 (for diagnostic applications), and of oxidizing short chain alkanes (with possible bioenergy applications).[94,95]

From a biomedical perspective, several cytochrome P450 inhibitors have been produced over the years, both as tools for investigating cytochrome P450 function and as specific CYP isoform inhibitors for drug treatments (Figure 8.8). The best known are the azole series of drugs, which inactivate (by heme iron coordination) the CYP51 lanosterol 14α-demethylase of yeasts and fungi. Azoles such as clotrimazole, fluconazole, econazole and ketoconazole are used worldwide to treat topical and systemic infections by pathogens such as *Candida albicans*.[96] The pyridine ring-containing drug metyrapone (Figure 8.8a) was one of the first clinically used cytochrome P450 inhibitors. The target enzyme was CYP11B1 (11-deoxycortisol-11-hydroxylase, the final step in cortisol biosynthesis) for the treatment of hypercortisolism (Cushing's disease).[97] While the azoles and metyrapone ligate the heme iron and achieve greater inhibitory potency through favourable interactions between the remaining parts of their molecules and their target cytochrome P450's active site residues, other cytochrome P450 inhibitory mechanisms have been exploited clinically. Examples include acetylenic sterol-derived drugs such as the oral contraceptive norgestrel (13-ethyl-17-hydroxyl-18, 19-dinor-17α-pregn-4-en-20-yn-3-one), and danazol (pregna-2,4-dien-20-yno[2,3-*d*]isoxazol-17-ol) used mainly for treatment of endometriosis. The effectiveness of these molecules relies on their substrate-like nature and the oxidative attack of the target cytochrome P450 compound I species on the carbon–carbon triple bound groups in these inhibitors, leading to reactive derivatives that may covalently attach to a heme pyrrole nitrogen and/or modify active site amino acid side chains.[41] Several diseases (*e.g.* hypercortisolism, hypercholesterolemia and type I vitamin D-dependent rickets) are also associated with mutations to specific

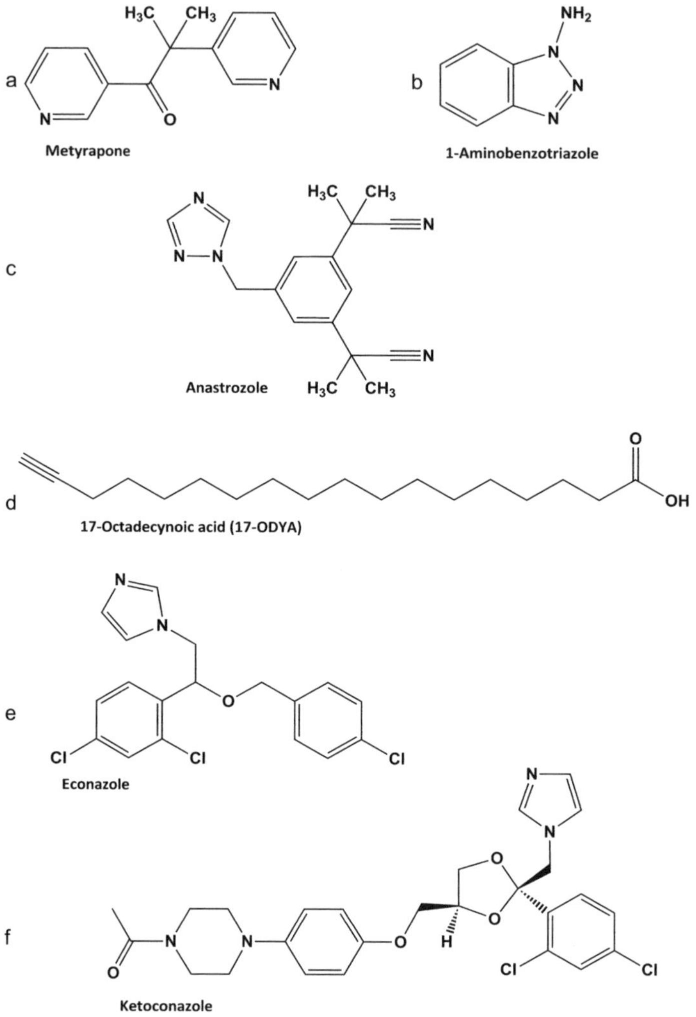

Figure 8.8 Selected inhibitors of cytochrome P450 reactions. (a) Metyrapone – inhibits by pyridine nitrogen coordination of heme iron. Inhibits CYP11B1 and has been used for hypercortisolism (Cushing's disease) and treatment of other hormonal disorders. (b) 1-Aminobenzotriazole – a mechanism-based suicide inhibitor, relatively unselective between cytochromes P450. (c) Anastrozole – a breast cancer drug, acting as a competitive inhibitor of aromatase (CYP19A1). (d) 17-Octadecynoic acid (ODYA) – a mechanism based (suicide) inhibitor of fatty acid ω-hydroxylase cytochromes P450, resulting in heme and/or protein covalent modification. (e) Econazole – imidazole-based azole inhibitor of fungal CYP51 (lanosterol demethylase). Shown to be highly effective in clearing *M. tuberculosis* infection in mice.[105] (f) Ketoconazole – another imidazole-based azole that competitively inhibits fungal CYP51 by heme coordination, but also inhibits human CYP3A4 and CYP3A5.

human *CYP* genes, and certain cytochromes P450 are also responsible for activation of carcinogens (*e.g.* benzo[*a*]pyrene and other polycyclic aromatic hydrocarbons by CYP1A1), providing further potential targets for therapeutic intervention.

8.10   Conclusions and Future Prospects

There has been enormous progress in the analysis of cytochrome P450 enzymes in the $\sim$50 years since Omura and Sato's first characterization of P450 as a cytochrome enzyme.[10] The huge number of cytochromes P450 among organisms throughout all the domains of life is now fully appreciated, as is their diversity of chemical reactivity, and range of substrate and redox partner recognition. Only very recently has the proposed reactive iron(IV)-oxo species (compound I) implicated in substrate oxidation been conclusively characterized, and even now controversy still rages regarding the oxidative potency of different reactive intermediates in the cytochrome P450 catalytic cycle, and their relative abilities to catalyse reactions such as, for example, C–H bond oxidation, C=C bond epoxidation and sulfoxidation. Further studies in this area should be complemented by research aimed at rationalizing cytochrome P450 conformational dynamics, since this would be important for understanding the structural properties of those states important for substrate entry and channelling to the active site. Such work could provide new perspectives on cytochrome P450 function, *e.g.* to enable therapeutic intervention or protein engineering to enhance activity or alter specificity. Similarly, personal genomics revealing CYP polymorphisms associated with metabolic defects may provide the basis for tailored treatments, with novel drugs developed using new structural information for the target cytochromes P450. From a biotechnological perspective, the promise of engineered cytochrome P450 catalysts with improved stability and activity, as well as novel cytochrome P450 reactivities, could bring the cytochromes P450 centre stage in efforts to compete with or supplant synthetic chemistry approaches to manufacture of industrially desirable molecules, such as chiral synthons for synthetic processes and drug metabolites for molecular diagnostics. An exciting future lies ahead for cytochrome P450 science with the emphasis shifting towards biotechnological applications for a class of enzymes that one of the field's great luminaries, the late Professor Judd Coon, labelled "Nature's most versatile biological catalyst".[98]

References

1. A. W. Munro, H. M. Girvan and K. J. McLean, *Nat. Prod. Rep.*, 2007, **24**, 585.
2. I. G. Denisov, T. M. Makris, S. G. Sligar and I. Schlichting, *Chem. Rev.*, 2005, **105**, 2253.
3. D. R. Nelson, *Biochim. Biophys. Acta*, 2011, **1814**, 14.

4. R. W. Estabrook, *Drug. Metab. Dispos.*, 2003, **31**, 1461.
5. G. C. Mueller and J. A. Miller, *J. Biol. Chem.*, 1949, **180**, 1125.
6. G. C. Mueller and J. A. Miller, *J. Biol. Chem.*, 1953, **202**, 579.
7. J. Axelrod, *J. Pharmacol. Exp. Ther.*, 1955, **114**, 430.
8. M. Klingenberg, *Arch. Biochem. Biophys.*, 1958, **75**, 376.
9. D. Y. Cooper, R. W. Estabrook and O. Rosenthal, *Fed. Proc.*, 1963, **22**, 587.
10. T. Omura and R. Sato, *J. Biol. Chem.*, 1962, **237**, 1375.
11. H. Remmer and J. H. Merker, *Ann. NY Acad. Sci.*, 1965, **123**, 79.
12. D. Y. Cooper, S. Levin, S. Narasimhulu, O. Rosenthal and R. W. Estabrook, *Science*, 1965, **147**, 400.
13. M. Katagiri, G. N. Ganguli and I. C. Gunsalus, *J. Biol. Chem.*, 1968, **243**, 3543.
14. A. Y. Lu and M. J. Coon, *J. Biol. Chem.*, 1968, **243**, 1331.
15. T. Kimura and K. Suzuki, *Biochem. Biophys. Res. Commun.*, 1965, **20**, 373.
16. K. Suzuki and T. Kimura, *Biochem. Biophys. Res. Commun.*, 1965, **19**, 340.
17. K. J. McLean, M. Sabri, K. R. Marshall, R. J. Lawson, D. G. Lewis, D. Clift, P. R. Balding, A. J. Dunford, A. J. Warman, J. P. McVey, A. M. Quinn, M. J. Sutcliffe, N. S. Scrutton and A. W. Munro, *Biochem. Soc. Trans.*, 2005, **33**, 796.
18. R. Perera, M. Sono, J. A. Sigman, T. D. Pfister, Y. Lu and J. H. Dawson, *Proc. Natl. Acad. Sci. USA*, 2003, **100**, 3641.
19. J. H. Dawson, *Science*, 1988, **240**, 433.
20. M. A. Noble, C. S. Miles, S. K. Chapman, D. A. Lysek, A. C. Mackay, G. A. Reid, R. P. Hanzlik and A. W. Munro, *Biochem. J.*, 1999, **339**, 371.
21. S. N. Daff, S. K. Chapman, K. L. Turner, R. A. Holt, S. Govindaraj, T. L. Poulos and A. W. Munro, *Biochemistry*, 1997, **36**, 13816.
22. S. G. Sligar and I. C. Gunsalus, *Proc. Natl. Acad. Sci. USA*, 1976, **73**, 1078.
23. H. Ouellet, P. M. Kells, P. R. Ortiz de Montellano and L. M. Podust, *Biorg. Med. Chem. Lett.*, 2001, **21**, 332.
24. L. M. Podust, T. L. Poulos and M. R. Waterman, *Proc. Natl. Acad. Sci. USA*, 2001, **98**, 3068.
25. V. Mougdal and J. Sobel, *Exp. Opin. Pharmacother.*, 2010, **11**, 2037.
26. K. J. McLean and A. W. Munro, *Drug Metab. Rev.*, 2008, **40**, 427.
27. K. J. McLean, A. J. Warman, H. E. Seward, K. J. Marshall, H. M. Girvan, M. R. Cheesman, M. R. Waterman and A. W. Munro, *Biochemistry*, 2006, **45**, 8427.
28. H. Ogura, C. R. Nishida, U. R. Hoch, R. Perera, J. H. Dawson and P. R. Ortiz de Montellano, *Biochemistry*, 2004, **43**, 14712.
29. M. D. Driscoll, K. J. McLean, C. Levy, N. Mast, I. A. Pikuleva, P. Lafite, S. E. Rigby, D. Leys and A. W. Munro, *J. Biol. Chem.*, 2010, **285**, 38270.
30. H. M. Girvan, K. R. Marshall, R. J. Lawson, D. Leys, M. G. Joyce, J. Clarkson, W. E. Smith, M. R. Cheesman and A. W. Munro, *J. Biol. Chem.*, 2004, **279**, 23274.

31. J. S. Miles, A. W. Munro, B. N. Rospendowski, W. E. Smith, J. McKnight and A. J. Thomson, *Biochem. J.*, 1992, **288**, 503.
32. T. G. Spiro, The resonance Raman spectroscopy of metalloporphyrins and heme proteins, in *Iron Porphyrins*, ed. A. B. P. Lever and H. B. Gray, Addison-Wesley Publishers: Reading, Massachusetts, 1983, pp. 89–159.
33. J. H. Dawson and S. P. Cramer, *FEBS Lett.*, 1978, **88**, 127.
34. J. H. Dawson, L. A. Andersson and M. Sono, *J. Biol. Chem.*, 1982, **257**, 3606.
35. D. F. V. Lewis, *Drug Metab. Rev.*, 1986, **17**, 1.
36. J. Rittle and M. T. Green, *Science*, 2010, **330**, 933.
37. R. J. Lawson, D. Leys, M. J. Sutcliffe, C. A. Kemp, M. R. Cheesman, S. J. Smith, J. Clarkson, W. E. Smith, I. Haq, J. B. Perkins and A. W. Munro, *Biochemistry*, 2004, **43**, 12410.
38. M. Sharrock, P. G. Debrunner, C. Schulz, J. D. Lipscomb, V. Marshall and I. C. Gunsalus, *Biochim. Biophys. Acta*, 1976, **420**, 8.
39. O. Bangcharoenpaurpong, A. K. Rizos, P. M. Champion, D. Jollie and S. G. Sligar, *J. Biol. Chem.*, 1986, **261**, 8089.
40. D. L. Harris and G. H. Loew, *J. Am. Chem. Soc.*, 1994, **116**, 11671.
41. P. R. Ortiz de Montellano and J. J. De Voss, Substrate oxidation by cytochrome P450 enzymes, in *Cytochrome P450: Structure, Mechanism and Biochemistry*, 3rd edn, ed. P. R. Ortiz de Montellano, Kluwer Academic/Plenum Publishers, New York, 2005, pp. 183–245.
42. R. Renneberg, J. Capdevila, N. Chacos, R. W. Estabrook and R. A. Prough, *Biochem. Pharmacol.*, 1981, **30**, 843.
43. A. D. N. Vaz, D. F. McGinnity and M. J. Coon, *Proc. Natl. Acad. Sci. USA*, 1998, **95**, 3555.
44. F. Ogliaro, S. P. de Visser, S. Cohen, P. K. Sharma and S. Shaik, *J. Am. Chem. Soc.*, 2002, **124**, 2806.
45. M. Altarsha, T. Benighaus, D. Kumar and W. Thiel, *J. Am. Chem. Soc.*, 2009, **124**, 4755.
46. S. Shaik, H. Hirao and D. Kumar, *Nat. Prod. Rep.*, 2007, **24**, 533.
47. S. Shaik and S. P. de Visser, Computational approaches to cytochrome P450 function, in *Cytochrome P450: Structure, Mechanism and Biochemistry*, 3rd edn, ed. P. R. Ortiz de Montellano, Kluwer Academic/Plenum Publishers, New York, 2005, pp. 45–85.
48. T. J. Volz, D. A. Rock and J. P. Jones, *J. Am. Chem. Soc.*, 2002, **124**, 9724.
49. P. K. Sharma, S. P. de Visser and S. Shaik, *J. Am. Chem. Soc.*, 2003, **125**, 8698.
50. R. Davydov, R. Perera, S. Jin, T.-C. Yang, T. A. Bryson, M. Sono, J. H. Dawson and B. M. Hoffman, *J. Am. Chem. Soc.*, 2005, **127**, 1403.
51. D. R. Nelson, *Hum. Genomics*, 2009, **4**, 59.
52. D. R. Nelson, L. Koymans, T. Kamataki, J. J. Stegeman, R. Feyereisen, D. J. Waxman, M. R. Waterman, O. Gotoh, M. J. Coon, R. W. Estabrook, I. C. Gunsalus and D. W. Nebert, *Pharmacogenetics*, 1996, **6**, 1.
53. Y. J. Chun, T. Shimada, R. Sanchez-Ponce, M. V. Martin, L. Lei, B. Zhao, S. L. Kelly, M. R. Waterman, D. C. Lamb and F. P. Guengerich, *J. Biol. Chem.*, 2007, **282**, 17486.

54. F. P. Guengerich, Human cytochrome P450 enzymes, in *Cytochrome P450: Structure, Mechanism and Biochemistry*, 3rd edn, ed. P. R. Ortiz de Montellano, Kluwer Academic/Plenum Publishers, New York, 2005, pp. 377–530.
55. K. J. McLean, A. J. Dunford, R. Neeli, M. D. Driscoll and A. W. Munro, *Arch. Biochem. Biophys.*, 2007, **464**, 228.
56. H. Ouellet, J. B. Johnston and P. R. Ortiz de Montellano, *Arch. Biochem. Biophys.*, 2010, **493**, 82.
57. C. M. Jenkins and M. R. Waterman, *J. Biol. Chem.*, 1994, **269**, 27401.
58. M. Wang, D. L. Roberts, R. Paschke, T. M. Shea, B. S. Masters and J. J. Kim, *Proc. Natl. Acad. Sci. USA*, 1997, **94**, 8411.
59. J. A. Peterson, M. C. Lorence and N. Amarneh, *J. Biol. Chem.*, 1990, **265**, 6066.
60. A. Bellamine, A. T. Mangla, W. D. Nes and M. R. Waterman, *Proc. Natl. Acad. Sci. USA*, 1999, **96**, 8937.
61. A. J. Green, A. W. Munro, M. R. Cheesman, G. A. Reid, C. von Wachenfeldt and S. K. Chapman, *J. Inorg. Biochem.*, 2003, **93**, 92.
62. D. B. Hawkes, K. E. Slessor, P. V. Bernhardt and J. J. De Voss, *ChemBioChem*, 2010, **11**, 1107.
63. A. V. Grinberg, F. Hannemann, B. Schiffler, J. Müller, U. Heinemann and R. Bernhardt, *Proteins*, 2000, **40**, 590.
64. A. Gutierrez, A. Grunau, M. J. Paine, A. W. Munro, C. R. Wolf, G. C. Roberts and N. S. Scrutton, *Biochem. Soc. Trans.*, 2003, **31**, 497.
65. L. O. Narhi and A. J. Fulco, *J. Biol. Chem.*, 1987, **262**, 6683.
66. S. D. Black and S. T. Martin, *Biochemistry*, 1994, **33**, 12056.
67. R. Neeli, H. M. Girvan, A. Lawrence, M. J. Warren, D. Leys, N. S. Scrutton and A. W. Munro, *FEBS Lett.*, 2005, **579**, 5582.
68. M. C. Gustafsson, O. Roitel, K. R. Marshall, M. A. Noble, S. K. Chapman, A. Pessegueiro, A. J. Fulco, M. R. Cheesman, C. von Wachenfeldt and A. W. Munro, *Biochemistry*, 2004, **43**, 5474.
69. R. De Mot and H. Parret, *Trends Microbiol.*, 2002, **10**, 502.
70. G. A. Roberts, A. Celik, D. J. Hunter, T. W. Ost, J. H. White, S. K. Chapman, N. J. Turner and S. L. Flitsch, *J. Biol. Chem.*, 2003, **278**, 48914.
71. E. L. Rylott, R. G. Jackson, J. Edwards, G. L. Womack, H. M. Seth-Smith, D. A. Rathbone, S. E. Strand and N. C. Bruce, *Nat. Biotechnol.*, 2006, **24**, 216.
72. C. J. Jackson, D. C. Lamb, T. H. Marczylo, A. G. Warrilow, N. J. Manning, D. J. Lowe, D. E. Kelly and S. L. Kelly, *J. Biol. Chem.*, 2002, **277**, 46959.
73. F. Brodhun, C. Göbel, E. Hornung and I. Feussner, *J. Biol. Chem.*, 2009, **284**, 11792.
74. A. Daiber, H. Shoun and V. Ullrich, *J. Inorg. Biochem.*, 2005, **99**, 185.
75. M. Hecker and V. Ullrich, *J. Biol. Chem.*, 1989, **264**, 141.
76. T. L. Poulos, B. C. Finzel, I. C. Gunsalus, G. C. Wagner and J. Kraut, *J. Biol. Chem.*, 1985, **260**, 16122.
77. T. L. Poulos, B. C. Finzel and A. J. Howard, *J. Mol. Biol.*, 1987, **195**, 687.

78. K. G. Ravichandran, S. S. Boddupalli, C. A. Hasermann, J. A. Peterson and J. Deisenhofer, *Science*, 1993, **261**, 731.
79. D. Leys, C. G. Mowat, K. J. McLean, A. Richmond, S. K. Chapman, M. D. Walkinshaw and A. W. Munro, *J. Biol. Chem.*, 2003, **278**, 5141.
80. T. L. Poulos and E. F. Johnson, Structures of cytochrome P450 enzymes, in *Cytochrome P450: Structure, Mechanism and Biochemistry*, 3rd edn, ed. P. R. Ortiz de Montellano, Kluwer Academic/Plenum Publishers, New York, 2005, pp. 87–114.
81. T. W. Ost, C. S. Miles, A. W. Munro, J. Murdoch, G. A. Reid and S. K. Chapman, *Biochemistry*, 2001, **40**, 13421.
82. S. Rupasinghe, M. A. Schuler, N. Kagawa, H. Yuan, L. Lei, B. Zhao, S. L. Kelly, M. R. Waterman and D. C. Lamb, *FEBS Lett.*, 2006, **580**, 6338.
83. P. A. Williams, J. Cosme, V. Sridhar, E. F. Johnson and D. E. McRee, *Mol. Cell*, 2000, **5**, 121.
84. M. Ekroos and T. Sjögren, *Proc. Natl. Acad. Sci. USA*, 2006, **103**, 13682.
85. H. Li and T. L. Poulos, *Nat. Struct. Biol.*, 1997, **4**, 140.
86. M. A. Schuler and D. Werck-Reichhart, *Annu. Rev. Plant Biol.*, 2003, **54**, 629.
87. K. A. Neilsen and B. L. Møller, Cytochrome P450s in plants, in *Cytochrome P450: Structure, Mechanism and Biochemistry*, 3rd edn, ed. P. R. Ortiz de Montellano, Kluwer Academic/Plenum Publishers, New York, 2005, pp. 553–583.
88. R. Feyereisen, *Annu. Rev. Entomol.*, 1999, **44**, 507.
89. R. Bernhardt, *J. Biotechnol.*, 2006, **124**, 128.
90. M. Sono, M. P. Roach, E. D. Coulter and J. H. Dawson, *Chem. Rev.*, 1996, **96**, 2841.
91. K. J. McLean, P. Lafite, C. Levy, M. R. Cheesman, N. Mast, I. A. Pikuleva, D. Leys and A. W. Munro, *J. Biol. Chem.*, 2009, **284**, 35524.
92. C. Duport, R. Spagnoli, E. Degryse and D. Pompon, *Nat. Biotechnol.*, 1998, **16**, 186.
93. J.-W. Park, J.-K. Lee, T.-J. Kwon, D.-H. Yi, Y.-J. Kim, S.-H. Moon, H.-H. Suh, S.-M. Kang and Y.-I. Park, *Biotechnol. Lett.*, 2003, **25**, 1827.
94. A. M. Sawayama, M. M. Chen, P. Kulanthaivel, M. S. Kuo, H. Hemmerle and F. H. Arnold, *Chem. Eur. J.*, 2009, **15**, 11723.
95. R. Fasan, Y. T. Meharenna, C. D. Snow, T. L. Poulos and F. H. Arnold, *J. Mol. Biol.*, 2008, **383**, 1069.
96. G. Girmenia, *Expert Opin. Investig. Drugs*, 2009, **18**, 1279.
97. T. E. Temple and G. W. Liddle, *Annu. Rev. Pharmacol.*, 1970, **10**, 199.
98. M. J. Coon, *Annu. Rev. Pharmacol. Toxicol.*, 2005, **45**, 1.
99. K. J. McLean, M. R. Cheesman, S. L. Rivers, A. Richmond, D. Leys, S. K. Chapman, G. A. Reid, N. C. Price, S. M. Kelly, J. Clarkson, W. E. Smith and A. W. Munro, *J. Inorg. Biochem.*, 2002, **91**, 527.
100. P. Belin, M. H. Le Du, A. Fielding, O. Lequin, M. Jacquet, J. B. Charbonnier, A. Lecoq, R. Thai, M. Courçon, C. Masson, C. Dugave, R. Genet, J. L. Pernodet and M. Gondry, *Proc. Natl. Acad. Sci. USA*, 2009, **106**, 7426.

101. D. J. Waxman and T. K. H. Chang, *Methods Mol. Biol.*, 2006, **320**, 153.
102. N. Mast, M. A. White, I. Bjorkhem, E. F. Johnson, C. D. Stout and I. A. Pikuleva, *Proc. Natl. Acad. Sci. USA*, 2008, **105**, 9546.
103. H. Ouellet, S. Guan, J. B. Johnston, E. D. Chow, P. M. Kells, A. L. Burlingame, J. S. Cox, L. M. Podust and P. R. Ortiz de Montellano, *Mol. Microbiol.*, 2010, **77**, 730.
104. N. Strushkevich, S. A. Usanov, A. N. Plotnikov, G. Jones and H. W. Park, *J. Mol. Biol.*, 2008, **380**, 95.
105. Z. Ahmad, S. Sharma and G. K. Khuller, *FEMS Microbiol. Lett.*, 2006, **261**, 181.

Drug Metabolism by Cytochrome P450: A Tale of Multistate Reactivity

DEVESH KUMAR

Molecular Modelling Group, Indian Institute of Chemical Technology, Hyderabad (A. P.) 500-607, India

9.1 Introduction

Cytochromes P450 (CYP 450 or P450) are heme based monoxygenases that form an iron(IV)-oxo heme cation radical in their catalytic cycle that is its active oxidant. They are named after the absorption band at 450 nm of the carbon-monoxide bound structure and essentially are a superfamily of cysteine thiolate-ligated heme enzymes that are responsible for most of phase I drug metabolism reactions. These enzymes are widespread in nature and have been identified in prokaryotic as well as eukaryotic cells.[1–7] The activity of CYP 450 enzymes determines the drug distribution in the human body, and as a consequence affects their pharmacological as well as toxic effects. There are many different isoforms of CYP 450 and these oxidize a wide range of chemical compounds although the substrate specificity does not correlate with their taxonomical classification. The CYP 450s catalyze various oxidation reactions, including aromatic and aliphatic hydroxylation, N-dealkylation, N-demethylation, O-dealkylation, O-demethylation, ester bond hydrolysis, dehalogenation, *etc.* A selection of substrates and products is given in Figure 9.1.[5] The CYP 450 enzymes play a central role in drug metabolism, which is likely evolved as an extension of the function of these enzymes in the metabolism of

Iron-Containing Enzymes: Versatile Catalysts of Hydroxylation Reactions in Nature
Edited by Sam P de Visser and Devesh Kumar
© Royal Society of Chemistry 2011
Published by the Royal Society of Chemistry, www.rsc.org

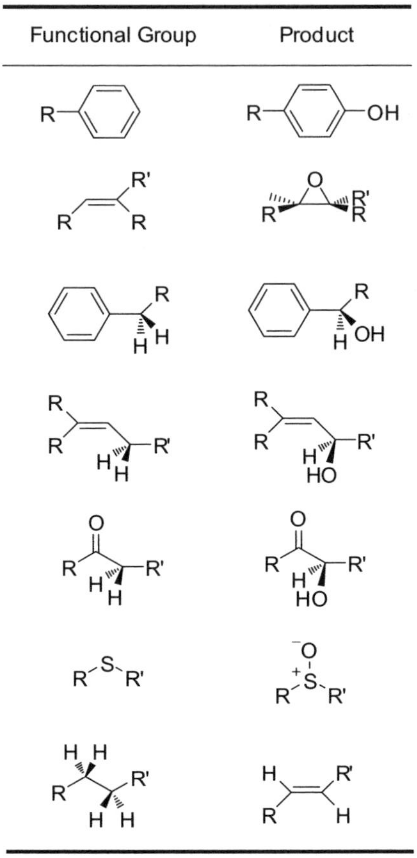

Figure 9.1 Reaction types catalyzed by CYP 450 enzymes.

endogenous substrates. The natural substrates, identified as endogenous compounds, for CYP 450 are classified as xenobiotic metabolizers.[8]

9.2 Nomenclature of Cytochrome P450 Enzymes

The Nomenclature Committee[9] has established a standardized system to name and assign individual genes of the CYP 450 superfamily into families and subfamilies. This is based on the level of amino acid sequence identity, phylogenetic association and gene organization as determined by the CYP 450. An italicized CYP is named for the root of all cytochrome P450 genomic and cDNA sequence, an Arabic numeral designates the individual family followed by the subfamily with letter. A pseudogene is a defective gene that often is a relic of gene duplications but does not produce a functional protein. The pseudogenes are designated by addition of the letter "P" immediately following the Arabic numeral family designation under the nomenclature rules for the cytochrome P450 superfamily.

With a large number of exceptions, CYP 450 sequences that have more than 40% amino acid identity are placed in the same family. Sequences that are more than 55% identical are placed in the same subfamily and those that are more than 97% identical are considered to represent alleles, unless there is evidence to the contrary.[2,10,11] On the basis of these rules, the first officially-named cytochrome P450 was CYP1A1. Since cytochrome P450 sequences are found throughout all domains of living organisms, the blocks of numbers that are reserved for the classification of new cytochrome P450 families, *i.e.*, CYP1–CYP49 and CYP301–CYP499 represent animal CYP 450s, whereas the CYP51–CYP69 and CYP501–CYP699 are CYP 450s found in lower eukaryotes, CYP71–CYP99 and CYP701–CYP999 are plant CYP 450s and finally the CYP101–CYP299 are bacterial CYP 450s.

With the aim to further organize the nomenclature of cytochrome P450s to reflect evolutionary history as well as better organization, the nomenclature committee has proposed to bundle the existing families into phylogenetically-relevant higher ordered clusters called "clans". A "clan" is defined as a group of gene families that evidently diverged from a single common ancestor based on commonalities among multiple phylogenetic assemblies of cytochrome P450 sequences.[12] These clans are then named either for the family with the lowest numerical CYP designation in the cluster or for the best-known family in the cluster. For example, the CYP3 (mammals), CYP5 (mammals), CYP6 (insects) and CYP9 (insects) families are part of the CYP3 clan. Enzyme derived from different gene is termed an isoform. Figure 9.2 presents the family tree of 60 sequenced genes of 18 families and 44 subfamilies of human cytochrome P450.[9]

9.3 Types of Drug Interactions

In the human body, the liver has been recognized as the major site for drug metabolism and, by the 1960s, the cytochrome P450 microsomal enzymes were characterized as key drug metabolizing enzymes. Adverse drug reactions happen if compounds that act as inducers or inhibitors of the enzymes disrupt the normal enzyme function. Enzyme induction can result in accelerated enzyme synthesis, faster drug metabolism and sub-therapeutic drug concentrations. Enzyme inhibition can lead to accumulation of a drug, in particular if the drug has a narrow therapeutic window, and serious toxicity may develop in a short time. Fortunately, because many drugs can be metabolized by more than one enzyme, inhibition of one pathway often brings an alternative pathway into play.[13]

9.3.1 Induction

Enzyme induction usually represents an absolute increase in enzyme synthesis, but phenobarbital, a well-known enzyme inducer, increases blood flow to the liver, and its use results in an increase of up to 50% in liver size.[14] Currently there are less than a dozen clinically important CYP 450 inducers that are

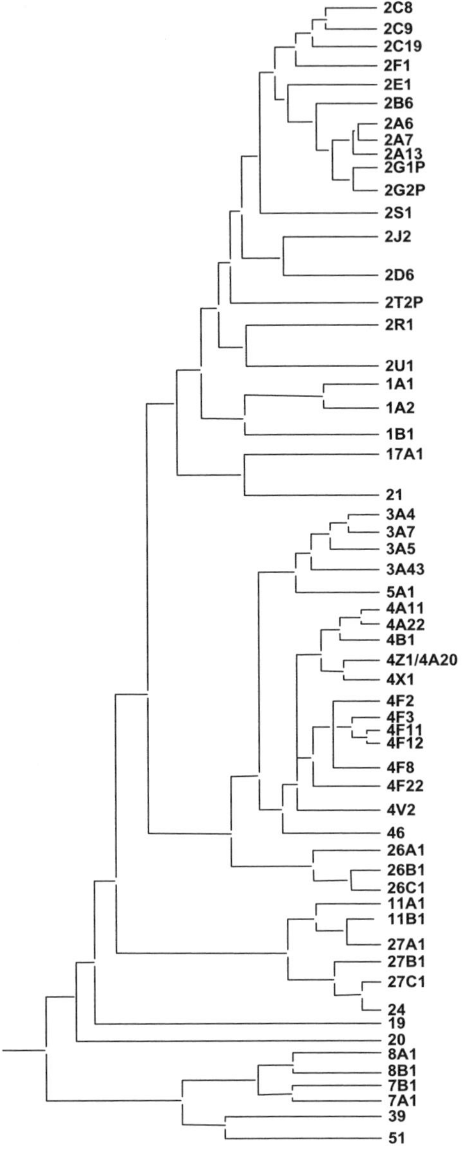

Figure 9.2 Family tree of human CYP 450 enzymes. All the names in the figure include the prefix CYP.

drugs. These include phenobarbital, other anticonvulsants and rifampin. One of the more serious recent examples is the induction of CYP3A4 by rifampin in HIV patients using protease inhibitors.[15] The onset and duration of the interaction depend both on the kinetics of the drug and of the half-life of the CYP enzyme, which ranges from 1–6 days.[14] Important non-pharmacologic inducers include the polycyclic aromatic hydrocarbons in cigarette smoke, which cause the induction of CYP1A2 and result in higher dosing requirements of drugs like theophylline in smokers.[14] Chronic alcohol use induces CYP2E1, but its effects on metabolism of CYP2E1 substrates like acetaminophen are less predictable.

9.3.2 Inhibition

Enzyme inhibition usually represents an absolute decrease in enzyme metabolic activity because of the binding of an inhibitor can stop a substrate from entering into the active or binding site of enzyme. A good example is the inhibition of the metabolism of tricyclic antidepressants like desipramine by fluoxetine.[16] The substrate, desipramine, is metabolized by CYP2D6, and its metabolism is strongly inhibited by binding of fluoxetine to the same isozyme. Fluoxetine at a dose of 20 mg daily can produce a four-to-five-fold increase in the plasma concentration of desipramine, so that a decrease in desipramine dose is necessary to avoid unpleasant or toxic side effects. The duration of the interaction depends on the half-life of the inhibitor.[14] A drug can be both a substrate and an inhibitor for a particular isozyme. A drug that is not a substrate can also inhibit the activity of a non-substrate isozyme.[13] Quinidine is metabolized by CYP3A3/4, but powerfully inhibits CYP2D6.

Sometimes, detailed knowledge of metabolic pathways can predict the clinical significance of an interaction. For example, the potential drug interactions of tertiary tricyclic antidepressants like imipramine by inhibitors of CYP2D6 are limited, because oxidative metabolism can still go on through secondary pathways catalyzed by CYP1A2, CYP2C19, and CYP3A3/4.[13] Although it is helpful to know the extent of the theoretical possibilities, only clinical experience is able to prove which interactions are clinically significant.[17]

9.4 Important Isoforms of Human CYP

There are about 18 families and 43 subfamilies of CYP genes in humans, which comprise up to 60 cytochrome P450s and 58 pseudogenes. In terms of drug metabolism these enzymes mainly belong to the CYP1, CYP2 and CYP3 families. Table 9.1 summarizes the human CYP family with their primary function. In recent years significant numbers of crystal structures of various CYP 450 isoforms have been reported. As an example, Figure 9.3 shows the active site of three isoforms, CYP1A2, CYP2D6 and CYP3A4, as taken from the crystal structure coordinates from 2HI4,[18] 2F9Q,[19] and 3NXU[20] protein databank (pdb) files, respectively.

9.4.1 CYP1A2 Isoform

CYP1A2 mostly metabolizes chemicals and environmental toxins[21] and drug substrates for this isoform are caffeine, propranolol and theophylline[14] The drugs imetidine, fluvoxamine, erythromycin, clarithromycin, enoxacin,

Table 9.1   Diversity and function of human cytochrome P450 enzymes.

CYP family	*Primary function*	*Subfamily, genes, pseudogenes*
CYP1	Drug metabolism	3 Subfamilies, 3 genes, 1 pseudogene
CYP2	Drug and steroid metabolism	13 Subfamilies, 16 genes 16 pseudogenes
CYP3	Drug metabolism	1 Subfamily, 4 genes, 2 pseudogenes
CYP4	Arachidonic acid or fatty acid metabolisms	5 Subfamilies, 11 genes, 10 pseudogenes
CYP5	Thromboxane A2 synthase	1 Subfamily, 1 gene
CYP7A	Bile acid biosynthesis 7-α hydroxylase of steroid nucleus	1 Subfamily, 1 gene
CYP7B	Brain specific form of 7-α hydroxylase	1 Subfamily, 1 gene
CYP8A	Prostacyclin synthase	1 Subfamily, 1 gene
CYP8B	Bile acid biosynthesis	1 Subfamily, 1 gene
CYP11	Steroid biosynthesis	2 Subfamilies, 3 genes
CYP17	Steroid biosynthesis (17-α hydroxylase)	1 Subfamily, 1gene
CYP19	Steroid biosynthesis (aromatase forms estrogens)	1 Subfamily, 1 gene
CYP20	Unknown function	1 Subfamily, 1 gene
CYP21	Steroid biosynthesis	1 Subfamily, 1 gene, 1 pseudogene
CYP24	Vitamin D degradation	1 Subfamily, 1 gene
CYP26	Retinoic acid hydroxylase	3 Subfamily, 3 gene
CYP27A	Bile acid biosynthesis	1 Subfamily, 1 gene
CYP27B	Vitamin D3, 1-α hydroxylase activates vitamin D3	1 Subfamily, 1 gene
CYP27C	Unknown function	1 Subfamily, 1 gene
CYP39	7-α Hydroxylation of 24 hydroxy cholesterol	1 Subfamily, 1 gene
CYP46	Cholesterol 24-hydroxylase	1 Subfamily, 1 gene
CYP51	Cholesterol biosynthesis, lanosterol 14-apha demethylase	1 Subfamily, 1 gene, 3 pseudogenes

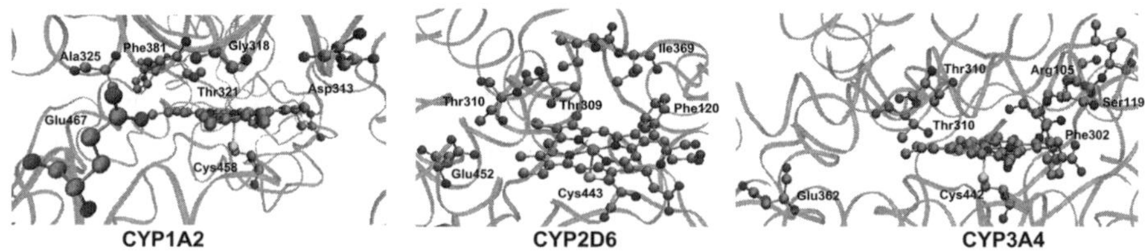

Figure 9.3   Active site of three typical CYP 450 isoforms: (a) CYP1A2 as taken from the 2HI4 pdb file; (b) CYP2D6 as taken from the 2F9Q pdb file; (c) CYP3A4 as taken from the 3NXU pdb file.

ciprofloxacin, isoniazid *etc.* are inhibitor whereas polycyclic aromatic hydrocarbons (cigarette smoke) and dioxin are inducers for CYP1A2.

9.4.2   CYP2C8, CYP2C9 and CYP2C19 Isoforms

CYP2C8, CYP2C9 and CYP2C19 are the most clinically important isoforms among the 2C subfamily. CYP2C8 is well known to catalyze the paclitaxel biosynthesis[22] and the bioconversion of arachidonic acid into epoxyeicosatrienoic acid in liver and kidney tissue.[23] CYP2C9 is highly expressed in hepatic tissue and displays strong metabolic activity towards numerous drugs, including the anticoagulant warfarin, the antidiabetic agent tolbutamide, and sedatives such as hexobarbital, ibuprofen, diclofen and several non-steroidal anti-inflammatory drugs. Drugs like phenobarbital and rifampin are its inducers, whereas other drugs such as amiodarone, fluconazole, fluoxetine, ticlopidine operate as inhibitors. CYP2C19 is known for the metabolism of the anticonvulsant agent mephenytoin and also shows significant activity towards various antidepressants such as diazepam and also towards anti-malarial drugs.

9.4.3   CYP2D6 Isoform

CYP2D6 can metabolize over 80 different clinically used drugs,[24] including most antidepressants, with the exception of fluvoxamine, nefazodone, citalopram and bupropion, which are either inhibitors of CYP2D6 or metabolized by CYP2D6.[25] The CYP2D6 inhibitors include drugs such as quinidine, paroxetine, fluoxetine, norfluoxetine, sertraline, desmethylsertraline, fluvoxamine, nefazodone, venlafaxine, erythromycin and ketoconazole. On the other hand, rifampin is induced by this CYP 450 isozyme.[14]

9.4.4   CYP3A4 Isoform

CYP3A4 metabolizes more than 150 different drugs and drug types, including most calcium channel blockers, benzodiazepines, cyclosporine and tacrolimus.[14] The CYP3A4 isoform does not exhibit polymorphism, but it shows large individual variations in isozyme levels[14,24] and this complicates predictions of drug interactions. About 70% of CYP3A4 isozymes are found in the gut, where they are involved in metabolism of substrates before they can reach the liver. Therefore, CYP3A4 is a major agent of the first-phase of drug metabolism. Commonly known drugs like rifampin, rifabutin and rifapentine act as inducer for CYP3A4 whereas drugs like cimetidine, erythromycin, grapefruit juice, ketoconazole and related species are well-known inhibitors.

Figure 9.4 shows contributions of different isoforms of CYP families towards drug metabolism of the 200 most frequently prescribed drugs in the United States of America. Obviously, from Figure 9.4, most (87%) of these drugs are metabolized by CYP families 1, 2 and 3, with major contributions from

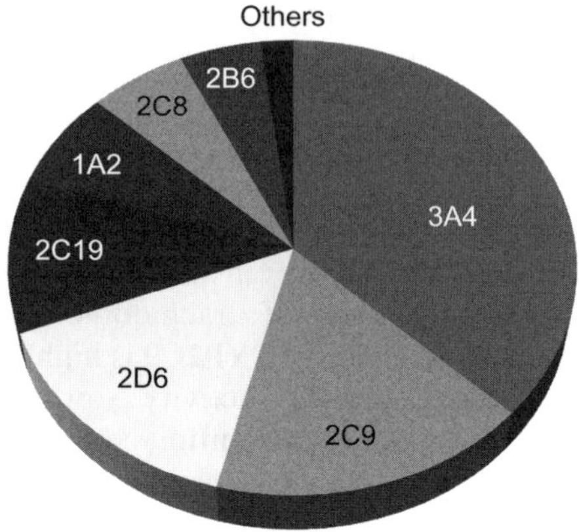

Figure 9.4   Contributions of CYP isoforms towards drug metabolism of the 200 most frequently prescribed drugs in the United States of America.

CYP3A4/5 (37% of drugs) followed by CYP2C9 (17%), CYP2D6 (15%), CYP2C19 (10%), CYP1A2 (9%), CYP2C8 (6%) and CYP2B6 (4%).[26]

9.5   Examples of Generation of Various Metabolites from a Single CYP 450

Several studies were carried out to identify which of the CYP 450 isoforms is responsible for the formation of the primary metabolites of drugs *in vitro* using human liver microsomes. Generally, only one CYP 450 isoform is responsible for formation of the different metabolites, while occasionally one other CYP 450 isoform generates one of the metabolites. For example, three CYP 450-mediated metabolites, namely, ziprasidone sulfoxide, ziprasidone sulfone and oxindole acetic acid, were identified using mass spectroscopy for ziprasidone (a FDA approved drug for the treatment of schizophrenia) *in vitro* using human liver microsomes (Figure 9.5). It was found that CYP3A4 is solely responsible for the formation of ziprasidone metabolites and that ziprasidone is not even a substrate for the other isoforms.[27–29]

Another example of different metabolites of a drug that are cleared by only one CYP 450 isoform is fluvastatin (lescol), which is a 3-hydroxy-3-methylglutaryl coenzyme A reductase inhibitor, and helps to lower the low density lipoprotein cholesterol levels by 20–30% at a daily dose of 20–40 mg.[30–32] Fluvastatin is metabolized by human liver microsomes into three metabolites, namely 5-hydroxy-, 6-hydroxy- and *N*-deisopropyl-fluvastatin (Figure 9.6). The HPLC and mass spectroscopy studies showed that although five CYP 450 isoforms can metabolize fluvastatin, actually all three metabolites are formed by CYP2C9 only, whereas CYP1A1, CYP2C8, CYP2D6 and CYP3A4 only produce 5-hydroxy-fluvastatin.[33]

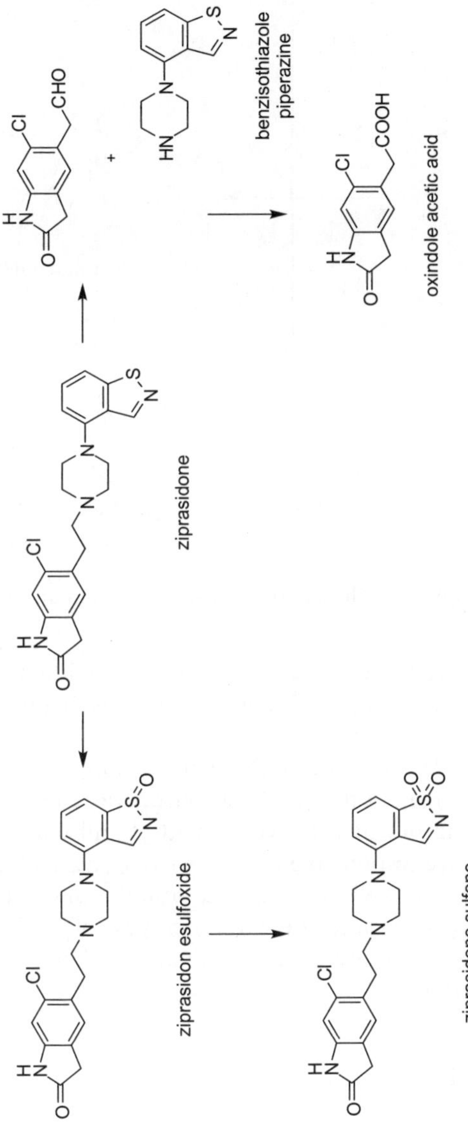

Figure 9.5 *In vitro* metabolic pathways of ziprasidone by human liver microsomes.

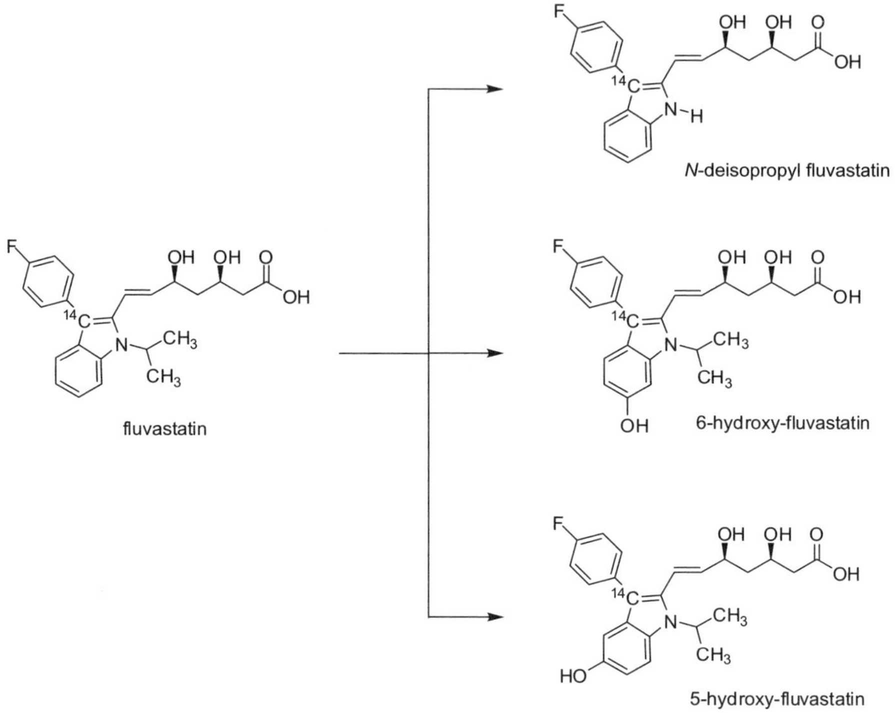

Figure 9.6 *In vitro* metabolic pathways of fluvastatin by human liver microsomes.

There are numerous drug compounds, such as rosiglitazone, thiazolidine-dione, rifampicin, paracetamol, *etc.*,[34–36] whose various metabolites are cleared by only one specific CYP 450 isoform *via* various oxidation reactions such as aromatic and aliphatic hydroxylation, N-dealkylation, O-dealkylation, ester bond hydrolysis, dehalogenation, *etc.* These large varieties of catalytic reactions, however, seem to implicate the involvement of multiple oxidant species in CYP 450 enzymes that are responsible for these reactions. Detailed mechanistic studies, experimentally as well as computationally, have established that actually only one active species in the catalytic cycle of CYP 450 enzymes is responsible for all oxygen mediated reactions, namely Compound I (which will be discussed in detail later).

9.6 CYP 450 Structure

In 1985, Poulos and coworkers reported the first X-ray crystal structure of a CYP 450 isozyme using the soluble bacterial CYP 450$_{cam}$ isolated from the bacterium *Pseudomonas putida*.[37] Now-a-days a range of different crystal structure coordinates are available for catalytic cycle intermediates, which have been characterized using spectroscopic techniques. More than 300 different crystal structures of various CYP 450 isozymes can be found in the protein

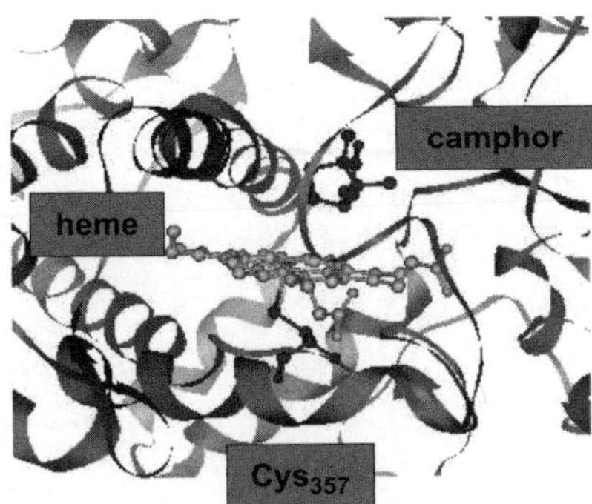

Figure 9.7   Active site of CYP 450$_{cam}$ as taken from the 1DZ9 crystal structure, with axial cysteinate and substrate highlighted.

databank.[38] These isozymes show considerable homology and all contain a central heme *b* group (iron protoporphyrin IX), where the catalysis takes place. Generally, the heme is linked to a cysteine (Cys) amino acid *via* thiolate bonding to the metal, which is often termed the axial ligand of the metal.[39]

Figure 9.7 shows active site of CYP 450$_{cam}$ (a bacterial enzyme that hydroxylates camphor) with the substrate molecule placed over the active site of the enzyme. In all CYP 450 isoforms, the active species of the enzyme is an iron ion ligated to a protoporphyrin IX macrocycle with two additional ligands: one labeled axial that is usually a thiolate group from a cysteinate residue (Cys$_{357}$ in CYP 450$_{cam}$) of the protein and the other labeled distal, which is a variable ligand and the site of molecular oxygen binding. This distal ligand changes during the catalytic cycle of the enzyme and thereby activates the enzyme's main function.

9.7 Catalytic Cycle of CYP 450

The catalytic cycle of CYP 450 enzymes is schematically depicted in Figure 9.8 and starts from a resting state (**1**) where a distal water molecule is ligated to the sixth ligand position of a low spin ferric state. Entry of substrate into substrate binding pocket expels this distal water molecule and also triggers a spin state crossing from low-spin to high-spin[40] resulting in a five-coordinate high spin ferric state (**2**). In most CYP 450 isozymes, the substrate binds tightly to the substrate binding pocket in a locked position through electrostatic and hydrogen bonding interactions.[41] Density functional theory and combined quantum mechanics/molecular mechanics studies confirmed that a spin state crossing occurs with a different spin state ordering for **1** and **2**.[42–44] The change in spin state triggers an electron transfer process from the reductase domain and reduces the ferric complex to a ferrous complex (**3**).[40] Molecular oxygen then binds to **3** to form the ferric-superoxo heme complex (**4**, Fe^{III}-O$_2$$^-$). Upon addition of a second electron to the heme, **4** converts into **5**, which is assumed to be the rate-limiting step in the catalytic cycle,[45] and yields a ferric-peroxo species (**5**, Fe^{III}-O$_2$$^{2-}$).

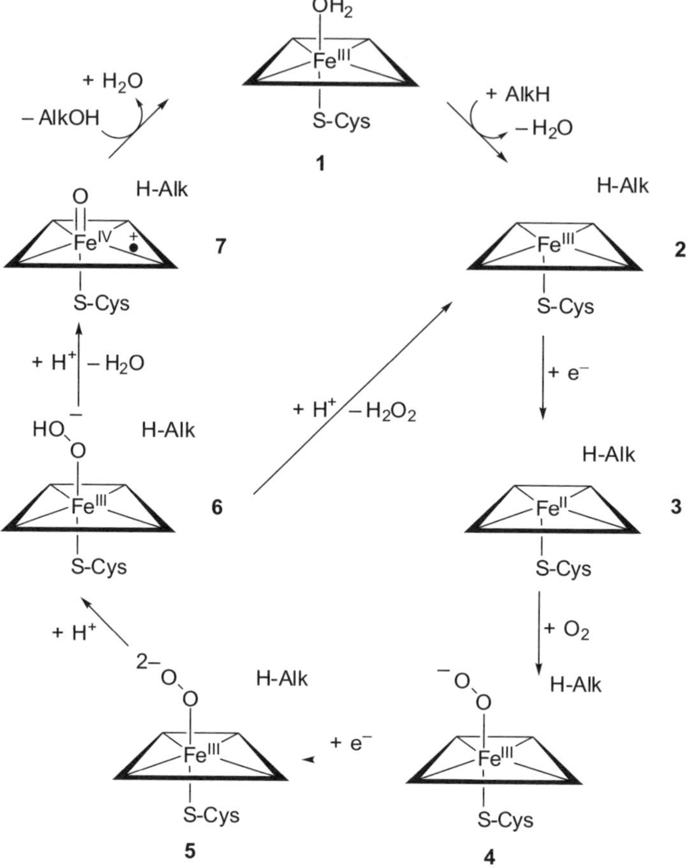

Figure 9.8 Catalytic cycle of cytochrome P450 enzymes.

A subsequent protonation step of **5** gives a ferric-hydroperoxo complex, **6**, also known as Compound 0 (Cpd 0). The delivery of the second proton onto the proximal oxygen atom of Cpd 0 will result in release of a hydrogen peroxide (H_2O_2) molecule to yield **2**, whereas delivery onto the distal oxygen atom of Cpd 0 results in heterolytic O–O bond cleavage to form a water molecule and generate the putative iron(IV)-oxo porphyrin π-cation radical intermediate (**7**). The latter is commonly known as Compound I (Cpd I) and is equivalent to Cpd I of horseradish peroxidase (HRP). The Cpd I (**7**) monoxygenates the substrate, for example, an alkane is converted into an alcohol and released from the binding pocket. The vacant binding site of the heme is then reoccupied with a water molecule that has re-entered the active site and restores the system to the resting state (**1**).

9.8 Compound I of CYP 450: The Active Species

Recently, Rittle and Green reported[46] spectroscopic and kinetic evidence of the existence of Cpd I in CYP119. In their study they found that the electronic structure of P450 Cpd I has a triplet iron(IV)-oxo group ($S=1$) coupled to an $S=\frac{1}{2}$ ligand-base radical. The UV/visible and Mössbauer spectra of CYP119 Cpd I are similar to that found for Cpd I of chloroperoxidase, although the EPR signatures are different. They also investigated the rate constants and kinetic isotope effects (KIE) of substrate hydroxylation reactions for hexanoic acid, octanoic acid, dodecanoic acid and camphor. These studies confirm the existence of Cpd I of CYP 450 and also the capability of Cpd I to hydroxylate unactivated hydrocarbons.[46] Earlier studies inferred its presence from cryogenic and electronic absorption studies[47,48] as well as from cryogenic EPR/ENDOR spectroscopy of the alcohol product of camphor hydroxylation.[49–51] The analogous reactivity (similar product distribution and stereochemical scrambling) of synthetic Cpd I species and CYP 450 towards the same substrates provides strong but indirect support for the participation of Cpd I in the reaction mechanism.[52] Owing to problems in detecting and characterizing Cpd I, many suggestions of alternative oxidant species were made such as Cpd 0,[53–55] and Cpd II,[56–58] which is the one-electron reduced species of Cpd I. However, detailed biomimetic experimental studies showed that Cpd 0 is less reactive than Cpd I and, therefore, ruled out the ferric-hydroperoxo species as a possible oxidant. So far all studies (computational as well as experimental) agree that only one oxidant, namely Cpd I, is active in CYP 450 enzymes.[7,59–61] Since experimental studies failed to characterize Cpd I and were unable to resolve the uncertainty of the status as the primary oxidant of CYP 450 enzymes, theoretical studies have been performed with the main focus of establishing the active species of this enzyme.[62–78] It is generally agreed that Cpd I has two low-lying degenerate spin states, doublet and quartet, that can masquerade the activity of the oxidant and lead to the seeming involvement of multiple oxidants.

Figure 9.9(a) and (b) presents the geometry of Cpd I with two different axial ligands, SH$^-$ and SMe$^-$, respectively, that mimic the cysteinate axial ligand in the enzyme. Also given are group spin densities (ρ) as obtained from DFT and DFT/MM.[79] These values are compared with those obtained from experimental EXAFS[80] and EPR-ENDOR[81] studies.

Figure 9.9(d) shows the high lying occupied and low lying virtual orbitals as well as the electronic configuration of Cpd I (**7**). In particular, the five metal 3d-type orbitals are shown and a high-lying mixed porphyrin-thiolate based orbital, labeled as a$_{2u}$. Since thiolate is a potent electron donor, the sulfur-porphyrin mixing in a$_{2u}$ is significant,[71] but for weaker electron-donor ligands, such as imidazole,[82,83] this mixing is negligible and a$_{2u}$ is a pure porphyrin orbital. The lowest three 3d-block orbitals are, in increasing order of energy, the nonbonding $\delta(3d_{x2-y2})$ orbital and the two π^*_{FeO} (π^*_{xz} and π^*_{yz}) orbitals; the other two involve antibonding interactions between the metal $3d_{xz,yz}$ and the oxygen $2p_{x,y}$ atomic orbitals and they are high-lying virtual orbitals of Cpd I (σ^* orbitals); the σ^*_{xy} orbital describes the Fe–N antibonding interactions whereas the σ^*_{z2} describes the interactions along the O–Fe–S axis. Their energy ordering is dominated by the length and consequently the strength of the Fe–O bond, and when the distance is short the σ^*_{z2} orbital rises above the σ^*_{xy}

(a) B3LYP/MM; (Por, SH)   (b) B3LYP/MM; (Por, SMe)   (c) EXAFS Geometry

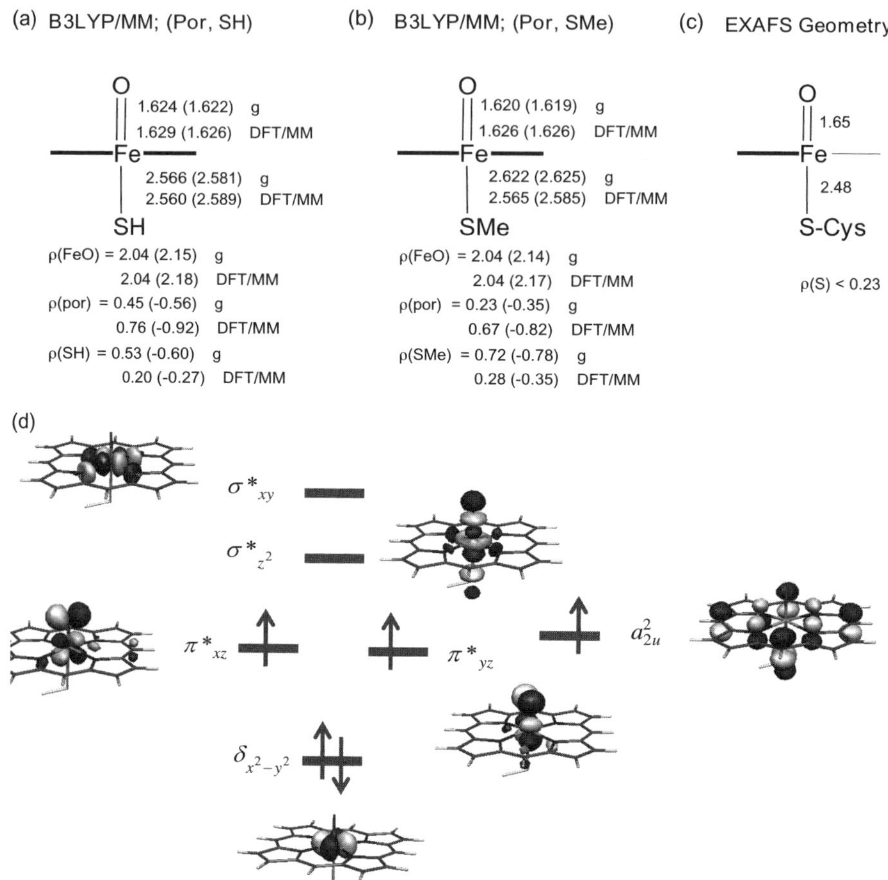

Figure 9.9 Structure of Cpd I (Å) with axial ligand (a) SH$^-$ and (b) SMe$^-$, and group spin densities.[79] (c) Geometry (Å) as obtained from experimental EXAFS[80] and EPR-ENDOR spin densities.[81] (d) Key molecular orbitals of Cpd I. Data refer to ^{4}Cpd I (^{2}Cpd I). Here g stands for gas-phase and ρ is spin density.

orbital. In Cpd I with the Por$^{+\bullet}$Fe^{IV} oxidation state, the 3d-block orbitals have four electrons whereas the porphyrin-based a_{2u} orbital is singly occupied. This occupation generates a pair of almost degenerate spin states: one quartet spin state ($S=\frac{3}{2}$) with occupancy $\delta^{\uparrow\downarrow} \pi^*_{xz}{}^{\uparrow} \pi^*_{yz}{}^{\uparrow} a_{2u}{}^{\uparrow}$ and a doublet spin state ($S=\frac{1}{2}$) with occupancy $\delta^{\uparrow\downarrow} \pi^*_{xz}{}^{\uparrow} \pi^*_{yz}{}^{\uparrow} a_{2u}{}^{\downarrow}$. Obviously, from Figure 9.9(a), in the gas phase the a_{2u} spin density is equally distributed over the porphyrin and sulfur groups whereas in a QM/MM calculated system more spin density is found on the porphyrin. There exists also some other possible spin states ($S=\frac{5}{2}$ and $\frac{3}{2}$) which are also relatively low lying and generated due to one-electron promotion from an electron in the δ orbital to the σ^*_{xy} orbital. These electronic states and their occupancy will be discussed below.[68,84]

9.8.1 Axial Ligand Effect of Compound I

Heme enzymes are common enzymes in nature, but not all heme enzymes have the heme linked to the protein *via* a thiolate bridge of a cysteinate residue. Thus, peroxidases have a central heme group, where the metal is bound to the imidazole ring of a histidine residue, while catalase is a heme enzyme with a tyrosinate axial ligand.[85–87] The differences in heme binding have been correlated with the functional differences of these enzymes, whereby peroxidases detoxify hydrogen peroxide to two water molecules and catalases convert hydrogen peroxide into molecular oxygen and water. The axial ligand of heme enzymes, therefore, seems to affect the function of the heme and it has been suggested that anionic ligands, such as cysteinate have electron-donating properties, *i.e.* entice a "push"-effect on the heme, whereas a neutral histidine ligand will be electron-withdrawing, *i.e.* has a "pull"-effect on the heme.[88,89] Since many computational studies addressed this "push" and "pull" effects, let us include a short summary of studies on peroxidases and catalases here.

Figure 9.10 displays the catalytic cycle of a typical peroxidase, namely cytochrome *c* peroxidase (C*c*P). As mentioned, the heme is linked to the protein *via* a histidine axial ligand.[90] This histidine ligand is part of a hydrogen bonded triad that includes a carboxylic acid group of an aspartic acid (Asp$_{235}$) and an axially located tryptophan group (Trp$_{191}$). On the distal site, substrate hydrogen peroxide binds, which is involved in hydrogen bonding interactions with a histidine group (His$_{52}$) and an arginine side chain (Arg$_{48}$). In the first step of the catalytic cycle, the substrate donates a proton to His$_{52}$ to form an iron(III)-hydroperoxo complex, similar to Cpd 0 in the catalytic cycle of CYP 450. This complex abstracts a proton from His$_{52}$ again and splits off a water molecule to form an iron(IV)-oxo Trp($^{+•}$) species, designated as Cpd I(C*c*P). In most other peroxidases, however, Cpd I has an electronic state similar to that found for CYP 450 with a heme cation radical rather than a tryptophan cation radical. Cpd I(C*c*P) after two reduction and two protonation steps releases a water molecule and returns back to the resting state.

Detailed computational studies using the QM/MM methodology on the enzyme horseradish peroxidase confirmed the first step in the catalysis as a proton relay where His$_{52}$ initially abstracts the proximal proton of hydrogen peroxide that is relayed to the distal oxygen atom in a subsequent step to give Cpd I and a water molecule.[91]

C*c*P was one of the first heme peroxidases for which a crystal structure was determined and Cpd I was characterized with electron paramagnetic resonance (EPR) studies and as such it has become the template of peroxidases in general.[92–94] This is despite the formation of a Cpd I structure with an unusual radical site. Cpd I of CYP 450, designated as [Cpd I(P450)], as well as Cpd I of catalase, or [Cpd I(Cat)], and many heme-peroxidases all have the same electronic state with an iron(IV)-oxo group embedded into a singly-oxidized heme group, *i.e.* [Fe^{IV}=O(heme$^{+•}$)L] with L being the axial ligand. By contrast, the electronic structure of Cpd I(C*c*P) is [Fe^{IV}=O(heme)(Trp$^{+•}$)], where essentially an electron transfer from the axial tryptophan residue to the heme has

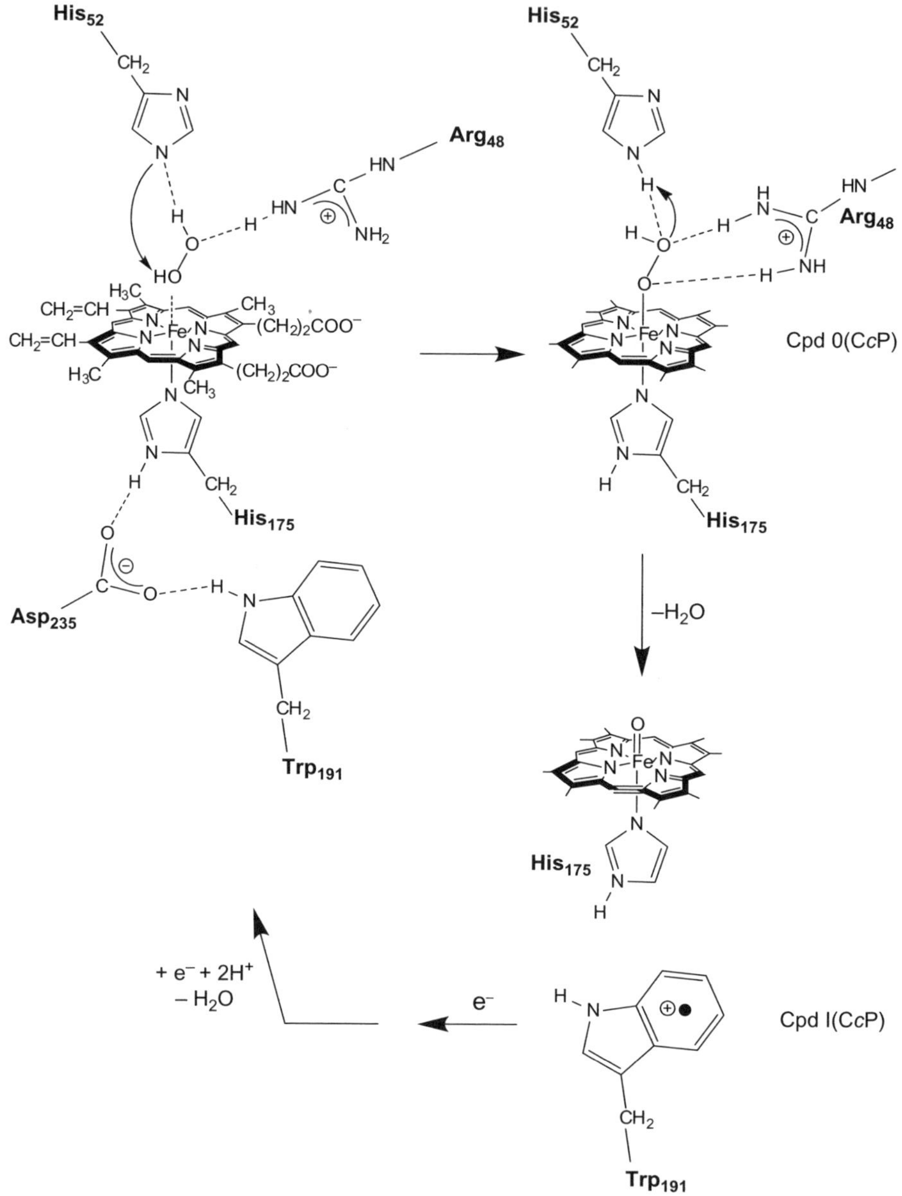

Figure 9.10 Catalytic cycle of cytochrome *c* peroxidase enzymes.

occurred. Structurally, the active site of C*c*P is not dramatically different from that obtained for, for example, horseradish peroxidase (HRP) and ascorbate peroxidase (APX).[95–97] In HRP and APX the heme is also linked to a histidine axial ligand of the protein that is hydrogen bonded to a carboxylic acid group of an aspartic acid residue. APX also has an axial tryptophan residue within hydrogen bonding distance to the axial histidine, but it is missing in HRP and replaced by a phenylalanine residue. Nevertheless, Cpd I in both HRP and APX are characterized as [Fe^{IV}=O(heme$^{+\bullet}$)His].

A series of computational studies were performed to determine the origin of the differences in electronic configuration of Cpd I(CcP) and Cpd I(APX).[98–101] Figure 9.11 displays the optimized geometries of a Cpd I(CcP) model that is described as an iron(IV)-oxo group in protoporphyrin IX (without side chains) and bound to an axial imidazole group. In addition, Asp$_{235}$ and Trp$_{191}$ are included as acetate and indole groups, respectively. Crystal structure coordinates of the active site of CcP and APX reveal a cation binding site in APX that accommodates a K$^+$ ion.[94] Density functional theory studies included a point charge in the model in the cation binding site with a charge $Q = 0$, $+1$ or -1.[100] As follows from the group spin densities depicted in Figure 9.11, the FeO group has a total spin density (ρ) of about 2, implicating two unpaired electrons in FeO type orbitals. These can be assigned to the π^*_{FeO} (π^*_{xz} and π^*_{yz}) orbitals similar to those shown in Figure 9.9(d) for Cpd I(P450). The third unpaired electron spin

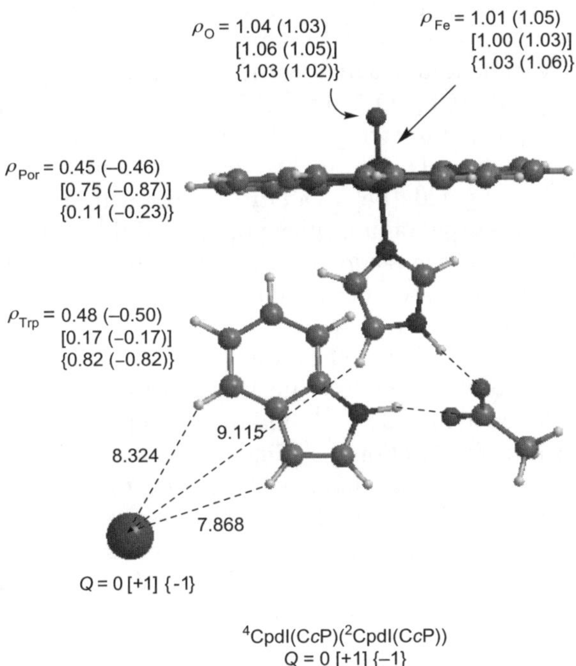

Figure 9.11　Group spin densities of quartet and doublet Cpd I(CcP) under the influence of an added point charge with magnitude $Q = 0$, $+1$ or -1. Group spin densities out of parentheses correspond to quartet and those in parentheses correspond to doublet spin state with $Q = 0$. Group spin densities in square brackets correspond to $Q = +1$ whereas those in curly brackets correspond to $Q = -1$ in the above fashion. Also given are distances (Å) between the point charge and atoms of the chemical model.

density is distributed over the heme (ρ_{Por}) and the tryptophan ligand (ρ_{Trp}). Without a point charge there is approximately equal spin density, *i.e.* radical character, on the heme and tryptophan units. This is in agreement with early studies of the Siegbahn group on peroxidases,[102] where also spin density on both groups was found. However, adding a point charge to the system dramatically changes the radical character of the heme and tryptophan groups, whereby a positive charge gives an [Fe^{IV}=O(heme$^{+\bullet}$)(Trp)] conformation, while a negative point charge an [Fe^{IV}=O(heme)(Trp$^{+\bullet}$)] situation. These studies, therefore, confirm the experimental observation of charge separation in the various Cpd I species and show the intricate balance charges in Cpd I. More recent QM/MM studies on the difference of CcP and APX enzymes revealed an induced electric field effect of the protein to affect the orbital energy levels of Cpd I, and hence the relative energies of [Fe^{IV}=O(heme$^{+\bullet}$)(Trp)] *versus* [Fe^{IV}=O(heme)(Trp$^{+\bullet}$)].[103,104]

Clearly, the axial ligand plays a vital role in controlling the electron affinity of Cpd I due to subtle charge/spin density relay from the heme to other groups in the vicinity. As a consequence, there have been many efforts to quantify the axial ligand effect of hemes. The first such study was performed by Gross and coworkers and relates to an experimental investigation using [Fe^{IV}=O(TMP$^{+\bullet}$)L] with L = F$^-$, Cl$^-$, CH$_3$COO$^-$, CF$_3$SO$_3^-$ and ClO$_4^-$ (TMP = tetramesitylporphyrin).[105–107] Thus, ligating anionic ligands (F$^-$, Cl$^-$, CH$_3$COO$^-$) gave enhanced reactivities in styrene epoxidation as compared to nonligating ligands (CF$_3$SO$_3^-$ and ClO$_4^-$). In addition, a *trans*-influence on the spectroscopic parameters was identified, whereby the systems with strong "push"-effect ligands gave higher Fe–O vibrational frequencies. Nam and coworkers investigated [Fe^{IV}=O(TPFPP$^{+\bullet}$)L] with L = Cl$^-$ or NCCH$_3$ [TPFPP = *meso*-tetrakis(pentafluorophenyl)porphyrin].[108] The axial ligand effect gave differences in *cis versus trans* product ratios in olefin epoxidation, differences in kinetic isotope effects and even regioselectivity changes from aliphatic to aromatic hydroxylation.

To understand the axial ligand effect on the catalysis of substrate activation by Cpd I models, many studies have investigated the electronic properties of Cpd I with variable axial ligand,[109–116] and the subsequent reactivity differences in oxygen atom transfer reactions.[117] Figure 9.12 displays a selection of calculated Cpd I type species that differ by the ligand *trans* to the oxo group as taken from refs 73, 83, 100, 113 and 118. These systems represent a CYP 450 Cpd I mimic, Cpd I(SH), where the cysteinate ligand is replaced by thiolate. It also has mimics of Cpd I of catalase, Cpd I(Cat), and cytochrome *c* peroxidase, Cpd I(CcP), where the axial ligands undergo hydrogen bonding interactions with nearby residues. Finally, Figure 9.12 displays a biomimetic model system with a chloride axial ligand, Cpd I(Cl). As follows from the optimized geometries in Figure 9.12, the average Fe–N distance falls within a narrow range of 2.017–2.022 Å. By contrast, much larger deviations are found for the Fe–O distances, which range from 1.648 Å for ^{2}Cpd I(SH) to 1.665 Å for ^{4}Cpd I(CcP). This will directly affect the stretch vibration for the Fe–O mode, but also implies that the bond is stronger for Cpd I(SH) as compared to Cpd I(CcP), for instance.

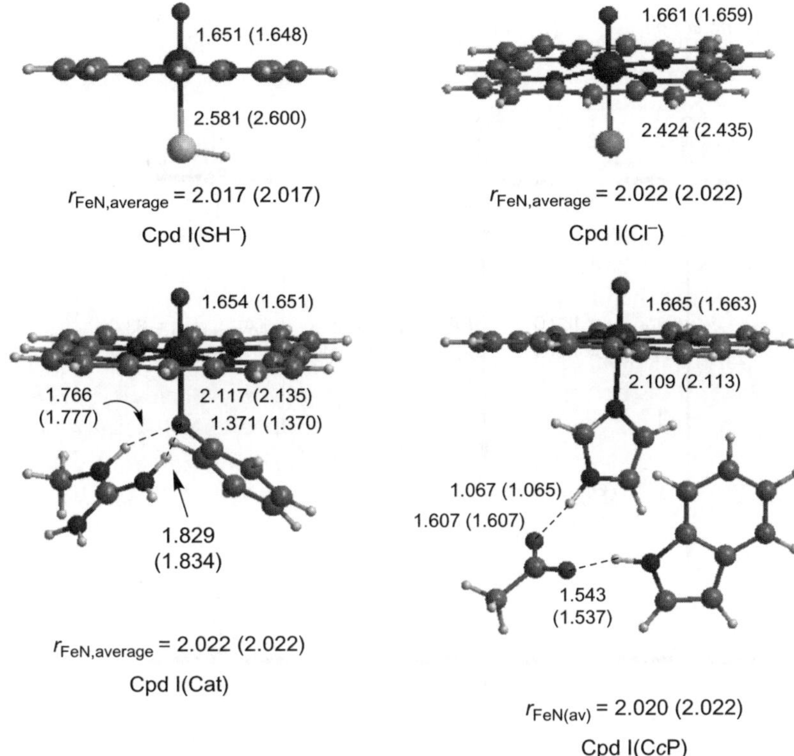

Figure 9.12   Optimized geometries of various Cpd I models with different axial ligand descriptions. Bond lengths are in ångströms.

Shaik, de Visser and coworkers estimated the axial ligand effect through investigating the electron affinity of Cpd I ($EA_{\text{Cpd I}}$) and compared the result with the electron affinity of a Cpd I system without axial ligand (EA_{NL}) and a model where the axial ligand is replaced by a point charge (EA_{PC}).[112,114] Figure 9.13 displays the calculated electron affinities for CYP 450 Cpd I and catalase Cpd I. A full geometry optimization of Cpd I(SH) and Cpd I(Cat) and their one-electron reduced forms, Compound II or Cpd II, was performed and subsequently single point calculations were carried out for the structures without axial ligand or with a point charge in the position of the atom that is bound to the metal. From these calculations, the effect of the axial ligand can be divided into an electric field effect (ΔE_{field}) and a quantum mechanical effect (ΔE_{QM}). The electric field effect induced by the axial ligand is estimated from the energy difference between EA_{NL} and EA_{PC}, whereas the quantum mechanical effect is estimated from the difference between EA_{PC} and $EA_{\text{Cpd I}}$. That way an electric field effect contribution of 76.7 kcal mol⁻¹ is calculated for Cpd I(Cat) and 54.0 kcal mol⁻¹ for Cpd I(SH). On the other hand, values of $\Delta E_{\text{QM}} = 1.8$ and 38.8 kcal mol⁻¹ are found for Cpd I(Cat) and Cpd I(SH), respectively. Thus, in the case of catalase the dominant contribution is the

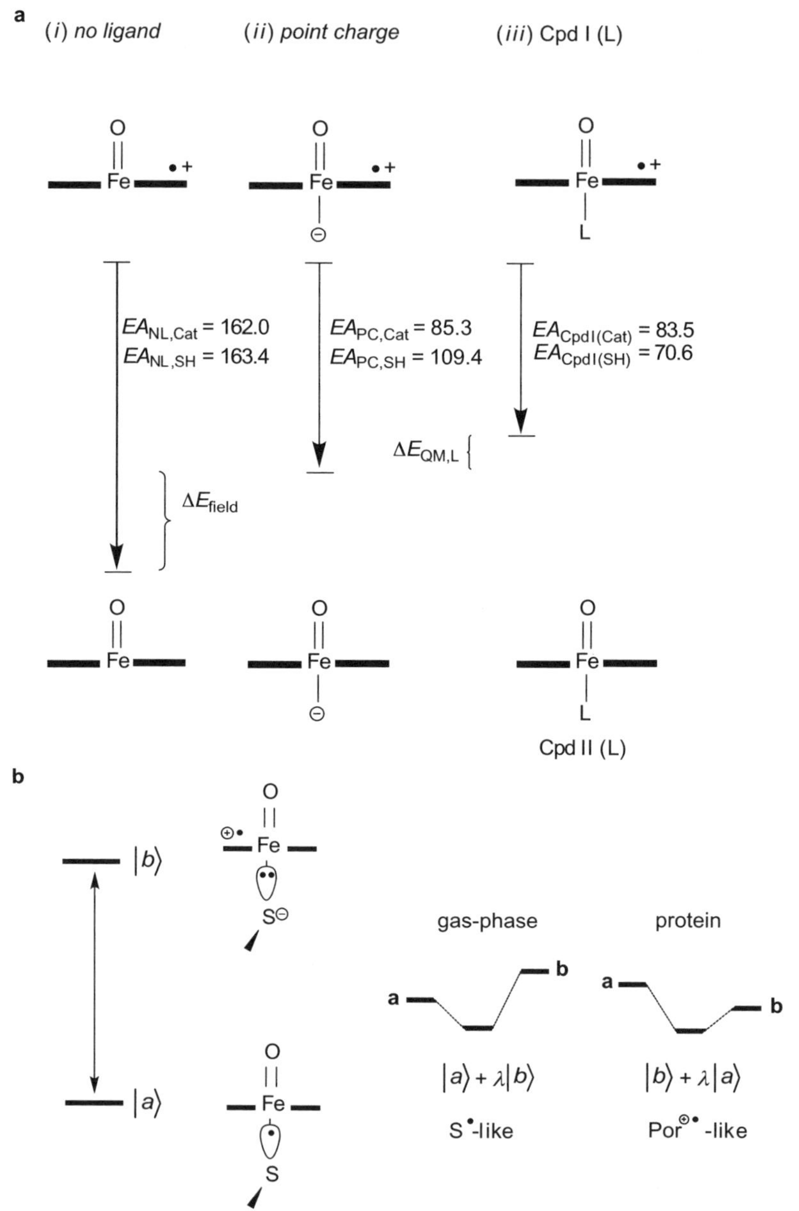

Figure 9.13 (a) Estimate of the "push"-effect strength from quantum chemical calculations, with electron affinities in kcal mol^{-1}. (b) Orbital mixing of porphyrin and thiolate orbitals in Cpd I.

electric field effect, due to the short distance between the metal and the axial ligand: $r_{Fe-OTyr} = 2.117$ Å while r_{Fe-S} is 2.581 Å. In Cpd I(SH) the quantum mechanical effect is large as a result of orbital interactions between a porphyrin type orbital with a lone pair on the axial ligand. Figure 9.13(b) explains the orbital mixing in Cpd I(SH) and the corresponding effect of this on the charge and spin density distribution. Cpd I has two high-lying molecular orbitals that are close in energy and mix, namely, a lone-pair on the thiolate axial ligand (π_S) and the porphyrin type a_{2u} orbital discussed and shown above in Figure 9.9(d). These orbitals are occupied by three electrons, which means a purely singly occupied π_S orbital will give a Por S$^\bullet$ electronic configuration, state $|a\rangle$, whereas a purely singly occupied a_{2u} orbital will result in a Por$^{+\bullet}$ S$^-$ configuration, state $|b\rangle$. In Cpd I(SH) these two states mix and partial radical character on both porphyrin and axial ligand is found. However, it has been shown that hydrogen bonding interactions toward the thiolate group, for instance mimicked by two ammonia molecules, already shift the orbital mixing to a more dominant porphyrin cation radical situation.[67,71] Subsequent, quantum mechanics/molecular mechanics calculations indeed showed that Cpd I has a large radical component on the porphyrin and lesser on the axial ligand.[79]

9.9   Reactivity of Compound I

As discussed above, Compound I is an iron(IV)-oxo heme cation radical species with three unpaired electrons in π^*_{xz}, π^*_{yz} and a_{2u} orbitals (Figure 9.9). Owing to the small spin–orbit coupling between the unpaired electrons there are two electromers (doublet and quartet) that are close in energy. In the quartet spin state all spins are aligned (ferromagnetically coupled), whereas in the doublet spin state the a_{2u} electron has opposite spin to the two π^* electrons. In the following we give the spin multiplicity in superscript on the left-hand-side of the labels. The reactant states of Compound I, ^{2,4}R, involve the two low lying spin states ($S = \frac{1}{2}, \frac{3}{2}$) with two excited states ($S = \frac{3}{2}, \frac{5}{2}$), ^{4,6}R', slightly higher in energy. The latter two states contain an excitation from the nonbonding δ_{x2-y2} orbital to the antibonding σ^*_{xy} orbital, which are two orbitals in the plane of the heme. In addition all states have a doubly occupied orbital, ϕ_{CH}, that represents the electron pair in the C–H bond. Technically, Cpd I can be assigned as an Fe^{IV} center with a hole in the porphyrin (*i.e.*, Por$^{+\bullet}$).

A substrate hydroxylation reaction gives products ^{2,4}P and ^{4,6}P' that are described by an Fe^{III} center and a closed-shell porphyrin. Thus, any reaction catalyzed by Cpd I of CYP 450 will involve two formal electron "transfer" reactions from the substrate undergoing oxidation. Figure 9.14 shows these two electron transfer processes from reactants *via* intermediates leading to products. Each of these steps involves a single electron transfer, one into the a_{2u} orbital and the second into the π^*_{xz}(FeO) orbital, in the quartet and doublet spin states for tri-radicaloid state. On the other hand, starting from the ^{4,6}R' states the first electron transfer is into the a_{2u} orbital, while the second electron transfer fills the σ^*_{xy} orbital with one electron. For the tri-radicaloid state, the first electron shift generates an Fe^{IV}-type intermediate with a closed-shell porphyrin, whereas for the penta-radicaloid state, the electron shift generates an Fe^{III}-type intermediate with a radical cationic situation on the porphyrin ring. In all cases, the substrate

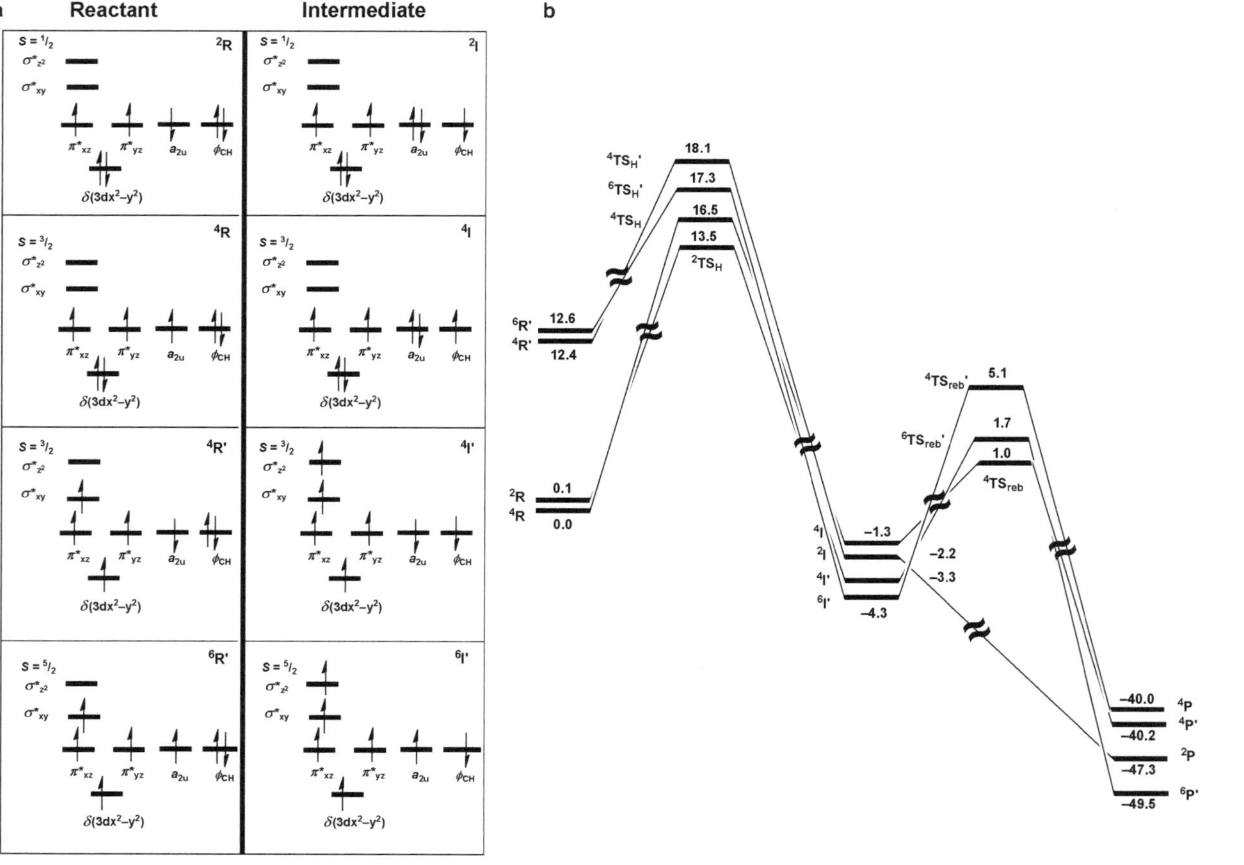

Figure 9.14 (a) Orbital occupancy evolution during substrate metabolism by CYP 450 Cpd I. (b) Reaction pathways for C–H hydroxylation of propene as obtained by the B3LYP/LACV3P+*//LACVP level of theory. The model includes two NH_3–S hydrogen bonds that mimic the interactions in the protein and a polarization effect through the addition of a medium with dielectric constant $\varepsilon = 5.7$. Relative energies are in kcal mol^{-1} and contain zero-point corrections calculated at B3LYP/LACVP.

maintains a radical center with a singly occupied orbital (ϕ_{CH}). Therefore, we have four closely packed radical intermediates during the reactions.[119] The second electron is transferred from the substrate to fill either the a_{2u} orbital or one of the d-orbitals on the iron. Looking at Figure 9.14 it is obvious that there are several other possible electronic states, which can be obtained by redistribution of the electrons in this dense orbital manifold.[120–125] Therefore, the reaction catalyzed by Cpd I of P450 will involve at least two spin states, *i.e.* will proceed *via* two-state reactivity (TSR), and possibly involves many electromeric states or multistate reactivity (MSR).[126–128] The most intriguing feature of TSR and MSR is that the different spin states and/or electromers may lead to reaction *via* different mechanisms and thereby produce different intermediates and products. This spin-dependent mechanistic phenomenon has been observed during oxidation processes ranging from the simple reaction of FeO^+ [129,130] to the complex mechanisms of enzymes and synthetic catalysts.[120,129–134] Sometimes, significant energy differences between the LS and HS activation barriers lead to a spin-selective single-state reactivity (SSR).

9.10　Aliphatic C–H Hydroxylation by Compound I of CYP 450

Aliphatic hydroxylation is the most common reaction catalyzed by CYP 450 enzymes and it is known to perform this reaction on a broad range of chemical substrates. Since this is an important reaction in catalysis many efforts have been made to rationalize substrate hydroxylation reactions and predict the rate constant of substrate hydroxylation.[135–137] Generally, aliphatic substrates are activated by Cpd I *via* an initial hydrogen atom abstraction, which leads to a radical intermediate.[138] Alcohol products are formed after radical rebound of the hydroxyl group to the radical rest group. The hydrogen atom abstraction from a substrate RH by Cpd I to form the intermediate complex I_H is described in Equation (9.1) and the reaction enthalpy for this reaction (ΔH_{HA}) is dependent on the relative energies of reactants and products:

$$[Fe^{IV}=O(Por^{+\bullet})Cys] + RH \rightarrow [Fe^{IV}(OH)(Por)Cys] + R^{\bullet} + \Delta H_{HA} \qquad (9.1)$$

The reaction enthalpy (ΔH_{HA}) for the above reaction can be written as a function of the C–H bond dissociation energy (BDE_{C-H}) of the substrate and the BDE_{O-H} for the O–H bond dissociation energy of $[Fe^{IV}(OH)(Por)Cys]$ through combination of Equations (9.1) and (9.2) to give Equation (9.3):

$$RH \rightarrow R^{\bullet} + H^{\bullet} + BDE_{C-H} \qquad (9.2)$$

$$\Delta H_{HA} = BDE_{C-H} - BDE_{O-H} \qquad (9.3)$$

In the case a series of calculations is performed on substrate hydroxylation by $[Fe^{IV}=O(Por^{+\bullet})SH]$ as a mimic of Cpd I of CYP 450 this implies that the

BDE_{O-H} value will be the same for all reaction mechanisms studied. Therefore, the exo- or endothermicity of formation of the intermediate complexes ($^{4,2}I_H$) in the reaction should correlate linearly with the BDE_{C-H} value of the substrates RH. Indeed a linear correlation of reaction energy or driving force with BDE_{C-H} was found with $R^2 = 0.99$.[135] The standard deviation from linearity for this driving force is small and consequently the DFT calculated reaction energies are within a few kcal mol^{-1} error. These values are much smaller than the systematic error generally reported for DFT calculations of 3–5 kcal mol^{-1} with respect to experimentally determined enthalpies.[139–141]

Subsequently, the relationship of hydrogen abstraction barrier height with BDE_{C-H} was investigated and the obtained correlation is shown in Figure 9.15 for aliphatic hydroxylation of the following substrates by [Fe^{IV}=O(Por$^{+•}$)SH]: methane, ethane, propane, propene, toluene, ethylbenzene, *trans*-methylphenylcyclopropane, *trans-i*-propylphenylcyclopropane, *N,N*-dimethylaniline and camphor. The trend depicted in Figure 9.15(b) represents the raw data of calculated barrier height *versus* BDE_{C-H}, while the one displayed in Figure 9.15(c) includes a correction factor for the resonance energy ($RE_{Substrate}$). The resonance energy represents the geometric distortion of the radical with respect to the reactant conformation.

A valence bond (VB) curve crossing diagram has been used to explain the substrate hydroxylation trends and confirms the correlation between hydrogen abstraction barrier height and BDE_{C-H}. Figure 9.16 displays the VB curve crossing diagram from reactants (Cpd I + substrate, RH) *via* a radical intermediate leading to alcohol products. The curve crossing diagram starts bottom-left with the structure of Cpd I. In principle, the π-system in the Fe–O bond is occupied with six electrons ($\pi_{xz}^2 \pi_{yz}^2 \pi^*_{xz}{}^1 \pi^*_{yz}{}^1$) each of those electrons is identified with a dot next to the atom in Figure 9.16. Since the π^*_{xz} and π^*_{yz} orbitals are singly occupied orbitals on the Fe–O bond, each of these orbitals gives about 50% spin density on the iron and 50% spin density on the oxygen atom. This results in the two electromers K_1 and K_2 depicted in Figure 9.16. Apart from the two unpaired electrons in the π^*_{xz} and π^*_{yz} orbitals there is also an unpaired electron in the porphyrin based a_{2u} orbital. The alcohol products by contrast have the metal in oxidation state Fe^{III} coupled to a closed-shell heme. The orbital occupation of ^{2}P is $\delta^2 \pi^*_{xz}{}^2 \pi^*_{yz}{}^1 a_{2u}{}^2$, whereas that for ^{4}P is $\delta^2 \pi^*_{xz}{}^1 \pi^*_{yz}{}^1 \sigma^*_{z^2}{}^1 a_{2u}{}^2$. In a VB diagram the electronic configuration of the reactants (with wave function Ψ_r) connects to an excited state in the products (with wave function Ψ_p^*). Similarly, the product electronic configuration (with wave function Ψ_p) connects to an excited state in the reactants (with wave function Ψ_r^*). At the point where the two curves cross, there will be an avoided crossing and a transition state leading from reactants to products. The direct barrier, *i.e.* concerted barrier, from reactants to products is high in energy,[142] but a lowering of the barrier is achieved thanks to the interception of an intermediate curve. This intermediate leads to the radical intermediate in the reaction mechanism, in this case with electronic configuration [Fe^{IV}(OH)(Por)Cys---R$^•$]. In addition, there also will be an intermediate curve leading to the [Fe^{III}(OH)(Por$^+$)Cys---R$^•$] intermediate, but in most gas-phase studies this is high in energy.

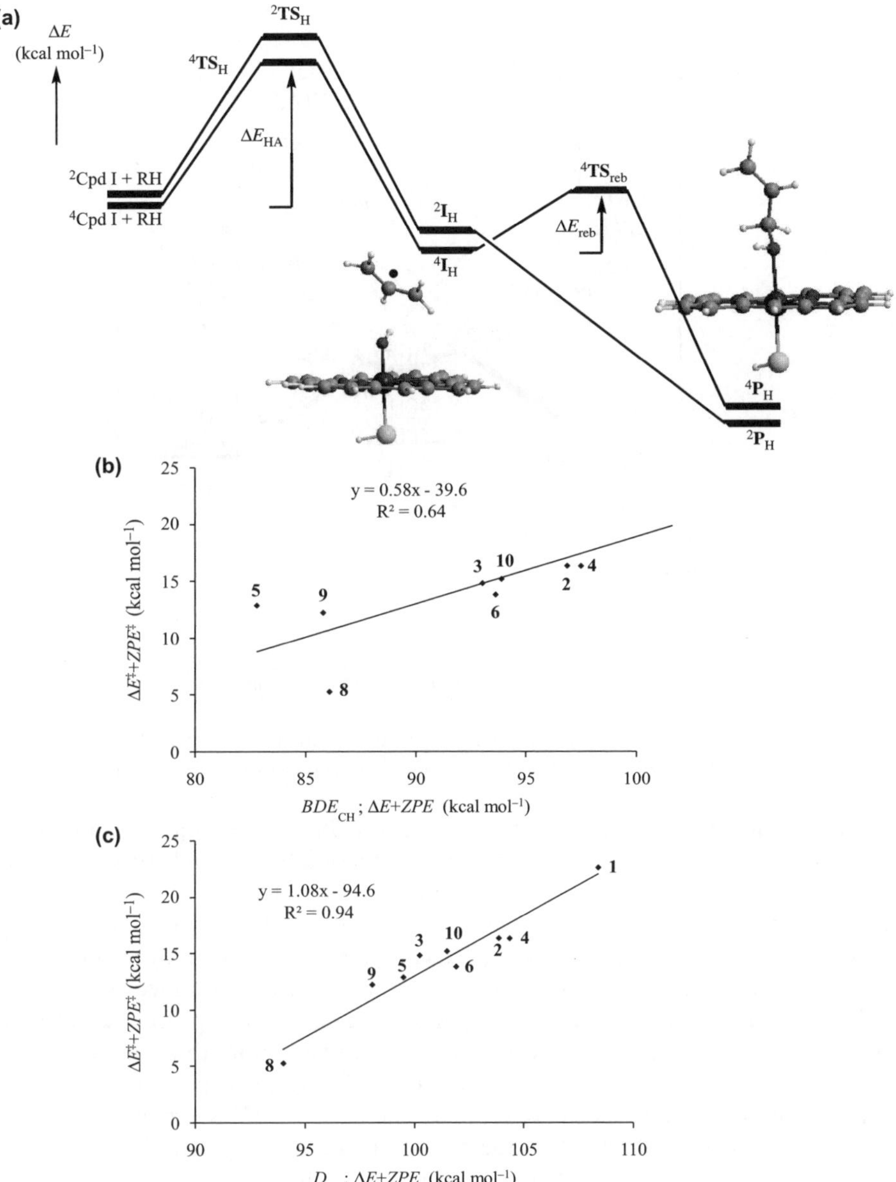

Figure 9.15 (a) General mechanism of substrate hydroxylation by Cpd I of CYP 450 in TSR. (b) Plot of the average barriers $\Delta E^‡ + ZPE^‡$ (UB3LYP/LACV3P+*) for the quartet and doublet hydrogen abstraction steps *versus* BDE$_{C-H}$ (UB3LYP/LACV3P+*). (c) Plot of the average barriers $\Delta E^‡ + ZPE^‡$ (UB3LYP/LACV3P+*) for the quartet and doublet hydrogen abstraction steps *versus* D_{CH} (D_{CH} = BDE$_{C-H}$ + RE$_{Substrate}$˙) as calculated with UB3LYP/LACV3P + *. Substrates are methane (**1**), ethane (**2**), propane (**3,4**), *trans*-methylphenylcyclopropane (**5**), *trans-i*-propylphenylcyclopropane (**6**), *N,N*-dimethylaniline (**7**), toluene (**8**), ethylbenzene (**9**) and camphor (**10**). In the case of propane, there is a separate transition state for hydrogen abstraction from the primary (**3**) and secondary (**4**) carbon atoms, respectively.

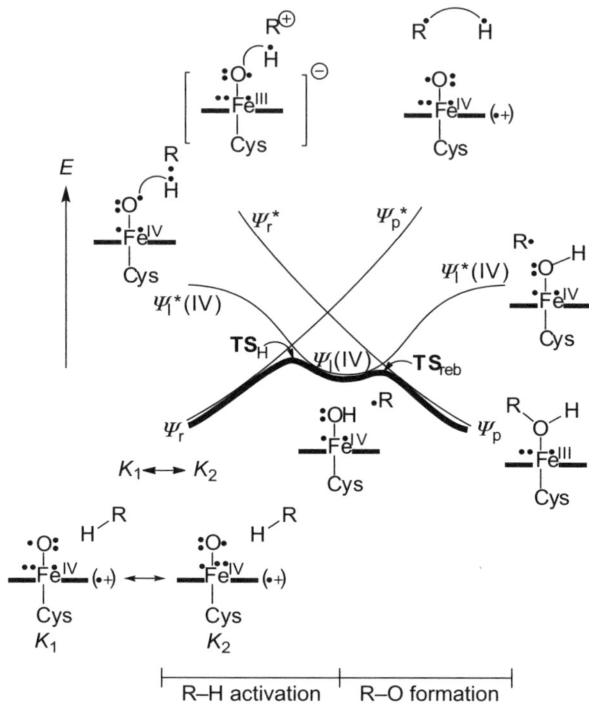

Figure 9.16 Valence bond curve crossing diagram for substrate hydroxylation by Cpd I of CYP 450 that takes place *via* an [Fe^{IV}(OH)(Por)Cys---R$^•$] intermediate.

The intermediate curve (with wave function Ψ_I) crosses the reactant wave function and gives rise to a hydrogen atom abstraction barrier and crosses with the product wave function to generate a radical rebound barrier. This VB curve diagram, therefore, predicts a stepwise mechanism for substrate hydroxylation by CYP 450 enzymes that proceed *via* a radical intermediate. Moreover, it explains the electron transfer processes that proceed during the reaction mechanism.

The VB diagram not only is a quantitative way to explain the reaction mechanism, but can also be used to predict hydrogen atom abstraction barriers from empirical values. Thus, the barrier height $\Delta E^{\ddagger}$ for a hydrogen atom abstraction is a factor B below the curve crossing energy (ΔE_c).[143] Moreover, the curve crossing energy is a fraction (f) of the excitation energy (or promotion gap, $G_{H,r}$) between the ground state and the electronic state that represents the products. Consequently, the barrier height $\Delta E^{\ddagger}$ can be written as follows using Equation (9.4):

$$\Delta E^{\ddagger} = \Delta E_c - B = fG_H - B \tag{9.4}$$

As follows from Figure 9.17, the excitation energy or promotion gap $G_{H,r}$ refers to a singlet to triplet excitation in the C–H bond of the substrate plus an electron transfer from the oxygen atom into the a_{2u} orbital. A singlet–triplet excitation in a single bond essentially equals the bond dissociation energy for

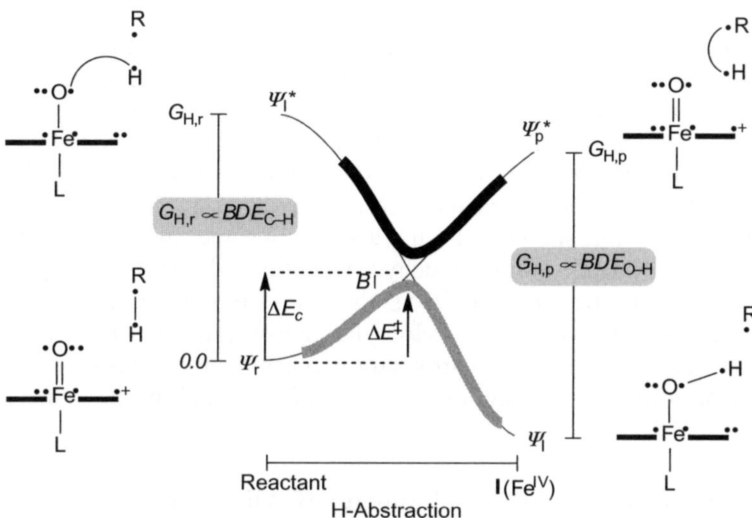

Figure 9.17　Rationalization of the correlation of barrier height with either BDE_{C-H} or BDE_{O-H}.

that bond, and hence is proportional to BDE_{C-H}. This VB curve crossing diagram, therefore, proves that the barrier height of substrate hydroxylation correlates linearly with BDE_{C-H}.

Recent studies also focused on the correlation of hydrogen atom abstraction barrier of propene with a range of different oxidants.[116] According to Equation (9.3), the reaction energy should correlate with BDE_{C-H} as well as with BDE_{O-H}. Changing the oxidant but keeping the substrate constant, therefore, will vary BDE_{O-H} and keep BDE_{C-H} constant. Indeed, linear correlations similar to those displayed in Figure 9.15 are found. The VB curve crossing diagram shown in Figure 9.17 explains the observed trends. Now, the relevant reaction mechanism is the backward reaction starting in products (the radical intermediates) and the barrier leading to Cpd I and substrate is investigated. For this process the barrier is proportional to an excitation energy ($G_{H,p}$) that is dependent on the singlet to triplet excitation in the O–H bond of the iron(IV)-hydroxo complex. Of course, the reverse barrier height is equal to the forward barrier height minus the driving force for the reaction. Consequently, both BDE_{C-H} and BDE_{O-H} connect to the hydrogen abstraction barrier.

Although the first hydrogen atom abstraction reactions by CYP 450 Cpd I model systems were studied using methane as a substrate,[144–146] in the past ten years many more realistic substrates have been studied as well. Special emphasis has always been on camphor hydroxylation by CYP 450_{cam}, because it is the natural substrate for hydroxylation by this CYP 450 isozyme. Initial studies used DFT methods but later also full QM/MM techniques were applied to investigate the complete enzyme.[147,148] CYP 450_{cam} regioselectively hydroxylates the C^5 position of camphor mostly because camphor is tightly bound to the substrate binding pocket and only the C^5 position approaches the heme close enough to enable substrate hydroxylation. All these studies predict the same mechanistic trends, namely, starting with an initial hydrogen atom abstraction followed by radical rebound to form the alcohol products.

Table 9.2 summarizes substrate hydroxylation reactions calculated with DFT or QM/MM methods for Cpd I(L) with axial ligand $L = SH^-$, SMe^-.[136,137,148–155] Essentially the general mechanism of substrate hydroxylation as supported by all calculations is the one shown in Figure 9.15(a). In some cases, the rebound step leads to either N-dealkylation or dehydrogenation.[156] Nevertheless, the initial hydrogen atom abstraction barrier is the rate-determining step as well. On average the low-spin barriers are below the high-spin barriers by 0.9 ± 1.3 kcal mol^{-1}. Obviously, substrate hydroxylation phenomenon follows a two-state-reactivity pathway with competing spin state surfaces and reaction barriers.

9.10.1 Rearrangement Mechanisms of Aliphatic Hydroxylation Reactions

Owing to the transient lifetime of Cpd I and the failure to identify and characterize it as the active species, many experimental studies have been performed to find indirect evidence of its activity. In one of those experiments, so-called, radical-clocks substrates were used.[157–159] These radical clock substrates react with an oxidant *via* partial rearrangement processes, as displayed by the reaction in Figure 9.18, where an initial hydrogen atom abstraction gives rebound to form the unrearranged products (U) or the obtained radical (Alk$^•$) can rearrange with rearrangement rate constant k_r to form a rearranged radical (Ãlk$^•$) that rebounds the hydroxyl to form rearranged products (R). Based on the ratio of [U/R] Newcomb and coworkers[53,159] estimated the lifetime of the radicals. However, the obtained radical lifetimes of the radicals in the reaction mechanisms did not correlate with known lifetimes of these types of chemical systems. This made Newcomb and coworkers suggest that a "second-oxidant" is involved in the reaction that catalyzes the substrate hydroxylation. In particular, the precursor of Cpd I in the catalytic cycle was suggested to act as an oxidant either side-by-side with Cpd I or replacing Cpd I. This hypothesis has prompted many experimental and computational studies in the field and therefore we will dedicate this section to the mechanism of radical clock substrate and the involvement of Cpd 0 in substrate hydroxylation.

The potential energy profile for the reaction mechanism of *trans*-methyl-phenylcyclopropane hydroxylation by [Fe^{IV}=O(Por$^{+•}$)SH] was calculated using DFT models leading to unrearranged and rearranged products;[120,121] a summary of the results is given in Figure 9.18. The calculations predict Cpd I(SH) to exist in close-lying doublet and quartet spin states and as a consequence the reaction proceeds *via* two-state reactivity on competing doublet and quartet spin state surfaces. On both spin state surfaces there is a hydrogen atom abstraction *via* a transition state TS_H leading to an [Fe^{IV}(OH)(Por)(SH)---Alk$^•$] radical intermediate (I_U). Radical rebound to give the unrearranged products

Table 9.2 DFT calculated aliphatic hydrogen atom abstraction barriers (kcal mol^{-1}) by Cpd I models. The description of the axial cysteinate ligand in the calculations is given as L.

L	*Substrate*	$\Delta E_{HA,HS}^{\ddagger}$	$\Delta E_{HA,LS}^{\ddagger}$	*Ref.*
SH$^-$	Methane	22.9	22.3	147
SH$^-$	Methane	22.9	24.7	170
SH$^-$	Ethane	17.4	15.3	147
SH$^-$	Ethanol: from C^1	11.7	10.3	175
SH$^-$	Ethanol: from OH	12.3	12.7	175
SH$^-$	Acetaldehyde	8.2	10	171
SH$^-$	Propane, 2° carbon	15.8	13.9	147
SH$^-$	Propane, 1° carbon	17.5	15.2	147
SH$^-$	Propane	20.6	19.5	167
SH$^-$	Propene	13	12.8	147
SH$^-$	*N,N*-Dimethylaniline	5.5	5	147
SH$^-$	Toluene	12.4	12.1	147
SH$^-$	Ethylbenzene	12.6	11.5	147
SH$^-$	Camphor	14.5	15.9	147
SH$^-$	*trans*-2-Phenyl-methylcyclopropane	14.5	13.1	147
SH$^-$	*trans*-2-Phenyl-isopropylcyclopropane	13.4	12.3	147
SH$^-$	Arginine	10.5	4.9	176
SH$^-$	*N*-Cyclopropyl-*N*-methylaniline	11.4	9.6	180
SH$^-$	*N*-Cyclopropyl-*N*-methylaniline	11.2	9.2	180
SH$^-$	Testosterone-β_1	13	11.4	177
SH$^-$	Testosterone-β_2	8.2	9.7	177
SH$^-$	Testosterone-β_6	6.1	5.3	177
SH$^-$	Testosterone-β_{15}	11.6	9.7	177
SMe$^-$	Trimethylamine	9.6	8.3	178
SMe$^-$	Morpholine	<1	<1	179
SCys$^-$	Camphor	21.8	21.1	167
SMe$^-$	Methane	20.7		160
SMe$^-$	Propane, 1° carbon	17.7		160
SMe$^-$	Propane, 2° carbon	14.8		160
SMe$^-$	Isobutane	14.3		160
SMe$^-$	Propene	12.9		160
SMe$^-$	Propionaldehyde	11.4		160
SMe$^-$	Toluene	13		160
SMe$^-$	Ethylbenzene, 2°	12.1		160
SMe$^-$	1-Methylethyl-benzene	13.3		160
SMe$^-$	Dimethyl ether	12.2		160
SMe$^-$	Dimethylsulfane	11		160
SMe$^-$	Methyl(phenyl)sulfane	10.8		160
SMe$^-$	Dimethylamine	7.6		160
SMe$^-$	Trimethylamine	6.7		160
SMe$^-$	Fluoroethane, 2°	18.4		160
SMe$^-$	Fluoroethane, 1°	14.7		160
SMe$^-$	Ethylbenzene, 1°	17.3		160
SMe$^-$	2-Fluoroprop-1-ene	13.2		160
SMe$^-$	Prop-1-en-2-ol	11.7		160
SMe$^-$	*p*-Xylene	12.7		160
SMe$^-$	1-Methyl-4-nitrosobenzene	11.8		160
SMe$^-$	Methoxybenzene	13		160
SMe$^-$	*N*-Methylaniline	7.6		160
SMe$^-$	*N,N*-Dimethylaniline	6.9		160

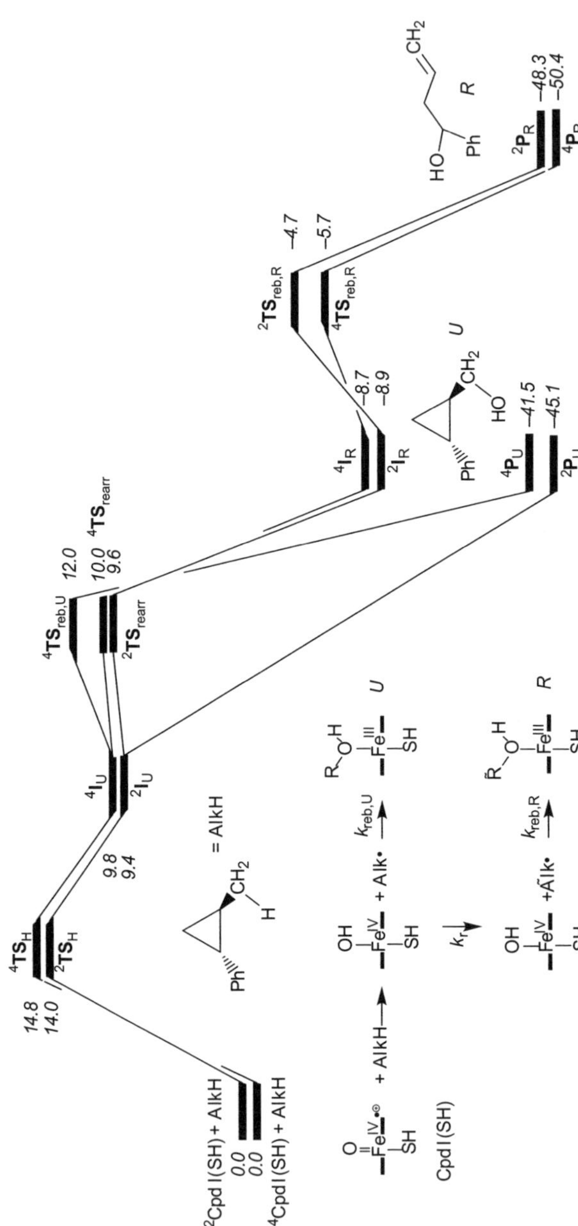

Figure 9.18 Calculated potential energy profile of *trans*-1-methyl-2-phenylcyclopropane hydroxylation by [Fe^{IV}=O(Por$^{+•}$)SH] with relative energies in kcal mol^{-1}. Also given is the reaction mechanism leading to unrearranged products (U) and rearranged products (R).

($\mathbf{P_U}$) encounters a barrier ($\mathbf{TS_{reb,U}}$) of 2.2 kcal mol^{-1} on the high-spin surface and is virtually barrier-free on the low-spin surface. An alternative mechanism leads to rearrangement within the radical intermediate to form the rearranged intermediate ($\mathbf{I_R}$). This reaction step has a small barrier ($\mathbf{TS_{rearr}}$) of about 0.2 kcal mol^{-1} on both spin state surfaces and large exothermicity, which implies that it will be an irreversible step. A final radical rebound barrier ($\mathbf{TS_{reb,R}}$) then leads to the rearranged product complexes ($\mathbf{P_R}$). As follows from the calculations depicted in Figure 9.18, on the low-spin surface $\mathbf{TS_{reb,U}} < \mathbf{TS_{rearr}}$, therefore, little rearranged product may be expected. On the other hand, on the high-spin surface $\mathbf{TS_{reb,U}} > \mathbf{TS_{rearr}}$, which indicates a dominant rearrangement process. Consequently, the reaction mechanism displayed in Figure 9.18 indicates spin-selective product formation, whereby the high-spin gives rearrangement and the low-spin no rearranged products.

The spin-selective product formation will also have a crucial effect on kinetic isotope effect in the experiments. To test this, Kumar and coworkers[120,121] calculated product isotope effects (PIE) from kinetic isotope effects (KIE). They reasoned that the PIE ratio for products U/R, Equation (9.5), can be estimated from the KIE values obtained for the LS and HS surfaces only:

$$\text{PIE}(U/R) = \text{KIE}(U)/\text{KIE}(R) = \text{KIE}(\text{LS})/\text{KIE}(\text{HS}) \qquad (9.5)$$

Kinetic isotope effects were calculated from the semi-classical Eyring equation ($\text{KIE}_{\text{Eyring}}$), Equation (9.6), which is dependent on the free energy of activation ($\Delta G^{\ddagger}$) of the reaction of Cpd I with *trans*-methylphenylcyclopropane and that where one or more of the methyl hydrogen atoms are replaced by deuterium atoms. In Equation (9.6) the temperature is given by T and R is the gas constant:

$$\text{KIE}_{\text{Eyring}} = \exp[(\Delta G^{\dagger}_{\text{D}} - \Delta G^{\dagger}_{\text{H}})/RT] \qquad (9.6)$$

More advanced KIE calculations include the effect of tunneling as calculated with either the Wigner ($\text{KIE}_{\text{Wigner}}$) or Bell ($\text{KIE}_{\text{Bell}}$) models, Equations (9.7–9.9). These models take the shape of the potential energy curve around the transition state into effect through the value of the imaginary frequency in the TS, namely, v. In Equations (9.7–9.9) the Boltzmann constant is given as k_{B} and Planck's constant as h:

$$\text{KIE}_{\text{tunneling}} = \text{KIE}_{\text{Eyring}} \times Q_{t,H}/Q_{t,D} \qquad (9.7)$$

$$Q_{t,\text{Wigner}} = 1 + u_t^2/24 \text{ with } u_t = hv/k_{\text{B}}T \qquad (9.8)$$

$$Q_{t,\text{Bell}} = \frac{u_t}{2\sin(0.5u_t)} - \sum_{n=1}^{\infty}(-1)^n \frac{\exp[(u_t - 2n\pi)\Delta E/u_t]}{(u_t - 2n\pi)/u_t} \qquad (9.9)$$

For *trans*-methylphenylcyclopropane *versus* *trans*-d_1-methylphenylcyclopropane a product isotope ratio $\text{PIE}_{\text{Eyring}} = 1.03$, $\text{PIE}_{\text{Wigner}} = 1.07$ and $\text{PIE}_{\text{Bell}} = 1.20$ at

298 K was calculated,[120,121] which compares favorably with the experimental value of 1.14.[160] The calculations, therefore, showed that the experimental product distributions and radical clock lifetimes can be explained with the two-state reactivity model, whereby Cpd I is the sole oxidant of substrate hydroxylation. The seeming observance of two active oxidants in the experimental studies, therefore, arises from the two spin states of Cpd I. Those two spin states of Cpd I lead to the masquerade of individual spin state processes and reactivities that are different on each spin state surface.

Further calculations on the relative activity of Cpd I and Cpd 0 in substrate epoxidation and sulfoxidation reactions gave evidence that Cpd 0 is a sluggish oxidant as compared to Cpd I and will not be able to activate substrates.[161–164] In support of this, Nam and coworkers carried out experimental studies in which they compared the reactivity of synthetic iron(IV)-oxo oxidants with iron(III)-hydroperoxo analogs.[165] No products were observed in the reaction of the iron(III)-hydroperoxo species with thioanisole and cyclohexene, therefore, Cpd 0 is unlikely to be an oxidant of aliphatic hydroxylation, epoxidation and sulfoxidation reactions.

9.11   C=C Epoxidation by Compound I of CYP 450

Substrate activation by CYP 450 enzymes commonly leads to oxygen atom transfer processes. When the substrate contains a double bond, the oxygen atom transfer will give epoxide product complexes. Figure 9.19 displays the common reaction products observed in double bond activation by CYP 450 enzymes. Thus, the dominant products will be epoxides, although stereochemical scrambling is possible, *i.e.* the production of *cis*-olefins from *trans*-alkenes. In addition, epoxidation reactions suffer from reactions leading to by-products. One type of by-products are the so-called suicidal complexes, where the olefin has formed bridging bonds with the oxo group of Cpd I and one of the nitrogen atoms of the porphyrin ring to form a five-membered ring.[166–169] This is a dead-end product and consequently terminates the activity of the enzyme. In addition, aldehyde by-products have been observed in the reaction mechanism. To gain insight into the mechanisms of olefin epoxidation by CYP 450 Cpd I and the formation of by-products (suicidal complexes and aldehydes), a series of computational studies were performed, which we briefly summarize here.[123,124,128,170–172]

A typical potential energy profile for substrate epoxidation is shown in Figure 9.20 and displays the results for ethene, propene and styrene as taken from ref. 124, 128 and 171. As follows, the reaction mechanism resembles that found for aliphatic hydroxylation closely, whereby an initial C–O bond formation leads to a radical intermediate prior to a ring-closure to form epoxide products.[172] DFT studies also investigated a possible concerted reaction mechanism where the oxo group inserts into the double bond and directly forms epoxide products. The latter mechanism, however, was found to proceed *via* a second-order transition state that is much higher in energy than the

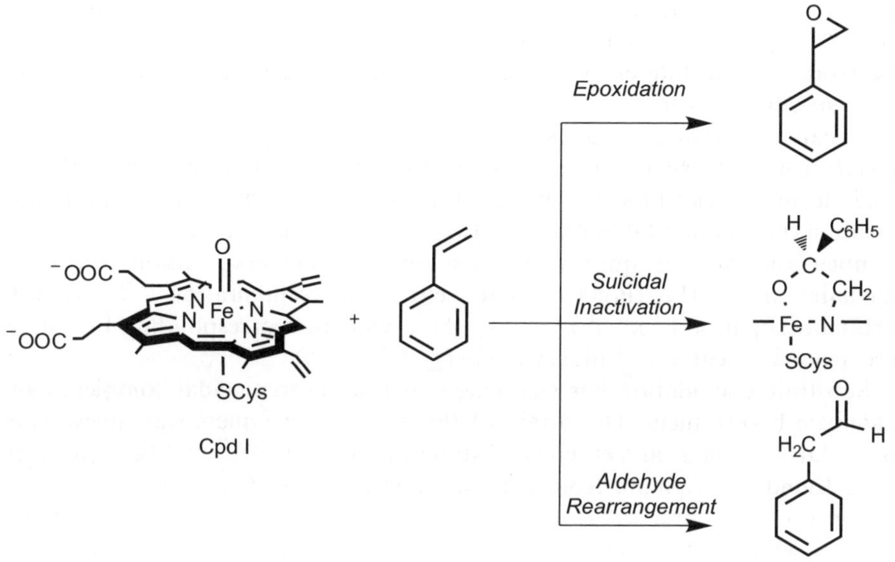

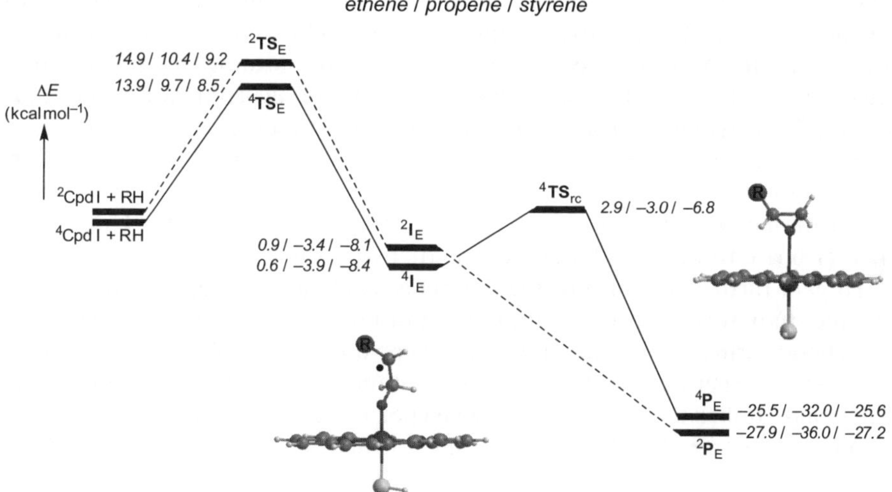

Figure 9.19   Olefin epoxidation by CYP 450 Cpd I leading to epoxide products, suicidal complexes and aldehyde by-products. Styrene is used as an example here.

Figure 9.20   Potential energy profile of substrate (ethene/propene/styrene) epoxidation by Cpd I(SH). All energies are in kcal mol^{-1} relative to isolated reactants and taken from refs 124, 128 and 171.

stepwise barriers, hence substrate epoxidation will takes place *via* a stepwise mechanism.[173] Interestingly, if one starts with pure *cis* or *trans* olefins one often ends up with a mixture of *cis* and *trans* epoxides.[52] The DFT studies on five different olefins (ethene, propene, cyclohexene, camphene and styrene)[124,125,128,174] revealed a multi-state reactivity (MSR) scenario in which electromers with different spin state take part in the product formation in a state specific manner.

As can be seen from Figure 9.20, the rate-determining step in the epoxidation mechanism is the first C–O bond formation step (*via* $\mathbf{TS_E}$) and generally the ring closure barrier ($\mathbf{TS_{rc}}$) is much lower. Thus, similarly to substrate hydroxylation the reaction takes place *via* two-state reactivity patterns on competing doublet and quartet spin states. Also in substrate epoxidation, the intermediates are of $[Fe^{IV}(Por)O–CH_2R^•]$ electronic configuration $^{4,2}\mathbf{I}_{rad,IV}$ with orbital occupation $\delta^2 \pi^*_{xz}{}^1 \pi^*_{yz}{}^1 a_{2u}{}^2 \phi_R{}^1$, while the corresponding Fe^{III} states are around 4 kcal mol^{-1} higher in energy.[171]

Substrate epoxidation has side reactions leading to suicidal complexes and aldehyde by-products. The origin of those two by-products was investigated using DFT methods and ethene and styrene as a substrate.[124,175] Both products were found to originate from cationic intermediates $^4\mathbf{I}_{cat,xy}$ and $^4\mathbf{I}_{cat,z2}$ with either a $\delta^2 \pi^*_{xz}{}^1 \pi^*_{yz}{}^1 \sigma^*_{xy}{}^1 a_{2u}{}^2 \phi_R{}^0$ or $\delta^2 \pi^*_{xz}{}^1 \pi^*_{yz}{}^1 \sigma^*_{z2}{}^1 a_{2u}{}^2 \phi_R{}^0$ configuration with orbitals defined as in Figure 9.9(d) and ϕ_R the radical on the substrate. Thus, in substrate epoxidation the mechanism is stepwise *via* a radical intermediate $^{4,2}\mathbf{I}_{rad,IV}$, which results in a C–O bond formation and a formal one-electron transfer from the substrate to oxidant. On the high-spin surface the radical intermediate has a finite lifetime due to the existence of a ring-closure transition state, but on the LS surface the radical lifetime will be short and quickly collapse to the epoxide product complexes. During the lifetime of the radical, the radical can rearrange leading to, for instance, stereochemical scrambling and the production of *cis*-epoxides from *trans*-olefins. This stereochemical scrambling, therefore, is likely on the HS surface and unlikely on the LS surface. In a similar vein, during the lifetime of the radical intermediate, the system can undergo a surface crossing from a radical to a cationic state giving either $^4\mathbf{I}_{cat,xy}$ or $^4\mathbf{I}_{cat,z2}$ intermediates. These states then give rise to the suicidal and aldehyde by-products.

To gain further insight into the mechanisms of substrate epoxidation, a systematic study was performed with the aim of rationalizing the rate constant, *i.e.* C–O bond formation barrier, in substrate epoxidation.[172] Thus, two sets of calculations were performed: firstly the epoxidation of seven different olefins by Cpd I(SH) was studied and secondly propene epoxidation by a range of iron(IV)-oxo porphyrin cation radical oxidants was investigated. The first studies used ethene, propene, 1-butene, *trans*-2-butene, 1,3-cyclohexadiene, 1,4-cyclohexadiene and styrene (Figure 9.21). The latter studies used a series of Cpd I models where the axial ligand was varied and a biomimetic nonheme iron(IV)-oxo complex and the iron(IV)-oxo active species of taurine/α-ketoglutarate dioxygenase.[176,177]

Since the highest occupied molecular orbital (HOMO) of an olefin is generally a π-orbital along a double bond, it was reasoned that a correlation between substrate epoxidation barrier height with double bond ionization could be expected. Indeed the trend of epoxidation barriers of Cpd I(SH) with these substrates corresponded to the strength of the double bond of the olefin,

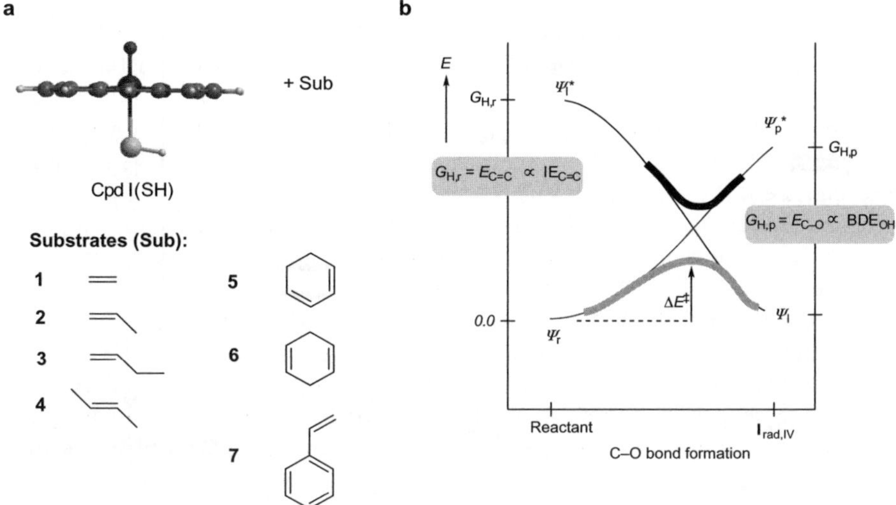

Figure 9.21　Systematic study of substrate epoxidation by Cpd I(SH). (a) Olefins investigated; (b) valence bond curve crossing diagram for C–O bond formation.

and hence ionization energy (IE) of the substrate. In fact the plot of barrier height with IE gave a correlation coefficient $R^2 = 0.80$. Subsequent studies using one substrate (propene) and a range of different oxidants showed that the formation of the O–C bond in substrate epoxidation can be modeled with the formation of an O–H bond and, consequently, a linear correlation between barrier height and BDE_{O-H} was found.

These observations of barrier height with either IE or BDE_{O-H} were rationalized with a valence bond curve crossing diagram as depicted in Figure 9.21(b). Thus, the forward barrier (from reactants to radical intermediates) is dependent on the singlet–triplet excitation energy in the double bond of the olefin that is broken, and can be estimated from IE values. On the other hand, using one substrate and multiple oxidants gives a correlation with BDE_{O-H}. The reverse epoxidation barrier, *i.e.* from radical intermediates to reactants, has a promotion gap that represents the singlet–triplet excitation in the C–O bond, which can be estimated from the excitation in an O–H bond and therefore is proportional to BDE_{O-H}. The valence bond curve crossing diagram, therefore, predicts that the mechanism is stepwise *via* a radical intermediate and confirms that the epoxidation barrier correlates with IE and BDE_{O-H}. This knowledge can also be used to predict barrier heights and rate constants of substrates where IE is known.

9.12　Sulfoxidation Reaction by Compound I of CYP 450

Sulfoxidation reactions are common in drug metabolism as well as substrate detoxification processes in the liver. For example, the metabolism of the

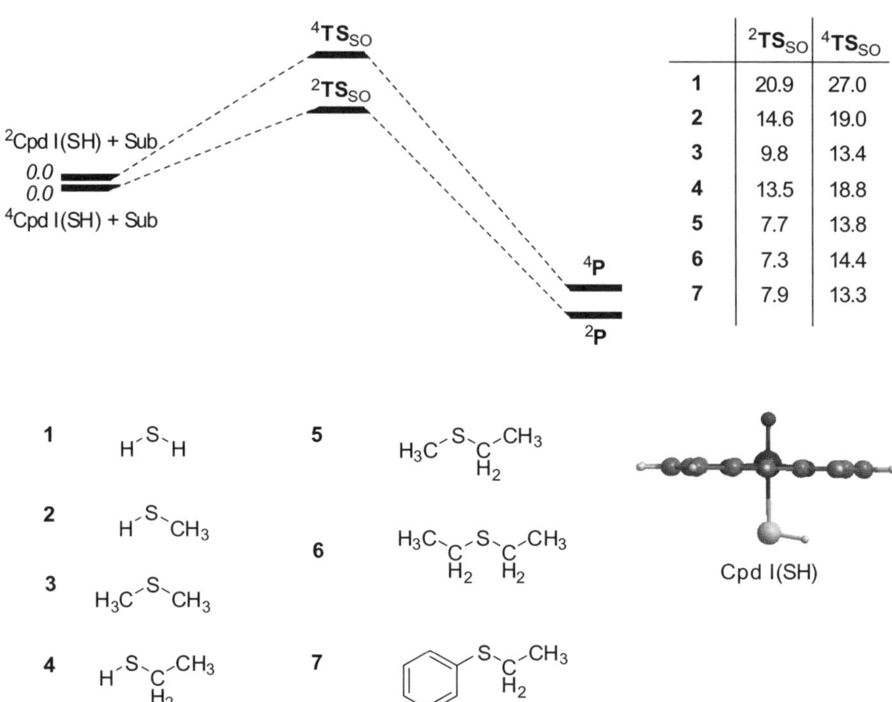

Figure 9.22 Reaction mechanism of substrate sulfoxidation by Cpd I(SH). Substrates studied are: H_2S (**1**), methyl sulfide (**2**), dimethyl sulfide (**3**), ethyl sulfide (**4**), methyl ethyl sulfide (**5**), diethyl sulfide (**6**) and ethyl phenyl sulfide (**7**). The barrier heights are given in kcal mol^{-1}.

pesticide aldicarb and the biodegradation of neuroleptic drugs in the liver is associated with substrate sulfoxidation by CYP 450 isozymes.[178–183] Detailed reaction mechanism studies on CYP 450 enzymes and biomimetic complexes for substrate sulfoxidation reactions have been carried out by several groups and the consensus mechanism for substrate sulfoxidation by CYP 450 enzymes is a concerted reaction mechanism as shown in Figure 9.22.[184–188]

The experimental observations are consistent with those obtained from density functional theory (DFT) and quantum mechanics/molecular mechanics (QM/MM) studies on substrate sulfoxidation by Cpd I of P450.[153,162,163,189,190] These studies confirmed a concerted reaction mechanism for substrate sulfoxidation in contrast to the stepwise mechanisms found for aliphatic hydroxylation and epoxidation reactions.

Experimental studies of the Watanabe group found a correlation between the rate constant of substrate sulfoxidation by CYP 450 enzymes and the ionization energy (IE) of the substrate.[191–193] Shaik and coworkers and Kumar and coworkers calculated substrate sulfoxidation by Cpd I of CYP 450 using a range of substrates. The obtained trends were explained with a valence bond curve crossing diagram including electron transfer processes.[194,195]

Using the series of substrates displayed in Figure 9.22, the barrier height of S–O bond formation was found to correlate linearly with the IE of the substrate. The correlation coefficient ($R^2 = 0.91$) supports the hypothesis of this relationship. Contrary to substrate epoxidation and aliphatic C–H hydroxylation by Cpd I of CYP 450, however, the inclusion of the reorganization energy (RE) on the obtained trends has very little effect and the correlation coefficient is virtually the same ($R^2 = 0.92$). The reason for this is that the RE values are very small, *i.e.* between 0.04 and 1.05 kcal mol^{-1} due to the attack of the oxo-group of Cpd I on a lone pair orbital on the sulfur atom of the substrate. Consequently, the geometry in the transition state will have distorted minimally with respect to the reactant conformation, resulting in a low value for the RE. Of course, thermodynamically the rate constant is connected to the free energy of activation; therefore, the rate constant of substrate sulfoxidation should connect with IE as well, as indeed observed by the Watanabe group.[191–193] Interestingly, since both substrate epoxidation and sulfoxidation barriers connect to the IE value of the substrate, this should imply that the regioselectivity of substrate epoxidation *versus* sulfoxidation will be dependent on the nature of the substrate only and not on the properties of the oxidant.

9.13 Aromatic Hydroxylation Reaction by Compound I of CYP 450

The final reaction mechanism for substrate activation by CYP 450 Cpd I that we will discuss is the aromatic hydroxylation. The mechanism of aromatic hydroxylation has long remained controversial and leads to the production of phenols often with ketones and/or epoxides as by-products. The production of ketones proceeds *via* the so-called NIH shift, where one of the hydrogen atoms from the aromatic ring undergoes a 1,2-shift.

Many computational studies on aromatic hydroxylation by CYP 450 Cpd I have been reported.[196–201] Figure 9.23 displays the computationally recommended reaction mechanism of aromatic hydroxylation by Cpd I of CYP 450 enzymes leading to phenol and cyclohexa-1,3-dienone products. Thus, the reaction starts with attack of the oxo group on one of the carbon atoms of the aromatic ring to form an intermediate complex *via* a transition state **TS$_{\mathrm{CO}}$**. In principle, C–O bond formation can lead to single electron transfer from the substrate to the heme to give a radical intermediate (**I$_{\mathrm{rad}}$**) or a double electron transfer to produce a cationic intermediate (**I$_{\mathrm{cat}}$**). Generally, these intermediates are close in energy and in the case of ethylbenzene hydroxylation the low-spin pathway gives a cationic intermediate, whereas the quartet spin state surface gives a radical intermediate instead. As a consequence, there are considerable energy differences between the doublet and quartet C–O bond formation barriers, which are in favor of the low-spin surface. Therefore, aromatic hydroxylation will take place on a dominant low-spin surface *via* single-state reactivity, in contrast to the two- or multi-state reactivity found for aliphatic hydroxylation and epoxidation.

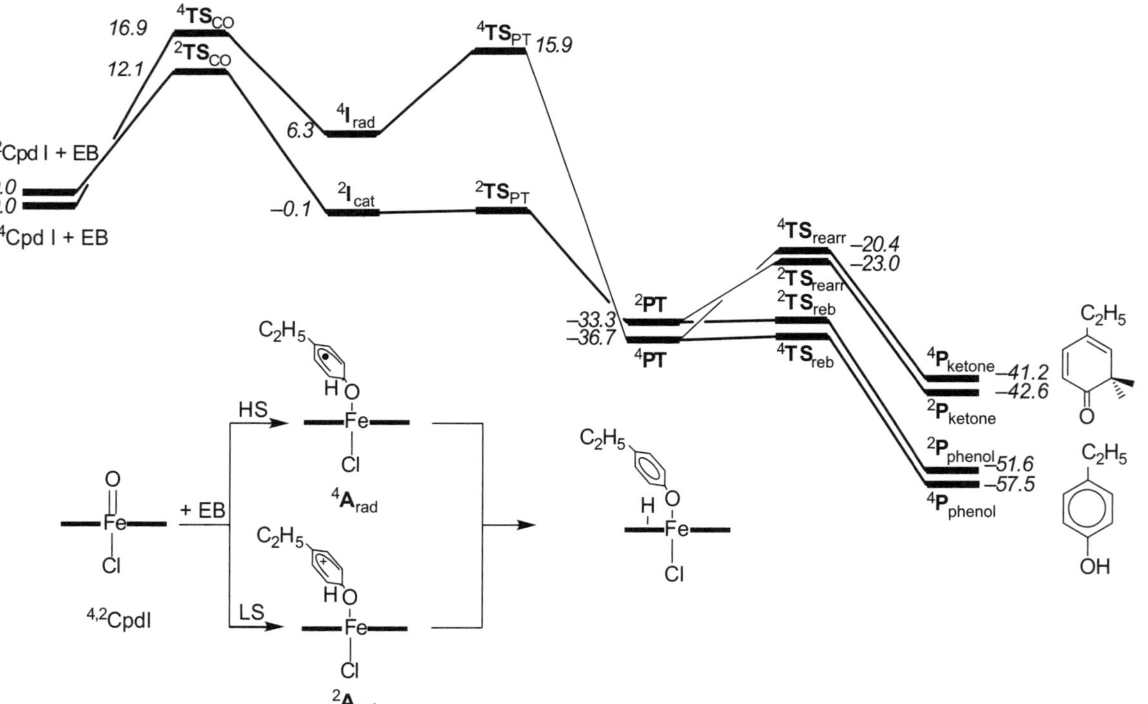

Figure 9.23 Aromatic hydroxylation of ethylbenzene (EB) at the *para*-position by Cpd I(Cl), passing through radical and cationic intermediates. Mechanisms leading to phenol and ketone products are given. All energies are in kcal mol^{-1} relative to isolated reactants in the quartet spin state.

In the next step of the reaction mechanism, the *ipso*-proton is transferred to one of the nitrogen atoms of the porphyrin ring to form the proton-transfer intermediate (**PT**). This essentially brings the aromaticity back in the benzene ring of the substrate and creates a protonated porphyrin group with one pyrrole group tilted out of the plane. This barrier is small and the process encounters a large exothermicity and, hence, will be an irreversible process. The proton from the protonation porphyrin group is transferred to the oxygen atom to form phenol products *via* rebound barrier (**TS**$_{reb}$) or alternatively is reshuttled to the C^2 carbon atom to give cyclohexa-1,3-dienone products *via* the rearrangement barrier (**TS**$_{rearr}$). In addition there is a mechanism leading to epoxide products that starts from the radical or cationic intermediates *via* a ring-closure barrier. For all calculations this ring-closure barrier was found to be larger than the proton-shuttle barrier to form **PT**.[196]

The mechanism displayed in Figure 9.23 explains the experimentally observed NIH effect and the by-product formation of ketones and epoxides. Studies on the regioselectivity of aliphatic *versus* aromatic hydroxylation of ethylbenzene by Cpd I(Cl) showed that the free energies of activation of the aromatic hydroxylation processes are just 1–2 kcal mol^{-1} higher in energy than those found for aliphatic hydroxylation. Since replacement of the hydrogen atoms of ethylbenzene by deuterium atoms raises the aliphatic hydroxylation barriers by several kcal mol^{-1} but not the aromatic hydroxylation barriers, this results in a regioselectivity switch from aliphatic hydroxylation for ethylbenzene-h_{10} to aromatic hydroxylation for ethylbenzene-d_{10}.[117]

9.14   Role of Water Molecule as Biocatalyst

Water is the medium in which the chemistry takes place in a biological system. Irrespective of its main role as bulk solvent, water also directly participates in a biochemical reaction. In the catalytic cycle of CYP 450, Cpd I is formed by release of a water molecule after protonation of its precursor Cpd 0. This water molecule, however, forms a hydrogen bond with oxo atom of Fe(IV)-oxo moiety and may linger in the substrate binding pocket during catalysis.[202] A systematic QM/MM investigation by the Thiel group discovered the possible catalytic role of this water molecule.[203,204] In their study they found that increase in the hydrogen bond strength between this water molecule and the oxo atom during the process of the hydrogen abstraction in CYP 450$_{cam}$ is because of a more negatively charged oxo atom in the transition state. The magnitude of lowering of the hydrogen abstraction barrier in CYP 450$_{cam}$ is found to be about 4 kcal mol^{-1} at the B3LYP/CHARMM level of theory. On the basis of a classical electrostatic model (with interaction energies calculated using Mulliken charges), they concluded that the catalytic effect of the water molecule arises due to more favorable electrostatic interactions in the transition state.[203,204] They further extended their work with the aim of exploring the generality of this effect by studying the hydrogen atom abstraction of nine substrates by Cpd I with a total of 12 different modes of abstraction, both in the doublet and quartet state using the DFT method. Table 9.3 summarizes the hydrogen atom abstraction barriers of these nine substrates in the presence and in the absence of this water molecule. Clearly, the magnitude in barrier lowering is correlated with the donor bond strength of the substrate and with the amount of charge transfer in the transition state.[205]

Table 9.3 B3LYP/LACVP barriers (kcal mol^{-1}) for hydrogen atom abstraction from substrates **1–9** by a Cpd I model system in the gas phase.

Substrate	Label[a]	Spin state[b]	B3LYP/LACVP barrier		
			Without water molecule in model	With water molecule in model	Difference
Ethane	**1**	D	20.32	18.13	2.19
		Q	21.28	18.71	2.57
Propane	**2n**	D	20.47	19.06	1.41
		Q	21.62	20.33	1.29
Propane	**2i**	D	18.02	16.61	1.41
		Q	19.26	16.63	2.63
Ethylbenzene	**3n**	D	19.96	17.70	2.26
		Q	21.16	18.30	2.86
Ethylbenzene	**3b**	D	14.30	10.53	3.77
		Q	14.57	10.81	3.76
*trans-*Methylphenylcyclopropane	**4**	D	17.43	14.85	2.58
		Q	18.46	14.20	4.26
Camphor	**5**	D	17.08	13.63	3.45
		Q	18.08	14.78	3.30
Propene	**6**	D	15.01	13.60	1.37
		Q	15.33	13.21	1.80
Toluene	**7**	D	15.78	12.66	2.67
		Q	15.15	12.05	3.01
*trans-i-*Propylphenylcyclopropane	**8**	D	13.68	13.39	0.29
		Q	14.33	12.92	1.41
N,N-Dimethylaniline	**9**	D	6.33	−1.57	7.90
		Q	7.79	2.27	5.52

[a]Substrates labeled as: ethane **1**, propane (abstraction of primary hydrogen atom) **2n**, propane (abstraction of secondary hydrogen atom) **2i**, ethylbenzene (abstraction of methyl hydrogen atom) **3n**, ethylbenzene (abstraction of benzyl hydrogen atom) **3b**, *trans*-methylphenylcyclopropane **4**, camphor **5**, propene **6**, toluene **7**, *trans-i*-propylphenylcyclopropane **8**, and *N,N*-dimethylaniline **9**.
[b]D, doublet spin state; Q, quartet spin state.

9.15 Conclusion

This chapter reviews the importance of heme enzymes CYP 450, which can be coupled to derive important structural and functional information on intermediates. Out of five subfamilies (1A, 2B, 2C, 2D, 3A) eight human cytochrome P450 enzymes are responsible for the vast majority of oxidative drug metabolisms of the most important clinical drugs. Out of the 200 most often prescribed drugs in the USA, about 80% are cleared primarily by these CYPs. The chapter also reviews the complexities that arise at the level of basic mechanisms such as substrate-dependent effects, linkage *etc.* Several metabolites of drugs like ziprasidone, fluvastatin, *etc.* are cleared by only one specific CYP 450 isozyme, such as CYP3A4, CYP2C9, respectively. This suggests that several alternative oxidant species of CYP 450, such as Cpd 0, Cpd II or Cpd I. On the basis of recent experimental as well theoretical studies this review suggests that Cpd 0 is less reactive than Cpd I. The chapter also focuses on the TSR/MSR concept, its occurrence in the reactions of high-valent metal-oxo reagents, and explains important features of TSR/MSR, that is, the different spin states and the different electromers that prefer different mechanisms and therefore produce different intermediates and products. Sometimes the reactivity is reduced to a spin-selective single-state reactivity (SSR) phenomenon when the low-spin state activation barrier is sufficiently lower than that for the high-spin state. This chapter clearly emphasizes that the enormous value of the pharmaceutical applications that can arise from *theoretical/computational* modeling of human drug metabolizing enzymes promises a drive toward another generation of research into the structure and catalytic activity of cytochrome P450s – one of the most ancient and versatile of all protein super families. The computational tools used to predicting drug metabolism will be crucial in rapidly growing drug discovery. Considerable challenges remain for drug metabolism prediction using structure-based methods, but apparently the opportunities for accurate prediction of drug metabolism using computational tools will become a mainstream component of drug design and will be an achievable goal in the near future.

Acknowledgements

D.K. is the Ramanujan Fellow of the Department of Science and Technology (New Delhi). The helpful scientific discussions with, and suggestions by, Dr Sam de Visser, University of Manchester, Manchester, United Kingdom to bring this chapter to its final shape are heartily acknowledged. The encouragement and support from Dr J. S. Yadav, Director, IICT, and scientific discussions with Dr G. N. Sastry, MMG, IICT, are also acknowledged.

References

1. J. H. Dawson and M. Sono, *Chem. Rev.*, 1987, **87**, 1255.
2. D. R. Nelson, T. Kamataki, D. J. Waxman, F. P. Guengerich, R. W. Estabrook, R. Feyereisen, F. J. Gonzalez, M. J. Coon, I. C. Gunsalus, O. Gotoh, K. Okuda and D. W. Nebert, *DNA Cell Biol*, 1993, **12**, 1.

3. M. Sono, M. P. Roach, E. D. Coulter and J. H. Dawson, *Chem. Rev.*, 1996, **96**, 2841.

4. K. M. Kadish, K. M. Smith and R. Guilard (eds.), *The Porphyrin Handbook*, Academic Press, San Diego, 2000, vol. 4.

5. F. P. Guengerich, *Chem Res. Toxicol.*, 2001, **14**, 611.

6. P. R. Ortiz de Montellano (ed.), *Cytochrome P450: Structure, Mechanism and Biochemistry*, 3rd edn, Kluwer Academic/Plenum Publishers, New York, 2004.

7. Y. Watanabe, H. Nakajima and T. Ueno, *Acc. Chem. Res.*, 2007, **40**, 554.

8. P. B. Danielson, *Curr. Drug. Metab.*, 2002, **3**, 561.

9. http://drnelson.uthsc.edu/cytochromeP450.html (accessed 16 March 2011).

10. D. W. Nebert and D. R. Nelson, in *Methods in Enzymology. Cytochrome P450*, vol. 206, ed. M. R. Waterman and E. F. Johnson, Academic Press, New York, 1991, pp. 3–11.

11. M. R. Waterman and E. F. Johnson (eds.), *Methods in Enzymology. Cytochrome P450*, vol. 206, Academic Press, New York, 1991.

12. D. R. Nelson, *Comp. Biochem. Physiol. C. Pharmacol. Toxicol. Endocrinol.*, 1998, **121**, 15.

13. L. Ereshefsky, C. Riesenman and Y. W. Lam, *J. Clin. Psychiatry*, 1996, **57**, 17.

14. E. L. Michalets, *Pharmacotherapy*, 1998, **18**, 84.

15. V. Dixit, N. Hariparsad, F. Li, P. Desai, K. E. Thummel and J. D. Unadkat, *Drug Metab Dispos*, 2007, **35**, 1853.

16. M. L. Catterson, S. H. Preskorn and R. L. Martin, *Psychiatr. Clin. North Am.*, 1997, **20**, 205.

17. L. Goshman, J. Fish and K. Roller, *J. Pharmacy Soc. Wisconsin 1999*, May/June, 23.

18. S. Sansen, J. K. Yano, R. L. Reynald, G. A. Schoch, K. J. Griffin, C. D. Stout and E. F. Johnson, *J. Biol. Chem.*, 2007, **282**, 14348.

19. P. Rowland, F. E. Blaney, M. G. Smyth, J. J. Jones, V. R. Leydon, A. K. Oxbrow, C. J. Lewis, M. G. Tennant, S. Modi, D. S. Eggleston, R. J. Chenery and A. M. Bridges, *J. Biol. Chem.*, 2006, **281**, 7614.

20. I. F. Sevrioukova and T. L. Poulos, *Proc. Natl. Acad. Sci. USA*, 2010, **107**, 18422.

21. K. E. Ketter, D. A. Flockhart and R. M. Post, *J. Clin. Psychopharmacol.*, 1995, **15**, 387.

22. J. W. Harris, A. Rahman, B. R. Kim, F. P. Guengerich and J. M. Collins, *Cancer Res.*, 1994, **54**, 4026.

23. M. A. Leo, J. M. Lasker, J. L. Raucy, C. I. Kim, M. Black and C. S. Lieber, *Arch. Biochem. Biophys.*, 1989, **269**, 305.

24. D. J. Waxman, *Adv. Mol. Cell Biol.*, 1996, **14**, 341.

25. B. G. Pollock, R. A. Sweet, M. Kirschner and C. F. Reynolds, *Ther. Drug. Monit.*, 1996, **18**, 581.

26. U. M. Zanger, M. Turpeinen, K. Klein and M. Schwab, *Anal. Bioanal. Chem.*, 2008, **392**, 1093.

27. C. Prakash, A. Kamel, D. Cui, R. D. Whalen, J. J. Miceli and D. Tweedie, *J. Clin. Pharmacol.*, 2000, **49**(Suppl. 1), 35S.

28. C. Prakash, A. Kamel, W. Anderson and H. Howard, *Drug Metab. Dispos.*, 1997, **25**, 206.
29. B. Testa, in *The Metabolism of Drugs and Other Xenobiotics*, ed. B. Testa, J. Caldwell, Academic Press, New York, 1995, pp. 203–234.
30. R. I. Levy, A. J. Troendle and J. M. Fattu, *Circulation*, 1993, **87**(Suppl. 4), III45.
31. T. K. Peters, J. Jewitt-Harris, M. Mehra and E. N. Muratti, *Am. J. Hypertens.*, 1993, **6**, S346.
32. M. H. Davidson, *Am. J. Med.*, 1994, **96**, S41.
33. V. Fischer, L. Johanson, F. Heitz, R. Tullman, E. Graham, J.-P. Baldeck and W. T. Robinson, *Drug Metabol. Dispos.*, 1999, **27**, 410.
34. P. J. Cox, D. A. Ryan, F. J. Hollis, A.-M. Harris, A. K. Miller, M. Vousden and H. Cowley, *Drug Metabol. Dispos.*, 2000, **28**, 772.
35. V. B. G. Reddy, B. V. Karanam, W. L. Gruber, M. A. Wallace, S. H. Vincent, R. B. Franklin and T. A. Baillie, *Chem. Res. Toxicol.*, 2005, **18**, 880.
36. B. Prasad and S. Singh, *J. Pharm. Biomed. Anal.*, 2009, **50**, 475.
37. T. L. Poulos, B. C. Finzel, I. C. Gunsalus, G. C. Wagner and J. Kraut, *J. Biol. Chem.*, 1985, **260**, 16122.
38. H. M. Berman, J. Westbrook, Z. Feng, G. Gilliland, T. N. Bhat, H. Wiessig, I. N. Shindylov and P. E. Bourne, *Nucleic Acids Res.*, 2000, **28**, 235.
39. I. G. Denisov, T. M. Makris, S. G. Sligar and I. Schlichting, *Chem. Rev.*, 2005, **105**, 2253.
40. S. G. Sligar, *Biochemistry*, 1976, **15**, 5399.
41. M. Swart, A. R. Groenhof, A. W. Ehlers and K. Lammertsma, *Chem. Phys. Lett.*, 2005, **403**, 35.
42. A. Altun and W. Thiel, *J. Phys. Chem. A*, 2005, **109**, 1268.
43. A. R. Groenhof, M. Swart, A. W. Ehlers and K. Lammertsma, *J. Phys. Chem. A*, 2005, **109**, 3411.
44. P. R. Balding, C. S. Porro, K. J. McLean, M. J. Sutcliffe, J.-D. Maréchal, A. W. Munro and S. P. de Visser, *J. Phys. Chem. A*, 2008, **112**, 12911.
45. C. B. Brewer and J. A. Peterson, *J. Biol. Chem.*, 1988, **263**, 791.
46. J. Rittle and M. T. Green, *Science*, 2010, **330**, 933.
47. T. Hishiki, H. Shimada, S. Nagano, T. Egawa, Y. Kanamori, R. Makino, S.-Y. Park, S. Adachi, Y. Shiro and Y. Ishimura, *J. Biochem. (Tokyo)*, 2000, **128**, 965.
48. T. Egawa, H. Shimada and Y. Ishimura, *Biochem. Biophys. Res. Commun.*, 1994, **201**, 1464.
49. I. G. Denisov, T. M. Makris and S. G. Sligar, *J. Biol. Chem.*, 2001, **276**, 11648.
50. D. G. Kellner, S. C. Hung, K. E. Weiss and S. G. Sligar, *J. Biol. Chem.*, 2002, **277**, 9641.
51. R. Davydov, T. M. Makris, V. Kofman, D. E. Werst, S. G. Sligar and B. M. Hoffman, *J. Am. Chem. Soc.*, 2001, **123**, 1403.
52. J. T. Groves, in *Cytochrome P450: Structure, Mechanism, and Biochemistry*, 3rd edn, ed. P. R. Ortiz de Montellano, Kluwer Academic/Plenum Publishers, New York, 2005, ch. 1.

53. M. Newcomb and P. H. Toy, *Acc. Chem. Res.*, 2000, **33**, 449.

54. S. Jin, T. M. Makris, T. A. Bryson, S. G. Sligar and J. H. Dawson, *J. Am. Chem. Soc.*, 2003, **125**, 3406.

55. A. D. N. Vaz, D. F. McGinnity and M. J. Coon, *Proc. Natl. Acad. Sci. USA*, 1998, **95**, 3555.

56. W. Nam, H. J. Lee, S.-Y. Oh, C. Kim and H. G. Jang, *J. Inorg. Biochem.*, 2000, **80**, 219.

57. W. Nam, S.-E. Park, I. K. Lim, M. H. Lim, J. K. Hong and J. Kim, *J. Am. Chem. Soc.*, 2003, **125**, 14674.

58. J. Kaizer, E. J. Klinker, N. Y. Oh, J.-U. Rohde, W. J. Song, A. Stubna, J. Kim, E. Münck, W. Nam and L. Que Jr., *J. Am. Chem. Soc.*, 2004, **126**, 472.

59. Y. Ichikawa, H. Nakajima and Y. Watanabe, *ChemBioChem*, 2006, **7**, 1582.

60. A.-R. Han, Y. J. Jeong, Y. Kang, J. Y. Lee, M. S. Seo and W. Nam, *Chem. Commun.*, 2008, 1076.

61. A. Franke, C. Fertinger and R. van Eldik, *Angew. Chem., Int. Ed.*, 2008, **47**, 5238.

62. G. H. Loew and D. L. Harris, *Chem. Rev.*, 2000, **100**, 407.

63. M. T. Green, *J. Am. Chem. Soc.*, 1999, **121**, 7939.

64. M. T. Green, *J. Am. Chem. Soc.*, 2000, **122**, 9495.

65. M. T. Green, *J. Am. Chem. Soc.*, 2001, **123**, 9218.

66. T. Ohta, K. Matsuura, K. Yoshizawa and I. Morishima, *J. Inorg. Biochem.*, 2000, **82**, 141.

67. F. Ogliaro, S. Cohen, S. P. de Visser and S. Shaik, *J. Am. Chem. Soc.*, 2000, **122**, 12892.

68. J. C. Schöneboom, F. Neese and W. Thiel, *J. Am. Chem. Soc.*, 2005, **127**, 5840.

69. A. Altun, D. Kumar, F. Neese and W. Thiel, *J. Phys. Chem. A*, 2008, **112**, 12904.

70. P. Rydberg, E. Sigfridsson and U. Ryde, *J. Biol. Inorg. Chem.*, 2004, **9**, 203.

71. F. Ogliaro, S. P. de Visser, S. Cohen, J. Kaneti and S. Shaik, *ChemBioChem*, 2001, **2**, 848.

72. F. Ogliaro, S. Cohen, M. Filatov, N. Harris and S. Shaik, *Angew. Chem., Int. Ed.*, 2000, **39**, 3851.

73. F. Ogliaro, S. P. de Visser, J. T. Groves and S. Shaik, *Angew. Chem., Int. Ed.*, 2001, **40**, 2874.

74. D. L. Harris, G. H. Loew and L. Waskell, *J. Inorg. Biochem.*, 2001, **83**, 309.

75. M. Filatov, N. Harris and S. Shaik, *Angew. Chem., Int. Ed.*, 1999, **38**, 3510.

76. M. Filatov, N. Harris and S. Shaik, *J. Chem. Soc., Perkin Trans. 2*, 1999, 399.

77. J. Antony, M. Grodzicki and A. X. Trautwein, *J. Phys. Chem. A*, 1997, **101**, 2692.

78. M. Radoń and E. Broclawik, *J. Chem. Theory Comput.*, 2007, **3**, 728.
79. J. C. Schöneboom, H. Lin, N. Reuter, W. Thiel, S. Cohen, F. Ogliaro and S Shaik, *J. Am. Chem. Soc.*, 2002, **124**, 8142.
80. K. L. Stone, R. K. Behan and M. T. Green, *Proc. Natl. Acad. Sci. USA*, 2005, **102**, 16563.
81. S. H. Kim, R. Perera, L. P. Hager, J. H. Dawson and B. M. Hoffman, *J. Am. Chem. Soc.*, 2006, **128**, 5598.
82. E. Derat, S. Cohen, S. Shaik, A. Altun and W. Thiel, *J. Am. Chem. Soc.*, 2005, **127**, 13611.
83. S. P. de Visser, S. Shaik, P. K. Sharma, D. Kumar and W. Thiel, *J. Am. Chem. Soc.*, 2003, **125**, 15779.
84. H. Hirao, D. Kumar, W. Thiel and S. Shaik, *J. Am. Chem. Soc.*, 2005, **127**, 13007.
85. P. R. Ortiz de Montellano, *Annu. Rev. Pharmacol. Toxicol.*, 1992, **32**, 89.
86. P. Nicholls, I. Fita and P. C. Loewen, *Adv. Inorg. Chem.*, 2000, **51**, 51.
87. N. G. Veitch and A. T. Smith, *Adv. Inorg. Chem.*, 2000, **51**, 107.
88. J. H. Dawson, R. H. Holm, J. R. Trudell, G. Barth, R. E. Linder, E. Bunnenberg, C. Djerasi and S. C. Tang, *J. Am. Chem. Soc.*, 1976, **98**, 3707.
89. T. L. Poulos, *J. Biol. Inorg. Chem.*, 1996, **1**, 356.
90. D. B. Goodin and D. E. McRee, *Biochemistry*, 1993, **32**, 3313.
91. E. Derat, S. Shaik, C. Rovira, P. Vidossich and M. Alfonso-Prieto, *J. Am. Chem. Soc.*, 2007, **129**, 6346.
92. M. Sivaraja, D. B. Goodin, M. Smith and B. M. Hoffman, *Science*, 1989, **245**, 738.
93. J. E. Huyett, P. E. Doan, R. Gurbriel, A. L. P. Houseman, M. Sivaraja, D. B. Goodin and B. M. Hoffman, *J. Am. Chem. Soc.*, 1995, **117**, 9033.
94. C. A. Bonagura, B. Bhaskar, H. Shimizu, H. Li, M. Sundaramoorthy, D. E. McRee, D. B. Goodin and T. L. Poulos, *Biochemistry*, 2003, **42**, 5600.
95. G. I. Berglund, G. H. Carlsson, A. T. Smith, H. Szöke, A. Henriksen and J. Hajdu, *Nature*, 2002, **417**, 463.
96. B. Bhaskar, C. A. Bonagura, H. Li and T. L. Poulos, *Biochemistry*, **41**, 2684.
97. E. L. Raven, *Nat. Prod. Rep.*, 2003, **20**, 367.
98. G. M. Jensen, S. W. Bunte, A. Warshel and D. B. Goodin, *J. Phys. Chem. B*, 1998, **102**, 8221.
99. D. K. Menyhárd and G. Náray-Szabó, *J. Phys. Chem. B*, 1999, **103**, 227.
100. S. P. de Visser, *J. Phys. Chem. A*, 2005, **109**, 11050.
101. S. P. de Visser, The axial ligand effect on substrate monoxygenation by the oxo-iron active species of heme enzymes. How does cytochrome c peroxidase compare to cytochrome P450?, in *Inorganic Biochemistry: Research Progress*, ed. J. G Hughes and A. J. Robinson, Nova Science Publishers, Inc., 2008, pp. 197–224.
102. M. Wirstam, M. R. A. Blomberg and P. E. M. Siegbahn, *J. Am. Chem. Soc.*, 1999, **121**, 10178.
103. C. M. Bathelt, A. J. Mulholland and J. N. Harvey, *Dalton Trans.*, 2005, 3470.

104. J. N. Harvey, C. M. Bathelt and A. J. Mulholland, *J. Comput. Chem.*, 2006, **27**, 1352.
105. Z. Gross and S. Nimri, *Inorg. Chem.*, 1994, **33**, 1731.
106. K. Czarnecki, S. Nimri, Z. Gross, L. M. Proniewicz and J. R. Kincaid, *J. Am. Chem. Soc.*, 1996, **118**, 2929.
107. Z. Gross, *J. Biol. Inorg. Chem.*, 1996, **1**, 368.
108. W. J. Song, Y. O. Ryu, R. Song and W. Nam, *J. Biol. Inorg. Chem.*, 2005, **10**, 294.
109. W. Nam, *Acc. Chem. Res.*, 2007, **40**, 522.
110. I. M. C. M. Rietjens, A. M. Osman, C. Veeger, O. Zakharieva, J. Antony, M. Grodzicki and A. X. Trautwein, *J. Biol. Inorg. Chem.*, 1996, **1**, 372.
111. H. Kuramochi, L. Noodleman and D. A. Case, *J. Am. Chem. Soc.*, 1997, **119**, 11442.
112. F. Ogliaro, S. P. de Visser and S. Shaik, *J. Inorg. Biochem.*, 2002, **91**, 554.
113. S. P. de Visser, *J. Biol. Inorg. Chem.*, 2006, **11**, 168.
114. R. Wang and S. P. de Visser, *J. Inorg. Biochem.*, 2007, **101**, 1464.
115. S. P. de Visser, L. Tahsini and W. Nam, *Chem. Eur. J.*, 2009, **15**, 5577.
116. S. P. de Visser, *J. Am. Chem. Soc.*, 2010, **132**, 1087.
117. S. P. de Visser, *Chem. Eur. J.*, 2006, **12**, 8168.
118. S. P. de Visser, *Inorg. Chem.*, 2006, **45**, 9551.
119. S. Shaik, H. Hirao and D. Kumar, *Acc. Chem. Res.*, 2007, **40**, 532.
120. D. Kumar, S. P. de Visser, P. K. Sharma, S. Cohen and S. Shaik, *J. Am. Chem. Soc.*, 2004, **126**, 1907.
121. D. Kumar, S. P. de Visser and S. Shaik, *J. Am. Chem. Soc.*, 2003, **125**, 13024.
122. D. Kumar, S. P. de Visser and S. Shaik, *J. Am. Chem. Soc.*, 2004, **126**, 5072.
123. E. Derat, D. Kumar, H. Hirao and S. Shaik, *J. Am. Chem. Soc.*, 2006, **128**, 473.
124. S. P. de Visser, D. Kumar and S. Shaik, *J. Inorg. Biochem.*, 2004, **98**, 1183.
125. D. Kumar, S. P. de Visser and S. Shaik, *Chem.Eur. J.*, 2005, **11**, 2825.
126. H. Hirao, D. Kumar and S. Shaik, *J. Inorg. Biochem.*, 2006, **100**, 2054.
127. S. Shaik, S. P. de Visser, F. Ogliaro, H. Schwarz and D. Schröder, *Curr. Opin. Chem. Biol.*, 2002, **6**, 556.
128. S. Shaik, D. Kumar, S. P. de Visser, A. Altun and W. Thiel, *Chem. Rev.*, 2005, **105**, 2279.
129. S. P. de Visser, F. Ogliaro, N. Harris and S. Shaik, *J. Am. Chem. Soc.*, 2001, **123**, 3037.
130. M. Filatov and S. Shaik, *J. Phys. Chem. A*, 1998, **102**, 3835.
131. D. Schröder, H. Schwarz, D. E. Clemmer, Y. Chen, P. B. Armentrout, V. I. Baranov and D. K. Bohme, *Int. J. Mass Spectrom. Ion Processes*, 1997, **161**, 175.
132. E. Derat and S. Shaik, *J. Am. Chem. Soc.*, 2006, **128**, 13940.
133. H. Hirao, D. Kumar, L. Que Jr. and S. Shaik, *J. Am. Chem. Soc.*, 2006, **128**, 8590.

134. A. M. Khenkin, D. Kumar, S. Shaik and R. Neumann, *J. Am. Chem. Soc.*, 2006, **128**, 15451.
135. S. Shaik, S. Cohen, Y. Wang, H. Chen, D. Kumar and W. Thiel, *Chem. Rev.*, 2010, **110**, 949.
136. S. P. de Visser, D. Kumar, S. Cohen, R. Shacham and S. Shaik, *J. Am. Chem. Soc.*, 2004, **126**, 8362.
137. L. Olsen, P. Rydberg, T. H. Rod and U. Ryde, *J. Med. Chem.*, 2006, **49**, 6489.
138. S. Shaik, D. Kumar and S. P. de Visser, *J. Am. Chem. Soc.*, 2008, **130**, 10128.
139. J. T. Groves and G. A. McClusky, *J. Am. Chem. Soc.*, 1976, **98**, 859.
140. A. D. Becke, *J. Chem. Phys.*, 1993, **98**, 5648.
141. M. Güell, J. M. Luis, M. Solà and M. Swart, *J. Phys. Chem. A*, 2008, **112**, 6384.
142. F. Neese, *Coord. Chem. Rev.*, 2009, **253**, 526.
143. S. J. Kim, R. Latifi, H. Y. Kan, W. Nam and S. P. de Visser, *Chem. Commun.*, 2009, 1562.
144. S. S. Shaik, *J. Am. Chem. Soc.*, 1981, **103**, 3692.
145. F. Ogliaro, N. Harris, S. Cohen, M. Filatov, S. P. de Visser and S. Shaik, *J. Am. Chem. Soc.*, 2000, **122**, 8977.
146. K. Yoshizawa, Y. Kagawa and Y. Shiota, *J. Phys. Chem. B*, 2000, **104**, 12365.
147. N. Harris, S. Cohen, M. Filatov, F. Ogliaro and S. Shaik, *Angew. Chem., Int. Ed.*, 2000, **39**, 2003.
148. T. Kamachi and K. Yoshizawa, *J. Am. Chem. Soc.*, 2003, **125**, 4652.
149. J. C. Schöneboom, S. Cohen, H. Lin, S. Shaik and W. Thiel, *J. Am. Chem. Soc.*, 2004, **126**, 4017.
150. Y. Wang, H. Wang, C. Yang, L. Yang and K. J. Han, *J. Phys. Chem. B*, 2006, **110**, 6154.
151. Y. Wang, C. Yang, H. Wang, K. Han and S. Shaik, *ChemBioChem*, 2007, **8**, 277.
152. S. P. de Visser and L. S. Tan, *J. Am. Chem. Soc.*, 2008, **130**, 12961.
153. Y. Zhang, P. Morisetti, J. Kim, L. Smith and H. Lin, *Theor. Chem. Acc.*, 2008, **121**, 313.
154. P. Rydberg, U. Ryde and L. Olsen, *J. Chem. Theory Comput.*, 2008, **4**, 1369.
155. A. R. Shaikh, R. Sahnoun, E. Broclawik, M. Koyama, H. Tsuboi, N. Hatakeyama, A. Endou, H. Takaba, M. Kubo, C. A. Del Carpio and A. Miyamoto, *J. Inorg. Biochem.*, 2009, **103**, 20.
156. D. Li, Y. Wang, C. Yang and K. Han, *Dalton Trans.*, 2009, 291.
157. D. Kumar, L. Tahsini, S. P. de Visser, H. Y. Kang, S. J. Kim and W. Nam, *J. Phys. Chem. A*, 2009, **113**, 11713.
158. P. R. Ortiz de Montellano and R. A. Stearns, *J. Am. Chem. Soc.*, 1987, **109**, 3415.
159. M. Newcomb, R. Shen, S.-Y. Choi, P. H. Toy, P. F. Hollenberg, A. D. N. Vaz and M. J. Coon, *J. Am. Chem. Soc.*, 2000, **122**, 2677.

160. M. Newcomb, D. Aebisher, R. Shen, R. Esala, P. Chandrasena, P. F. Hollenberg and M. J. Coon, *J. Am. Chem. Soc.*, 2003, **125**, 6064.
161. F. Ogliaro, S. P. de Visser, S. Cohen, P. K. Sharma and S. Shaik, *J. Am. Chem. Soc.*, 2002, **124**, 2806.
162. T. Kamachi, Y. Shiota, T. Ohta and K. Yoshizawa, *Bull. Chem. Soc. Jpn.*, 2003, **76**, 721.
163. P. K. Sharma, S. P. de Visser and S. Shaik, *J. Am. Chem. Soc.*, 2003, **125**, 8698.
164. C. S. Porro, M. J. Sutcliffe and S. P. de Visser, *J. Phys. Chem. A*, 2009, **113**, 11635.
165. M. J. Park, J. Lee, Y. Suh, J. Kim and W. Nam, *J. Am. Chem. Soc.*, 2006, **128**, 2630.
166. P. R. Ortiz de Montellano, H. S. Beilan, K. L. Kunze and B. A. Mico, *J. Biol. Chem.*, 1981, **256**, 4395.
167. K. L. Kunze, B. L. K. Mangold, C. Wheeler, H. S. Beilan and P. R. Ortiz de Montellano, *J. Biol. Chem.*, 1983, **258**, 4202.
168. P. R. Ortiz de Montellano, B. L. K. Mangold, C. Wheeler, K. L. Kunze and N. O. Reich, *J. Biol. Chem.*, 1983, **258**, 4208.
169. Z.-Q. Tian, J. Richards and T. G. Traylor, *J. Am. Chem. Soc.*, 1995, **117**, 21.
170. S. P. de Visser, F. Ogliaro, P. K. Sharma and S. Shaik, *Angew. Chem., Int. Ed.*, 2002, **41**, 1947.
171. S. P. de Visser, F. Ogliaro, P. K. Sharma and S. Shaik, *J. Am. Chem. Soc.*, 2002, **124**, 11809.
172. D. Kumar, B. Karamzadeh, G. N. Sastry and S. P. de Visser, *J. Am. Chem. Soc.*, 2010, **132**, 7656.
173. S. P. de Visser, F. Ogliaro and S. Shaik, *Chem. Commun.*, 2001, 2322.
174. S. Shaik, S. P. de Visser and D. Kumar, *J. Biol. Inorg. Chem.*, 2004, **9**, 661.
175. S. P. de Visser, F. Ogliaro and S. Shaik, *Angew. Chem., Int. Ed.*, 2001, **40**, 2871.
176. S. P. de Visser, *J. Am. Chem. Soc.*, 2006, **128**, 9813.
177. S. P. de Visser, *J. Am. Chem. Soc.*, 2006, **128**, 15809.
178. M. Werner, G. Birner and W. Dekant, *Chem. Res. Toxicol.*, 1996, **9**, 41.
179. A. Madan, A. Parkinson and M. D. Faiman, *Drug Metabol. Dispos.*, 1995, **23**, 1153.
180. K. A. Usmani, E. D. Karoly, E. Hodgson and R. L. Rose, *Drug Metabol. Dispos.*, 2004, **32**, 333.
181. B. Furnes and D. Schlenk, *Drug Metabol. Dispos.*, 2005, **33**, 214.
182. E. J. Perkins, A. El-Alfy and D. Schlenk, *Toxicol. Sci.*, 1999, **48**, 67.
183. J. Wójcikowski, P. Maurel and W. A. Daniel, *Drug Metabol. Dispos.*, 2006, **34**, 471.
184. Y. Goto, T. Matsui, S.-i. Ozaki, Y. Watanabe and S. Fukuzumi, *J. Am. Chem. Soc.*, 1999, **121**, 9497.
185. S. Kato, H.-J. Yang, T. Ueno, S.-i. Ozaki, G. N. Phillips Jr., S. Fukuzumi and Y. Watanabe, *J. Am. Chem. Soc.*, 2002, **124**, 8506.

186. T. J. Volz, D. A. Rock and J. P. Jones, *J. Am. Chem. Soc.*, 2002, **124**, 9724.
187. K.-B. Cho, Y. Moreau, D. Kumar, D. A. Rock, J. P. Jones and S. Shaik, *Chem. Eur. J.*, 2007, **13**, 4103.
188. J. Benet-Buchholz, P. Comba, A. Llobet, S. Roeser, P. Vadivelu and S. Wiesner, *Dalton Trans.*, 2010, **39**, 3315.
189. D. Kumar, S. P. de Visser, P. K. Sharma, H. Hirao and S. Shaik, *Biochemistry*, 2005, **44**, 8148.
190. C. Li, L. Zhang, C. Zhang, H. Hirao, W. Wu and S. Shaik, *Angew. Chem., Int. Ed.*, 2007, **46**, 8168.
191. Y. Watanabe, T. Iyanagi and S. Oae, *Tetrahedron Lett.*, 1980, **21**, 3685.
192. Y. Watanabe, T. Numata, T. Iyanagi and S. Oae, *Bull. Chem. Soc. Jpn.*, 1981, **54**, 1163.
193. Y. Watanabe, T. Iyanagi and S. Oae, *Tetrahedron Lett.*, 1982, **23**, 533.
194. S. Shaik, Y. Wang, H. Chen, J. Song and R. Meir, *Faraday Discuss.*, 2010, **145**, 49.
195. D. Kumar, G. N. Sastry and S. P. de Visser, *Chem. Eur. J.*, 2011, in press, DOI: 10.1002/chem.201003187.
196. S. P. de Visser and S. Shaik, *J. Am. Chem. Soc.*, 2003, **125**, 7413.
197. C. Hazan, D. Kumar, S. P. de Visser and S. Shaik, *Eur. J. Inorg. Chem.*, 2007, 2966.
198. S. P. de Visser, L. Tahsini and W. Nam, *Chem. Eur. J.*, 2009, 5577.
199. C. M. Bathelt, L. Ridder, A. J. Mulholland and J. N. Harvey, *J. Am. Chem. Soc.*, 2003, **125**, 15004.
200. C. M. Bathelt, L. Ridder, A. J. Mulholland and J. N. Harvey, *Org. Biomol. Chem.*, 2004, **2**, 2998.
201. C. M. Bathelt, A. J. Mulholland and J. N. Harvey, *J. Phys. Chem. A*, 2008, **112**, 13149.
202. J. Zheng, D. Wang, W. Thiel and S. Shaik, *J. Am. Chem. Soc.*, 2006, **128**, 13204.
203. A. Altun, V. Guallar, R. A. Friesner, S. Shaik and W. Thiel, *J. Am. Chem. Soc.*, 2006, **128**, 3924.
204. A. Altun, S. Shaik and W. Thiel, *J. Comput. Chem.*, 2006, **27**, 1324.
205. D. Kumar, A. Altun, S. Shaik and W. Thiel, *Faraday Discuss.*, 2011, **148**, 373.

Oxidation of Unnatural Substrates by Engineered Cytochrome P450$_{cam}$

SAPTASWA SEN, SOUMEN KANTI MANNA AND
SHYAMALAVA MAZUMDAR*

Department of Chemical Sciences, Tata Institute of Fundamental Research,
Homi Bhabha Road, Colaba, Mumbai 400005, India

10.1 Introduction

Cytochrome P450s (CYP 450s) refer to a superfamily of heme-containing
enzymes involved in activation of molecular oxygen to carry out numerous
metabolic processes.[1-3] The CYP 450 enzymes play key roles in drug metabolism
and xenobiotic detoxification as well as in steroid hormone biosynthesis in
mammalian tissues. These enzymes are also involved in catalyzing the synthesis
of a large number of secondary metabolites in plants and other organisms. The
first microbial CYP 450 was discovered by Gunsalus and coworkers[4-6] from the
soil bacterium *Pseudomonas putida*,[5,7] which is involved in the metabolism of
(1R)-camphor to form 5-*exo* hydroxycamphor.[4-6,8-10] This enzyme, commonly
known as CYP 450$_{cam}$ or CYP101, is one of the most extensively studied
members of the CYP 450 superfamily.[6] CYP 450$_{cam}$ shows very high substrate
specificity and product selectivity, and hence became a favoured model enzyme
to understand the steroid biosynthesis by mammalian CYP 450 enzymes.[11]

CYP 450$_{cam}$ (molecular weight $\sim$47 kDa) has 414 amino acid residues
containing a single ferric protoporphyrin-IX (Heme-b)[1,2] as the prosthetic
group buried deep inside the protein matrix with the ferric ion axially

Iron-Containing Enzymes: Versatile Catalysts of Hydroxylation Reactions in Nature
Edited by Sam P de Visser and Devesh Kumar

Published by the Royal Society of Chemistry, www.rsc.org

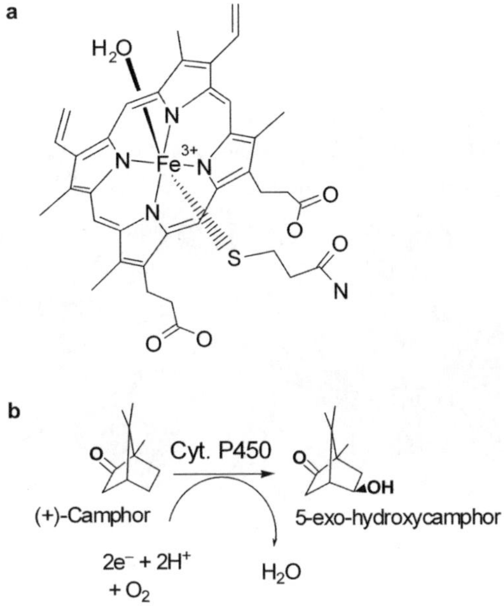

Figure 10.1　(a) Prosthetic group of CYP 450$_{cam}$, an iron(III) protoporphyrin-IX linked with a proximal cysteine-357 residue. (b) Stereospecific hydroxylation of the *exo* C–H bond at the C^5 position of camphor by CYP 450$_{cam}$.

coordinated to a Cys$_{357}$ residue (Figure 10.1).[12,13] This cysteine residue has been found to be critical for the catalytic activity of the enzyme and replacement of this residue leads to gross inactivation of the enzyme.[14–16] The heme propionate groups in the enzyme form salt bridges and H-bond interactions with His$_{355}$, Arg$_{299}$ residues[13] that also provide stability to the active site. The heme propionates have also been shown to be involved in a water network at the distal pocket of the enzyme.[13] The resting state of the substrate-free enzyme consists of a low-spin ferric heme with distally bound water that has been proposed to be a part of this water network.[6] The crystal structure of the enzyme in the absence of the substrate (PDB code: 1PHC) also shows the presence of water cluster at the distal pocket near the heme.

10.2　Binding of the Substrate

The natural substrate [(1R)-camphor] and the oxygen binding sites are located at the distal side of the heme. Although there is no clear opening for the substrate entrance, the substrate is believed to enter through the space between the F-G loop and B′-helix (Figure 10.2).[6,17] The binding of substrate removes these water molecules and the substrate is held in the active site through non-covalent interactions. The substrate binding pocket consists mostly of hydrophobic residues such as Phe$_{87}$, Phe$_{98}$, Leu$_{244}$, Val$_{247}$, Val$_{295}$, Thr$_{185}$, Ile$_{395}$, and Val$_{396}$; the latter holds the hydrophobic substrate in position through van der Waals

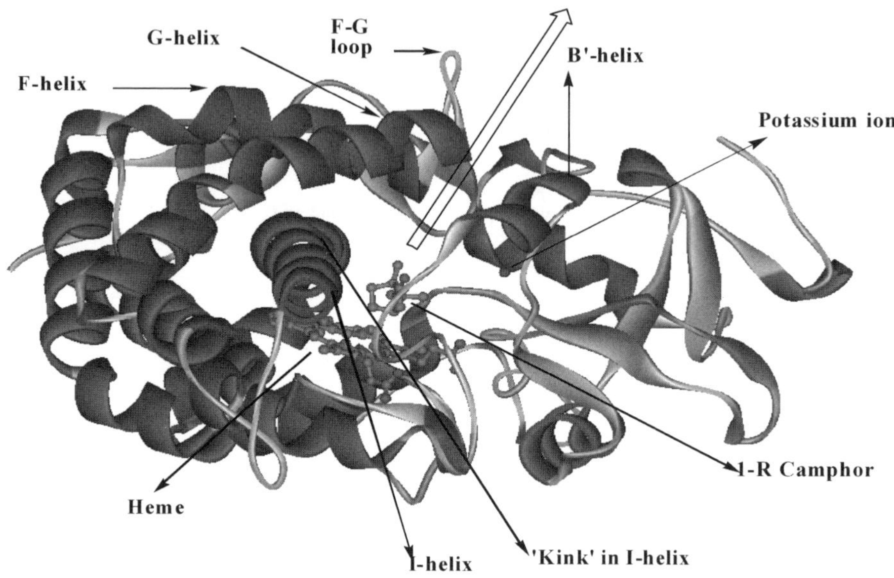

Figure 10.2   Structure of (1*R*)-camphor bound CYP 450$_{cam}$. The open arrow indicates the proposed substrate access channel; also indicated is the potassium bound enzyme near the B'-helix.

forces.[6,13] These residues can be thought of as defining a barrier about the substrate, which allows 1–2 Å movement in any direction. For example, Thr$_{185}$ and Phe$_{87}$ residues form hydrophobic contacts with the C^{10} methyl group of camphor, while the methyl groups of the Val$_{295}$ residue are involved in horn locking interactions with the C^8 and C^9 methyl groups of camphor.[6,13,18] On the other hand, the polar residues (Tyr$_{96}$, Thr$_{252}$, and Asp$_{297}$) in the distal pocket have been found to play an important role in the catalytic cycle of the enzyme. The side chain of the Tyr$_{96}$ residue forms an H-bond with the C^2 carbonyl group of (1*R*)-camphor. This H-bond has been found to influence the substrate binding and most importantly the spin equilibrium of this enzyme.[6] It has been recently revealed that the enzyme also contains a potassium ion binding site in the B-C loop and B' helix region next to the substrate binding pocket (Figure 10.2).[18,19] The oxygen binding to the heme in an "end on" orientation is stabilized by a small distortion (kink) of the "I-helix" (Figure 10.2)[6,18] from Gly$_{248}$ to Thr$_{252}$ residue. The Thr$_{252}$ residue has been implicated in the activation of the molecular oxygen by forming a H-bond network in the distal pocket during catalysis.[20,21] The salt bridges formed by the side chains of Asp$_{251}$ with Lys$_{178}$ and Arg$_{186}$ residues have been shown to stabilize the aforementioned kink in the I-helix.[6,18] The Asp$_{251}$ residue also forms a proton delivery channel through the distal H-bond network that extends to the surface of the enzyme and involves other charged residues, such as Lys$_{178}$, Asp$_{182}$, and Arg$_{186}$.[17] A disruption of this H-bond network has been found to result in uncoupling and concomitant production of H_2O_2 and water.[6,17,22] Therefore, the H-bond network has been implicated to be crucial in controlled proton delivery to the Fe-bound molecular oxygen and in coupling of the electron transfer to the product formation. Till now structures of CYP 450$_{cam}$ with several alternative substrates have also been solved.

The most variable regions of the CYP 450s are the substrate recognition sites (SRSs) that spatially form substrate access/exit channels and the substrate binding site near the heme active centre. This accounts for the different substrate specificity manifested by different CYP 450 isoforms. In general, at the time of entry or exit, the substrate has to make contact with different hydrophobic or hydrophilic residues depending on its nature.[23] This gating mechanism has been well addressed by several researchers through experiments or molecular dynamics simulation on CYP 450s.[24] It should be mentioned that there are distinctive differences between mammalian and bacterial CYP 450s in substrate/product entry/exit pathways.[25] In summary, we can say that substrate recognition/binding is not only influenced by the interactions at the heme active site but also by the properties of the access routes to the binding site such as size, shape, and physicochemical properties of the enzyme as well as for the substrate.[26–28]

CYP 450s are divided into two major classes (Class I and Class II) according to the different types of electron transfer systems they use.[29,30] CYP 450$_{cam}$ is a part of the Class I family, which includes bacterial and mitochondrial CYP 450s, and uses a two-component shuttle system consisting of an iron-sulfur protein (ferredoxin) and a ferredoxin reductase. In this particular isozyme, the electron is transferred from NADH to the enzyme (CYP 450$_{cam}$) through its electron transfer partners, putidaredoxin (PdX) and putidaredoxin reductase (PdR). The enzyme forms a part of the energy harvesting machinery of the bacterium, and has evolved an excellent strategy to optimize the efficiency of the oxygenase system. The redox potential of putidaredoxin ($E^0 = -240$ mV) is such that no electron transfer can take place in absence of the substrate.[6,31,32] The substrate binding induces a spin state change and increases the redox potential of this enzyme from –300 mV (substrate free, low spin) to –170 mV (substrate bound, high spin form), so that the electron is transferred from NADH to substrate bound CYP 450$_{cam}$ through the putidaredoxin.[6] Elegant theoretical studies by Harris and Loew have suggested[33] that nature has utilized this kind of sophisticated application of electrostatic stabilization to poise the ferric $S = \frac{5}{2}$ to $S = \frac{1}{2}$ spin transition such that simple water ligation can alter the dynamic spin state equilibrium significantly. Through the detailed analysis of kinetic processes, we now know that the spin state of CYP 450 heme iron is regulated by the accessibility of water in the sixth coordination position. Thus, substrate binding acts as a switch to turn on the catalytic cycle. The putidaredoxin binds to the surface of this enzyme at the proximal side of heme through H-bond/salt bridge interactions of the Asp$_{34}$ and Asp$_{38}$ residues of putidaredoxin with the Arg$_{109}$ and Arg$_{112}$ residues of CYP 450$_{cam}$.[34,35] This perturbs the proximal H-bond network of CYP 450$_{cam}$ involving Arg$_{112}$, Leu$_{356}$, His$_{355}$, and the heme propionate group. Hence, the binding of putidaredoxin induces a conformational change on the active site of the enzyme, involving the substrate binding pocket, I-helix and heme, which helps to carry out regio- and stereospecific hydroxylation of camphor in tight coupling with NADH oxidation.[36–38]

10.3 CYP 450$_{cam}$ Reaction Cycle

For the last two to three decades, the proposed reaction cycle of CYP 450 has remained nearly invariant.[39] It is established by now that CYP enzymes operate by means of a generic electron transfer pathway (Figure 10.3).[40,41] The catalytic cycle in the CYP 450 enzymes is turned on by the spin-state change at the heme active site due to expulsion of the distal water molecule on substrate binding, which is followed by electron transfer from redox partners. The principal features of the catalytic mechanism of these enzymes are outlined in the Figure 10.4.

The hexa-coordinated Fe^{III} complex (**1**) exists largely in a spin doublet state (low spin), and as long as this is the case the enzyme will remain inactive:[18,41]

1. Binding of substrate (*e.g.* an alkane, RH) to the enzyme, by displacing the water molecule and leaving a penta-coordinated ferric porphyrin (**2**) accompanied by spin state change (high spin) of iron, affords an enzyme–substrate adduct.
2. Owing to the loss of the water ligand, the ferric complex (**2**) is now a slightly better electron acceptor than the resting state (by approximately 130–300 mV) and this is sufficient for the reduction of the ferric CYP 450 by the associated reductase (operating at 200 mV) with an NADPH-derived electron to the ferrous CYP 450 (**3**).
3. Binding of molecular oxygen to the ferrous heme produces a ferrous CYP 450-dioxygen complex (**4**), similar to the situation in oxymyoglobin; this is a good electron acceptor and has a singlet spin ground state.
4. A second one-electron reduction takes place, and is considered to be the rate-determining step (in most cases)[42] in the catalytic cycle;[6] the reduced dioxygen anion-complex (**5**) is a strong base that quickly is protonated and converts into the Fe(III)-hydroperoxo species known as Compound 0.
5. Compound 0 abstracts another proton and a water molecule splits off to generate (by heterolytic cleavage of the O–O bond) a reactive iron(IV)-oxo intermediate, called a ferryl species known as Compound I (**6**).

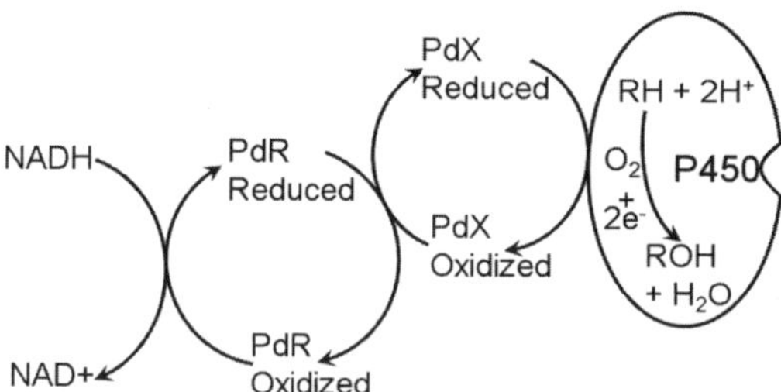

Figure 10.3   Electron transfer pathway in CYP 450$_{cam}$ enzymes system.

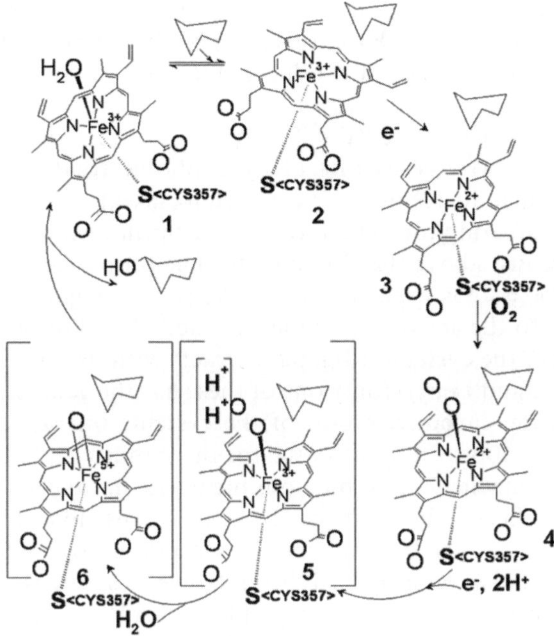

Figure 10.4   Schematic representation of CYP 450 catalytic cycle showing various intermediate states of the heme.

6. Finally, Compound I monoxygenates substrates, *i.e.* oxygen-atom transfer from this iron(IV)-oxo complex to the bound substrate to form the oxygenated product complex; the product exits from the pocket and thereby allows water molecules to re-enter and restore the resting state (**1**) in a phenomenon that completes the cycle.

Interestingly, a single water molecule is used to gate the catalytic cycle in Figure 10.4; however, the enzyme also requires a protonation machinery to complete the full cycle. The protonation machinery is believed to involve an acid–alcohol pair as in the case of the CYP 450$_{cam}$ isozyme (the Asp$_{251}$ and Thr$_{252}$ residues).[18,41,43] It is thought that these residues, and perhaps also Glu$_{366}$ in CYP 450$_{cam}$, shuttle the protons to the dioxygen moiety in those intermediates shown in Figure 10.4, *via* water molecules.[22] On the one hand, water molecules are important for protonation,[22] but the presence of water may aid formation of hydrogen peroxide and other uncoupling products that waste the oxygen and the reducing equivalents.[10,44,45] Extensive theoretical calculations have been reported on the issue of controlled and uncontrolled reaction products and those findings agreed very well with experimental results and thus confirmed the assumption that uncontrolled proton delivery by solvent water networks is responsible for the uncoupling reaction.[46,47] Thus, the enzyme has to function as a middle road: having water molecules in the heme region and at the same time denying them free access to the neighbourhood of the Fe-OO moiety, in order to be able to activate bonds and to produce oxidized products efficiently. The control of water is influenced in part by the tightness of substrate binding by the enzyme. For example, in the case of CYP 450$_{cam}$ (Figure 10.2), this is achieved by a substrate-binding cavity consisting of hydrophobic residues Val$_{295}$, Leu$_{244}$, Thr$_{101}$, and the phenolic residue Tyr$_{96}$, which forms a hydrogen bond with the carbonyl group of camphor. These residues help in anchoring the substrate steady at a place just above the FeO moiety. The propionate side chain of the heme in CYP 450$_{cam}$ has been shown to take part in regulating the access of water molecules to the active site of the enzyme.[48] In addition to this feature, the sulfur atom of the cysteinate ligand interacts with the three amidic groups of Gln$_{360}$, Gly$_{359}$, and Leu$_{358}$ (only one of these has the geometry of a classical H-bond), while the carbonyl group of the cysteine interacts with Gln$_{360}$.[18] These interactions appear to be essential for the stability and monoxygenation activity of the enzyme.[16] All the intermediates in oxygen activation by CYP 450 have been observed by several means like low-temperature EPR,[40] ENDOR, cryogenic X-ray crystallographic study and other spectroscopy.[49–51] But it has been proposed through several spectroscopic techniques that the ferryl species is the most responsible for CYP 450 catalysed oxidation or hydroxylation reactions. Activation of molecular oxygen can often be circumvented if peroxides are used as activated oxygen donors. Recently, extensive research has been undertaken on trapping those intermediates using different CYP 450 enzymes.[52–63]

10.4 Rational Design of the Active Site of CYP 450$_{cam}$

The CYP 450 enzyme that shows activity towards one class of substrates often is not active in reactions of other class substrates. The substrate specificity is particularly important in steroidogenic CYP 450s, and CYP 450$_{cam}$, although a bacterial CYP 450 shows a high degree of substrate specificity and product selectivity.[6] Modification of the active site of CYP 450$_{cam}$ has been shown to induce reactivity in the enzyme towards unnatural substrates. It is possible to produce new enzymes in recombinant systems, altering the amino acid sequence and, therefore, the properties through appropriate modifications at the DNA level. Numerous protein engineering experiments have demonstrated that changes in protein properties are brought about by the cumulative effects of many small adjustments, many of which are distributed or propagated over significant distances. There are relatively few examples where "rational" design has yielded useful enzymes, which do not alter the view that rational protein design is often a fruitless exercise.

The first step in the CYP 450-reaction cycle (see Figure 10.4) is the binding of substrate. This basic aspect of molecular recognition not only completely defines the regio- and stereochemistry of oxygenation but also provides the fit into the buried active site that excludes bulk water and ensures the efficient coupling of reducing equivalents from putidaredoxin. But, one of the main bottlenecks in the application of CYP 450$_{cam}$ for biotechnological purposes is the poor affinity of the enzyme for substrates, such as aromatic hydrocarbons, nitrogenous compounds, drugs, polyhalogenated hydrocarbons, flavonoid compounds, pesticides, *etc*. Poor substrate binding and incomplete spin-transition have been shown to decrease the catalytic efficiency of the enzyme.[6] Introduction of hydrophilic substituents like a hydroxyl group into these compounds is expected to increase the solubility and, hence, the bioavailability of these compounds for degradation. Moreover, biodegradation of many organic compounds has been shown to involve the oxygenation of the parent compound or the intermediate.[64,65]

Rational designs of CYP 450$_{cam}$ variants are largely based on the following three points:

1. increase of the active site space for bigger molecules;
2. decrease of the active site space for smaller and flatter molecules;
3. enhancement of the substrate binding affinity by complementary non-covalent interactions between the substrate and a suitable amino acid side chain in the active site.

Here, it should be mentioned that there are three primary important factors with respect to designing efficient CYP 450 biocatalysts: (i) the substrate must be oxidized at a significant rate, (ii) the regioselectivity must favour the desired product, and (iii) the enzyme must use the majority of reducing equivalents from NADH to produce products.

Site-specific mutagenesis or oligonucleotide-directed mutagenesis is a molecular biology technique in which a mutation is created at a defined site in a *c*DNA of the protein. The nucleotide sequence of a cloned DNA fragment may be changed at will by site-directed mutagenesis using synthetic oligo-nucleotides as mutation primers. The oligonucleotide primer is designed so that it is complementary to the relevant part of the parent DNA template but containing an internal mismatch to cause the mutation. In addition to single-point mutations, this approach may also be used to construct multiple mutations, insertions, and deletions. For plasmid manipulations, this tech-nique has been combined with the polymerase chain reaction (PCR) where a pair of complementary mutagenic primers is used to amplify the entire plasmid. It generates a nicked, circular DNA that can undergo repair by endogenous bacterial machinery. The template parent DNA is eliminated by enzymatic digestion with an enzyme specific for methylated DNA (DPN1). Site-directed mutagenesis has become an invaluable technique for studying protein structure–function relationships.[66,67] The mutagenesis of active site residues has been taken as the main approach to achieve the goal of altering substrate specificity of the enzyme. This has not only helped to enhance the binding and oxidation of unnatural substrates but also provided useful information to understand the determinants of substrate–protein interactions that can be exploited for better design of a catalyst.[68] Figure 10.5 shows

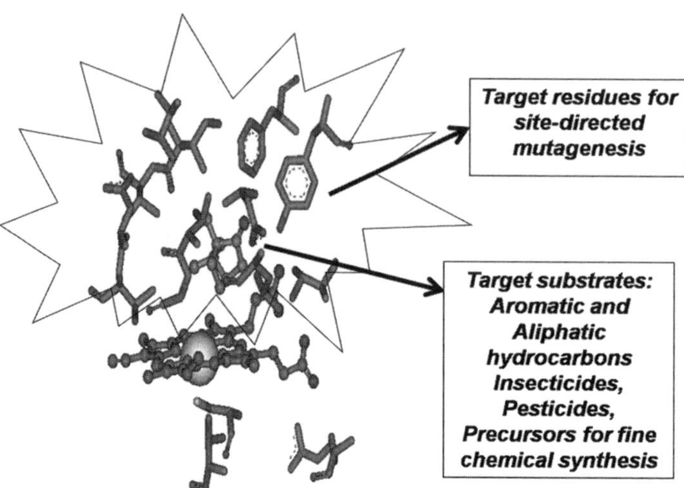

Figure 10.5 Active site of camphor bound wild-type CYP 450_{cam} and a schematic representation of the active site engineering strategy employed to catalyse the oxidation of unnatural substrate using CYP 450_{cam}.

the key residues in the active site to accommodate the unnatural substrate in CYP 450_{cam}.

10.5 Metabolism of Unnatural Substrates by CYP 450_{cam} Variants

The availability of the crystal structures (1DZ4)[13,69] of CYP 450_{cam} has led to several approaches to investigate the mechanism of camphor hydroxylation.[70,71] Site-directed mutagenesis for mutating the active site residues, isotopic labelling experiments, reconstitution of heme by other metal-porphyrins in the active site, intermediate trapping experiments,[72] electrochemistry,[73–75] NMR[76–79] and many other spectroscopic studies[80–82] have been performed on this enzyme system with camphor and camphor analogues like camphene, norcamphor, norborane, and thiocamphor[6,10] to understand the mechanistic aspects of the enzyme. Combinations of all those experiments have been also used to quantify the regiospecificity, reaction rate, and also the efficiency of the wild type as well as mutated enzymes.[83]

In efforts to make CYP 450 a "green" biocatalyst for synthetic chemistry, most research and developments have been made on the oxidation of new substrates and probing the intermediates resulting oxidized products by CYP 450_{cam}. Extensive work has been conducted in attempting to use variants of this enzyme to carry out environmental detoxification processes, which are an important component of bioremediation. Figure 10.6 briefly describes various single and higher order mutants of CYP 450_{cam} reported in the literature to date.

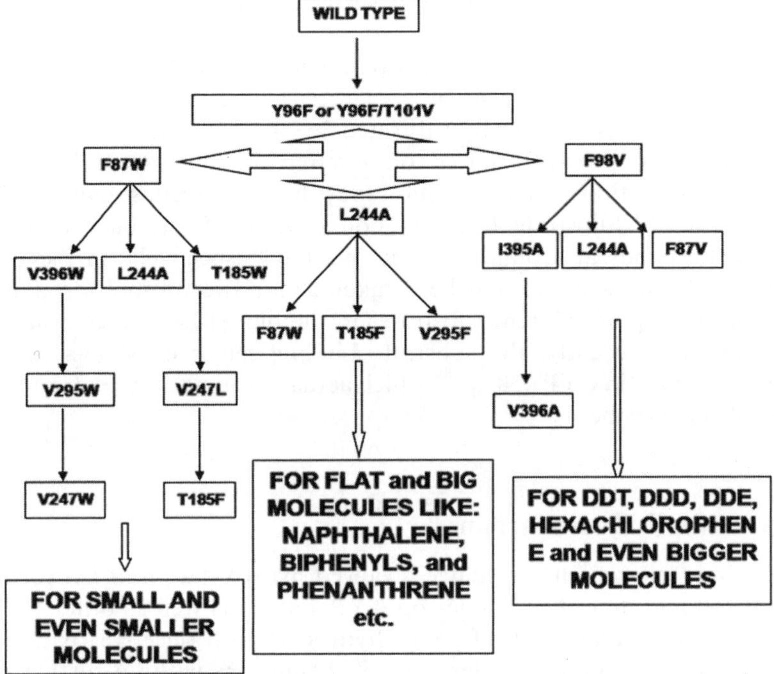

Figure 10.6   Schematic diagram of a possible protein (CYP101) engineering strategy for binding of unnatural substrates.

10.6   Binding of Unnatural Substrate, Hydroxylation, and Product Release

The stoichiometry of catalytic reaction by CYP 450$_{cam}$ requires one molecule of oxygen and two electrons from NADH to add one oxygen atom to the substrate. If the ratio of NADH consumed to product formed is greater than one, the enzyme is said to be uncoupled. Uncoupling occurs when (i) the ferrous dioxy complex reverts to the ferric state by dissociation of superoxide, (ii) a molecule of H_2O_2 dissociates from the ferric hydroperoxide complex, or (iii) two electrons are used to reduce the ferryl species to a molecule of water before it can be used in a reaction with the substrate. The parameters that govern uncoupling are till now unclear. There is very little experimental evidence concerning how substrates enter and exit from the buried active site of the heme.[24,27,84] The substrate association to the enzyme involves specific anchoring of the substrate molecule to the enzyme at a site close to the heme active site that is deeply buried inside the protein matrix. There are several key amino acid residues in the heme active site that have been identified to play important roles in monoxygenations of the substrate in the enzyme. Appropriate modification of the substrate binding site in this enzyme can allow altering of the substrate recognition properties of this enzyme and thus catalytic monoxygenation of certain unnatural substrates can be carried out by the modified enzyme.

Substrate concentration is a critical parameter in the design of the binding assay. The Soret band position shift is largest at the substrate concentration where the enzyme is completely saturated by the substrate. Again, enzyme-saturating levels of substrate in free solution may not always be achievable due to solubility limitations. Hence, caution must be taken against over-saturation of substrates.[85,86] Substrate free and bound CYP 450$_{cam}$ have two special characteristic bands at around 417 and 392 nm, respectively, in the UV–Vis region.[32,87,88] This signature has been taken as a powerful tool for characterization of binding of different substrates in the buried heme active site of the protein matrix. Recently, the substrate binding has been shown to induce structural changes in CYP 450$_{cam}$,[89] which actually provides the redox potential change of the enzyme.

10.6.1 Small Hydrocarbons

Ortiz de Montellano[90] has recently reviewed hydroxylation of hydrocarbons by CYP 450 and described studies on CYP 450$_{cam}$ as well as other CYP 450 enzymes in the metabolism of small hydrocarbons. Selective oxidation of simple alkanes to alcohols under ambient conditions without further oxidation to aldehydes/ketones or deep oxidation to carbon dioxide is of great industrial importance. The oxidation of methane to methanol is of course a prime objective of catalysis research, but also oxidation of other gaseous alkanes to alcohols could afford new feed-stocks for chemical synthesis. Solid state heterogeneous catalysts require elevated temperatures and pressures. In contrast, biological catalysts such as heme and non-heme monoxygenases catalyse alkane oxidation under mild conditions and, as such, are potential candidates for environmentally friendly, green chemistry alkane oxidation catalysts.

Several papers have been published on oxidation of various gaseous alkanes by CYP 450$_{cam}$,[91–93] and, broadly, the results proposed that specificity of this enzyme could be biased towards the linear alkanes against their branched isomers by reducing their active site volume by bulky amino acid substitution such as Val to Leu. Tables 10.1–10.3 summarize the NADH oxidation rate, coupling efficiency, and hydroxylation rates for ethane, propane, butane, and pentane using different mutants of the enzyme. The starting point of CYP 450$_{cam}$ engineering for gaseous alkanes was the Y96F/V247L double mutant, which favours the oxidation of hexane over 3-methylpentane. To constrain butane and propane close to the heme, Bell *et al.*[91] replaced Phe$_{87}$ by tryptophan and Thr$_{101}$ by leucine. Studies of butane and propane oxidation by several mutants like-Y96F, F87W/Y96F, Y96F/V247L, F87W/Y96F/T101L, F87W/Y96F/V247L, and F87W/Y96F/T101L/V247L were reported. They found that wild-type CYP 450$_{cam}$ and all the active site mutants oxidized butane to butan-2-ol, and propane to propan-2-ol, with no evidence for the formation of the

Table 10.1 Butane and propane oxidation[a] catalysed by wild-type CYP 450$_{cam}$ and active site mutants.[91–93]

CYP 450$_{cam}$ enzymes	*Butane*			*Propane*		
	NADH oxidation rate	*Butan-2-ol formation rate*	*Coupling efficiency (%)*	*NADH oxidation rate*	*Propan-2-ol formation rate*	*Coupling efficiency (%)*
Wild-type	10	0.4	5	9	0.5	0.9
Y96F	100	42	42	18	2.2	12
F87W/Y96F	200	132	66	60	12	20
Y96F/V247L	250	186	75	110	43	39
F87W/Y96F/ T101L	540	460	85	198	51	26
F87W/Y96F/ V247L	263	88	34	82	13.7	17
F87W/Y96F/ T101L/V247L	795	750	95	346	110	32
F87W/Y96F/ T101M/V247L	373	205	55	164	66	40
F87W/Y96F/ V247L/V295I	363	99	27.5	124	3.5	3
F87W/Y96F/ V247L/V396L	52	12.5	24	7.5	1.5	18
F87W/Y96F/ T101L/V247L/ D297M	385	243	63	214	56	26
F87W/Y96F/ T101L/V247L/ L244M	578	520	90	266	173	66

[a]Rates of NADH turnover and product formation are given in nmol (nmol CYP 450)$^{-1}$ min^{-1}. The coupling efficiency is the ratio of the total amount of products formed to the amount of NADH consumed.

Table 10.2 Pentane oxidation activity[a] of CYP 450$_{cam}$ mutants.[91–93]

CYP 450$_{cam}$ enzymes	*Pentane*		
	NADH oxidation rate	*Hydroxylated pentane formation rate*	*Coupling efficiency (%)*
Wild-type	6.4	0.07	1.1
Y96F	405	110	29.5
Y96F/V247A	152	31	22
Y96F/V247L	510	227	44.5
Y96F/F98W	32	0.4	1.3
F87W/Y96F	471	192	42.5
F87W/Y96F/F98W	367	26.5	7.2
F87W/Y96F/V247L	510	237	46.5
F87W/Y96W/V247L	67	–	–
F87W/Y96F/F98W/V247L	26	1.0	3.8
F87W/Y96F/V247L/V396L	76	1.7	2.3

[a]Rates of NADH turnover and product formation are given in nmol (nmol CYP 450)$^{-1}$ min^{-1}. The coupling efficiency is the ratio of the total amount of products formed to the amount of NADH consumed.

Table 10.3 Comparison of propane and ethane oxidation^a for CYP 450$_{cam}$ mutants.[91–93]

CYP 450$_{cam}$ enzymes	Propane				Ethane			
	NADH oxidation rate	*Product formation rate*	*Coupling efficiency (%)*	*H_2O_2 formation (%)*	*NADH oxidation rate*	*Product formation rate*	*Coupling efficiency (%)*	*H_2O_2 formation (%)*
F87W/Y96F/T101L/V247L/L244M (EB mutant)	266	176	66.2	25	<20	–	–	–
EB mutant/L294M	414	193	46.8	38	<20	–	–	–
EB mutant/L294M/T185M	302	228	75.6	23.4	45	–	–	–
EB mutant/L294M/T185M/L358P	464	379	81.9	12.4	269	10	4	30
EB mutant/L294M/T185M/L358P/ G248A	590	505	85.6	15	741	78.2	10.5	39.6

^{a}Rates of NADH turnover and product formation are given in nmol (nmol CYP 450)$^{-1}$ min^{-1}. The coupling efficiency is the ratio of the total amount of products formed to the amount of NADH consumed.

terminal alcohols or the further oxidation products butan-2-one and acetone. The F87W/Y96F/T101L/V247L quadruple mutant showed the highest activity towards butane (turnover rate 750 min^{-1}, 2000 times faster than the wild-type enzyme), which is quite comparable with camphor hydroxylation by wild-type enzyme (1200 min^{-1}) under identical conditions. While the propane turnover rates were lower; importantly, the activity trend for propane mirrors that for butane, suggesting that high propane oxidation activity may be achieved with additional mutations that further reduce the active site volume. We learned from this work that the attack took place exclusively at the secondary C–H bonds. The absence of attack at the primary C–H bonds reflected substrate binding orientation and mobility, which allowed the ferryl intermediate to select the more activated secondary C–H bonds. Primary C–H bond oxidation could only be observed if substrate was constrained such that primary the C–H bond was much closer to heme.

In 2005, Xu *et al.* reported the engineering of CYP 450$_{cam}$ for the oxidation of very small hydrocarbons, such as ethane to ethanol.[93] Their approach was to decrease the active site volume even more combined with directed evolution. They introduced more bulky amino acids in the heme active site to promote the binding and oxidation of those gaseous alkanes. The CYP 450$_{cam}$ active site is highly irregular in shape and the F87W/Y96F/T101L/L244M/V247L mutant, previously known for propane oxidation, was found to have some cavities between Leu$_{294}$ and Thr$_{252}$, Leu$_{247}$ and Gly$_{248}$, and in the Ala$_{296}$–Asp$_{297}$ regions. It also had been observed that the cavity between the Trp$_{87}$, Leu$_{247}$, and Val$_{396}$ side chains at the top of the active site was covered by Thr$_{185}$. Therefore, they introduced mutations like L294M, G248A, and T185M to force ethane to bind closer to the heme active centre. The previous mutant, which actually did show no activity towards ethane oxidation, showed a tremendous increase in NADH oxidation rates in the presence of ethane, along with the mutations like L294M/G248A/T185M. By altering the hydrogen bonding to the proximal side chain by another L358P mutation, a general tightening of the pocket was observed and it was also found that the heme was pushed towards the substrate binding pocket. This mutation along with the other aforementioned mutations showed a tremendous increase in propane oxidation and coupling (86%) with respect to simple mutants. This multiple mutant was also examined for ethane oxidation. They found >85% heme-spin shift and higher NADH oxidation activity than that for propane. They have also explained why ethane showed lower coupling efficiency than propane. Actually, the ethane molecule was not bound closer to the heme as is the case for propane and, hence, ferryl-oxo species competed *via* two pathways: substrate oxidation and peroxide uncoupling, where peroxide uncoupling was dominant; this also explained why simple mutants did not show readily detectable ethane oxidation activity. Overall, studies on this mutant had not only provided information about its interesting property for metabolizing small molecules like ethane, butane, or propane, but also afford novel examples on manipulation of its active site structure to provide new insight into the origin of heme spin state equilibrium in CYP 450 enzymes.

10.6.2 Alkyl Benzenes

To understand the molecular basis of enzymatic catalysis, CYP 450$_{cam}$-mediated oxidation of ethylbenzene and several closely related structures obtained by incrementally adding one, two, three, or four carbon atoms to the ethylbenzene structure was studied. For each compound, the ability of substrates to bind to CYP 450$_{cam}$, the extent of stimulation of catalytic turnover as measured by NADH consumption, the partitioning between the formation of organic and reduced oxygen metabolites, and the nature of the organic products have been assimilated. Similar studies have been carried out with the T185L and T185F mutants of CYP 450$_{cam}$, in which the size of the active site is incrementally decreased.[94] The results showed that no detectable correlation exists between the observed spin state change and any other parameter studied. For substrates of equal size, coupled and uncoupled turnover vary widely but both are maximized when the phenyl ring bears one large alkyl substituent or a methyl group in *ortho* position relative to the largest alkyl substituent. The presence of substituents in addition to these decreases activity. In the absence of other changes, coupled turnover is correlated with the size of the largest substituent, but no such correlation exists for uncoupled turnover. Decreasing the size of the active site cavity by a T185L mutation generally increases coupled turnover without altering the dependence on the alkyl group size. A T185F mutation causes too great an active site perturbation for structure–activity studies. Substrate oxidation occurs preferentially at 2° or 3° C–H bonds of the largest substituent or on the benzylic methyl-group located in the *ortho* position. Aromatic hydroxylation only competes with the oxidation of non-benzylic 1° C–H bonds. The extent of coupled turnover is a function of substrate shape, substrate size, and cavity size, but still elusive parameters control the extent of uncoupled turnover.

10.6.3 Polycyclic Aromatic Hydrocarbons (PAHs)

Polycyclic aromatic hydrocarbons (PAHs) are hazardous and persistent environmental contaminants, posing a grave threat to human health.[95–97] The persistence of those PAHs is basically due to low bioavailability and the lack of naturally occurring microbial enzymes for their efficient degradation. The hydroxylation of those aromatic compounds involves nucleophilic attack of the hydroxyl group on the aromatic ring. The nucleophilic substitutions on aromatic rings are difficult and generally carried out under harsh conditions.[98–100] Moreover, the tailor-made regioselectivity of aromatic hydroxylation is difficult to achieve *via* conventional organic synthesis as the substitution takes place preferentially at the *ortho/-para/-meta* positions depending on the existing functional group on the aromatic ring.[98–100]

Therefore, PAH bioremediation by microorganisms is of interest because all CYP 450s follow the same mechanism and the product formed mainly depends on the orientation of substrate binding. Hence, PAH oxidation by CYP 450$_{cam}$ could have significant implications in different areas. It has been shown that substitution of active site residue Tyr$_{96}$ by a hydrophobic amino acid greatly enhanced the activity of CYP 450$_{cam}$ towards small as well as large hydrophobic molecules. The first approach was oxidation of styrene by Y96F and Y96A, where it was found that, being slightly smaller than camphor and more planar, styrene might not bind to wild type, but removal of the tyrosine-OH group of Tyr$_{96}$ by mutations drastically increased the percentage of high-spin state of heme in the mutant enzymes (45% for Y96A and 55% for Y96F).[101] The same approach was also undertaken for phenanthrene, fluoranthene, pyrene, and benzo[a]pyrene. In these studies,[102,103] L.-L. Wong and coworkers oxidized the aforementioned polycyclic hydrocarbons with CYP 450$_{cam}$ by mutating two active site residues Phe$_{87}$ and Tyr$_{96}$. A single mutation at the 96th position by Gly, Ala, Val, or Phe alone could not facilitate coupling/NADH turnover. They prepared double mutants like F87A/Y96F and F87L/Y96F of CYP 450$_{cam}$ and then compared the oxidation of those hydrophobic compounds with wild-type CYP 450$_{cam}$ and these mutants. They found that mutation at Phe$_{87}$ and Tyr$_{96}$ positions at the active site dramatically increased the activity towards all those substrates while wild-type CYP 450$_{cam}$ had a low activity of less than 0.01 min^{-1}. Their rationale was that the smaller side chain of alanine and the more flexible side chain of leucine compared to a rigid phenylalanine residue at the 87th position with Y96F mutation might promote the entry and the binding of PAHs and also the binding orientations of these substrates so that larger amounts of products are formed. Phenanthrene was oxidized to 1-, 2-, 3-, and 4-phenanthrol while fluoranthene gave mainly 3-fluoranthol. Pyrene was oxidized to 1-pyrenol, and then to 1,6- and 1,8-pyrenequinone. Small amounts of 2-pyrenol was also formed with Y96A mutant. Only F87A/Y96F oxidized benzopyrene to its hydroxylated form.

10.6.4 2-Ethylhexanol

Jones and coworkers studied the oxidation of 2-ethylhexanol to 2-ethylhexanoic acid (Scheme 10.1) by CYP 450$_{cam}$.[104,105] The oxygen binding to the carbon already bound to an OH group possibly leads to formation of a vicinal diol intermediate, which in a subsequent reaction forms the acid. CYP 450$_{cam}$ catalyses several such multistep reactions analogous to that involved in steroid biosynthesis. New mutations were chosen to decrease the active site volume, increase the active site hydrophobicity, and improve stereoselectivity. They introduced a total of four active site mutations, namely, F87W, Y96W, T185F, and L244A. The F87W and Y96W mutations are located in the B'-helix and at the end of the B'-helix region, respectively, and improved the regioselectivity

Scheme 10.1 Conversion of 2-ethylhexanol into 2-ethylhexanoic acid catalysed by CYP 450 mutants.

significantly and give almost exclusively the desired product.[104,105] Substitution of these amino acids by a tryptophan residue in both cases decreased the active site volume by several fold, and thereby forced the substrate closer to the heme and increased the possibility of heme–substrate interaction in the active site. The T185F mutant showed the highest NADH coupling value for the formation of product. Being a part of the F-helix end, this mutation actually decreases hydrogen peroxide production and increases water production during 2-ethylhexanol catalysis and, interestingly, it indicates that water and hydrogen peroxide production are controlled by different features of the heme active site. On the other hand, L244A mutation on wild-type CYP 450$_{cam}$ altered the stereoselectivity of 2-ethylhexanoic acid production. The L244A mutation on the I-helix increases the active site volume by 11%. There is a possible change of nonbonding interactions around the ethyl side chain of 2-ethylhexanol and its mutated residue. 2-Ethylhexanol enantioselectivity assays have also been addressed and quantified by chiral GC-MS. All these studies with single mutants showed how active site mutation can affect the coupling of NADH with product formation.

10.6.5   Aromatic–Aliphatic Hydrocarbon, Phenylcyclohexane

Selective oxidation of saturated C–H bonds under mild conditions is the most difficult to achieve using conventional synthetic methodologies with a catalyst at room temperature. The homolytic cleavage of a benzene C–H bond is also not easy, due to a bond dissociation energy of 111 kcal mol^{-1}, which is well above the values for aliphatic alkanes (90–101 kcal mol^{-1}) and even methane (105 kcal mol^{-1}).[106] This is all the more difficult when the oxidant is molecular oxygen since customarily hydrogen peroxide or water is generated as a by-product. Consequently, research has been performed on the oxidation of a molecule like phenylcyclohexane by monoxygenation reactions.[107] It was found that CYP 450$_{cam}$ selectively oxidized the cyclohexane ring (Scheme 10.2), while any chemical systems would be expected to oxidize the unsaturated aromatic ring or a benzylic C–H bond. This observation strongly suggested that the cyclohexane ring of the molecule is highly mobile inside the CYP 450$_{cam}$ substrate binding pocket and the ring is possibly bound to the heme with C^3 and C^4 closest to the iron centre. It had been speculated that there are van der Waals interactions between the cyclohexane ring and the aliphatic side chains of Leu$_{244}$, Val$_{247}$,

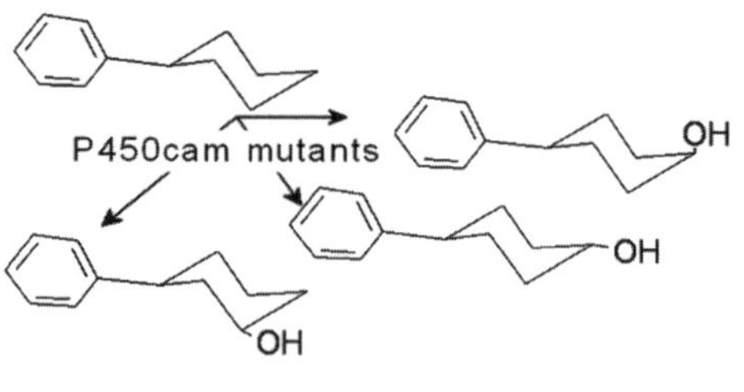

Scheme 10.2   Hydroxylation of phenylcyclohexane catalysed by CYP 450$_{cam}$ mutants.

and Val$_{295}$ and/or the stacking interactions between the phenyl groups of phenylcyclohexane and the aromatic side chains at positions 87, 96, and 98.

10.6.6 Diphenylmethane

Unlike cytochrome CYP 450$_{BM3}$, CYP 450$_{cam}$ requires two separate proteins (PdR, PdX) to mediate electron transfer. The feasibility of CYP 450 enzymes in biotransformations depends on maximizing the catalyst lifetime whilst minimizing or removing the need to use cofactors. *in vitro,* several artificial cofactors are available and preparative scales have already been reported,[108–111] including cofactor recycling and electrochemically driven turnover. Another alternative strategy is a whole cell system in which all the proteins in the CYP 450 enzyme system are expressed in a single heterologous host to constitute a catalytically competent *in vivo* substrate oxidation system. Bell and Wong reported construction of a tricistronic plasmid containing three genes of a CYP 450$_{cam}$ system.[112] *Escherichia coli* transformed with the tricistronic plasmid expressing Y96F/V247L, Y96F/I395G mutants readily and gave significant quantities of 4-hydroxydiphenylmethane from diphenylmethane and 1-phenylethanol in 24–72 h. The Y96F/V247L mutant oxidized 1 mM styrene and ethylbenzene within 12 h. Consequently, this new *in vivo* system cannot only be used for preparative larger scale reactions for product characterization but also will allow rapid screening and directed evolution of CYP 450$_{cam}$ mutants.

10.6.7 Valporic Acid

If the crystal structure of the enzyme is known, then the interaction between the enzyme and its ligand can be probed directly by *in silico* docking analyses. Docking methods have been widely used for determining the protein–ligand interaction from known protein structures, especially as an application in computer-aided drug design and risk assessments. DOCK, AUTODOCK, FlexX, GRID, GOLD, and related docking programs have been used to fit potential substrates within the CYP 450$_{cam}$ active site. One of the oldest approaches of these involved the interaction of a flexible ligand such as valporic acid (VPA), an anti-epileptic agent, in its free state and its behaviour as a ligand in the binding site of this oxidative metabolizing enzyme.[113] These studies aimed to develop strategies for the prediction of the product distribution of flexible ligands. From the docking studies, two relatively different initial docking orientations of VPA in the CYP 450$_{cam}$ binding site were obtained. One was obtained by a rigid rotation of the same conformer by 180°. Both orientations allow an electrostatic recognition between the C=O group of VPA with the OH group of Tyr$_{96}$. This particular 180° rotated conformer was used to predict the regio- and stereoselective product distribution of VPA for each of the CYP 450$_{cam}$-VPA complexes. By contrast, experimental studies found C^4-OH VPA as the predominant hydroxylation product with a small amount of C^5-OH VPA.

This type of study was proceeded further with the work of substrate docking algorithms and prediction of the substrate specificity of CYP 450$_{cam}$ and its L244A mutant,[114] where 16 different types of compounds were evaluated as CYP 450$_{cam}$ substrates by measuring (i) the binding of the substrate to the enzyme, (ii) stimulation of NADH and O$_2$ consumption, (iii) enhancement of H$_2$O$_2$ production, and (iv) formation of organic metabolites. Some of the compounds were predicted to fit into the active site, while others were found to be poor/non-substrates. DOCK, a receptor-constrained three-dimensional screening program has been successfully used to identify lead molecules for the development of enzyme inhibitors and receptor ligands.

10.6.8   Terpenoids

Terpenes are a good starting material for the synthesis of many fine chemicals due to their similar carbon skeleton. The terpenes are secondary metabolites of plants that are produced, in part, as a defence against microorganisms and insects but, in addition, also have pollinator-attractive properties.[115] Members of this class of chemicals have carbon structures that can be decomposed into isoprene (C$_5$H$_8$) residues and are classified, based on the number of carbons in the molecule, as monoterpenes, sesquiterpenes, diterpenes, triterpenes, and tetraterpenes or carotenes.[116] The simpler terpenes (mono- and sesquiterpenes) are the major constituents of essential oils and are widely used in the perfume industry, while di- and triterpenes are less volatile and are obtained from plant gums and resins.[117] Carotenes are synthesized by bacteria, algae, fungi, and green plants and comprise more than 500 known structures.[118] Recently, a review paper on the bio-oxidation of terpenes by different enzymes was published.[119]

10.6.8.1   α-Pinene

Oxygenated derivatives of the monoterpene (+)-α-pinene are found in plant essential oils and used in fragrances and flavourings because of their distinctive, pleasant odours and taste notes.[120] This compound is structurally related to camphor, the natural substrate of CYP 450$_{cam}$. Combinations of Y96F, F87A, F87W, F87L, and V247L mutations were considered as active site mutations for α-pinene.[121] All tested CYP 450$_{cam}$ mutations showed enhanced substrate binding and an increased rate of oxidation for α-pinene (Scheme 10.3). Some mutants showed tighter α-pinene binding than camphor binding by wild-type enzyme. The most active mutant was Y96F/V247L with (+)-α-pinene oxidation of 270 nmol (nmol CYP 450)$^{-1}$ min^{-1}, which was 70% of camphor oxidation by wild-type enzyme. If the gem dimethyl groups of pinene occupied positions similar to those found for camphor, *cis*-verbenol would be the dominant product. All CYP 450$_{cam}$ enzymes gave *cis*-verbenol products but with much reduced selectivity compared to camphor oxidation by wild-type enzyme. (+)-Verbenone, (+)-myrtenol, and the (+)-(*R*)-pinene epoxides were among the minor products. The crystal structure of the Y96F/F87W/V247L mutant, the most selective of the CYP 450$_{cam}$ mutants initially examined, was determined to provide further insight into the CYP 450$_{cam}$ substrate binding and catalysis. (+)-(*R*)-Pinene was bound in two orientations that were related by a rotation of the molecule. One of these orientations was similar to that of camphor in the wild-type enzyme while the other was significantly different. Analysis of the enzyme–substrate contacts suggested rationalizations of the product distribution. In particular, competition rather than cooperativity between the F87W and V247L mutations and substrate movement during catalysis were proposed to be the major factors. The crystal structure led to the introduction of the L244A mutation to increase the selectivity of pinene oxidation by further biasing the binding orientation toward that of camphor in the wild-type structure. Tables 10.4 and 10.5 show the results of binding and activity studies on various mutants of CYP 450$_{cam}$ for the metabolism of α-pinene, indicating that the mutations of the active site do indeed play an important role in imparting stereoselectivity to the product formation.

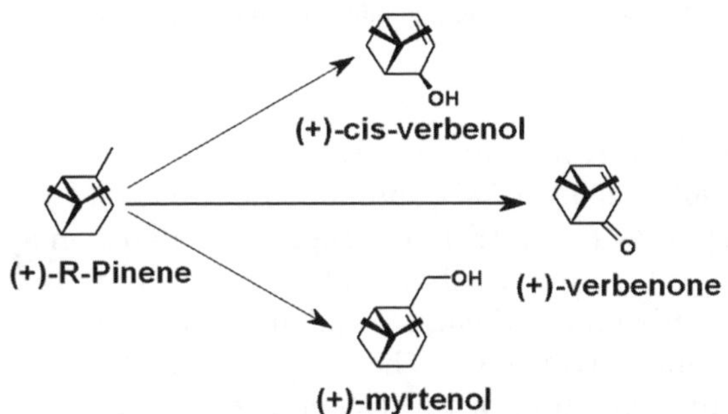

Scheme 10.3   Hydroxylation of (+)-(*R*)-pinene catalysed by CYP 450$_{cam}$ mutants.[121]

Table 10.4   Binding and oxidation of (+)-(*R*)-pinene by wild-type CYP 450$_{cam}$ and its mutants (ref. 121)[a]

CYP 450$_{cam}$ enzymes	% High spin	K$_d$ (µM)	NADH oxidation rate	Product formation rate	Coupling efficiency (%)
WT	85	1.1	82	18.6	23
Y96F	95	0.15	147	56	38
F87A/Y96F	70	0.12	210	150	72
F87L/Y96F	80	0.1	163	88	54
F87W/Y96F	95	0.08	171	96	56
Y96F/V247L	95	0.14	98	271	91
F87W/Y96F/V247L	75	0.30	129	65	51
F87W/Y96F/L244A/V247L	85	0.24	233	156	67
F87W/Y96F/L244A	65	0.26	122	120	95
Y96F/L244A/V247L	90	0.15	223	129	58

[a]Rates of NADH turnover and product formation are given in nmol (nmol CYP 450)$^{-1}$ min^{-1}. The coupling efficiency is the ratio of the total amount of products formed to the amount of NADH consumed.

Table 10.5 Distribution of products (%) from (+)-α-pinene oxidation catalysed by wild-type and mutants of CYP 450$_{cam}$. (ref. 121)[a]

CYP 450$_{cam}$ enzymes	(+)-(R)-Pinene epoxides (%)	(+)-cis-Verbenol (%)	(+)-Myr-tenol (%)	(+)-verbenone (%)	Others (%)
WT	4	31	4	10	51
Y96F	3	46	10	12	29
F87A/Y96F	22	33	7	8	30
F87L/Y96F	12	38	7	10	33
F87W/Y96F	7	66	3	7	17
Y96F/V247L	12	55	7	7	19
F87W/Y96F/V247L	3	70	4	8	15
F87W/Y96F/L244A/V247L	12	44	–	13	31
F87W/Y96F/L244A	2	86	2	5	5
Y96F/L244A/V247L	1	55	–	32	12

[a]Rates of NADH turnover and product formation are given in nmol (nmol CYP 450)$^{-1}$ min^{-1}. The coupling efficiency is the ratio of the total amount of products formed to the amount of NADH consumed.

The F87W/Y96F/L244A mutant gave 86% (+)-*cis*-verbenol and 5% (+)-verbenone. The Y96F/L244A/V247L mutant gave 55% (+)-*cis*-verbenol, but interestingly also 32% (+)-verbenone, suggesting that it may be possible to engineer a CYP 450$_{cam}$ mutant that could oxidize (+)-(R)-pinene directly to (+)-verbenone. Verbenol, verbenone, and myrtenol are naturally occurring plant fragrance and flavourings. The preparation of these compounds by selective enzymatic oxidation of (+)-(R)-pinene, which is readily available in large quantities, could have applications in synthesis. The results also show that the protein engineering of CYP 450$_{cam}$ for high selectivity of substrate oxidation is more difficult than achieving high substrate turnover rates because of the subtle and dynamic nature of enzyme–substrate interactions.

10.6.8.2 Valencene

Another large class of terpenoid compounds is the sesquiterpenes, which are common constituents of plant essential oils.[121] Their derivatives are biologically active or precursors to metabolites with biological functions.[122] For example, the phytoalexin capsidiol is derived from the sesquiterpene 5-epi-aristolochene,[123] and amorpha-4, 11-diene is the precursor to the potent antimalarial compound artemisinin.[124,125] Valencene is a sesquiterpene that is an aroma component of citrus fruit and citrus-derived odorants. Its ketone derivative (+)-nootkatone is a sought after fragrance compound found in grapefruit juice.[126] The valencene synthase gene from orange has been cloned and functionally expressed in *E. coli*.[127,128] This makes the oxidation of (+)-valencene an attractive synthetic route to the high added-value compound (+)-nootkatone by CYP 450$_{cam}$. While wild-type enzyme did not show any activity towards valencene, because of its large structure and more hydrophobicity, the CYP 450$_{cam}$ variants F87A/Y96F, F87L/Y96F, F87A/Y96F/V247L, F87A/Y96F/L244A, F87V/Y96F/L244A, and F87A/Y96F/L244A/V247L showed higher NADH consumption rates as well as product formation rates (Table 10.6). The main product of oxygenation of valencene is schematically shown in Scheme 10.4. Here, a comparison between CYP 450$_{BM3}$ and CYP 450$_{cam}$ has been investigated. Wild-type CYP 450$_{BM3}$ and mutants were found to have higher activities (up to 43 min^{-1}) than CYP 450$_{cam}$ but were much less selective. The selectivity patterns suggested that (+)-valencene has one binding orientation in CYP 450$_{cam}$ but multiple orientations in CYP 450$_{BM3}$. Their hydroxylated and epoxidized products were investigated from whole-cell reactions as well as by several other techniques.

Table 10.6 Binding and oxidation of (+)-valencene by wild-type CYP 450$_{cam}$ and its mutants.[121]

CYP 450$_{cam}$ Enzymes	%High spin heme	K$_d$ (μM)	NADH oxidation rate	Product formation rate	Overall yield (%)	(+)-trans-Nootkatol (%)	(+)-Nootkatone (%)
Wild type	<5	–	–	–	–	–	–
F87A/Y96F	80	3.09 ± 0.26	36 ± 5.3	7.2 ± 2.0	20	70	12
F87L/Y96F	85	4.23 ± 0.31	22 ± 2.4	2.8 ± 1.0	12.7	65	20
F87A/Y96F/V247L	25	1.88 ± 0.30	30 ± 1.8	5.7 ± 0.9	19	78	6.7
F87A/Y96F/L244A	80	2.70 ± 0.16	63 ± 13	9.8 ± 1.6	15.6	60	12
F87V/Y96F/L244A	65	3.40 ± 0.45	48 ± 5.4	9.7 ± 1.7	20.2	38	47
F87A/Y96F/L244A/V247L	100	2.76 ± 0.35	27 ± 6.7	6.9 ± 0.9	25.6	86	4.2

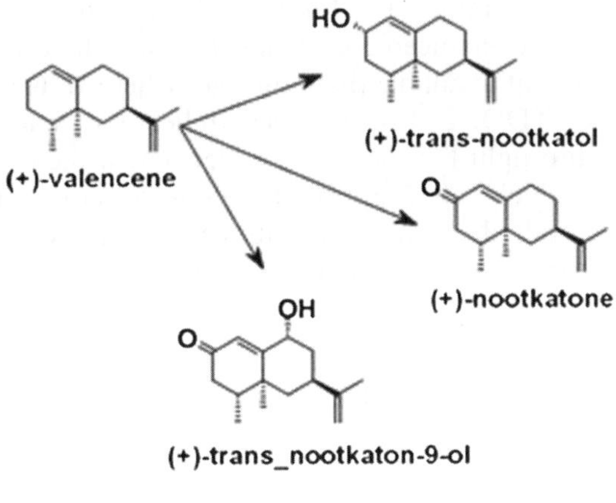

Scheme 10.4 Main products of oxidation of (+)-valencene by CYP 450$_{cam}$ mutants.[121]

10.6.9 Fused Benzene–Cycloalkane Compounds

The oxygenated benzo-cycloarene family [benzo-cyclobutene, benzo-cyclopentene (indane), benzo-cyclohexene (tetralin), benzo-cycloheptene] contains useful reactants in asymmetric synthesis.[129] Moreover, these compounds are useful for the preparation of a photosensitive polymer. It was already known from ethylhexanol experiments that CYP 450 catalyses monoxygenation only on the aliphatic ring. Hence, this work differentiated the catalytic functionality of CYP 450_{cam} from the dioxygenases that have been applied to this class of substrates.[130–133]

Some experimental and theoretical studies have been performed for the hydroxylation of those compounds by wild type and Y96F mutant of CYP 450_{cam} to understand the factors affecting product distribution, catalytic rate, and cofactor utilization.[134] The products of all reactions, except those of benzo-cycloheptene, were regiospecifically hydroxylated at the 1-position. In fact, there are significant differences and similarities with the products formed by dioxygenase catalysis and CYP 450_{cam} for these compounds. Quantum mechanical calculations carried out using B3LYP/6-31G comparing the energetics of all possible radical intermediates predicted hydroxylation at the 1- and 3-positions of benzo-cycloheptene, and at the 1-position for the other three compounds. However, the fact that the ratio of 1-alcohol to 3-alcohol changes significantly between wild type and Y96F mutant CYP 450_{cam} indicates that active site geometry and composition also play a significant role in determining benzo-cycloheptene product regiospecificity. The (*R*)-enantiomer (88% and 95%) was predominantly formed for indane and tetralin. Steric constraints of the active site were confirmed by molecular dynamics calculations to control the enantiomer distribution for tetralin hydroxylation. NADH coupling, measured at saturating benzo-cycloarene concentration, was found to be much higher for Y96F CYP 450_{cam} than that for wild-type protein (70% for Y96F compared to 12% for wild type of benzo-cycloheptene), indicating that removal of the hydroxyl group of a tyrosine amino acid residue made the enzyme better suitable for hydrophobic molecules. Binding affinities and catalytic rates were greatly increased for the engineered enzymes. Because the toluene dioxygenase (TDO) structure was not available then, this work helped us to understand the mechanistic cycle of TDO. Based on these studies it was speculated that the TDO active site is too tight for the larger molecules, while being too loose for benzo-cyclobutene, which can adopt multiple conformations in the active site and hence is subject to hydroxylation at both rings. This work rationalized also that a free radical can form at any point of a substrate.

10.6.10 Nitrogenous Compounds

10.6.10.1 Indole

One of the most commonly used dyes since ancient times is indigo.[135–137] This dye is heavily used in textile industry[136,137] and also as additives for food colour in the form of disulfonate derivatives (indigo carmine).[138] Indigo carmine is also used as a pH indicator as well as for diagnostic purposes.[139–141] Until the nineteenth century this dye was extracted from plant sources, mainly from *Indigofera tinctoria*.[142] However, with the advent of organic chemistry, the industrial production of the dye has replaced agricultural production by adopting the synthetic routes reported by Heumann in 1890 and later on by Pfleger in 1901.[143,144] In both cases, the synthesis not only involves harmful corrosive chemicals like NaOH or KOH at high temperatures but also is carried out in organic solvents. In addition, *N*-phenylglycine, the precursor for the industrial indigo production is synthesized from aniline using cyanide.[144]

The ancient as well as industrial production of this dye involves the synthesis of 3-hydroxyindole (indoxyl), which subsequently is oxidized by aerial oxygen to indigo. Interestingly, this intermediate may, potentially, be produced by selective oxidation of indole by CYP 450$_{cam}$. In view of this fact, we explored the ability of the wild type and the engineered CYP 450$_{cam}$ variants to oxidize and finally produce indigo from indole. The previous docking studies have shown that the distal pocket of the heme consisting of the Tyr$_{96}$, Thr$_{101}$, Phe$_{87}$, and Leu$_{244}$ residues could be suitably mutated to change the substrate specificity of the enzyme. Scheme 10.5 gives a schematic representation of various methods of indigo synthesis. We found that the mutant enzymes could catalyse oxygenation of indole to produce indigo. While Y96F was found to be several times better as a catalyst for conversion of indole into indigo, the double mutant Y96F/L244A showed the highest NADH oxidation rate as well as yield of indigo. The oxidative catalysis using H$_2$O$_2$ as the oxygen source was found to produce a higher purity of indigo, and lesser or no formation of indirubin was detected.

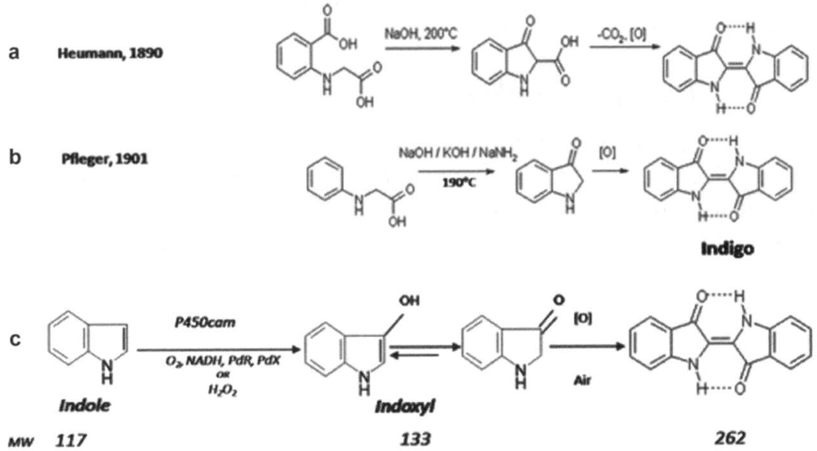

Scheme 10.5   Schematic representation of indigo synthesis through (a) Heumann's method, (b) Pfleger's method, and (c) CYP 450$_{cam}$ catalysed pathway.

A high-throughput screening of CYP 450_{cam} libraries was performed,[145,146] because of their ability to oxidize indole to indigo and indirubin. This resulted in the identification of CYP 450_{cam} variants with activity towards the structurally unrelated substrate diphenylmethane. Random mutation of Tyr_{96} and Phe_{98} to different amino acids resulted good activities towards indole and diphenylmethane. This study also emphasized the importance of maintaining a significant degree of hydrophobicity in the active site.

10.6.10.2   Indoline, Coumarin, Indene, and Quinoline

The addition of aromatic compounds like indoline, quinoline and coumarin were found to have no significant effect on the UV–Vis absorption spectra of wild-type CYP 450_{cam}. However, the absorption spectra of the mutants were found to undergo drastic changes on addition of these compounds. While addition of indoline and quinoline induced a redshift (Type II spectra) of the absorption spectrum of the Soret maxima of the mutant enzymes, coumarin and indene showed a blue-shift, *i.e.* a Type I difference spectrum (peak around 385 nm and trough at about 420 nm). A Type II change indicates the formation of direct Fe–N coordination in the enzyme–substrate complex (trough around 390–410 nm and a peak around 425–435nm). Compared to indole where the lone pair of nitrogen resides in a conjugated 'p' orbital as a part of the Hückel $(4n+2)$ aromatic 'π' electron system, the lone pair of indoline resides in a sp^3 orbital and is available for easy electron donation to the iron centre. Experimentally, indene binding constants with the mutants were found to be considerably higher than those for coumarin. This may be attributed to the fact that indene is a more hydrophobic molecule than coumarin and thereby prefers to reside in the hydrophobic environment of the active site of CYP 450_{cam} mutants more than coumarin. Nonetheless coumarin was selected as a representative of these substrates, based on the fact that hydroxyl-coumarin derivatives are used as drugs, dyes/sensors, sunscreen agents, and also for the assay of microsomal CYP 450s. Thus, selective hydroxylation may find potential industrial application.

10.6.10.3   Tetrahydrocarboline

CYP 450_{cam} also oxidizes 1,2,3,4-tetrahydro-β-carboline, an indolic alkaloid. To rationalize the experimental results of product distributions, *ab initio* quantum-chemistry calculations of the product stabilities and molecular-dynamics calculations of the substrate-binding mode in the active site have been performed by the Jankowski group.[147] The calculations suggest that the product distribution is influenced by both the relative intrinsic gas-phase stabilities of the monohydroxy products from that alkaloid and by conformational rearrangement of the heme active site on substrate binding.

10.6.10.4   Nicotine

Some crystallographic as well as spectroscopic studies were performed on the non-native substrates such as nicotine, imidazole *etc.*, for the same heme-protein to check the binding behaviour. Despite the existence of theoretical models, crystallographic studies showed that the binding mode of nicotine is non-productive, and an indication of Type II spectra was found. Nicotine, (*S*)-*N*-methyl-2-phenylpyrrolidine, imidazole, and 1-methylimidazole all were found to coordinate through one of their nitrogen atoms with the heme-iron, resulting in a low-spin 420 nm species.[148,149] Experimental binding constants with wild type and mutants were also found to be very low compared to camphor. These types of studies also support the idea that most of the heterocyclic compounds generally give rise to Type II difference spectra due to direct Fe–N coordination on substrate binding.[150–153] Nicotine, imidazole, and 1-methylimidazole were found to have low binding constants compared to camphor with the wild-type CYP 450$_{cam}$.

10.6.11 Halogenated Compounds

Polyhalogenated compounds are used for several domestic, agricultural, and industrial purposes as pesticides, herbicides, solvents, *etc.*[154–158] Most of the organohalide compounds are environmentally hazardous and many of them have been found to be potential carcinogens or mutagens.[154,157,159,160] One of the most problematic issues with these substrates is that they are highly insoluble and inert in nature and, therefore, accumulate in fatty tissues of plants, *i.e.* animal food sources.[154] Thus, they enter the human body through the food chain and pose a potential hazard to human health. Even breast milk has been found to be contaminated with organochlorine pesticides.[154,156]

10.6.11.1 Aliphatic and Aromatic Small Polyhalogenated Compounds

Chlorine chemistry has been at the forefront of public concern, since organochlorine compounds constitute a large group of environmental pollutants that adversely affect human health and life.[161] More than 2000 halogenated chemicals are known to be released into the biosphere by various marine organisms, higher plants, insects, bacteria, fungi, and mammals.[162,163] Some organisms are thought to use organohalogen compounds as feeding deterrents, irritants, or pesticides, *i.e.* as chemical defence. Several bacteria[164] and microbes[165–169] are known to have evolved for the enzyme-catalysed dechlorination and degradation of both haloaliphatic and haloaromatic compounds. Several review articles deal with such aspects.[155,167,168,170,171]

For biotechnological purposes, degradation of chlorinated compounds by CYP 450$_{cam}$ has been one of the major research fields in the last few decades. CYP 450$_{cam}$ under oxygen-rich conditions catalyses dehalogenation reactions,[172] whereas under low oxygen levels a reductive dechlorination takes place.[173,174] The ability of heme proteins, which generally function as oxidases, to catalyse the reduction of bromochloromethane and related compounds like CCl_4 and NO_2CCl_3 has been shown by Castro and coworkers.[175]

This reductase capacity of heme proteins arises from the Fe(II)-porphyrin form of the protein isolated from mammalians and microbial sources. In 1985, biodehalogenation reactions were carried out for bromotrichloromethane (BTM) and 1,2-dibromo, 3-chloropropane (DBCP) at ppm levels using soil bacterium CYP 450$_{cam}$ at high concentration in aerobic media for the first time.[175] Whole cell experiments with CYP 450$_{cam}$ and a homogeneous solution of that isolated enzyme with BTM have been shown to give $CHCl_3$ as degraded product. Since, the C–Br bond is weaker than the C–Cl bond, some amounts of bromodichloromethane were observed as dechlorinated product. Subsequently, the authors proposed that further reduction to mono-halomethanes was not possible because of the unfavourable resonance stability of the intermediate halomethane radicals generated during the approach of a chlorine group to the Fe-active site. However, for DBCP, the mechanism for degradation was difficult to predict.

10.6.11.2 Pesticides

Although pesticides are designed to be selective, that is, more harmful to organisms than to man, they are beneficial to other organisms and their selectivity is not always complete. Most pesticides are degraded chemically and biologically in nature very slowly. Their chemical degradation in water and soil is influenced by several factors like temperature, moisture, sunlight, pH, and coexisting organic and inorganic compounds. Figure 10.7 shows examples of organochlorine compounds. Because of their persistence in the environment, the use of organochlorine pesticides has been slowly minimized in developed as well as developing countries. Because of their persistence and widespread use in industries, several fractions of them have been detected in crops, soil, rain water, the atmosphere, surface water, and also the food chain and drinking water. As stated earlier, several bacteria and microbes can metabolize halogen-containing compounds and pesticides.

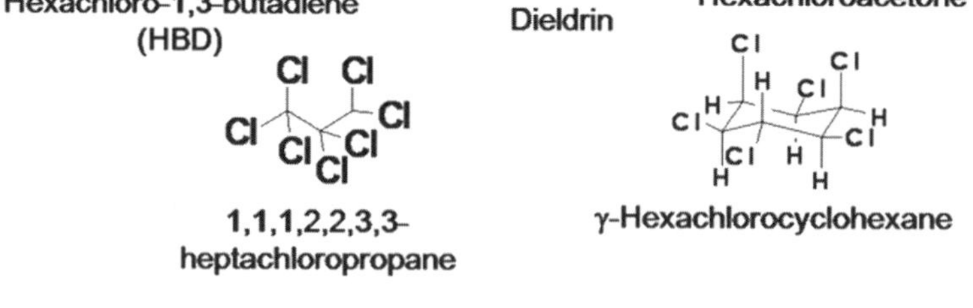

Figure 10.7 Commonly used halogenated pesticides.

We have found that protein engineering in the active site indeed helps to increase the binding affinities significantly of those aforementioned unnatural substrates. Aldrin, dieldrin, and endrin are bulkier molecules and much more hydrophobic substrates than the natural substrate (camphor) of the enzyme. Hence, they showed weak or no binding to the wild-type CYP 450$_{cam}$. The substrate binding pocket had to be modified to make more space and/or make it more hydrophobic to enhance the binding affinity for insecticides.

From the equilibrium binding studies in our laboratory and the results obtained from the crystal structures, pesticides binding by mutants of CYP 450$_{cam}$ showed interesting variations. The hydrophobic substitution Y96F with larger phenyl side chain was the only one to strengthen the pesticide binding. This observation is obviously due to the more hydrophobic nature of insecticides compared to camphor.

10.7 Summary

The conformational properties of the residues at the active site of CYP 450$_{cam}$ are the main key in regulating the substrate specificity of the enzyme. The architecture of the substrate binding pocket of the wild-type enzyme is such that it can anchor the natural substrate, camphor, extremely efficiently and small variations in the substrate structure alter the affinity of the enzyme for the substrate. In recent years an enormous effort has been made to modify the active site of the CYP 450$_{cam}$ with the aim of altering the substrate specificity and also the product selectivity of the enzyme. The present chapter gives a brief overview of the reports that deal with alteration of the active site of CYP 450$_{cam}$ in terms of tuning the activity of the enzyme. Reactions of a large number of unnatural substrates have been shown to be catalysed by specific mutants of CYP 450$_{cam}$. These studies demonstrate that the rational design of the active sites of enzymes has many shortcomings, but this approach has led to enormous success in our area of research. No doubt, the next few years will uncover novel aspects of CYP 450 function and will lead to a deeper and more sophisticated understanding of the catalytic mechanisms of the amazing family of CYP 450 enzymes.

The diversity and complexity of reactions catalysed by the CYP 450 enzymes, however, still continues to surprise researchers. In the past decade, there has been major progress in characterizing the intermediates that are formed as molecular oxygen is activated to the final oxidizing species. It is now well established that the general structures of CYP 450s are similar, including the mode of interaction with substrates, electron donors. However, membranes vary widely between different isoforms of the enzyme. Understanding the ligand interactions with the active site and recognition of the substrate by the specific CYP 450 isoform is of critical importance to determine the molecular basis of the substrate specificity of the enzymes. The specific role of water molecules on ligand dependent conformational change or in 'bridging' protein–ligand contacts also is not resolved unambiguously. Theoretical and computational methods have contributed greatly to this area. CYP 450_{cam} is one of the simplest prototypes of this complex superfamily of enzymes – most experimental and theoretical studies have been directed on the CYP 450_{cam} and its mutants.

Apart from understanding the fundamental aspects of the enzymatic functions, the CYP 450 enzymes have been shown to have potential for biotechnological applications. Efforts are being made to create viable green biocatalysts using recombinant mutants of CYP 450 enzymes. The present chapter has discussed reports on studies of CYP 450_{cam} as a green biocatalyst. However, the requirement of the reductases and NADH for efficient catalysis by this enzyme has remained a bottleneck in making it a successful biocatalyst. Recently, efforts have been made to develop electrochemically driven catalysis of CYP 450_{cam} by designing suitable electrode systems, which would help in making this enzyme a good candidate for industrial applications.

References

1. D. R. Nelson, T. Kamataki, D. J. Waxman, F. P. Guengerich, R. W. Estabrook, R. Feyereisen, F. J. Gonzalez, M. J. Coon, I. C. Gunsalus and O. Gotoh, *DNA Cell Biol.*, 1993, **12**, 1.
2. D. R. Nelson, D. C. Zeldin, S. M. Hoffman, L. J. Maltais, H. M. Wain and D. W. Nebert, *Pharmacogenetics*, 2004, **14**, 1.
3. O. Hayaishi, M. Katagiri and S. Rothberg, *J. Am. Chem. Soc.*, 1955, **77**, 5450.
4. H. E. Conrad, R. Dubus and I. C. Gunsalus, *Biochem. Biophys. Res. Commun.*, 1961, **6**, 293.
5. J. Hedegaard and I. C. Gunsalus, *J. Biol. Chem.*, 1965, **240**, 4038.
6. E. J. Muller, P. J. Loida and S. G. Sligar, in *Cytochrome P450: Structure, Mechanism and Biochemistry*, ed, P. R. Ortiz de Montellano, 2nd edn, Plenum Press, New York, 1995, pp. 83–124.
7. Y. Anzai, H. Kim, J. Y. Park, H. Wakabayashi and H. Oyaizu, *Int. J. Syst. Evol. Microbiol.*, 2000, **50**, 1563.
8. H. J. Ougham, D. G. Taylor and P. W. Trudgill, *J. Bacteriol.*, 1983, **153**, 140.
9. M. Katagiri, B. N. Ganguli and I. C. Gunsalus, *J. Biol. Chem.*, 1968, **243**, 3543.
10. R. Raag and T. L. Poulos, *Biochemistry*, 1991, **30**, 2674.
11. P. R. Ortiz de Montellano, (ed.), *Cytochrome P450: Structure, Mechanism and Biochemistry*, 3rd edn, Kluwer Academic/Plenum Publisher, New York, 2005.
12. V. B. Urlacher, S. Lutz-Wahl and R. D. Schmid, *Appl. Microbiol. Biotechnol.*, 2004, **64**, 317.
13. T. L. Poulos, B. C. Finzel and A. J. Howard, *J. Mol. Biol.*, 1987, **195**, 687.
14. R. Murugan and S. Mazumdar, *ChemBioChem*, 2005, **6**, 1204.
15. S. Yoshioka, S. Takahashi, H. Hori, K. Ishimori and I. Morishima, *Eur. J. Biochem.*, 2001, **268**, 252.

16. S. Yoshioka, S. Takahashi, K. Ishimori and I. Morishima, *J. Inorg. Biochem.*, 2000, **81**, 141.

17. L. L. Wong, A. C. G. Westlake and D. P. Nickerson, *Metal Sites in Proteins and Models: Redox Centers*, Springer, Berlin, 1999.

18. I. Schlichting, J. Berendzen, K. Chu, A. M. Stock, S. A. Maves, D. E. Benson, R. M. Sweet, D. Ringe, G. A. Petsko and S. G. Sligar, *Science*, 2000, **287**, 1615.

19. S. K. Manna and S. Mazumdar, *J. Inorg. Biochem.*, 2008, **102**, 1312.

20. M. Imai, H. Shimada, Y. Watanabe, Y. Matsushima-Hibiya, R. Makino, H. Koga, T. Horiuchi and Y. Ishimura, *Proc. Natl. Acad. Sci. USA*, 1989, **86**, 7823.

21. S. Nagano and T. L. Poulos, *J. Biol. Chem.*, 2005, **280**, 31659.

22. M. Vidakovic, S. G. Sligar, H. Li and T. L. Poulos, *Biochemistry*, 1998, **37**, 9211.

23. R. K. Behera and S. Mazumdar, *Biophys. Chem.*, 2008, **135**, 1.

24. S. K. Lüdemann, V. Lounnas and R. C. Wade, *J. Mol. Biol.*, 2000, **303**, 797.

25. K. Schleinkofer, Sudarko, P. J. Winn, S. K. Ludemann and R. C. Wade, *EMBO Rep.*, 2005, **6**, 584.

26. P. Rowland, F. E. Blaney, M. G. Smyth, J. J. Jones, V. R. Leydon, A. K. Oxbrow, C. J. Lewis, M. G. Tennant, S. Modi, D. S. Eggleston, R. J. Chenery and A. M. Bridges, *J. Biol. Chem.*, 2006, **281**, 7614.

27. D. Fishelovitch, S. Shaik, H. J. Wolfson and R. Nussinov, *J. Phys. Chem. B*, 2009, **113**, 13 018.

28. P. A. Williams, J. Cosme, A. Ward, H. C. Angove, D. Matak Vinkovic and H. Jhoti, *Nature*, 2003, **424**, 464.

29. G. A. Roberts, G. Grogan, A. Greter, S. L. Flitsch and N. J. Turner, *J. Bacteriol.*, 2002, **184**, 3898.

30. F. Hannemann, A. Bichet, K. M. Ewen and R. Bernhardt, *Biochim. Biophys. Acta (BBA)-Gen. Subjects*, 2007, **1770**, 330.

31. S. G. Sligar, P. G. Debrunner, J. D. Lipscomb, M. J. Namtvedt and I. C. Gunsalus, *Proc. Natl. Acad. Sci. USA*, 1974, **71**, 3906.

32. S. G. Sligar and I. C. Gunsalus, *Proc. Natl. Acad. Sci. USA*, 1976, **73**, 1078.

33. D. Harris and G. H. Loew, *J. Am. Chem. Soc.*, 1993, **115**, 8775.

34. H. Shimada, S. Nagano, H. Hori and Y. Ishimura, *J. Inorg. Biochem.*, 2001, **83**, 255.

35. M. Unno, J. F. Christian, T. Sjodin, D. E. Benson, I. D. Macdonald, S. G. Sligar and P. M. Champion, *J. Biol. Chem.*, 2002, **277**, 2547.

36. S. Nagano, T. Tosha, K. Ishimori, I. Morishima and T. L. Poulos, *J. Biol. Chem.*, 2004, **279**, 42844.

37. T. Tosha, S. Yoshioka, K. Ishimori and I. Morishima, *J. Biol. Chem.*, 2004, **279**, 42836.

38. T. Tosha, S. Yoshioka, S. Takahashi, K. Ishimori, H. Shimada and I. Morishima, *J. Biol. Chem.*, 2003, **278**, 39809.

39. C. A. Tyson, J. D. Lipscomb and I. C. Gunsalus, *J. Biol. Chem.*, 1972, **247**, 5777.

40. R. Davydov, T. M. Makris, V. Kofman, D. E. Werst, S. G. Sligar and B. M. Hoffman, *J. Am. Chem. Soc.*, 2001, **123**, 1403.
41. I. G. Denisov, T. M. Makris, S. G. Sligar and I. Schlichting, *Chem. Rev.*, 2005, **105**, 2253.
42. F. P. Guengerich, G. P. Miller, I. H. Hanna, H. Sato and M. V. Martin, *J. Biol. Chem.*, 2002, **277**, 33711.
43. K. Von König and I. Schlichting, in *Metal Ions in Life Sciences*, Eds. S. Sigel, H. Sigel, R. K. O. Sigel, John Wiley & Sons, Ltd., Chichester, 2007, Vol. 3, Chapter 8, pp. 235–266.
44. S. G. Bell, R. J. Sowden and L.-L. Wong, *Chem. Commun.*, 2001, 635.
45. C. Jung, in *Metal Ions in Life Sciences*, Eds. S. Sigel, H. Sigel, R. K. O. Sigel, John Wiley & Sons, Ltd., Chichester, 2007, Vol. 3, Chapter 7, pp. 187–234.
46. M. Altarsha, T. Benighaus, D. Kumar and W. Thiel, *J. Am. Chem. Soc.*, 2009, **131**, 4755.
47. D. Kumar, A. Altun, S. Shaik and W. Thiel, *Faraday Discuss.*, 2011, **148**, 373.
48. T. Hayashi, K. Harada, K. Sakurai, H. Shimada and S. Hirota, *J. Am. Chem. Soc.*, 2009, **131**, 1398.
49. S. H. Kim, T. C. Yang, R. Perera, S. Jin, T. A. Bryson, M. Sono, R. Davydov, J. H. Dawson and B. M. Hoffman, *Dalton Trans.*, 2005, 3464.
50. P. M. Champion, I. C. Gunsalus and G. C. Wagner, *J. Am. Chem. Soc.*, 1978, **100**, 3743.
51. M. Sharrock, P. G. Debrunner, C. Schulz, J. D. Lipscomb, V. Marshall and I. C. Gunsalus, *Biochim. Biophys. Acta*, 1976, **420**, 8.
52. J. C. Hackett, T. T. Sanan and C. M. Hadad, *Biochemistry*, 2007, **46**, 5924.
53. J. N. Harvey, C. M. Bathelt and A. J. Mulholland, *J. Comput. Chem.*, 2006, **27**, 1352.
54. D. G. Kellner, S.-C. Hung, K. E. Weiss and S. G. Sligar, *J. Biol. Chem.*, 2002, **277**, 9641.
55. M. Newcomb, R. Zhang, R. E. P. Chandrasena, J. A. Halgrimson, J. H. Horner, T. M. Makris and S. G. Sligar, *J. Am. Chem. Soc.*, 2006, **128**, 4580.
56. S. Prasad and S. Mitra, *Biochem. Biophys. Res. Commun.*, 2004, **314**, 610.
57. K. L. Stone, R. K. Behan and M. T. Green, *Proc. Natl. Acad. Sci. USA*, 2005, **102**, 16563.
58. S. P. de Visser, F. Ogliaro and S. Shaik, *Chem. Commun.*, 2001, 2322.
59. T. Kamachi and K. Yoshizawa, *J. Am. Chem. Soc.*, 2003, **125**, 4652.
60. T. Spolitak, J. H. Dawson and D. P. Ballou, *J. Inorg. Biochem.*, 2006, **100**, 2034.
61. R. Lonsdale, J. N. Harvey and A. J. Mulholland, *J. Phys. Chem. B*, 2009, **114**, 1156.
62. M. T. Green, J. H. Dawson and H. B. Gray, *Science*, 2004, **304**, 1653.
63. T. L. Poulos, *Phil. Trans. Roy. Soc. A: Math., Phys. Eng. Sci.*, 2005, **363**, 793.

64. M. Fukuda, Y. Yasukochi, Y. Kikuchi, Y. Nagata, K. Kimbara, H. Horiuchi, M. Takagi and K. Yano, *Biochem. Biophys. Res. Commun.*, 1994, **202**, 850.
65. K. M. Weir, T. D. Sutherland, I. Horne, R. J. Russell and J. G. Oakeshott, *Appl. Environ. Microbiol.*, 2006, **72**, 3524.
66. J. A. Peterson, M. C. Lorence and B. Amarneh, *J. Biol. Chem.*, 1990, **265**, 6066.
67. T. L. Poulos, *Nature*, 1989, **339**, 580.
68. *U.S. Patent* 6117661, 2000.
69. T. L. Poulos, B. C. Finzel, I. C. Gunsalus, G. C. Wagner and J. Kraut, *J. Biol. Chem.*, 1985, **260**, 16122.
70. P. P. Tamburini, G. G. Gibson, W. L. Backes, S. G. Sligar and J. B. Schenkman, *Biochemistry*, 1984, **23**, 4526.
71. B. W. Griffin and J. A. Peterson, *Biochemistry*, 1972, **11**, 4740.
72. A. K. Udit, M. G. Hill and H. B. Gray, *J. Inorg. Biochem.*, 2006, **100**, 519.
73. V. Reipa, M. P. Mayhew, M. J. Holden and V. L. Vilker, *Chem. Commun.*, 2002, 318.
74. P. S. Stayton and S. G. Sligar, *Biochemistry*, 1990, **29**, 7381.
75. E. Deprez, N. C. Gerber, C. Di Primo, P. Douzou, S. G. Sligar and G. Hui Bon Hoa, *Biochemistry*, 1994, **33**, 14464.
76. S. Modi, W. U. Primrose, J. M. Boyle, C. F. Gibson, L. Y. Lian and G. C. Roberts, *Biochemistry*, 1995, **34**, 8982.
77. C. R. McCullough, P. K. Pullela, S.-C. Im, L. Waskell and D. S. Sem, *J. Biomol. NMR*, 2009, **43**, 171.
78. W. Zhang, S. S. Pochapsky, T. C. Pochapsky and N. U. Jain, *J. Mol. Biol.*, 2008, **384**, 349.
79. B. OuYang, S. S. Pochapsky, M. Dang and T. C. Pochapsky, *Structure*, 2008, **16**, 916.
80. G. Niaura, V. Reipa, M. P. Mayhew, M. Holden and V. L. Vilker, *Arch. Biochem. Biophys.*, 2003, **409**, 102.
81. C. Jung, H. Schulze and E. Deprez, *Biochemistry*, 1996, **35**, 15088.
82. N. Bistolas, A. Christenson, T. Ruzgas, C. Jung, F. W. Scheller and U. Wollenberger, *Biochem. Biophys. Res. Commun.*, 2004, **314**, 810.
83. *U.S. Patent*, 6 117 661, 2000.
84. V. Helms and R. C. Wade, *Proteins*, 1998, **32**, 381.
85. M. C. Marden and G. Hui Bon Hoa, *Arch. Biochem. Biophys.*, 1987, **253**, 100.
86. S. Narasimhulu, L. M. Havran, P. H. Axelsen and J. D. Winkler, *Arch. Biochem. Biophys.*, 1998, **353**, 228.
87. R. Tsai, C. A. Yu, I. C. Gunsalus, J. Peisach, W. Blumberg, W. H. Orme-Johnson and H. Beinert, *Proc. Natl. Acad. Sci. USA*, 1970, **66**, 1157.
88. G. H. Loew and D. L. Harris, *Chem. Rev.*, 2000, **100**, 407.
89. K. Sakurai, H. Shimada, T. Hayashi and T. Tsukihara, *Acta Crystallogr. Sect. F*, 2009, **65**, 80.
90. P. R. Ortiz de Montellano, *Chem. Rev.*, 2009, **109**, 932.

91. S. G. Bell, E. Orton, H. Boyd, J.-A. Stevenson, A. Riddle, S. Campbell and L.-L. Wong, *Dalton Trans.*, 2003, 2133.

92. S. G. Bell, J. A. Stevenson, H. D. Boyd, S. Campbell, A. D. Riddle, E. L. Orton and L. L. Wong, *Chem. Commun.*, 2002, 490.

93. F. Xu, S. G. Bell, J. Lednik, A. Insley, Z. Rao and L. L. Wong, *Angew. Chem., Int. Ed.*, 2005, **44**, 4029.

94. O. Sibbesen, Z. Zhang and P. R. Ortiz de Montellano, *Arch. Biochem. Biophys.*, 1998, **353**, 285.

95. P. L. Carmichael, M. Ni She, A. Hewer, J. Jacob, G. Grimmer and D. H. Phillips, *Cancer Lett.*, 1992, **64**, 137.

96. R. D. Harvey, *Polycyclic Hydrocarbons and Carcinogenesis*, American Chemical Society, Washington DC, 1985.

97. R. K. Hewstone, *Sci. Total Environ.*, 1994, **156**, 255.

98. F. A. Carey and R. J. Sundberg, *Advanced Organic Chemistry Part B: Reaction and Synthesis*, 4th edn, Kluwer Academic/Plenum Publishers, New York, 2001.

99. J. Clayden, N. Greeves, S. Warren, P. Wothers, *Organic Chemistry*, Oxford University Press, Oxford, 2000.

100. J. March, *Advanced Organic Chemistry: Reaction, Mechanism and Structure*, 4th edn, John Wiley & Sons, Noida, Delhi, 2005.

101. D. P. Nickerson, C. F. Harford-Cross, S. R. Fulcher and L. L. Wong, *FEBS Lett.*, 1997, **405**, 153.

102. C. F. Harford-Cross, A. B. Carmichael, F. K. Allan, P. A. England, D. A. Rouch and L. L. Wong, *Protein Eng.*, 2000, **13**, 121.

103. P. A. England, C. F. Harford-Cross, J. A. Stevenson, D. A. Rouch and L. L. Wong, *FEBS Lett.*, 1998, **424**, 271.

104. K. J. French, D. A. Rock, D. A. Rock, J. I. Manchester, B. M. Goldstein and J. P. Jones, *Arch. Biochem. Biophys.*, 2002, **398**, 188.

105. K. J. French, M. D. Strickler, D. A. Rock, D. A. Rock, G. A. Bennett, J. L. Wahlstrom, B. M. Goldstein and J. P. Jones, *Biochemistry*, 2001, **40**, 9532.

106. J. Mater, *Biomimetic Oxidations Catalysed by Transition Metal Complexes*, Imperial College Press, Imperial College, London, 2000.

107. P. A. England, D. A. Rouch, A. C. G. Westlake, S. G. Bell, D. P. Nickerson, M. Webberley, S. L. Flitsch and L. L. Wong, *Chem. Commun.*, 1996, 357.

108. J. Kazlauskaite, A. C. G. Westlake, L.-L. Wong and H. A. O. Hill, *Chem. Commun.*, 1996, 2189.

109. R. W. Estabrook, M. S. Shet, K. Faulkner and C. W. Fisher, *Endocr. Res.*, 1996, **22**, 665.

110. Y. M. Lvov, Z. Lu, J. B. Schenkman, X. Zu and J. F. Rusling, *J. Am. Chem. Soc.*, 1998, **120**, 4073.

111. R. W. Estabrook, *Methods in Enzymology*, 1996.

112. S. G. Bell, C. F. Harford-Cross and L. L. Wong, *Protein Eng.*, 2001, **14**, 797.

113. C. Yan-Tyng, H. L. Gilda, E. R. Allan, A. B. Thomas, R. S. Pamela and R. O. D. M. Paul, *Int. J. Quantum Chem.*, 1993, **48**, 161.
114. J. J. De Voss, O. Sibbesen, Z. Zhang and P. R. Ortiz de Montellano, *J. Am. Chem. Soc.*, 1997, **119**, 5489.
115. J. Gershenzon and N. Dudareva, *Nat. Chem. Biol.*, 2007, **3**, 408.
116. P. J. Teisseire, P. A. Cadby, *Chemistry of Fragrant Substances*, VCH, New York, 1994.
117. P. W. Trudgill, Terpenoid metabolism by *Pseudomonas*, in *Bacteria: A Treatise on Structure and Function*, Eds. I. C. Gunsalus, J. R. Sokatch, L. N. Ornston, Academic Press, Inc., San Diego, 1986, pp. 483–525.
118. G. Sandmann, *Arch. Biochem. Biophys.*, 2001, **385**, 4.
119. J. L. Bicas, A. P. Dionilosio and G. M. Pastore, *Chem. Rev.*, 2009, **109**, 4518.
120. S. Arctander, *Perfume and Flavor Materials of Natural Origin*, Allured Publ. Co., Wheaton, IL, 1960.
121. R. J. Sowden, S. Yasmin, N. H. Rees, S. G. Bell and L. L. Wong, *Org. Biomol. Chem.*, 2005, **3**, 57.
122. B. M. Fraga, *Nat. Prod. Rep.*, 2004, **21**, 669.
123. L. Ralston, S. T. Kwon, M. Schoenbeck, J. Ralston, D. J. Schenk, R. M. Coates and J. Chappell, *Arch. Biochem. Biophys.*, 2001, **393**, 222.
124. Y. J. Chang, S. H. Song, S. H. Park and S. U. Kim, *Arch. Biochem. Biophys.*, 2000, **383**, 178.
125. D. L. Klayman, *Science*, 1985, **228**, 1049.
126. M. Furusawa, T. Hashimoto, Y. Noma and Y. Asakawa, *Chem. Pharm. Bull.*, 2005, **53**, 1513.
127. J. Chappell, *Trends Plant Sci.*, 2004, **9**, 266.
128. L. Sharon-Asa, M. Shalit, A. Frydman, E. Bar, D. Holland, E. Or, U. Lavi, E. Lewinsohn and Y. Eyal, *Plant J.*, 2003, **36**, 664.
129. H. L. Holland, in *Stereoselective Biocatalysis*, Marcel Dekker, Inc., New York, 2000.
130. D. T. Gibson, S. M. Resnick, K. Lee, J. M. Brand, D. S. Torok, L. P. Wackett, M. J. Schocken and B. E. Haigler, *J. Bacteriol.*, 1995, **177**, 2615.
131. L. P. Wackett, L. D. Kwart and D. T. Gibson, *Biochemistry*, 1988, **27**, 1360.
132. D. A. Grayson and V. L. Vilker, *J. Mol. Catal. B: Enzymatic*, 1999, **6**, 533.
133. D. R. Boyd, N. D. Sharma, T. A. Evans, M. Groocock, J. F. Malone, P. J. Stevenson and H. Dalton, *J. Chem. Soc., Perkin Trans.*, 1997, **1**, 1879.
134. M. P. Mayhew, A. E. Roitberg, Y. Tewari, M. J. Holden, D. J. Vanderah and V. L. Vilker, *New J. Chem.*, 2002, **26**, 35.
135. A. O. Kareem, *Ann. Chim.*, 2007, **97**, 527.
136. K. Pawlak, M. Puchalska, A. Miszczak, E. Rosloniec and M. Jarosz, *J. Mass Spectrom.*, 2006, **41**, 613.
137. E. S. Ferreira, A. N. Hulme, H. McNab and A. Quye, *Chem. Soc. Rev.*, 2004, **33**, 329.
138. H. Y. Huang, C. L. Chuang, C. W. Chiu and M. C. Chung, *Electrophoresis*, 2005, **26**, 867.

139. M. D. Rutter, B. P. Saunders, G. Schofield, A. Forbes, A. B. Price and I. C. Talbot, *Gut*, 2004, **53**, 256.

140. Y. Sakai, R. Eto, J. Kasanuki, F. Kondo, K. Kato, M. Arai, T. Suzuki, M. Kobayashi, T. Matsumura, D. Bekku, K. Ito, S. Nakamoto, T. Tanaka and O. Yokosuka, *Gastrointest. Endosc.*, 2008, **68**, 635.

141. H. Yamashita, J. Kitayama, H. Ishigami, J. Yamada, H. Miyato, S. Kaisaki and H. Nagawa, *Dig. Liver Dis.*, 2007, **39**, 389.

142. N. Chanayath, S. Lhieochaiphant and S. Phutrakul, *Chiang Mai Univ. J.*, 2002, **1**, 149.

143. S. A. Lawrence, *Amines: Synthesis, Properties and Applications*, Cambridge University Press, Cambridge, 2004.

144. H. Zollinger, *Colour Chemistry: Synthesis, Properties and Applications of Organic Dyes and Pigments*, 3rd edn, Wiley-VCH Verlag, Weinheim, 2003.

145. A. Celik, R. E. Speight and N. J. Turner, *Chem. Commun.*, 2005, 3652.

146. R. E. Speight, F. E. Hancock, C. Winkel, H. S. Bevinakatti, M. Sarkar, S. L. Flitsch and N. J. Turner, *Tetrahedron: Asymmetry*, 2004, **15**, 2829.

147. J. S. Sopková-de Oliveira, J. C. Smith, M. Delaforge, H. Virelizier and C. K. Jankowski, *Eur. J. Biochem.*, 1998, **251**, 398.

148. M. Strickler, B. M. Goldstein, K. Maxfield, L. Shireman, G. Kim, D. S. Matteson and J. P. Jones, *Biochemistry*, 2003, **42**, 11943.

149. A. Verras, A. Alian and P. R. Ortiz de Montellano, *Protein Eng. Des. Select.*, 2006, **19**, 491.

150. F. Cabello-Hurtado, M. Taton, N. Forthoffer, R. Kahn, S. Bak, A. Rahier and D. Werck-Reichhart, *Eur. J. Biochem.*, 1999, **262**, 435.

151. C. A. Hitchcock, K. Dickinson, S. B. Brown, E. G. Evans and D. J. Adams, *Biochem. J.*, 1990, **266**, 475.

152. M. P. Pietila, P. K. Vohra, B. Sanyal, N. L. Wengenack, S. Raghava-kaimal and C. F. Thomas Jr, *Am. J. Respir. Cell Mol. Biol.*, 2006, **35**, 236.

153. L. M. Podust, J. P. von Kries, A. N. Eddine, Y. Kim, L. V. Yermalitskaya, R. Kuehne, H. Ouellet, T. Warrier, M. Altekoster, J. S. Lee, J. Rademann, H. Oschkinat, S. H. Kaufmann and M. R. Waterman, *Antimicrob. Agents Chemother.*, 2007, **51**, 3915.

154. E. E. Calle, H. Frumkin, S. J. Henley, D. A. Savitz and M. J. Thun, *CA Cancer J. Clin.*, 2002, **52**, 301.

155. S. Fetzner and F. Lingens, *Microbiol. Rev.*, 1994, **58**, 641.

156. K. Fytianos, G. Vasilikiotis, L. Weil, E. Kavlendis and N. Laskaridis, *Bull. Environ. Contam. Toxicol.*, 1985, **34**, 504.

157. D. McLean, N. Pearce, H. Langseth, P. Jappinen, I. Szadkowska-Stanczyk, B. Persson, P. Wild, R. Kishi, E. Lynge, P. Henneberger, M. Sala, K. Teschke, T. Kauppinen, D. Colin, M. Kogevinas and P. Boffetta, *Environ. Health Perspect.*, 2006, **114**, 1007.

158. G. K. Robinson and M. J. Lenn, *Biotechnol. Genet. Eng. Rev.*, 1994, **12**, 139.

159. Q. Huang, X. Wang, Y. Liao, L. Kong, S. Han and L. Wang, *Bull. Environ. Contam. Toxicol.*, 1995, **55**, 796.

160. P. K. Tithof, J. Olivero, K. Ruehle and P. E. Ganey, *Toxicol. Sci.*, 2000, **53**, 40.
161. B. Hileman, *Chem. Eng. News*, 1993, **71**, 11.
162. G. W. Gribble, *Am. J. Public Health*, 1994, **84**, 1183.
163. G. W. Gribble, *Fortschr. Chem. Org. Naturst.*, 1996, **68**, 1.
164. S. Fetzner, *Appl. Microbiol. Biotechnol.*, 1998, **50**, 633.
165. D. B. Janssen, J. E. Oppentocht and G. J. Poelarends, *Curr. Opin. Biotechnol.*, 2001, **12**, 254.
166. R. Mavoungou, R. Masse and M. Sylvestre, *Sci. Total Environ.*, 1991, **101**, 263.
167. W. W. Mohn and J. M. Tiedje, *Microbiol. Rev.*, 1992, **56**, 482.
168. J. H. Slater, A. T. Bull and D. J. Hardman, *Adv. Microb. Physiol.*, 1997, **38**, 133.
169. J. Wiegel and Q. Wu, *FEMS Microbiol. Ecol.*, 2000, **32**, 1.
170. S. D. Copley, *Chem. Biol.*, 1997, **4**, 169.
171. D. J. Hardman, *Crit. Rev. Biotechnol.*, 1991, **11**, 1.
172. G. S. Koe and V. L. Vilker, *Biotechnol. Prog.*, 2005, **21**, 1119.
173. S. Li and L. P. Wackett, *Biochemistry*, 1993, **32**, 9355.
174. M. E. Walsh, P. Kyritsis, N. A. Eady, H. A. Hill and L. L. Wong, *Eur. J. Biochem.*, 2000, **267**, 5815.
175. C. E. Castro, R. S. Wade and N. O. Belser, *Biochemistry*, 1985, **24**, 204.

QM/MM Studies of Cytochrome P450 Systems: Application to Drug Metabolism

RICHARD LONSDALE, JEREMY N. HARVEY AND
ADRIAN J. MULHOLLAND*

Centre for Computational Chemistry, University of Bristol, Cantock's Close,
Bristol, BS8 1TS, United Kingdom

11.1 Introduction

Cytochrome P450 (CYP 450) is the collective name for a family of over 1000
heme-containing monoxygenases that are found in almost all living organisms.
CYPs catalyze a wide variety of reactions including hydroxylation, epoxida-
tion, heteroatom dealkylation, heteroatom oxygenation and group migration.[1]
The general equation for CYP-catalysed oxidation of a substrate RH, to form
ROH, is shown in Equation (11.1):

$$RH + O_2 + 2H^+ + 2e^- \rightarrow ROH + 2H_2O \tag{11.1}$$

From genetic analysis, there are 57 different CYPs in the human body,[2]
known as isoforms, that carry out a large range of biochemical transformations. CYPs are typically membrane-associated, located either in the inner
membrane of mitochondria or in the endoplasmic reticulum of cells. They differ
in sequence and specificity. For example, several CYP 4F isoforms found in the
liver, together with CYP2J2 found in the lungs, are involved in the metabolism
and bioactivation of the essential fatty acid arachidonic acid.[3] Arachidonic acid

Iron-Containing Enzymes: Versatile Catalysts of Hydroxylation Reactions in Nature
Edited by Sam P de Visser and Devesh Kumar
© Royal Society of Chemistry 2011
Published by the Royal Society of Chemistry, www.rsc.org

is a precursor to the eicosanoids, a group of molecules involved in inter- and intracellular messaging.[4] CYP11A1 is involved in the initiation of steroid synthesis, catalyzing the conversion of cholesterol into pregnenolone.[5] Of great interest to the pharmaceutical industry are the drug-metabolizing CYPs, including the 3A4, 2C9, 2D6 and 1A2 isoforms found in the liver.[6,7] These CYPs are believed to have evolved as a defense mechanism against ingested foreign compounds, such as drugs and chemicals found in foodstuffs, by introducing polar groups into lipophilic compounds. This is known as Phase I xenobiotic metabolism (CYPs are the most important enzymes in Phase I metabolism, though other enzymes, such as epoxide hydrolase, are also involved). Phase II metabolism typically involves conjugation reactions of (usually oxidized drug molecules) with groups such as glutathione or sulfonates, which should increase the molecular weight and hence leave the molecule biologically inactive. Glutathione *S*-transferases are important enzymes in Phase II metabolism.[8]

The consensus catalytic cycle common to all CYPs is depicted in Figure 11.1.[9] The resting state, **11.1**, contains the hexacoordinated heme Fe(III) with an octahedral geometry. The four nitrogen atoms of the porphyrin bind at the four equatorial positions of the low-spin iron centre, while the cysteine thiolate and water molecule occupy the axial positions. The entrance of substrate (RH) into the active site displaces the bound water molecule to form the five-coordinate iron complex, **11.2**.[9] This reduction in coordination number causes the iron to adopt a high-spin electronic configuration, and leads to a corresponding increase in reduction potential,[10] thus facilitating the reduction to Fe(II). The Fe(II) complex, **11.3**, binds molecular oxygen to form the ferric superoxide complex **11.4**. This complex undergoes a further one-electron reduction to form the ferric peroxo complex **11.5**. NAD(P)H is the source of both electrons, which are transferred to the heme by a redox partner protein.[11] In eukaryotic CYPs this redox partner is NADPH cytochrome P450 reductase. In prokaryotic CYPs the redox partner is often a combination of ferredoxin and a ferredoxin reductase. Intermediate **11.5** is highly basic and can accept a proton from a nearby water molecule to form the hydroperoxo-ferric complex **11.6**. A hydrogen-bonded network of water molecules provides a proton relay to the active site.[12] In CYP101 (also known as CYP 450_{cam}), Asp_{251} and Thr_{252} have been identified as important to this proton-relay mechanism, and this "acid-alcohol pair" motif is observed in many CYPs.[12–17] The peroxo-species **11.6** accepts a further proton and with loss of a water molecule forms the iron(IV)-oxo complex **11.7**, often referred to as Compound I (Cpd I). Cpd I is believed to be the active oxygenating species of the CYPs, responsible for the addition of an oxygen atom to the substrate to form the product (ROH).[9,17] ROH is subsequently displaced by a water molecule and the resting state of the enzyme (**11.1**) is restored.

Intermediates **11.1–11.7** have been isolated and characterized spectroscopically using various techniques, including UV–visible absorption,[18] electron paramagnetic resonance,[19] magnetic circular dichroism,[20] resonance Raman,[21] nuclear magnetic resonance[22] and Mössbauer.[23] Many CYP

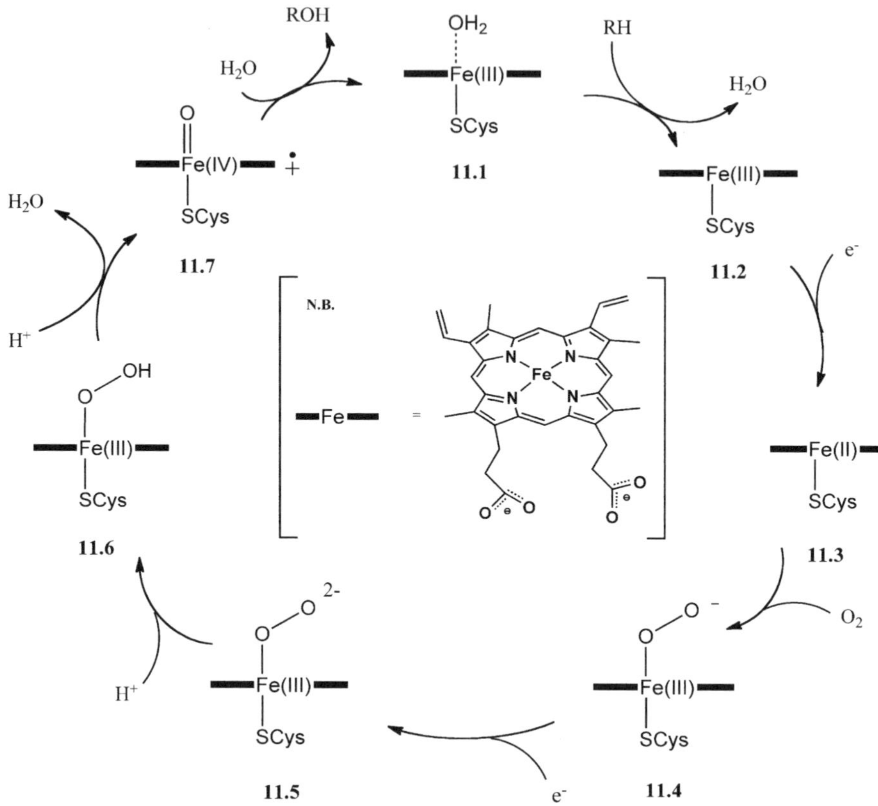

Figure 11.1 Catalytic cycle of the cytochrome P450 enzymes.

isoforms have now been characterized structurally by X-ray crystallography.[24] A significant amount of information regarding the steps of the cycle has also been obtained from biomimetic models and quantum chemical calculations. Biomimetic models typically involve metalloporphyrins with high-valent oxidants such as peracids.[25]

The ferryl-oxo π-cation porphyrin radical, Compound I (Cpd I or **11.7**) is widely believed to be the active oxygenating species for the CYPs.[9,26] Cpd I is electrophilic in nature and, until very recently, has remained experimentally elusive, presumably due to its fast formation and depletion. Cpd I of CYP119 has very recently been isolated and characterized spectroscopically.[27] The Mössbauer spectrum was found to resemble that of Cpd I of chloroperoxidase. When reacting with aliphatic hydrocarbon substrates, hydroxylation occurs with preference for tertiary C–H bonds followed by secondary and then primary sites.[28] Mono-substituted aromatic compounds are mainly hydroxylated at the *ortho* and *para* positions.[29]

The protein environment plays an important role in controlling the reactivity of CYPs.[30] In particular, the anionic sulfur atom of the cysteine thiolate group, bound to the heme iron at the axial position, is believed to play a significant role in CYP chemistry.[26] This thiolate is postulated to facilitate the breaking of the O–O bond in complex **11.6**, by donating electrons to the heme iron, and is referred to as the "push" effect.[31] Other heme-thiolate containing enzymes, such as chloroperoxidase, exist that do not display the same reactivity as CYPs, so the reactivity of CYPs must be due to more than just the presence of heme-thiolate ligation. Amino acids located around the heme-thiolate and in the substrate-binding cavity also play important roles. These residues allow the enzymes to function efficiently and help to fine-tune their selectivity. The residues surrounding the axial cysteinate follow a sequence that is conserved in many CYPs,[32] providing three amide nitrogens that hydrogen-bond to the thiolate sulfur, thereby stabilizing the negative charge. Several experimental and computational studies have been carried out in which the number of NH–S hydrogen-bonds was varied.[33–35] These studies have shown that decreasing the number of hydrogen-bonds to the thiolate lowers the reduction potential of the heme iron, and increases the electron donating ability of the cysteine.

11.2   CYPs and Drug Metabolism

Aside from efficacy, the suitability of drug candidates for clinical use is determined by five main criteria: administration, distribution, metabolism, elimination and toxicology (ADMET).[36] Metabolism issues are a significant cause of drug candidate failure at a late stage in the discovery process, and many are related to CYP activity.[6] Several drugs have been withdrawn from use, and drug candidates have failed (often in late stages of development) due to the reasons listed below.

Roughly 75% of drugs taken by humans are metabolized by CYPs, with five particular isoforms (1A2, 2C9, 2C19, 2D6, 3A4) accounting for around 95% of the CYP-mediated metabolism.[37] The highest concentration of xenobiotic-metabolizing CYPs in humans is in the endoplasmic reticulum of liver cells. The general action of the CYPs located in the liver is to increase the water solubility of lipophilic drug molecules, aiding their excretion by the kidneys. Other organs that are directly exposed to the environment, such as the nasal mucosa and lungs, have also been found to contain xenobiotic-metabolizing CYPs.[38]

CYPs are known to be associated with many adverse drug reactions (ADRs), the unwanted consequences associated with some medications. There are several general mechanisms for CYP mediated ADRs. One involves genetic polymorphism, where subgroups of populations possess a gene that results in production of a CYP isoform with decreased, zero or increased activity.[39] Examples of polymorphic CYP isoforms include 1A2, 2C9, 2C19 and 2D6.[39] Drug–drug interactions are another cause of ADRs, which are often associated with CYPs. There are two major classes of drug–drug interactions: induction and inhibition. When induction takes place, a drug affects the ability of a CYP to oxidize another drug. This is due to enhanced expression of the enzyme. An example of this is the enhanced oxidization by CYP3A4 with some drugs when co-administered with the antibiotic drug rifampicin.[40] During inhibition, a molecule binds so strongly to a CYP that the enzyme is rendered inactive and unable to oxidize further substrates (or inhibition may alter the preferred product or metabolic route). This can lead to a potentially lethal accumulation of an administered drug. Azoles are perhaps the best-known inhibitors of CYPs and are frequently employed in both inhibition studies and crystallography of CYPs.[41] Azoles typically coordinate to the ferric heme iron, preventing reactivity with dioxygen.[41]

The CYP-mediated metabolism of some drug molecules can lead to toxic products that bind to proteins covalently, or cause oxidative damage to cells. For example, paracetamol is metabolized *via* several pathways to the sulfate glucuronide and *via* CYPs 2E1, 1A2 and 3A4 to a reactive iminoquinone.[42–44] At low concentrations the iminoquinone reacts with glutathione and is thus detoxified. At high concentrations there is a deficiency of glutathione and the iminoquinone reacts with proteins, causing liver cell damage.

In some cases, CYPs catalyse the oxidation of drugs to their active forms. For example, codeine has relatively little pain-relieving activity in its ingested form; however, 15% of absorbed codeine is converted into the more potent pain-killer morphine by CYP2D6.[45,46]

The drug-metabolizing CYPs found in the human liver vary greatly in their selectivity, and react with a wide range of compounds. Drug molecules often contain several sites at which oxidation by a CYP could occur. The ability of the drug-metabolizing CYPs to react with many different compounds complicates the prediction of the site of oxidation, as their large active sites can often accommodate ligands in multiple orientations. Whilst oxidation at one particular site might deactivate a substrate, oxidation at another point on the same molecule might yield a toxic metabolite. An example of a substance that displays this type of reactivity is aflatoxin B1, a toxin produced by certain fungi and a substrate of CYP3A4.[47] Figure 11.2 shows the two possible pathways for the metabolism of aflatoxin B1. Hydroxylation at the 3α-position results in detoxification; however, epoxidation of the double bond at the 8,9 position results in a highly damaging epoxide. The prediction of formation of such metabolites from ingested compounds has the potential to help to avoid ADRs.

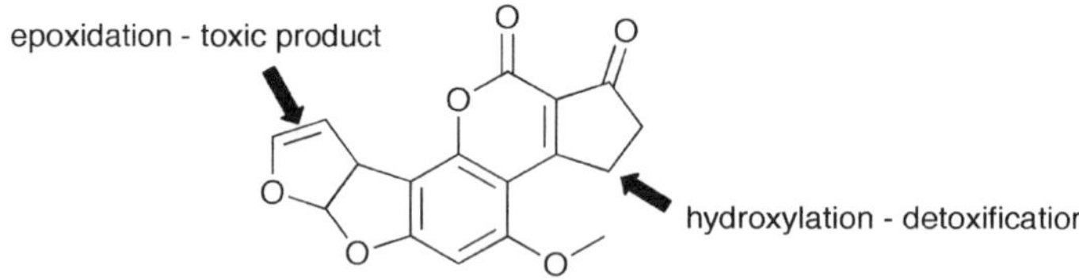

Figure 11.2   Sites of metabolism of aflatoxin B1 by CYP3A4. Epoxidation forms the toxic epoxide product, whereas hydroxylation results in detoxification.

Early stage *in vitro* testing of drug candidates in the discovery process for CYP interaction has reduced the number of drug withdrawals significantly.[48] Owing to the expense of these methods, *in silico* methods are of considerable interest for the pharmaceutical industry in the prediction of interactions (and potentially reactions) of CYPs with new chemical entities.[49,50] Several *in silico* methods are used for the prediction of CYP metabolism. These are generally split into four areas; ligand-based methods, structure-based methods, rule-based methods and reactivity-based methods, and differ by the amount of molecular detail involved and the theoretical approximations that are made.

Ligand-based methods include quantitative structure–activity relationships (QSAR) and pharmacophore models. QSAR methods aim to establish correlations between molecular structure, or other features, and observed properties. An example of such a correlation is the Hammett equation, a free energy relationship that relates equilibrium constants or rate constants with a set of parameters for many reactions involving benzene derivatives with different *meta-* and *para*-substituents.[51] A pharmacophore model is a set of features that a molecule requires to possess a certain biological activity. Pharmacophore development for a CYP isoform involves the determination of a set of features that are necessary for a substrate to be metabolized by that isoform. Typical features include hydrophobic regions, aromatic rings, hydrogen-bond donors/acceptors and cationic/anionic sites. Large databases of chemical compounds can be searched against a pharmacophore model for molecules that have the same features located a similar distance from each other.[52]

Rule-based methods make use of metabolic information for a large number of chemical entities that are compiled from the literature and stored in databases. By using computer-driven queries across these databases, the sites on a query molecule that are probably most prone to metabolic transformation can be identified.[53]

Structure-based methods make use of the structural data known about CYPs, such as X-ray crystal structures. Most of these use automated docking algorithms to place molecules in the active sites of CYPs. The binding affinities of the molecules can be calculated using an empirical scoring function.[54] These scoring functions typically ignore protein flexibility as well as the water molecules present in the active site. Simulation techniques, such as molecular dynamics, allow these factors to be taken into account.[55]

Reactivity-based methods predict the most likely site of metabolism of a substrate based on quantum chemical calculations. These calculations can provide valuable information on energy barriers involved in CYP catalysed reactions as well as geometries of intermediates and transition states.

11.3 Quantum Mechanical/Molecular Mechanical (QM/MM) Methods

Hybrid quantum mechanical/molecular mechanical (QM/MM) methods have become very important in the study of CYP reactivity in recent years.[55,56] QM/MM allows the exploitation of the advantages of both quantum mechanical (QM) and molecular mechanical (MM) approaches, and is applicable to large systems with thousands of atoms, such as enzymes. Several detailed reviews of QM/MM methods are available in the literature[57–60] and hence only a brief outline is provided here. The region where the chemical reaction takes place, along with a subset of the surrounding residues, is treated with a QM method, thus allowing calculations based on first principles to be carried out, at a high level of accuracy. For CYPs, the QM method of choice is usually density functional theory (DFT), and in particular the B3LYP functional, as it (generally) tends to reliably predict the electronic structure of key intermediates.[26,61,62] Semiempirical QM methods are currently not adequate enough to model these systems and correlated *ab initio* methods are too computationally expensive for optimization of geometries and calculation of reaction barriers. The good performance of B3LYP in QM/MM studies of CYPs has been confirmed by comparison with QM/MM single point calculations on Cpd I using *ab initio* multi-reference configuration interaction (MRCI) as the QM method.[63] An example of a typical QM/MM partitioning scheme for CYPs is shown in Figure 11.3. The remainder of the protein is modelled at the MM level of theory, providing appropriate structural restraints on the reaction region, as well as

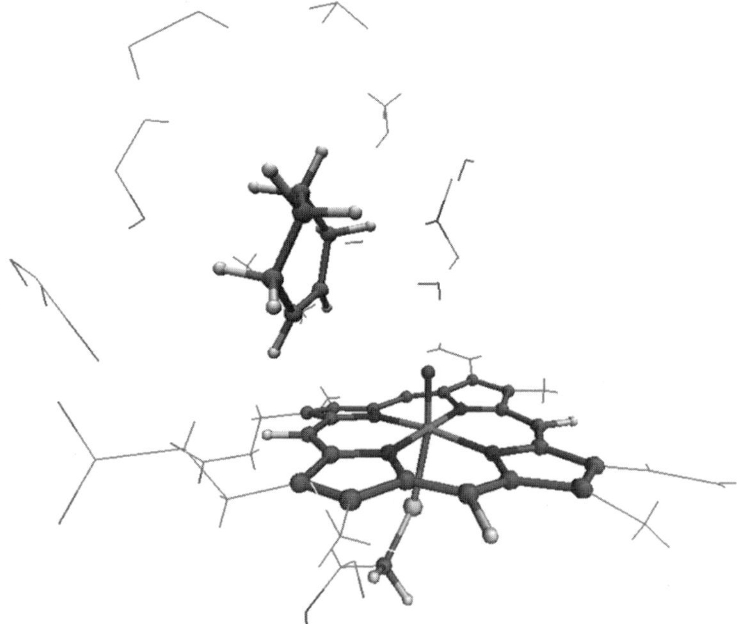

Figure 11.3 Typical division of the system into the QM (ball and stick) and MM (line) regions in a QM/MM calculation of CYP101 (CYP 450$_{cam}$). The majority of the enzyme has been omitted for clarity.

electrostatic and van der Waals interactions. The energy of the system can be calculated using the following equation:

$$E_{\text{Total}} = E_{\text{QM}} + E_{\text{MM}} + E_{\text{QM/MM}} \tag{11.2}$$

where E_{QM} and E_{MM} are the respective energies of the QM and MM subsystems and $E_{\text{QM/MM}}$ is the energy due to the interactions between the regions as described above.

Often, the division of a system into QM and MM regions requires that the boundary between the two regions intersects a covalent bond. There are three commonly used approaches to dealing with this issue. One approach is the frozen orbital method, where the continuity of electron density at the boundary is maintained by a frozen orbital along the bond between the QM and MM atoms.[64] The frozen orbital is derived from model compounds. The second method is a more generalized form of the frozen orbital approach, where a hybrid sp^2 orbital containing one electron is inserted at the QM/MM boundary.[65] The third method is to use "link atoms" to satisfy the valences.[66] These are typically hydrogen atoms, and some of the interactions between these link atoms and the MM region are usually neglected. Generally, the positions of the link atoms are chosen such that they do not cut across polar bonds, to avoid unrealistic effects. To avoid the introduction of unbalanced charge interactions at the boundary between the QM and MM regions, the charges of MM groups adjacent to QM groups are commonly set to zero.

The QM region typically consists of between 50 and 100 atoms. Geometry optimization of snapshots and points on potential energy surfaces is feasible, but MD simulations on CYP systems (typically with thousands of steps) are too computationally demanding with DFT-based QM/MM methods to be generally practical at present. When modelling a chemical reaction, whether in a protein, the gas phase or in solution, optimized structures are required for key intermediates and transition states. Optimization of minimum energy structures is relatively straightforward, but optimization of transition states is more difficult. Several algorithms exist for transition state optimization of small molecules in the gas phase; however, these are generally not ideally suited for large systems such as enzymes due to computational expense and convergence issues. One method used in modelling approximate reaction pathways is the "adiabatic mapping" approach.[67] A reaction coordinate is selected, for example, the distance between two atoms where the bond is being broken during the reaction step being considered. The energy of the system is minimized several times whilst holding this reaction coordinate at fixed values, and an energy profile is thereby obtained. Discontinuities in energy profiles can occur with this procedure, due to the large number of similar conformational states accessible to a protein. Where discontinuities occur, profiles must be followed in the forward and reverse directions until a converged energy profile is obtained. To sample the conformational space available to the enzyme, multiple profiles are commonly obtained, using different snapshots taken from MM molecular dynamics simulations, and average barriers are calculated.[68]

11.4   QM/MM Studies of CYPs

Owing to its importance in drug metabolism, CYP reactivity has been studied by several groups using QM/MM methods.[55,56,69] The remainder of this chapter will focus on applications of QM/MM methods in the study of CYP enzymes.

11.4.1   Catalytic Cycle of CYP101 (CYP 450$_{cam}$)

CYP101 (CYP 450$_{cam}$) was the first CYP to have its crystal structure solved, and although it is only found in bacteria, much of what has been learnt about reactivity in CYPs has come from studying this particular isoform. Intermediates at each step of the catalytic cycle of CYP 450$_{cam}$ (Figure 11.1) have been studied using QM/MM methods. Some of these studies are highlighted below. As crystal structures are available for most intermediates in the catalytic cycle of CYP 450$_{cam}$,[12] most of these studies have been performed using a different crystal structure as starting point for each step. Thiel *et al.* have since shown that the entire catalytic cycle of CYP 450$_{cam}$ can be modelled using a single, common crystal structure, because most of the changes that take place during the cycle are localized to the active site.[70] QM/MM studies of Cpd I, and of its role in the oxidation of substrates, will be covered in later sections.

11.4.1.1   Resting State (11.1)

The resting state of CYP 450$_{cam}$ has been studied using QM/MM calculations by Schöneboom and Thiel.[71] MM MD simulations were performed on the 1DZ4 crystal structure, with camphor replaced by six water molecules. Randomly chosen MD structural "snapshots" were optimized at the B3LYP/CHARMM22 level in the doublet, quartet and sextet spin states (Figure 11.4). The doublet spin state was found to be lowest in energy, in agreement with experiment.[20,72] An "upright" conformation of the distal water ligand was discovered to be preferred, rather than the "tilted" geometry previously suggested, in which the water protons form hydrogen bonds to the porphyrin nitrogen atoms (Figure 11.5). Comparison with QM-only calculations revealed

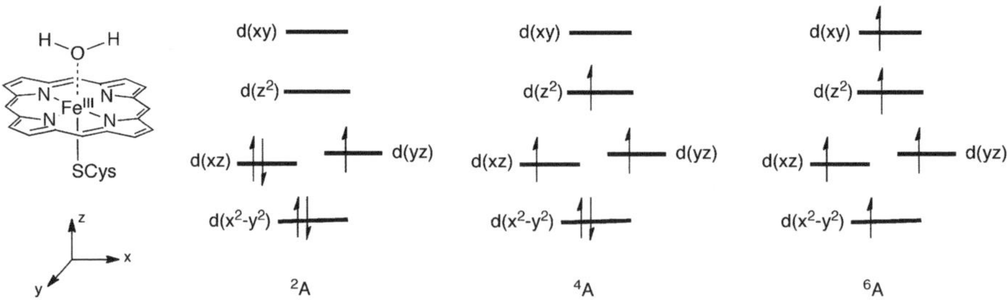

Figure 11.4   Orbital diagrams, indicative of the occupations in the doublet (^{2}A), quartet (^{4}A) and sextet (^{6}A) spin states, of the resting state (**11.1**) of CYP 450$_{cam}$ (shown on the left).

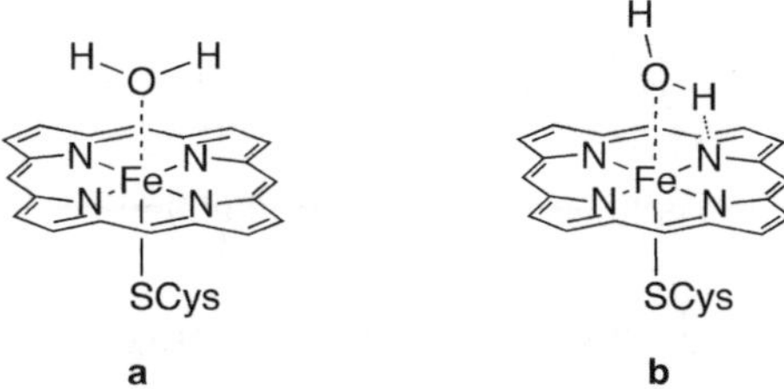

Figure 11.5 Upright (a) and tilted (b) orientations of the bound water molecule proposed for the resting state (**11.1**) in the CYP catalytic cycle. QM/MM calculations by Schöneboom *et al.* support the upright (a) orientation.[71]

that the protein and solvent environment reduces the doublet–quartet and quartet–sextet energy gaps.

11.4.1.2 Five-coordinate Fe(III) and Fe(II) Complexes (11.2 and 11.3)

The pentacoordinated ferric (**11.2**) and ferrous (**11.3**) intermediates for CYP 450_{cam} have both been identified experimentally as high-spin complexes.[9,17] Gas-phase Hartree–Fock calculations predict a ground state sextet spin state for the ferric complex,[73] but the finding at this level of a large doublet–sextet energy gap is inconsistent with the high-spin/low-spin equilibrium suggested by experiment. Friesner and Guallar have studied the ferric state (**11.2**) with B3LYP/OPLS QM/MM methods and found the quartet spin state to lie 2.0 and $5.6\,\mathrm{kcal\,mol^{-1}}$ lower than the sextet and doublet spin states, respectively.[74] Altun and Thiel modelled the ferric and ferrous complexes with B3LYP/ CHARMM.[75] The effect of the amount of exact Hartree–Fock exchange in the DFT functional on the relative energies of the spin states was investigated, by comparing B3LYP, BHLYP and BLYP energies. The B3LYP relative energies were found to be more consistent with the experimentally observed high-spin/ low-spin equilibria.

The binding of the substrate camphor in the active site of CYP 450_{cam} has been studied computationally by Swart *et al.*[76] A combination of gas-phase DFT calculations and QM/MM calculations was performed. In these calculations, a strong hydrogen bond was found between camphor and Tyr_{96}, and this interaction was proposed to explain the high degree of specificity in camphor hydroxylation. The geometry of camphor was found to be distorted in the protein, compared to the gas phase; however, this effect was found not to have a large impact on the strong hydrogen bonding interaction and has not been observed in other QM/MM studies with camphor bound in CYP 450_{cam}.[70,77–82]

11.4.1.3 Dioxygen Complexes (11.4 and 11.5)

The ferrous dioxygen (or ferric superoxo) complex (**11.4**) is diamagnetic and EPR silent, suggesting a singlet ground state.[83] There are two proposed models

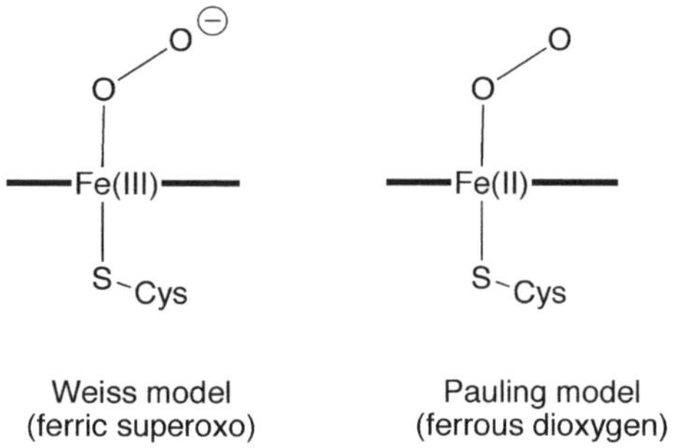

Figure 11.6 Weiss[85] (ferric superoxo) and Pauling[84] (ferrous dioxygen) models proposed for the electronic structure of **11.4**.

for the electronic configuration of **11.4**: the Pauling model[84] and the Weiss model.[85] The Pauling model consists of singlet Fe(II) bonded to singlet O_2. The Weiss model contains doublet Fe(III) bound to doublet superoxide $O_2^{-\bullet}$ (Figure 11.6). Previous DFT calculations predicted the Weiss model;[86] however, symmetry adapted cluster configuration interaction calculations predict the Pauling structure.[87] As these previous calculations were based on simplified models, and did not take the protein environment into account, it is possible that some essential features were not included. The oxyheme complexes of CYP 450_{cam} were subsequently studied by Wang and Thiel at the B3LYP/CHARMM level of theory, and their QM/MM calculations support a Weiss model.[88] The reduction potential for **11.4** was calculated with QM/MM methods and was found to vary between –1.93 and –2.28 V, depending on the size of the QM region (the latter energy corresponds to the inclusion of a distal water chain). The relative spin state energies support a singlet species for **11.4** and **11.5**, and higher spin states were observed not to be stable and dissociated, which is consistent with EPR experiments.

11.4.1.4 Compound 0 (*11.6*) and Formation of Compound I (*11.7*)

Zheng *et al.* have studied the formation of Cpd I (**11.7**) from Cpd 0 (**11.6**) using QM/MM methods.[89] Two mechanisms were proposed (Figure 11.7). In mechanism 1, a proton is transported to the distal oxygen atom *via* the hydrogen-bonding network provided by the protein. Heterolytic O–O bond cleavage then follows this step, to generate Cpd I and water. Mechanism 2 begins with O–O bond cleavage, to form an intermediate that undergoes proton-coupled electron transfer to form Cpd I and water. B3LYP/CHARMM calculations predict that mechanism 2 is favoured, and that Asp_{251} is the most likely proton source. Cpd I was found to be slightly lower in energy than Cpd 0. The rate-limiting step for Cpd I formation is believed to be the O–O bond cleavage step, with a barrier of *ca.* 14 kcal mol^{-1}.

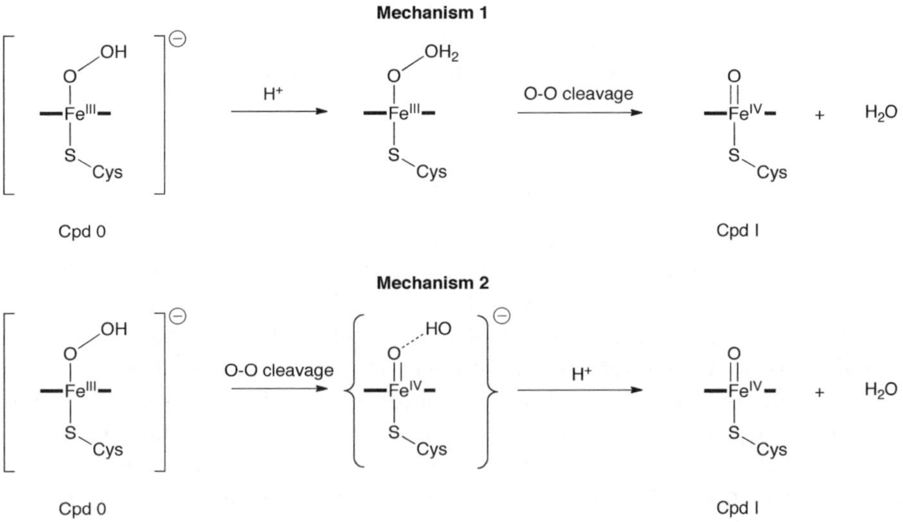

Mechanism 1

Mechanism 2

Figure 11.7 Proposed mechanisms for the two-step conversion of Cpd 0 into Cpd I studied by Zheng *et al.*[89] Mechanism 2 was found to be more energetically favourable at the B3LYP/CHARMM22 level of QM/MM theory.

11.4.1.5 Product Complex

Experimental evidence suggests that the product complex of CYP 450_{cam} consists of 5-*exo*-camphor coordinated to the heme iron atom, with a mixture of doublet and sextet spin-states.[17] Spectroscopic results identify that the immediate product formed after hydroxylation is transient, and that the heme pocket subsequently relaxes to allow the long Fe–O bond observed in the crystal structure (2.67 Å).[90] Schöneboom *et al.* modelled this latter species at the B3LYP/CHARMM22 level of theory, considering the doublet and quartet spin-states.[81] Their study supported a quartet ground state in which the Fe–O bond is broken (2.82 Å). The doublet state optimized to two minima, close in energy: one in which the Fe–O bond was long (2.82 Å) and the other in which it was short (2.26 Å). Lin *et al.* studied the product complex of CYP 450_{cam} at the B3LYP/CHARMM22 level of theory.[91] Potential energy surface scans were performed along the Fe–O bond distance for the doublet, quartet and sextet spin states. The first step in the product release was found to be the breaking of the Fe–O bond, before the product leaves the active site. Energies were calculated with the BLYP functional on the B3LYP/CHARMM22 geometries, as it was reported in this study that the B3 exchange functional has a tendency to overestimate the stability of high-spin states relative to low-spin states. B3LYP/CHARMM22 energies predict a quartet ground state, whereas BLYP/CHARMM22 energies correctly predict a doublet ground state. However, it should be stressed here that the effect is small and will not affect the overall mechanism for camphor hydroxylation. Two minima are observed on the doublet spin surface, corresponding to short and long Fe–O distances, and are linked by a flat potential energy surface with a very small barrier. Calculations were also performed in the absence of the protein environment and it was shown that the energy cost of Fe–O bond dissociation is lowered in the enzyme, compared to the gas phase.

11.4.1.6 *Electronic Structure of Compound I (11.7)*

The active oxidizing species of CYP enzymes, Cpd I, has only been isolated very recently.[27] Until this time, only indirect experimental evidence and theoretical studies supported its role in CYP chemistry.[92–95] Cpd I has three unpaired electrons located in closely-lying molecular orbitals (Figure 11.8). Gas-phase calculations on a small QM model of Cpd I revealed that two of the unpaired electrons lie in the π^*_{yz} and π^*_{xz} orbitals of the Fe–O moiety.[26,96] The remaining unpaired electron is distributed between the a_{2u} orbital on the porphyrin moiety and the lone pair $p_\pi(S)$ orbital of the cysteinate sulfur. The spin of this electron couples to the $S = 1$ spin of the Fe–O moiety, giving rise to closely lying doublet and quartet spin states.[26] The distribution of the third electron between the a_{2u} and $p_\pi(S)$ orbitals varies depending on the molecular model used to represent Cpd I. QM-only studies using a mercaptide or cysteinate anion in the absence of its hydrogen-bonding environment predict the third unpaired electron to remain exclusively on the sulfur atom, giving rise to $^{4,2}\Pi_S$ states. The $^{4,2}\Pi_S$ states are inconsistent with experimental studies of

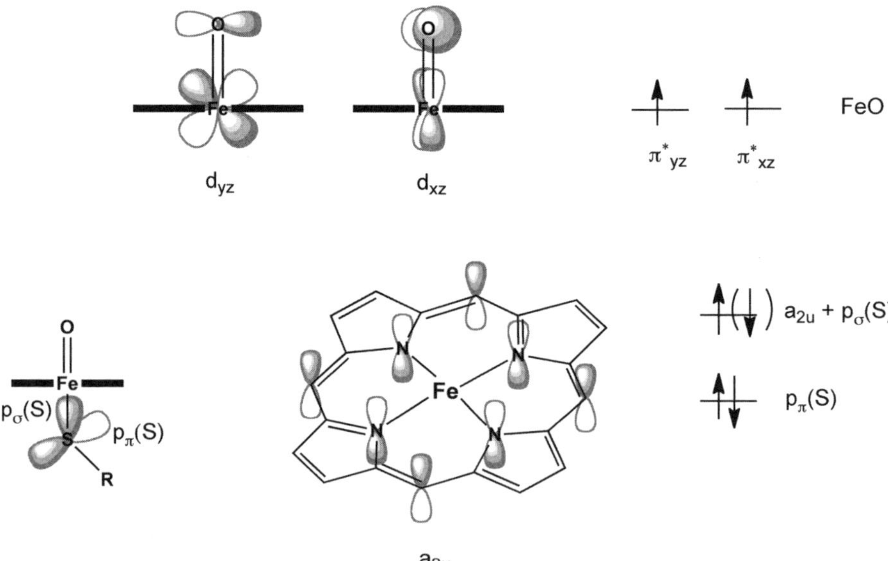

Figure 11.8 Frontier molecular orbitals for the quartet spin state of Cpd I. The heme substituents are omitted for clarity. The alternative electron spin corresponding to the doublet spin state is shown in parentheses.

Cpd I of chloroperoxidase, which is green in colour, consistent with $^{2,4}A_{2u}$ states.[97] By analogy with chloroperoxidase, which like CYPs is a heme-thiolate enzyme, Cpd I of CYPs is also believed to have a $^{2,4}A_{2u}$ ground state.[92]

The inclusion of NH–S hydrogen bonds in QM calculations on small models increases the amount of spin density distributed over the porphyrin ring. These added hydrogen bonded groups are designed to mimic the hydrogen bonding environment of the cysteinate ligand by the surrounding protein amino acids. The introduction of a polarizing field has a similar effect.[98] The Fe–S bond length is very sensitive to the environment and treatment of the cysteinate ligand and shortens by 0.1 Å when hydrogen bonding and polarization are included in DFT QM calculations on small models.[98] These observations first led to the suggestion that Cpd I acts as a chameleon species that changes its electronic structure according to its environment.[99]

As the electronic structure of Cpd I has been found to be so sensitive to the methodology used in the small model QM studies mentioned above, it follows that a theoretical treatment, such as QM/MM, that can include the electrostatic, steric and hydrogen bonding interactions of the enzyme, will provide a more definitive system setup. The electronic structure of Cpd I has been studied with QM/MM methods by several groups. Schöneboom *et al.* studied Cpd I in CYP 450$_{cam}$ at the B3LYP/CHARMM level of theory.[100] Three snapshots were chosen from an MD simulation derived from the 1DZ9 crystal structure.[12] This crystal structure was initially identified as that of Cpd I; however, further experimental studies have shed doubt on this claim. Schöneboom *et al.* compared the electronic structure of Cpd I in the QM/MM calculations with that in the gas phase and found the Fe–S bond length to be significantly shortened in the enzyme compared to the gas phase. The doublet and quartet spin states were found to lie within $0.2\,\text{kcal}\,\text{mol}^{-1}$ of each other in the enzyme, with the doublet the lowest in energy, in agreement with previous gas-phase calculations.[98] The unpaired spin density on the Fe–O unit was found to be insensitive to the MD snapshot chosen and the QM model used for the cysteinate ligand (SH^-, SCH_3^- and $SCys^- + $ backbone of Leu$_{356}$ and Leu$_{358}$). The Fe–S bond length and sulfur spin density were found to be sensitive to these perturbations. The third unpaired electron was distributed over both the porphyrin and cysteinate sulfur, in contrast to the gas-phase calculations. It is believed that the actual Cpd I species is a resonance form of two states, the first corresponding to a thiolyl radical and closed-shell porphyrin, with a long Fe–S distance. This state is the lowest in energy in the gas phase; however, in the protein environment, this species becomes close in energy to another state, in which an electron is transferred from the porphyrin to the sulfur, resulting in a negatively charged sulfur and a positively charged porphyrin. This latter state is stabilized by hydrogen bonding interactions to the sulfur. Schöneboom *et al.* proposed that Cpd I behaves as a "chameleon" species that adapts its electronic structural character to the specific environment.[100]

The electronic structure of Cpd I has also been modelled in human CYP isoforms. Bathelt *et al.* modelled Cpd I in three human drug-metabolizing CYPs (2C9, 2B4 and 3A4) at the B3LYP/CHARMM level, and also

CYP 450$_{cam}$ as a comparison.[101] The electronic structure of Cpd I was found to be remarkably similar in all isoforms, with the third unpaired electron located predominantly on the porphyrin ring. Calculations were performed with larger QM regions that incorporated more of the proximal cysteine and porphyrin side chains but little change was observed between the different model systems. No unpaired electron density was found on the heme propionates, in contrast to the findings of Guallar *et al.* in their QM/MM study of CYP 450$_{cam}$.[82] Cpd I was also shown not to be sensitive to other details of the QM/MM methodology, such as DFT functional and basis set. A small amount of variability in the spin density on the cysteinyl sulfur (26 to 50%) was observed due to the starting protein geometry. This conformational effect was shown to be larger than the difference between isoforms, leading to the conclusion that isoforms are unlikely to be distinguishable on electronic grounds. The most significant difference in Cpd I was between the human CYPs and the bacterial CYP 450$_{cam}$; however, two calculations on the same isoform using different MD snapshots were found to yield similar differences in Fe–S bond length and sulfur spin density. A small change was also found when comparing an isoform in the presence and absence of a substrate molecule in its active site cavity. It was suggested from this work that the differences in electronic structure and Fe–S bond length were due to thermal motions of the protein, substrate and solvent. These motions cause the hydrogen bonding environment and electric field surrounding Cpd I to fluctuate, but are expected to make no net difference between different isoforms.

The effect of the presence of substrate in the active site of CYPs on the electronic structure of Cpd I has also been investigated by others. Fishelovitch *et al.* modelled Cpd I in CYP3A4 using various models that were either substrate-free or contained one or two substrate molecules.[102] Three snapshots from (MM) MD were structurally optimized at the QMMM level for each system. The authors observed that in the absence of substrate the Fe–S bond tended to elongate, thus localizing the unpaired electron on the sulfur. In the presence of the substrate diazepam close to Cpd I, the NH–S interactions with the cysteinate ligand were found to be strengthened, shortening the Fe–S bond and maintaining the delocalization of the unpaired electron over the porphyrin ring and cysteinate sulfur.

11.4.2   Hydroxylation of Camphor by CYP 450$_{cam}$

Hydroxylation of camphor by CYP 450$_{cam}$ has become an important test case in the study of CYPs with QM/MM techniques. To utilize QM/MM methods for predictive purposes, *e.g.* potential formation of toxic metabolites during drug metabolism, it must first be established whether it can replicate existing experimental observations. CYP 450$_{cam}$ is found in the bacterium *Pseudomonas putida*, and is responsible for the stereospecific hydroxylation of camphor at the 5-*exo* position, and is part of the bacterium's energy cycle (Scheme 11.1). The two electrons required for oxidation are supplied by putidaredoxin, a protein

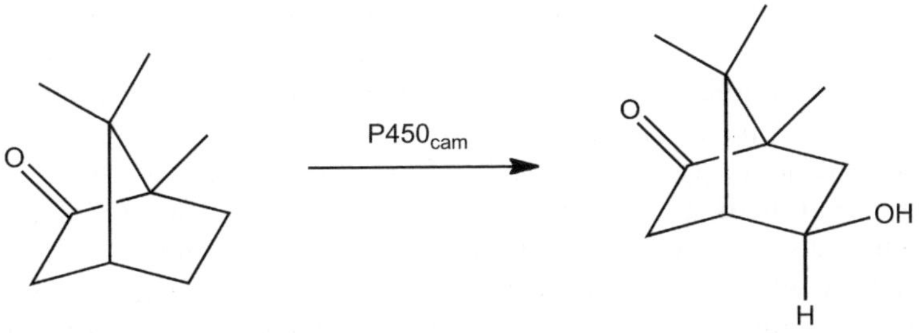

Scheme 11.1 Hydroxylation of camphor by CYP 450_{cam} at the 5-*exo* position.

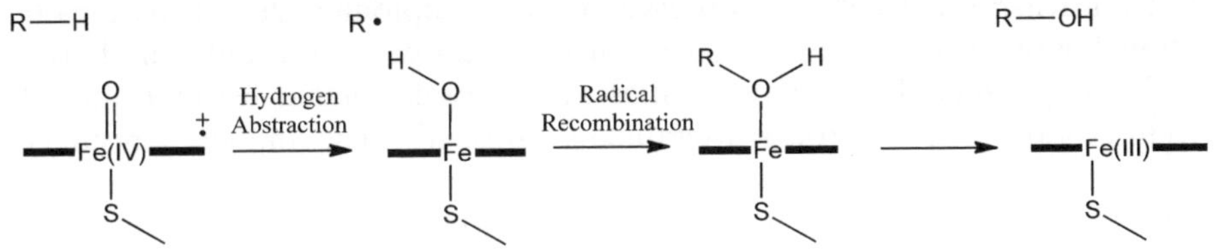

Figure 11.9 "Rebound" mechanism for alkane hydroxylation by CYPs.[103]

containing a 2Fe-2S cluster, and the flavoprotein putidaredoxin reductase. The selectivity of CYP 450_{cam} is controlled by the enzyme environment, as the substrate is anchored into an orientation favouring 5-*exo*-hydroxylation, by a hydrogen bond formed between the carbonyl oxygen of camphor and the hydroxyl group of Tyr_{96}.

Camphor hydroxylation follows the "rebound" mechanism proposed by Groves and McClusky (Figure 11.9).[103] This mechanism is supported by the kinetic isotope effects observed with deuterium labelled substrates, which indicate that the rate-limiting step involves breakage of a C–H bond. The rebound mechanism is further supported by the loss of stereochemistry observed in CYP-mediated hydroxylation, as a concerted process would result in retention of configuration about the carbon atom at which hydroxylation occurs.[9,104]

Calculations using the B3LYP density functional on a model system of ethane with Cpd I show that the initial hydrogen abstraction step is probably rate-limiting, with the doublet and quartet surfaces having similar barrier heights.[104] The doublet pathway proceeds through the subsequent radical recombination step with little or no energetic barrier, whereas on the quartet surface this step has a barrier of $4.9\,\text{kcal}\,\text{mol}^{-1}$. Hence the quartet pathway is thought to account for the loss of stereochemistry in some CYP-mediated hydroxylation reactions, as this barrier allows time for radical rearrangement to take place.

The first (and rate-limiting) step for camphor hydroxylation is hydrogen atom abstraction of the 5-*exo*-hydrogen by the Cpd I ferryl oxygen. In this step an electron is transferred from camphor to Cpd I, to form a radical on C^5 of camphor. The subsequent rebound step involves bond formation between the C^5 of camphor and the ferryl oxygen, forming the product complex.

Oxidation of camphor in CYP 450_{cam} has been modelled by several groups, using small gas-phase QM models and QM/MM calculations.[78–82,105] The hydrogen abstraction step was found to have a barrier of *ca.* $20\,\text{kcal}\,\text{mol}^{-1}$ in the gas phase at the B3LYP/6-31 + G** level of theory.[105] Several discrepancies have been observed between the QM/MM modelling studies. Guallar *et al.* observed that when the heme propionate groups were included in the QM region a considerable amount of unpaired electron density was found on one of the propionate side chains.[82] This side chain was considered to be partially oxidized, and hence postulated to play an important role in the reactivity of Cpd I. This observation was questioned, as such high amounts of unpaired electron density had not been observed on the propionates in calculations by Schöneboom *et al.*[81] Further uncertainty arose due to the different barrier heights and hence different reactivities that were calculated between the different groups. The two groups attempted to reconcile their conflicting results in a joint paper.[78]

Żurek *et al.* embarked on modelling the hydroxylation of camphor independently.[80] The key issue to address was whether the differences that were observed between the calculated results in the previous studies[81,82] were due to a difference between the methods or to some detail in the setup of the calculations. Calculations on enzymes are complicated – an incorrect assignment of a protonation state of an amino acid residue can sometimes lead to large inaccuracies in optimized geometries and hence energies. The side chain of Asp_{297} is close to one of the propionate chains of the heme group in the 1DZ9 crystal structure of CYP 450_{cam} (Figure 11.10), a feature that seems to be unique to this isoform. As X-ray crystal structures do not usually contain information about the location of protons, the protonation state of Asp_{297}, as in the case of

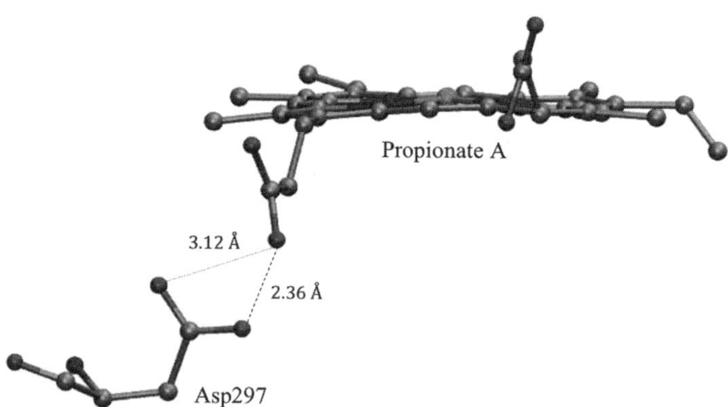

Figure 11.10   Proximity of Asp_{297} to the heme propionate observed in the crystal structure of CYP 450_{cam} (PDB code 1DZ9).[12] The distances shown are between the carboxylate oxygen atoms of Asp_{297} and the closest propionate oxygen. The short O–O distance indicates a hydrogen bond.[80]

other titratable amino acids, must be inferred from pK_a calculations, as well as inspection of local hydrogen bonding environments. Aspartic acid residues are usually ionized at physiological pH, and hence the automatic assignment would be to have Asp_{297} deprotonated at both carboxylate oxygen atoms. The close proximity of the heme propionate group, however, suggests that Asp_{297} should be protonated, resulting in a stabilizing hydrogen bond to the propionate. In the work of Guallar *et al.*,[82] Asp_{297} was ionized and not protonated.

Żurek *et al.* modelled the hydroxylation of camphor using two systems: one in which Asp_{297} was ionized and the other in which it was protonated. Initial MD simulations on the two systems revealed that in the ionized system the interaction between the two negatively charged groups lead to a distortion of the protein environment surrounding the propionate and Asp_{297}. In the system where Asp_{297} was protonated, the geometry of this region remained close to that of the crystal structure throughout the MD simulation. Three QM/MM reaction pathways were generated for the ionized Asp_{297} system, in which the propionates were included in the QM region. A small amount of electron density was observed on the propionates, as observed by Schöneboom *et al.*,[81] and a much smaller amount than observed by Guallar *et al.*[82] The setup protocol used by Żurek *et al.* involved deriving starting geometries for QM/MM from MD simulations, whereas Guallar *et al.* used the crystal structure directly. This difference may help to explain the difference in the observed amounts of propionate spin density between the two groups. No unpaired electron density was observed on the propionates in the model system in which Asp_{297} was protonated. The average calculated barriers for hydrogen abstraction were found to be 15.3 and 18.3 kcal mol^{-1} for the ionized and protonated models, respectively. This difference of 3.0 kcal mol^{-1} can probably be attributed to slight differences in the conformation of the enzymes, due to the difference in setup. This work is consistent with that of Schöneboom *et al.*,[81] showing that, even with different setup procedures, careful QM/MM work can lead to reproducible results.

11.4.3 Compound I Reactivity and Selectivity

CYPs catalyse a wide variety of reactions and can oxidize many different functional groups.[1] The prediction of which site in a drug candidate will undergo oxidation by CYP is an important problem to address, particularly in the area of drug metabolism. Oxidation at a particular site of a substrate may result in detoxification, whereas oxidation at an alternative position could result in a toxic response. Most models that are currently used to predict the selectivity of metabolism by CYPs are based on the relative accessibility of the different parts of the substrate molecule to the active species of Cpd I, rather than the reactivity of the different functional groups.[1] For example, the anti-inflammatory drug diclofenac undergoes hydroxylation by CYP2C9 at the 4′ position whereas metabolism of diclofenac by CYP3A4 occurs at the more reactive 5 position. This difference in selectivity has been suggested to be due to an arginine residue in the active site of CYP2C9 anchoring diclofenac in a

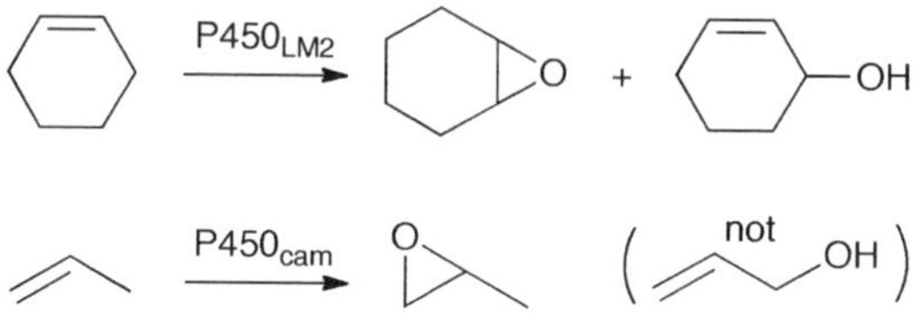

Scheme 11.2 Oxidation of cyclohexene and propene by the bacterial CYPs CYP 450_{LM2} and CYP 450_{cam}. Cyclohexene oxidation yields both epoxidation and hydroxylation products,[111–113] whereas propene oxidation only yields epoxidation products.[114]

position favouring oxidation by Cpd I at the 4' position.[106,107] There are numerous examples of where substrates can adopt multiple binding positions in the active sites of CYPs.[108] An example of this is the drug nicotine, which has been shown to display more than one binding position in the active site of CYP 450_{cam}.[108,109] For such cases the relative reactivity of the accessible functional groups of the substrate to Cpd I must be taken into account.

The oxidation of cyclohexene and propene by CYP 450_{cam} have been modelled using QM/MM methods.[68,110] The oxidation of cyclohexene by CYP 450_{LM2} results in formation of C=C epoxidation and allylic hydroxylation products with an approximate 2 : 1 preference for epoxidation.[111–113] In contrast, the oxidation of propene results in formation of propene-1-oxide, with no hydroxylation products observed (Scheme 11.2).[114] Whilst the experimental data correspond to two different isoforms, the relative size of both substrates is small with respect to the active site of CYP 450_{cam}, suggesting that the two enzymes will produce similar relative product yields.

Cases where selectivity effects are small, such as this, provide a stringent test for computational approaches. One would expect that the relative barriers obtained with QM/MM methods would reflect the experimentally observed selectivity, *i.e.* a lower barrier for epoxidation compared to hydroxylation in propene and roughly equal barriers in the case of cyclohexene. Cohen *et al.* modelled the epoxidation and hydroxylation of propene and cyclohexene by CYP 450_{cam}, and obtained barriers that did not agree with this expected result.[110] The authors obtained energy barriers that favoured the hydroxylation of propene and cyclohexene by 2.7 and 6.8 kcal mol^{-1} respectively, concluding that the relative energy barriers for oxidation of propene and cyclohexene obtained by QM/MM are not good predictors of the observed selectivity. The experimentally observed selectivity was instead attributed to the relative rates of product release. Cyclohexenol and cyclohexene oxide products were postulated to bind to the heme roughly equally strongly, which the authors suggested could account for the observed product ratio of C=C/C–H $\approx$ 1. For propene, however, the alcohol was calculated to bind more strongly to the heme than the epoxide and hence there is a preference for formation of propene-1-oxide. Since (for CYP 450_{cam}) the rate-limiting step in the catalytic cycle is the second reduction step (**11.4**$\rightarrow$ **11.5**), Cohen *et al.* postulated that the barrier that should be used to determine

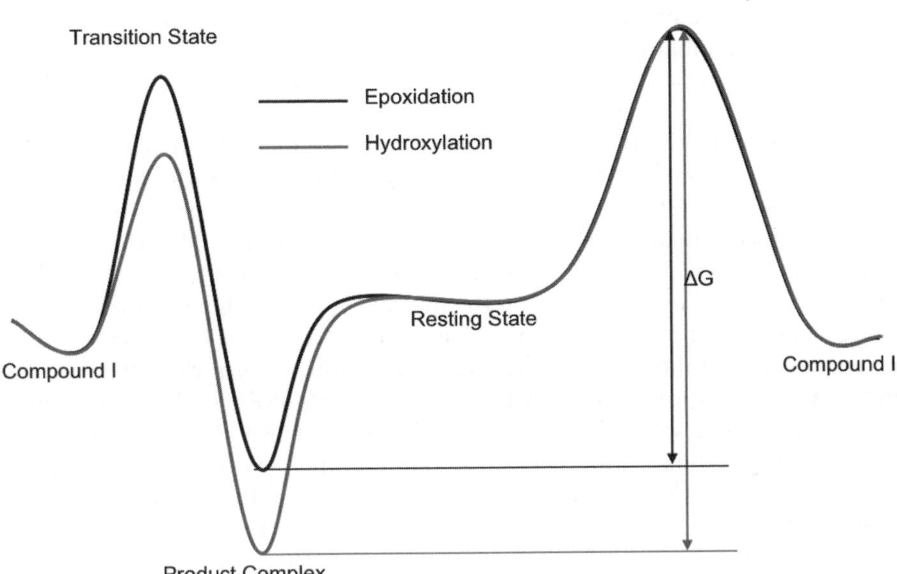

Figure 11.11   Free energy surfaces illustrating the proposal of Cohen *et al.*[110] for product release determining selectivity. Recent QM/MM calculations[68] indicate that this mechanism is not applicable to cyclohexene and propene oxidation in CYP 450$_{cam}$.

selectivity should be the difference in energy between the highest and lowest points on the free energy surface, *i.e.* between the electron transfer step (the same for both pathways) and the enzyme–product complex (Figure 11.11).

The theory surrounding the product-release effect on selectivity cannot be correct as it assumes that the two reaction cycles involving hydroxylation and epoxidation are disconnected, *i.e.* that a CYP releasing an epoxide can then only go on to form another epoxide (and similarly for hydroxylation). This assumption is not valid because the product leaves the active site cavity, making space for a water molecule which regenerates the resting state,[19] prior to entrance of the subsequent substrate molecule. The widely accepted view is that epoxidation and hydroxylation occur *via* a common intermediate when the substrate enters the protein active site.

An alternative explanation for the discrepancy between the QM/MM energy barriers obtained by Cohen *et al.* and the experimental product formation ratio is incomplete convergence of the computational results. In turn, this could be due to choice of starting structure from MD simulations for QM/MM optimization. Energy barriers that are calculated from starting snapshots in which the alkene is in an unfavourable position for reaction will naturally lead to higher barriers, as the two atoms between which a bond is being formed will have further to move to form a bond. If this positioning effect discriminates between the two types of reaction, it could account for the poor agreement with experiment. Propene and cyclohexene do not have any hydrogen bonding groups, and are small in comparison to the size of the active site of CYP 450_{cam}, and are hence expected to undergo free tumbling when in the active site, thereby being able to adopt structures favourable to both reactions. An independent study was hence performed by our group, in which cyclohexene and propene oxidation was modelled with QM/MM methods. Initial structures used to model both reactions were chosen from a subset of snapshots selected from MD simulations such that the substrates were in a position that resembled the transition state.[68] As expected, such positions were adopted frequently in the unconstrained MD simulations.

The mechanism for the hydroxylation of alkenes by Cpd I is the same "rebound" mechanism observed in the hydroxylation of alkanes (Figure 11.9). The first and rate-limiting step is the abstraction of an allylic hydrogen from the alkene, which is followed by radical recombination to form the allylic alcohol product. Epoxidation of alkenes by Cpd I proceeds *via* C–O bond formation between an alkene sp^2 carbon and the Cpd I ferryl oxygen, leaving a radical on the other sp^2 carbon atom. This is the rate-limiting step for the reaction; the barrier has been calculated to be around $12\,kcal\,mol^{-1}$ for propene in a small model DFT study by de Visser *et al.*[115] The subsequent step is ring formation between the carbon radical and the ferryl oxygen (Figure 11.12).

In our initial study of propene and cyclohexene oxidation, MD simulations were first performed to address whether free tumbling of cyclohexene and propene occurs in CYP 450_{cam}. Two 5 ns stochastic boundary MD simulations were performed, using the 1DZ9 crystal structure, with propene and cyclohexene docked by hand. Free tumbling of the substrate molecules was indeed observed. Angles and distances that are expected to correlate to pre-reactive conformations for hydrogen abstraction and C–O bond formation we monitored during MD, and were found to span a broad range of values. Snapshots were chosen for QM/MM adiabatic mapping profiles based on the fairly large subsets in which these angle and distance values were close to those observed in the relative transition state geometries. A large range of barrier heights was observed for each process, spanning values in some cases that range from 8 to $25\,kcal\,mol^{-1}$ despite the use of snapshot selection criteria. The lowest barriers that were observed are likely to be most representative of the actual enzyme reactivity; hence an averaging scheme incorporating the Boltzmann equation was used to provide a "best estimate" of the overall reaction barriers for both

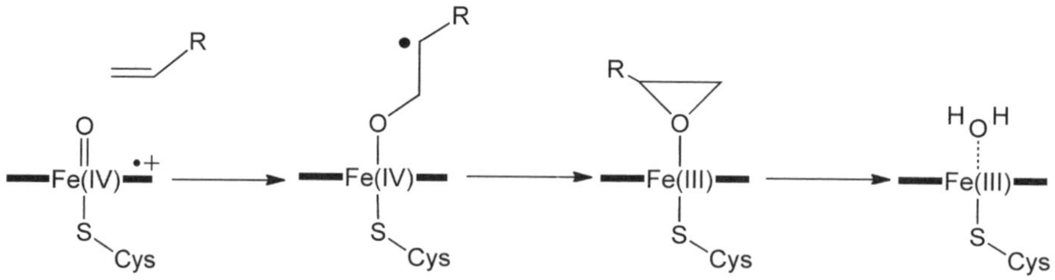

Figure 11.12 Mechanism for epoxidation of alkenes by Cpd I.[116]

processes. These barriers displayed a much-improved agreement with experiment compared with the previous QM/MM study.[110] This particular system highlights the complexity involved in QM/MM studies of enzymes, and stresses the need for conformational sampling during setup.

11.4.4   Aromatic Hydroxylation

The metabolic activity of CYPs has long been associated with liver dysfunction and carcinogenesis.[1] This toxicity is due to the ability of CYPs to activate various substrates to form reactive intermediates. Aromatic hydroxylation is one of the most problematic cases, where, in addition to phenol-type products, highly reactive intermediates such as arene oxides may be generated.[117] Aromatic rings in drug molecules typically contain multiple sites at which oxidation may occur, leading to different intermediates with varying reactivities. Some of these intermediates can form covalent bonds with macromolecules such as DNA or proteins, which may lead to events causing a toxicological response. An example of a compound in which the position of aromatic hydroxylation dictates the toxicity of the metabolite formed is bromobenzene. Hydroxylation at the 4-position is linked with toxicity, in contrast to the competing hydroxylation at the 2-position.[118] There are many examples of commonly administered drugs containing aromatic rings, such as the sedative diazepam, the stimulant ritalin, and many non-steroidal anti-inflammatory drugs such as ibuprofen, naproxen and codeine (Figure 11.13). As a

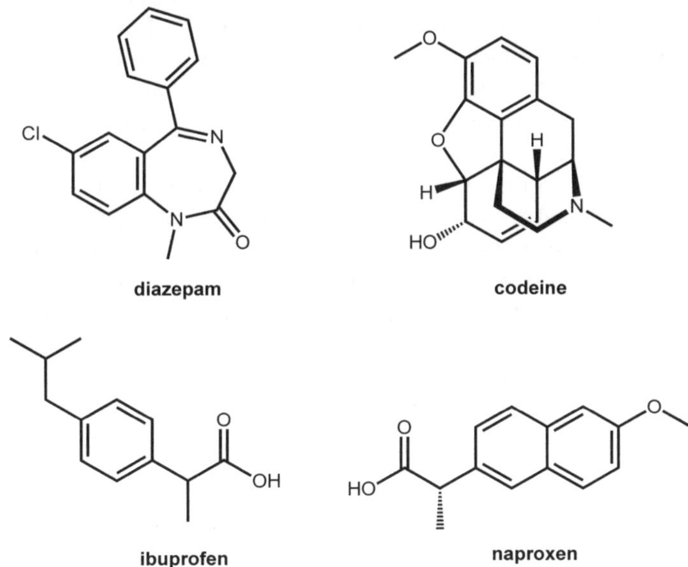

Figure 11.13   Examples of pharmaceutical compounds that contain aromatic rings. CYP-mediated oxidation is the main metabolism route for all of the compounds shown.

consequence, a better understanding of the mechanism of aromatic hydroxylation in drug molecules is useful for pharmaceutical research and drug development.

Aromatic hydroxylation of substrates by CYPs has been the subject of study by several groups.[118–120] As with aliphatic hydroxylation, Cpd I is the active oxidizing species responsible for aromatic hydroxylation.[26] Hydrogen atom migration from the site of hydroxylation to the adjacent carbon, the "NIH-shift", has been observed by means of deuterium labelled substrates.[121] A wide variety of substituents also migrate in this manner, including halogens and alkyl groups.[117,122] This observation is not consistent with a "rebound" type mechanism found for aliphatic hydroxylation, neither is it consistent with direct oxygen atom insertion into the C–H bond, as these mechanisms do not allow for migration of substituents.

The addition–rearrangement mechanism is the most commonly accepted mechanism for CYP-catalysed aromatic hydroxylation.[123,124] In this mechanism, hydroxylation proceeds *via* a tetrahedral σ-complex, formed upon addition of Cpd I to the aromatic carbon. Evidence for the addition–rearrangement mechanism originates from the isotope effects observed in the hydroxylation of a series of selectively deuterated chlorobenzenes,[124] which rule out initial epoxide formation as a possible mechanism for aromatic hydroxylation with these substrates.

As with most other theoretical studies on CYPs, prior to any QM/MM calculations, aromatic hydroxylation was modelled in the gas phase.[29,120,125] In the gas phase, several mechanisms for benzene hydroxylation were considered and the electrophilic addition pathway, to form an intermediate tetrahedral σ-complex, was found to be the most favourable.[120] Two-state reactivity is observed, *i.e.* reaction takes place on both the quartet and doublet spin surfaces, as found with aliphatic hydroxylation. The σ-complex can have radical or cationic character; however, a preference for a cationic σ-complex was observed (Figure 11.14). The σ-complex can rearrange *via* competitive pathways to form several products, including the epoxide, phenol or ketone. These results were confirmed in a subsequent independent study.[29,125] In the latter study, the orientation of the benzene ring relative to the porphyrin ring was found to have an effect on the barrier to electrophilic addition. Addition of benzene to Cpd I in a side-on orientation (with the benzene ring perpendicular to the porphyrin) was found to be more favourable than face-on addition (with the benzene ring parallel to the porphyrin – see Figure 11.15). Activation barriers were also calculated for a series of substituted benzenes, at the *meta*- and *para*-position. Structure–activity relationships were determined: calculated activation barriers were shown to be correlated to a combination of (experimentally derived) Hammett σ-constants and a theoretical scale based on bond dissociation energies of hydroxyl adducts of the substrates.[125]

The effect of the enzyme environment on aromatic hydroxylation was investigated by modelling the hydroxylation of benzene in CYP 2C9 with QM/MM methods.[126] Side-on and face-on modes[125] of benzene approach were modelled, in different conformations, generated from (MM) MD simulations.

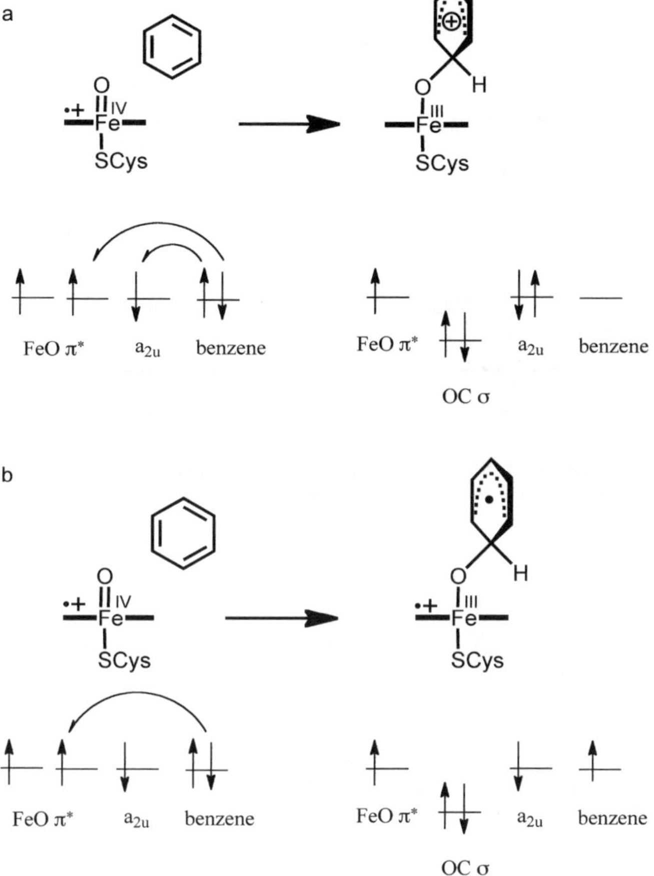

Figure 11.14   Mechanism for the addition of Cpd I to benzene to form a tetrahedral σ-adduct with (a) cationic and (b) radical character.[120,125,126]

For the QM/MM system, it was found that the side-on and face-on activation barriers are of comparable energy, suggesting that there is no preferential orientation in the protein environment. The QM/MM barrier heights were calculated in the range 18.0–21.7 kcal mol^{-1} and the optimized σ-complexes were found to be more radical-like than cation-like. The rearrangement pathways from the tetrahedral σ-intermediate to benzene epoxide, ketone and phenol were also modelled (Figure 11.16). In the face-on pathway, formation of the epoxide and ketone products had very similar barrier heights, suggesting that both epoxide and ketone products can form on this pathway. In the case of the side-on pathway, epoxide formation was found to be more favourable than ketone formation. An additional rearrangement pathway was found for the side-on orientation, in which a proton is transferred to a porphyrin nitrogen atom, which can be followed by a rearrangement to form the phenol product.

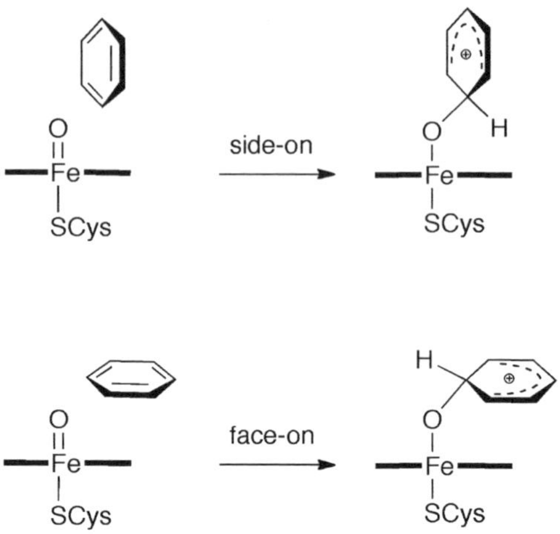

Figure 11.15 Schematic representation of the alternative side-on and face-on pathways for addition of benzene to Cpd I.[125,126]

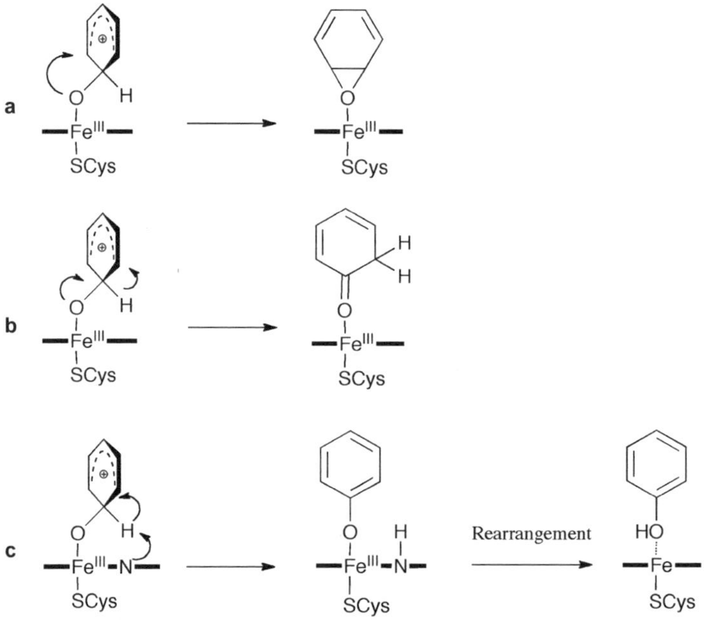

Figure 11.16 Rearrangement pathways from the tetrahedral σ-intermediate formed on addition of Cpd I to benzene for (a) epoxide formation, (b) ketone formation and (c) proton shuttle *via* a porphyrin nitrogen atom. These pathways were modelled in QM-only[125] and QM/MM[126] studies at the B3LYP and B3LYP/CHARMM22 levels of theory, respectively.

This work suggests that aromatic hydroxylation may proceed *via* several pathways in the enzyme environment, potentially resulting in several different products, including an epoxide.

11.4.5   Other QM/MM Studies of CYPs

11.4.5.1   CYP102A1 (CYP 450$_{BM3}$)

CYP102A1 (also known as CYP 450$_{BM3}$), isolated from the soil bacterium *Bacillus megaterium*, catalyses the hydroxylation of several long-chain fatty acids.[127] In contrast to CYP 450$_{cam}$, crystal structures of CYP 450$_{BM3}$ in complex with substrate do not contain the substrate in an orientation in which the site of oxidation is close to the heme.[128] It is possible that the active site residues are required to reorganize for the substrate to bind in a reactive position. "Induced-fit" computational docking studies have been carried out on CYP 450$_{BM3}$ with the substrate *N*-palmitoylglycine, revealing a gating mechanism in which Phe$_{87}$ flips upwards to allow the tail of the fatty acid substrate to access Cpd I.[129] Replica exchange MD simulations showed that this new conformation increases in population as the temperature is increased. Tian and Friesner performed QM/MM simulations on the hydrogen abstraction reaction of *N*-palmitoylglycine by CYP 450$_{BM3}$, using this induced-fit structure.[130] QM/MM calculations were performed at the B3LYP/OPLS level of theory. The hydrogen abstraction reaction was modelled using both QM/MM and gas-phase QM calculations. The barrier for hydrogen abstraction was similar in magnitude to that calculated previously for CYP 450$_{cam}$.[81] Both of these barriers are too high, as they are not consistent with the experimentally observed fast depletion of Cpd I, and the authors suggest that there is a possible problem with the methodology for CYP enzymes. The gas-phase calculations indicate that the substrate is subjected to steric strain in the enzyme binding site, imposing constraints on the geometry, but this does not have a significant effect on the activation barrier. The protein was shown to decrease the barrier by 3.3 kcal mol^{-1} in the quartet spin state, which is believed to be due to electrostatic interactions that stabilize spin density on the cysteinyl sulfur atom. A water molecule is observed in the crystal structure in a position close to the Cpd I ferryl oxygen.[128] The effect of this water molecule on the reaction barrier was investigated and it was found that the presence of this water molecule lowers the barrier to hydrogen abstraction by 2–3 kcal mol^{-1}.[130] A decrease in spin on the ferryl oxygen was observed in the presence of the water molecule and, because it was found that this interaction can be reproduced by point charges, it was deemed that this interaction is mainly electrostatic in nature. A similar effect has been observed in CYP 450$_{cam}$.[78,131]

11.4.5.2   CYP245A1 (CYP 450 StaP)

A very unusual mechanism for a CYP enzyme has been put forward for the conversion of chromopyrrolic acid (CPA) to form staurosporine by CYP245A1 (CYP 450 StaP) (Scheme 11.3). This CYP is of significant interest, as staurosporine is linked to anticancer activity.[132] CYP 450 StaP has been postulated to display reactivity closer to that observed in peroxidases, such as horseradish peroxidase (HRP) and cytochrome *c* peroxidase (C*c*P).[133] Peroxidases contain

Scheme 11.3 Conversion of chromopyrrolic acid (CPA) into staurosporine by CYP245A1 (CYP 450 StaP).[133]

a proximal histidine residue rather than the cysteine found in CYPs. This histidine imidazole is a superior electron acceptor to cysteine, and it results in oxidation of an active-site tryptophan residue in C*c*P, leading to the one-electron reduced form of Cpd I (Cpd II) and a tryptophan radical.

The structure of CYP 450 StaP complexed with CPA has been solved crystallographically.[133] In this structure, the CPA molecule is held in position by an array of hydrogen bonds, but this position is too far from the active site heme for direct oxidation by Cpd I to occur. The electrostatic potential surrounding the heme is strikingly similar to that observed in C*c*P, where the nearby Trp_{191} is oxidized and the active species is Cpd II.[134] It was, hence, postulated that CYP 450 StaP Cpd I initially forms Cpd II with a radical cation on the proximal indole ring of CPA.[133]

The mechanism of CYP 450 StaP has been modelled using QM/MM techniques by Wang *et al.*[135,136] The Cpd I species was studied and it was found that the electronic structure was sensitive to the protonation state of the CPA substrate. In the case where one of the carboxylate groups is ionized (CPA^-), a typical Cpd I electronic structure is obtained. When both carboxylate groups of the CPA substrate are ionized (CPA^{2-}), substantial electron transfer is observed from CPA to Cpd I, as observed in QM only calculations. Two of the three unpaired electrons are found to be located on the Fe–O unit, with the third shared between the porphyrin a_{2u} orbital and the proximal CPA indole ring. Altering the hydrogen bonding environment surrounding the active site was found to change the relative amounts of unpaired electron density found on the indole, porphyrin and cysteine fragments, in QM/MM calculations on C*c*P.[137] A hydrogen bonding pathway is proposed to form between the proximal indole N–H group and the Cpd I ferryl oxygen *via* His_{250} and two water molecules (Figure 11.17).

Three different initial steps were modelled in the QM/MM study by Wang *et al.* The proton coupled electron transfer pathway was found to be the most energetically favourable (as found with HRP),[138] whereby proton transfer from the proximal indole N–H, *via* the hydrogen-bonding network in Figure 11.17, occurs with simultaneous electron transfer from CPA to the Cpd I a_{2u} orbital.

The subsequent steps of this unusual CYP reaction have also been modelled with QM/MM methods.[136] The mechanism is believed to progress *via*

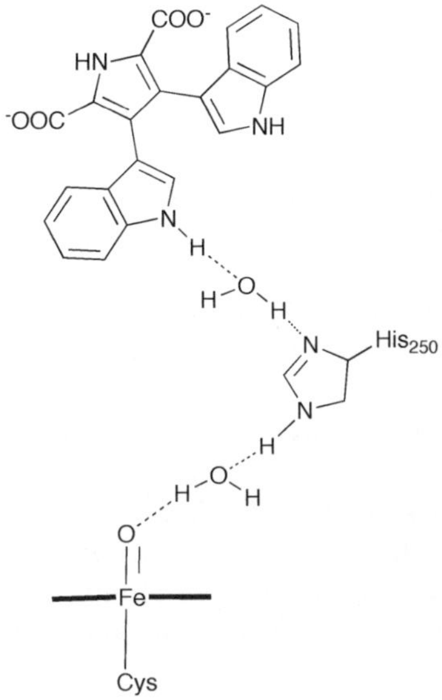

Figure 11.17   Hydrogen bonding network linking the proximal indole N–H moiety in CPA with the Cpd I ferryl oxygen *via* two water molecules and His_{250} in the active site of CYP 450 StaP, as modelled by Wang *et al.*[135,136]

water-mediated proton transfers and a C–C coupling step, which is coupled to the second electron transfer from CPA to the heme. An experimental study was performed in collaboration with the QM/MM modelling, in which oxidation of CPA was carried out by the H250F mutant of CYP 450 StaP. A marked decrease in activity was observed, but the enzyme was still able to turn over, supporting the catalytic role of water molecules in this enzyme.

11.5   Conclusions

QM/MM methods have provided detailed mechanistic insight into the reactivity of this important and interesting family of enzymes. This area is an excellent example of QM/MM calculations giving significant insight beyond analogous small model QM calculations. There are now many examples of the use of QM/MM calculations to provide detailed mechanistic understanding into the chemistry of CYP enzymes.[139] In many cases the type of insight obtained is beyond that which can be derived from current experimental approaches, or at least is complementary to it. Improvements to theoretical methods and (particularly) increasing computer power will no doubt lead to the ability to study more complex systems, such as inclusion of membrane effects. It is likely that some of the QM/MM studies and methods described here will be useful in future for the design and development of new pharmaceutical compounds. QM/MM methods also have the potential to contribute to the design of modified CYPs as practical catalysts for biotechnology.

References

1. F. P. Guengerich, *Chem. Res. Toxicol.*, 2001, **14**, 611.
2. D. R. Nelson, *Human Genomics*, 2009, **4**, 59.
3. P. Needleman, J. Turk, B. A. Jakschik, A. R. Morrison and J. B. Lefkowith, *Annu. Rev. Biochem.*, 1986, **55**, 69.
4. J. Capdevila, *Biochem. Biophys. Res. Commun.*, 2001, **285**, 571.
5. D. W. Nebert and D. W. Russell, *Lancet*, 2002, **360**, 1155.
6. F. P. Guengerich, in *The Ubiquitous Roles of Cytochrome P450 Proteins*, Metal Ions in Life Sciences, vol. 3, John Wiley & Sons Ltd., 2007, p. 561.
7. P. Anzenbacher and E. Anzenbacherova, *Cell. Mol. Life Sci.*, 2001, **58**, 737.
8. R. N. Armstrong, *Chem. Res. Toxicol.*, 1997, **10**, 2.
9. P. R. Ortiz de Montellano, (ed.), *Cytochrome P450: Structure, Mechanism and Biochemistry*, Kluwer Academic/Plenum Publishers, New York, 2004.
10. M. Sono, M. P. Roach, E. D. Coulter and J. H. Dawson, *Chem. Rev.*, 1996, **96**, 2841.
11. A. W. Munro, H. M. Girvan and K. J. McLean, *Nat. Prod. Rep.*, 2007, **24**, 585.
12. I. Schlichting, *Science*, 2000, **287**, 1615.
13. R. Raag, S. Martinis, S. G. Sligar and T. L. Poulos, *Biochemistry*, 1991, **30**, 11420.
14. J. Aikens and S. G. Sligar, *J. Am. Chem. Soc.*, 1994, **116**, 1143.
15. M. Vidakovic, S. G. Sligar, H. Li and T. L. Poulos, *Biochemistry*, 1998, **37**, 9211.
16. R. Davydov, I. MacDonald, T. M. Makris, S. G. Sligar and B. M. Hoffman, *J. Am. Chem. Soc.*, 1999, **121**, 10654.
17. R. Davydov, T. M. Makris, V. Kofman, C. R. Werst, S. G. Sligar and B. M. Hoffman, *J. Am. Chem. Soc.*, 2001, **123**, 1403.
18. J. A. Peterson, Y. Ishimura and B. W. Griffin, *Arch. Biochem. Biophys.*, 1972, **149**, 197.
19. S. G. Sligar, *Biochemistry*, 1976, **15**, 5399.
20. J. H. Dawson and M. Sono, *Chem. Rev.*, 1987, **87**, 1255.
21. S. Hu, A. Schneider and J. Kincaid, *J. Am. Chem. Soc.*, 1991, **113**, 4815.
22. T. Tosha, *J. Biol. Chem.*, 2003, **278**, 39809.
23. M. Sharrock, P. G. Debrunner, C. Schulz, J. D. Lipscomb, V. Marshall and I. C. Gunsalus, *Biochim. Biophys. Acta*, 1976, **420**, 8.
24. T. L. Poulos and Y. T. Meharenna, in *The Ubiquitous Roles of Cytochrome P450 Proteins*, Metal Ions in Life Sciences, vol. 3, John Wiley & Sons Ltd., Chichester, 2007, pp. 57–96.
25. W. Lohmann and U. Karst, *Anal. Bioanal. Chem.*, 2008, **391**, 79.
26. S. Shaik, D. Kumar, S. P. de Visser, A. Altun and W. Thiel, *Chem. Rev.*, 2005, **105**, 2279.
27. J. Rittle and M. T. Green, *Science*, 2010, **330**, 933.
28. R. E. White, *Pharmacol. Ther.*, 1991, **49**, 21.

29. C. M. Bathelt, L. Ridder, A. J. Mulholland and J. N. Harvey, *J. Am. Chem. Soc.*, 2003, **125**, 15004.
30. K. von König and I. Schlichting, in *The Ubiquitous Roles of Cytochrome P450 Proteins*, Metal Ions in Life Sciences, vol. 3, John Wiley & Sons Ltd., Chichester, 2007, pp. 235–265.
31. F. Ogliaro, S. P. de Visser and S. Shaik, *J. Inorg. Biochem.*, 2002, **91**, 554.
32. T. L. Poulos, *J. Biol. Inorg. Chem.*, 1996, **1**, 356.
33. T. Ueno, N. Nishikawa, S. Moriyama, S. Adachi, K. Lee, T. Okamura, N. Ueyama and A. Nakamura, *Inorg. Chem.*, 1999, **38**, 1199.
34. S. Yoshioka, S. Takahashi, K. Ishimori and I. Morishima, *J. Inorg. Biochem.*, 2000, **81**, 141.
35. N. Ueyama, N. Nishikawa, Y. Yamada, T. Okamura and A. Nakamura, *J. Am. Chem. Soc.*, 1996, **118**, 12826.
36. H. van de Waterbeemd and E. Gifford, *Nat. Rev. Drug Discovery*, 2003, **2**, 192.
37. J. A. Williams, *Drug Metab. Dispos.*, 2004, **32**, 1201.
38. X. Ding and L. S. Kaminsky, *Annu. Rev. Pharmacol. Toxicol.*, 2003, **43**, 149.
39. M. Ingelman-Sundberg, S. Sim, A. Gomez and C. Rodriguez-Antona, *Pharmacol. Ther.*, 2007, **116**, 496.
40. J. H. Lin, *Pharm. Res.*, 2006, **23**, 1089.
41. A. W. Munro, K. J. McLean and H. M. Girvan, in *The Ubiquitous Roles of Cytochrome P450 Proteins*, Metal Ions in Life Sciences, vol. 3, John Wiley & Sons Ltd., Chichester, 2007, pp. 285–317.
42. C. Patten, P. Thomas, R. Guy, M. Lee, F. Gonzalez, F. P. Guengerich and C. Yang, *Chem. Res. Toxicol.*, 1993, **6**, 511.
43. S. Lee, J. Buters, T. Pineau, P. Fernandez-Salguero and F. Gonzalez, *J. Biol. Chem.*, 1996, **271**, 12063.
44. B. H. Rumack, *Clin. Toxicol.*, 2002, **40**, 3.
45. T. Adler and F. Shaw, *J. Pharmacol. Exp. Ther.*, 1952, **104**, 1.
46. J. Desmeules, M. Gascon, P. Dayer and M. Magistris, *Eur. J. Clin. Pharmacol.*, 1991, **41**, 23.
47. L. K. Kamdem, I. Meineke, U. Gödtel-Armbrust, J. Brockmöller and L. Wojnowski, *Chem. Res. Toxicol.*, 2006, **19**, 577.
48. S. Asha and M. Vidyavathi, *Appl. Biochem. Biotechnol.*, 2010, **160**, 1699.
49. P. Crivori and I. Poggesi, *Eur. J. Med. Chem.*, 2006, **41**, 795.
50. C. de Graaf, N. Vermeulen and K. Feenstra, *J. Med. Chem.*, 2005, **48**, 2725.
51. L. Hammett, *J. Am. Chem. Soc.*, 1937, **59**, 96.
52. A. R. Leach, *Molecular Modelling: Principles and Applications*, 2nd edn, Prentice Hall, Harlow, 2001.
53. D. Korolev, K. V. Balakin, Y. Nikolsky, E. Kirillov, Y. A. Ivanenkov, N. P. Savchuk, A. A. Ivashchenko and T. Nikolskaya, *J. Med. Chem.*, 2003, **46**, 3631.
54. M. Eldridge, C. Murray, T. Auton, G. Paolini and R. Mee, *J. Comput.-Aided Mol. Des.*, 1997, **11**, 425.

55. R. Lonsdale, K. E. Ranaghan and A. J. Mulholland, *Chem. Commun.*, 2010, 2354.
56. S. Shaik, S. Cohen, Y. Wang, H. Chen, D. Kumar and W. Thiel, *Chem. Rev.*, 2010, **110**, 949.
57. J. Åqvist and A. Warshel, *Chem. Rev.*, 1993, **93**, 2523.
58. R. A. Friesner and V. Guallar, *Annu. Rev. Phys. Chem.*, 2005, **56**, 389.
59. H. M. Senn and W. Thiel, *Angew. Chem., Int. Ed.*, 2009, **48**, 1198.
60. K. E. Ranaghan and A. J. Mulholland, *Int. Rev. Phys. Chem.*, 2010, **29**, 65.
61. M. T. Green, *J. Am. Chem. Soc.*, 1998, **120**, 10772.
62. M. T. Green, *J. Am. Chem. Soc.*, 1999, **121**, 7939.
63. H. Chen, J. Song, W. Lai, W. Wu and S. Shaik, *J. Chem. Theory Comput.*, 2010, **6**, 940.
64. V. Thery, D. Rinaldi, J. Rivail, B. Maigret and G. Ferenczy, *J. Comput. Chem.*, 1994, **15**, 269.
65. A. Warshel and M. Levitt, *J. Mol. Biol.*, 1976, **103**, 227.
66. M. J. Field, P. A. Bash and M. Karplus, *J. Comput. Chem.*, 1990, **11**, 700.
67. K. Eurenius, D. Chatfield, B. R. Brooks and M. Hodoscek, *Int. J. Quantum Chem.*, 1996, **60**, 1189.
68. R. Lonsdale, J. N. Harvey and A. J. Mulholland, *J. Phys. Chem. B*, 2010, **114**, 1156.
69. V. Guallar and F. H. Wallrapp, *Biophys. Chem.*, 2010, **149**, 1.
70. J. Zheng, A. Altun and W. Thiel, *J. Comput. Chem.*, 2007, **28**, 2147.
71. J. Schöneboom and W. Thiel, *J. Phys. Chem. B*, 2004, **108**, 7468.
72. R. Tsai, C. A. Yu, I. C. Gunsalus, J. Peisach, W. Blumberg, W. H. Orme-Johnson and H. Beinert, *Proc. Natl. Acad. Sci. USA*, 1970, **66**, 1157.
73. M. J. de Groot, R. W. Havenith, H. M. Vinkers, R. Zwaans, N. P. Vermeulen and J. H. van Lenthe, *J. Comput.-Aided Mol. Des.*, 1998, **12**, 183.
74. V. Guallar and R. A. Friesner, *J. Am. Chem. Soc.*, 2004, **126**, 8501.
75. A. Altun and W. Thiel, *J. Phys. Chem. B*, 2005, **109**, 1268.
76. M. Swart, A. Groenhof, A. Ehlers and K. Lammertsma, *Chem. Phys. Lett.*, 2005, **403**, 35.
77. A. Altun, S. Shaik and W. Thiel, *J. Am. Chem. Soc.*, 2007, **129**, 8978.
78. A. Altun, V. Guallar, R. A. Friesner, S. Shaik and W. Thiel, *J. Am. Chem. Soc.*, 2006, **128**, 3924.
79. A. Altun, S. Shaik and W. Thiel, *J. Comput. Chem.*, 2006, **27**, 1324.
80. J. Żurek, N. Foloppe, J. N. Harvey and A. J. Mulholland, *Org. Biomol. Chem.*, 2006, **4**, 3931.
81. J. Schöneboom, S. Cohen, H. Lin, S. Shaik and W. Thiel, *J. Am. Chem. Soc.*, 2004, **126**, 4017.
82. V. Guallar, M. Baik, S. Lippard and R. A. Friesner, *Proc. Natl. Acad. Sci. USA*, 2003, **100**, 6998.
83. R. Davydov, R. Kappl, J. Hüttermann and J. A. Peterson, *FEBS Lett.*, 1991, **295**, 113.
84. L. Pauling, *Nature*, 1964, **203**, 182.
85. J. J. Weiss, *Nature*, 1964, **202**, 83.

86. D. Harris, G. H. Loew and L. Waskell, *J. Am. Chem. Soc.*, 1998, **120**, 4308.
87. H. Nakatsuji, J. Hasegawa, H. Ueda and M. Hada, *Chem. Phys. Lett.*, 1996, **250**, 379.
88. D. Wang and W. Thiel, *J. Mol. Struct. THEOCHEM*, 2009, **898**, 90.
89. J. Zheng, D. Wang, W. Thiel and S. Shaik, *J. Am. Chem. Soc.*, 2006, **128**, 13204.
90. H. Y. Li, S. Narasimhulu, L. M. Havran, J. D. Winkler and T. L. Poulos, *J. Am. Chem. Soc.*, 1995, **117**, 6297.
91. H. Lin, J. C. Schöneboom, S. Cohen, S. Shaik and W. Thiel, *J. Phys. Chem. B*, 2004, **108**, 10083.
92. T. Egawa, H. Shimada and Y. Ishimura, *Biochem. Biophys. Res. Commun.*, 1994, **201**, 1464.
93. D. G. Kellner, S.-C. Hung, K. E. Weiss and S. G. Sligar, *J. Biol. Chem.*, 2002, **277**, 9641.
94. T. Spolitak, J. H. Dawson and D. P. Ballou, *J. Biol. Chem.*, 2005, **280**, 20300.
95. I. G. Denisov, T. M. Makris, S. G. Sligar and I. Schlichting, *Chem. Rev.*, 2005, **105**, 2253.
96. G. H. Loew and D. Harris, *Chem. Rev.*, 2000, **100**, 407.
97. R. Rutter, L. P. Hager, H. Dhonau, M. Hendrich, M. Valentine and P. Debrunner, *Biochemistry*, 1984, **23**, 6809.
98. F. Ogliaro, S. Cohen, S. P. de Visser and S. Shaik, *J. Am. Chem. Soc.*, 2000, **122**, 12892.
99. S. P. de Visser, F. Ogliaro, P. K. Sharma and S. Shaik, *Angew. Chem., Int. Ed.*, 2002, **41**, 1947.
100. J. Schöneboom, H. Lin, N. Reuter, W. Thiel, S. Cohen, F. Ogliaro and S. Shaik, *J. Am. Chem. Soc.*, 2002, **124**, 8142.
101. C. M. Bathelt, J. Żurek, A. J. Mulholland and J. N. Harvey, *J. Am. Chem. Soc.*, 2005, **127**, 12900.
102. D. Fishelovitch, C. Hazan, H. Hirao, H. J. Wolfson, R. Nussinov and S. Shaik, *J. Phys. Chem. B*, 2007, **111**, 13822.
103. J. T. Groves and G. A. McClusky, *J. Am. Chem. Soc.*, 1976, **98**, 859.
104. S. Shaik, S. Cohen, S. P. de Visser, P. K. Sharma, D. Kumar, S. Kozuch, F. Ogliaro and D. Danovich, *Eur. J. Inorg. Chem.*, 2004, 207.
105. T. Kamachi and K. Yoshizawa, *J. Am. Chem. Soc.*, 2003, **125**, 4652.
106. A. Mancy, P. Broto, S. Dijols, P. Dansette and D. Mansuy, *Biochemistry*, 1995, **34**, 10365.
107. P. Watkins, S. Wrighton, P. Maurel, E. Schuetz, G. Mendez-Picon, G. Parker and P. Guzelian, *Proc. Natl. Acad. Sci. USA*, 1985, **82**, 6310.
108. M. Strickler, B. Goldstein, K. Maxfield, L. Shireman, G. Kim, D. Matteson and J. P. Jones, *Biochemistry*, 2003, **42**, 11943.
109. S. Modi, D. E. Gilham, M. J. Sutcliffe, L. Y. Lian, W. U. Primrose, C. R. Wolf and G. C. K. Roberts, *Biochemistry*, 1997, **36**, 4461.
110. S. Cohen, S. Kozuch, C. Hazan and S. Shaik, *J. Am. Chem. Soc.*, 2006, **128**, 11028.

111. J. T. Groves and D. V. Subramanian, *J. Am. Chem. Soc.*, 1984, **106**, 2177.

112. A. D. N. Vaz, D. McGinnity and M. J. Coon, *Proc. Natl. Acad. Sci. USA*, 1998, **95**, 3555.

113. R. E. White, J. T. Groves and G. A. McClusky, *Acta Biol. Med. Germ.*, 1979, **38**, 475.

114. J. T. Groves, G. E. Avarianeisser, K. M. Fish, M. Imachi and R. L. Kuczowski, *J. Am. Chem. Soc.*, 1986, **108**, 3837.

115. S. P. de Visser, F. Ogliaro, P. K. Sharma and S. Shaik, *J. Am. Chem. Soc.*, 2002, **124**, 11809.

116. B. Meunier, S. P. de Visser and S. Shaik, *Chem. Rev.*, 2004, **104**, 3947.

117. D. M. Jerina and J. Daly, *Science*, 1974, **185**, 573.

118. I. M. C. M. Rietjens, C. den Besten, R. Hanzlik and P. van Bladeren, *Chem. Res. Toxicol.*, 1997, **10**, 629.

119. K. R. Korzekwa, W. F. Trager, M. Gouterman, D. Spangler and G. H. Loew, *J. Am. Chem. Soc.*, 1985, **107**, 4273.

120. S. P. de Visser and S. Shaik, *J. Am. Chem. Soc.*, 2003, **125**, 7413.

121. G. Guroff, J. W. Daly, D. M. Jerina, J. Renson, B. Witkop and S. Udenfriend, *Science*, 1967, **157**, 1524.

122. D. M. Jerina, N. Kaubisch and J. W. Daly, *Proc. Natl. Acad. Sci. USA*, 1971, **68**, 2545.

123. J. E. Tomaszewski, D. M. Jerina and J. W. Daly, *Biochemistry*, 1975, **14**, 2024.

124. K. R. Korzekwa, D. C. Swinney and W. F. Trager, *Biochemistry*, 1989, **28**, 9019.

125. C. M. Bathelt, L. Ridder, A. J. Mulholland and J. N. Harvey, *Org. Biomol. Chem.*, 2004, **2**, 2998.

126. C. M. Bathelt, A. J. Mulholland and J. N. Harvey, *J. Phys. Chem. A*, 2008, **112**, 13149.

127. A. W. Munro, D. Leys, K. J. McLean, K. R. Marshall, T. W. B. Ost, S. Daff, C. S. Miles, S. K. Chapman, D. A. Lysek, C. C. Moser, C. C. Page and P. L. Dutton, *Trends Biochem. Sci.*, 2002, **27**, 250.

128. D. C. Haines, D. R. Tomchick, M. Machius and J. A. Peterson, *Biochemistry*, 2001, **40**, 13456.

129. T. Jovanovic, R. Farid, R. A. Friesner and A. E. McDermott, *J. Am. Chem. Soc.*, 2005, **127**, 13548.

130. L. Tian and R. A. Friesner, *J. Chem. Theory Comput.*, 2009, **5**, 1421.

131. D. Kumar, A. Altun, S. Shaik and W. Thiel, *Faraday Discuss.*, 2011, **148**, 373.

132. H. J. Chae, J. S. Kang, J. O. Byun, K. S. Han, D. U. Kim, S. M. Oh, H. M. Kim, S. W. Chae and H. R. Kim, *Pharmacol. Res.*, 2000, **42**, 373.

133. M. Makino, H. Sugimoto, Y. Shiro, S. Asamizu, H. Onaka and S. Nagano, *Proc. Natl. Acad. Sci. USA*, 2007, **104**, 11591.

134. J. Stubbe and W. A. van Der Donk, *Chem. Rev.*, 1998, **98**, 705.

135. Y. Wang, H. Hirao, H. Chen, H. Onaka, S. Nagano and S. Shaik, *J. Am. Chem. Soc.*, 2008, **130**, 7170.
136. Y. Wang, H. Chen, M. Makino, Y. Shiro, S. Nagano, S. Asamizu, H. Onaka and S. Shaik, *J. Am. Chem. Soc.*, 2009, **131**, 6748.
137. C. M. Bathelt, A. J. Mulholland and J. N. Harvey, *Dalton Trans.*, 2005, 3470.
138. E. Derat and S. Shaik, *J. Am. Chem. Soc.*, 2006, **128**, 13940.
139. J. Olah, A. J. Mulholland and J. N. Harvey, *Proc. Natl. Acad. Sci. USA*, 2011, **108**, 6050.

Mechanism and Function of Tryptophan and Indoleamine Dioxygenases

SARAH J. THACKRAY,[a] IGOR EFIMOV,[b]
EMMA LLOYD RAVEN[b] AND CHRISTOPHER G.
MOWAT*[a]

[a] EastChem School of Chemistry, University of Edinburgh, West Mains
Road, Edinburgh, EH9 3JJ, United Kingdom; [b] Department of Chemistry,
University of Leicester, University Road, Leicester, LE1 7RH,
United Kingdom

12.1 Introduction

The uses of heme groups in biology include processes such as oxygen transport
and storage, electron transport, and the oxygenation of substrates. This final
example may involve the insertion of either one or both atoms of molecular
oxygen into substrates and such reactions are catalysed by monoxygenases and
dioxygenases, respectively. Heme-dependent enzymes such as the cytochromes
P450 and nitric oxide synthase are classic examples of monoxygenases, effecting
hydroxylation and other chemical transformations by insertion of O between
two other atoms (*e.g.* into a C–H or N–H bond). Such hydroxylating proteins
are specific and powerful oxidation catalysts, enabling the otherwise spin-
forbidden addition of oxygen to relatively inert substrates. The mechanisms of
these enzymes usually utilize the ability of iron to form the highly reactive ferryl
iron-oxo species to activate oxygen and allow its insertion into substrates.

Iron-Containing Enzymes: Versatile Catalysts of Hydroxylation Reactions in Nature
Edited by Sam P de Visser and Devesh Kumar
© Royal Society of Chemistry 2011
Published by the Royal Society of Chemistry, www.rsc.org

However, in the case of dioxygenation reactions the enzymes typically employed are non-heme iron-dependent proteins such as the intradiol and extradiol catechol dioxygenases that catalyse the oxidative cleavage of catechol substrates.[1-3] The subjects of this chapter, indoleamine 2,3-dioxygenase (IDO) and tryptophan 2,3-dioxygenase (TDO), have properties in common with both the heme-dependent monoxygenases and the non-heme iron-dependent dioxygenases. This chapter seeks to describe the function and physiological roles of these enzymes, to summarize the structural data available, and to review their kinetic, ligand-binding and electrochemical properties. Furthermore, we will consider inhibition of these enzymes by small molecules and the potential for using TDO and/or IDO as drug targets. Finally, we will draw comparisons between the mechanisms of oxygen activation by the heme-dependent mono- and dioxygenases in the light of mechanistic and structural information available for both classes of enzyme. To date, the catalytic mechanism of this reaction is unclear but emerging evidence suggests the activating iron species may be the same as for the cytochromes P450.

Of all the many dioxygenase enzymes and heme-containing enzymes, there are only two known heme-dependent dioxygenases, both of which catalyse the same reaction. Here we ask why this should be the case. Why have these two dioxygenases evolved to utilize heme as a cofactor in their reaction mechanism, and what makes them different from other dioxygenases or heme-utilizing enzymes? In this respect, and given the similarities between them, the over-arching question is this: How do TDO and IDO catalyse dioxygenation rather than monooxygenation?

12.2 Biological and Physiological Function of Indoleamine Dioxygenase and Tryptophan Dioxygenase

IDO and TDO are members of a small family of heme-dependent enzymes that also includes PrnB, a pyrrolnitrin biosynthetic enzyme from *Pseudomonas fluorescens*, and SO4144, an IDO-like protein from *Shewanella oneidensis* (sIDO) (although neither PrnB or sIDO have been demonstrated to have dioxygenase activity).[4-6]

IDO was first identified in 1967 and is found in many eukaryotes where it is expressed ubiquitously throughout the body, except for in the liver.[7] Bacteria are believed to use members of this family for the aerobic metabolism of L-Trp *via* the kynurenine pathway, and some prokaryotic IDO-like proteins (such as sIDO) have been identified. These proteins bear little sequence similarity to eukaryotic IDO proteins, and homology is based on their predicted 3D structures. Recent crystallization of sIDO has shown its high structural homology to IDO.[5]

TDO was discovered in the 1930s and is found in both eukaryotes and prokaryotes. TDO expression is normally restricted to the liver, but it has been identified in the brain and epididymis of some species.[8]

Sequence similarity between family members is low, with typical identities of around 10%, and homology is only apparent from their 3D structures and the conservation of certain amino acids in their active sites. In the case of the same enzyme from different species the sequence conservation is often higher. For example, sequence identities between prokaryotic and eukaryotic TDOs are around 20–30%.

TDO and IDO activate molecular oxygen and catalyse its insertion into L-Trp to form *N*-formylkynurenine, in the first and rate-limiting step in the kynurenine pathway (Figure 12.1).[9] The kynurenine pathway processes over 90% of L-Trp utilized by humans. Both IDO and TDO play a central role in the physiological regulation of tryptophan in the human body, controlling the kynurenine pathway and influencing serotogenic regulation. The kynurenine pathway is important not only as a source of metabolites but also due to the effect of IDO and TDO activity on the local tryptophan concentration. The build up of pathway metabolites such as quinolinic acid and L-kynurenine can lead to numerous physiological and pathophysiological conditions, such as multiple sclerosis, AIDS related dementia and ischemic brain injury.[10–12] The concentrations of these metabolites increases with the severity of neurological dysfunction or brain injury under a wide range of inflammatory conditions. For example, the kynurenine pathway metabolites 3-hydroxyanthranilic acid and 3-hydroxykynurenine are UV filters that can bind to the lens proteins in the eye and have been implicated in cataract formation when present at high concentrations.[13] In addition, some kynurenine pathway metabolites are immunomodulatory and can consequently contribute to immunosuppressive pathways and suppress proliferation (and even cause apoptosis) of T-cells, although the mechanisms of action of such metabolites are unknown.[14–16]

The local depletion of tryptophan due to IDO or TDO activity is associated with an antimicrobial response by TDO or IDO, or with immune regulation by IDO.[17] Some pathogens are sensitive to tryptophan degradation, and this may be an effective mechanism for controlling their ability to proliferate.

Although both enzymes catalyse the same reaction, compartmentalization of IDO and TDO expression is thought to reflect their differing biological roles. IDO plays an active role in the human body's immune response, but there is little evidence to suggest that TDO plays a part in this process. Immune regulation by IDO is extremely complex and is influenced by many factors including tryptophan depletion, the accumulation of immunosuppressive metabolites, and interactions with signalling and immunoregulatory pathways.[18]

IDO is a normal effector of peripheral tolerance in the immunoregulatory pathway of tryptophan metabolism and helps create a balanced immune response between inflammation and tolerance, suppressing excessive immune activation.[19] This delicate balance between suppression and activation of the immune response is not well understood, and IDO plays only a small part. However, the effects observed when IDO activity is disturbed are widespread, suggesting that it plays an important role. Furthermore, IDO is now thought to play an essential role in acquired immune tolerance. Inhibition of IDO activity

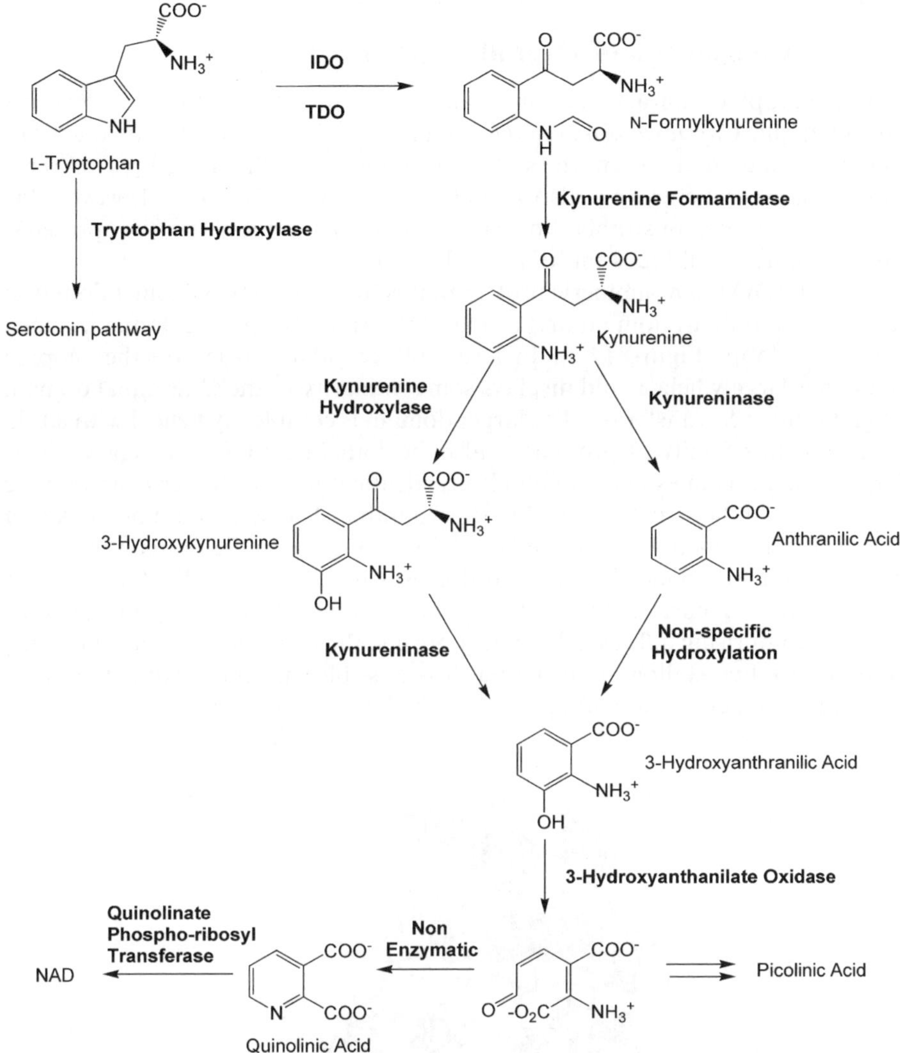

Figure 12.1 Schematic representation of the kynurenine pathway. The first step shows the insertion of dioxygen into L-Trp to form *N*-formylkynurenine. Over 95% of L-Trp is metabolized *via* the kynurenine pathway in mammals.

in mice has been shown to lead to allogenic fetal rejection, and increased IDO activity is thought to be responsible for renal allograft rejection.[20–22] In both of these examples the body must acquire immune tolerance. In addition, it has been proposed that IDO helps tumours induce tolerance from the host's immune system by interacting with T-cells and depleting L-Trp levels. Consequently, IDO has emerged as an attractive drug target in cancer and transplant treatments.

12.3   Structures of TDO and IDO

12.3.1   Comparison of Overall Structure

Crystallographic characterization of human IDO, *Xanthomonas campestris* (xcTDO) and *Cupriavidus metallidurans* TDO (rmTDO) has greatly aided our understanding of these enzymes.[4,5,23] In addition to the structures of these dioxygenases, structures are also available for PrnB and sIDO.[5,6] However, due to the lack of demonstrable dioxygenase activity for the latter two proteins, discussion here will focus on hIDO and TDO.

Human IDO is a monomeric protein that is folded into two distinct domains; a large (C-terminal) domain and a small (N-terminal) capping domain, joined by a long loop (Figure 12.2). In terms of secondary structure, the capping domain is largely helical and displays some similarity to the C-terminal domain of glutathione *S*-transferase. The larger domain is completely helical with a fold unique to this family of proteins, and inter-domain contact is extensive. This larger domain is in essence topologically identical to a protomer of tetrameric TDO, with all the structures readily superposable. The heme-binding pocket of IDO is located mainly within the larger domain, close to the domain interface. The loop region connecting the two domains lies above the distal face of the heme, providing part of the substrate-binding cavity. In the structure of IDO, residues comprising a flexible loop, just outside the heme pocket, have not been fully resolved crystallographically but it is possible that this flexible loop may play a role in substrate binding, as described below for TDO.

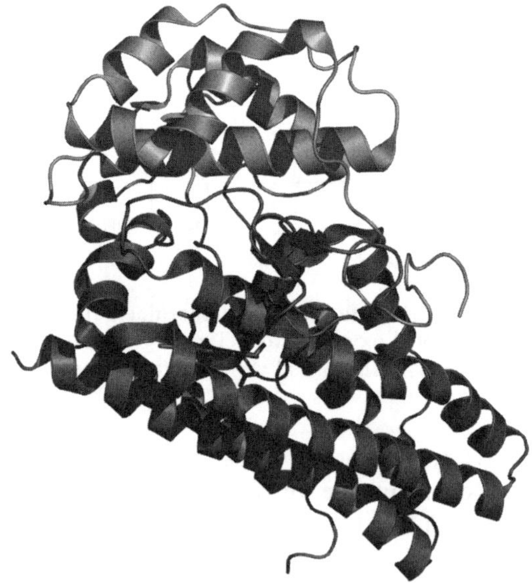

Figure 12.2   Overall structure of human IDO.[4] The smaller N-terminal domain is at the top of the figure.

The TDOs from *C. metallidurans* and *X. campestris* share 47% sequence identity, and their overall structures are essentially identical (Figure 12.3).[5,23] TDO is homotetrameric and is perhaps best described as a "dimer of dimers" because the N-terminal region (residues Arg_{21}–Leu_{40} of the *X. campestris* enzyme) of each monomer infiltrates the active site of the adjacent monomer and is crucial for substrate binding. The extent of the tetrameric interaction is illustrated by the fact that the TDO monomer has an average surface area of 14 500 $Å^2$ of which approximately 4500 $Å^2$ is buried within the tetramer interface.[5]

This interprotomer association results in similar interactions in the active site to those provided by the long loop from the capping domain in IDO. A flexible

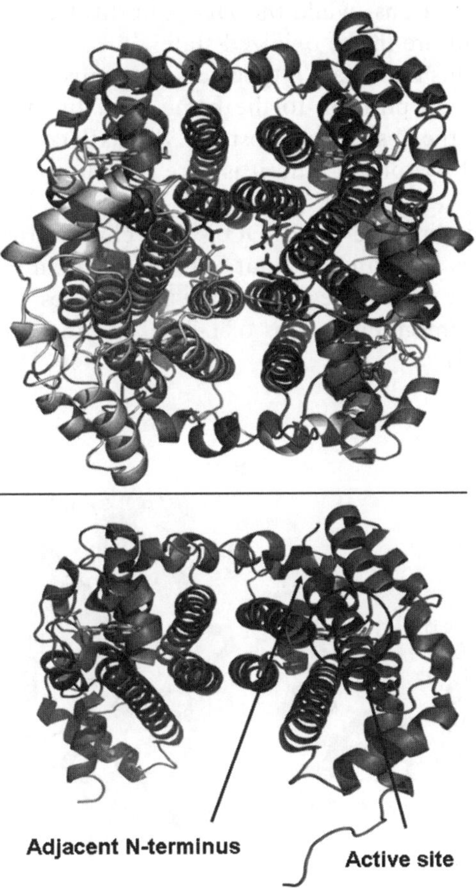

Figure 12.3 Overall structure of *Xanthomonas campestris* TDO.[5] Above: the tetrameric arrangement. Below: a dimer of xcTDO showing the intimate association of the N-terminus of one monomer in the active site of the adjacent one.

loop, just outside the heme pocket, has been fully resolved crystallographically and it is involved in binding both the ammonium and carboxylate moieties of L-Trp.

12.3.2 Active Site Environments

Currently, the only heme-dependent dioxygenase structures available with substrate bound at the active site are of xcTDO, these being the wild-type enzyme and the H55A and H55S variants.[5,24] Owing to a high similarity between the active sites of TDO and IDO it is perhaps reasonable to assume a similar substrate-binding mode in both. Of the available structures, only that of xcTDO (with either L-tryptophan or 6-fluoro-tryptophan bound) is in the catalytically active, ferrous state, making it the best available model of substrate binding interactions, while the His_{55}-substituted variant enzymes have substrate bound but are in the oxidized state.

All members of this family are *b*-type heme proteins with histidine occupying the axial coordination position to the heme iron. In the available structural models obtained in the absence of substrate, the sixth coordination site is either vacant or occupied by solvent, and it is here that dioxygen coordinates to the heme iron.

The substrate binding site of xcTDO is well characterized due to the availability of structures with and without L-Trp bound. Interactions of the substrate with the active site can be split into two categories – interactions with the carboxylate and ammonium groups of L-Trp, and those with the indole ring system (Figure 12.4).

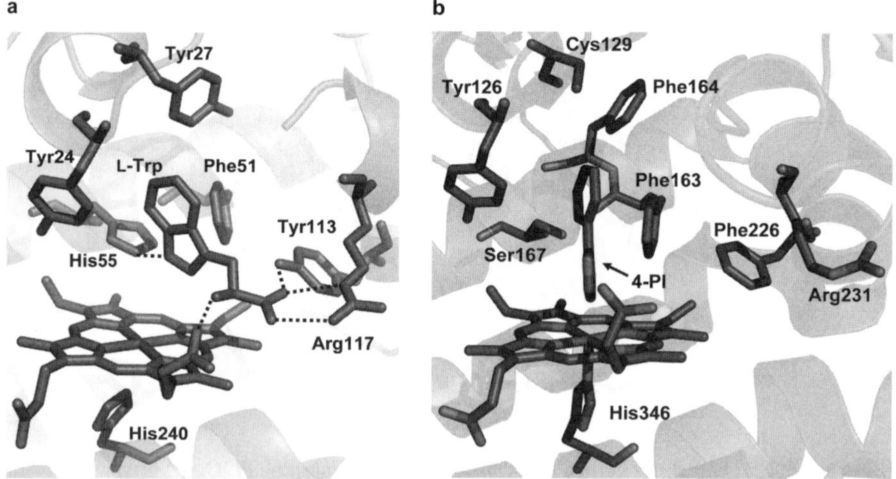

Figure 12.4 (a) Active site of xcTDO with L-Trp bound. (b) Active site of human IDO with the inhibitor 4-phenylimidazole (4-PI) bound.

Arg$_{117}$, Tyr$_{113}$ and Thr$_{254}$ all interact with the L-Trp carboxylate group in xcTDO, with the arginine forming an electrostatic interaction with the substrate carboxylate group that appears to be essential for substrate binding (the equivalent residue in hIDO is Arg$_{231}$). This arginine reorientates in the presence of substrate, coordinating the carboxylate group of L-Trp, with "open" and "closed" conformations observed.[5] The closed conformation is only observed in TDO crystals grown in the presence of tryptophan, although no bound L-Trp is observed in the crystal structure of rmTDO.[23] (This absence is probably due to a low affinity of L-Trp binding to oxidized rmTDO, similar to that observed for xcTDO [K_{d} (Fe^{III}) = 3.8 mM], whilst only 1 mM L-Trp was present in the crystallization solution.[5]) The ammonium group of the substrate is hydrogen-bonded to the side chain hydroxyl group of Thr$_{254}$, the 7-propionate group of the heme, and a water molecule. Thr$_{254}$ is involved in binding both the ammonium and carboxylate moieties of L-Trp and is part of a flexible loop (consisting of residues 250–260 of xcTDO), similar to that encoded in IDO. This loop is only ordered in the structures grown in the presence of L-Trp (xcTDO and rmTDO), as substrate binding appears to induce ordering.

The substrate indole ring binds into a hydrophobic pocket, interacting with the side chain of Phe$_{51}$ and several other hydrophobic residues including Tyr$_{24}$ and Tyr$_{27}$ from the adjacent monomer. A tyrosine/glutamate motif linking the N-terminal loop of the adjacent monomer to the active site in TDO, or between the capping and larger domains in IDO, is observed close to the active site pocket. Perhaps the most obvious difference between the active sites of xcTDO and IDO is the presence of a histidine residue in the active site of TDO (His$_{55}$ in xcTDO). The histidine side chain is hydrogen-bonded with the indole nitrogen atom of L-Trp, and a possible role for His$_{55}$ might be in controlling the orientation of substrate binding.[24] In comparison, the equivalent residue in IDO is a serine (Ser$_{167}$, Figure 12.4). It would appear that this serine is unlikely to form an analogous hydrogen bonding interaction and it is therefore doubtful that it fulfils any similar regulatory role in IDO. This extensive network of enzyme–substrate interactions serves to stabilize the active site and control substrate binding and catalysis. As yet there is no available crystal structure for IDO with the substrate bound, but a structure has been solved with the inhibitor 4-phenylimidazole (4-PI) bound in the active site.[4] However, this ligand binds to the heme iron and therefore does not provide any insight into the mode of L-tryptophan binding in IDO.

Finally, in the crystal structure of L-Trp-bound, reduced, xcTDO there is a second L-Trp binding site in the tetramer interface (Figure 12.5). The well-defined electron density shows four L-Trp molecules within the tetramer interface, interacting with residues Arg$_{85}$, Lys$_{92}$ and Asp$_{228}$ from one monomer and stabilized by Lys$_{86}$ from another. This interfacial binding site is also identified in the crystal structures of mutant enzymes H55A and H55S (PDB IDs 3BK9 and 3E08 respectively).[24] In the light of reports of allosteric activation of TDO by L-Trp it could be possible that this site represents an allosteric effector site. Unfortunately, no clear data are available on these interactions.

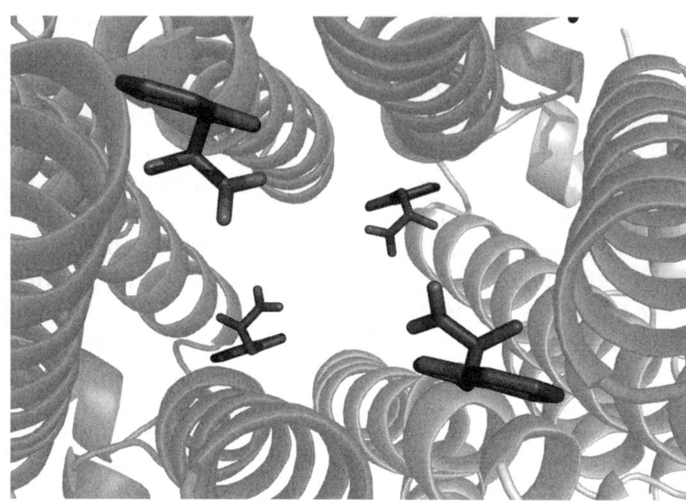

Figure 12.5   Interfacial L-Trp binding site in xcTDO. This interfacial site can also be
seen in the centre of the tetramer in Figure 12.3.

12.4   Turnover and Inhibition

12.4.1   Steady State Kinetics

IDO and TDO catalyse the dioxygenation of L-Trp, and in the case of IDO
some related indoleamine derivatives, with the absolute requirement that the
α-amino group of tryptophan must be present. More specifically, TDO has
been shown to catalyse the dioxygenation of L-Trp and derivatives substituted
with a halogen or a methyl group in the 5- or 6-positions of the indole
ring, whilst IDO displays much broader substrate selectivity, utilizing D-Trp,
5-hydroxy-tryptophan, tryptamine and serotonin (5-hydroxytryptamine) in
addition to those substrates utilized by TDO.[5,24,25]

In vitro assays must contain reducing equivalents to facilitate catalytic
activity: these mimic the *in vivo* maintenance of the ferrous state, a role that has
been suggested to be fulfilled by cytochrome b_5.[26,27] During the catalytic cycle
there is no net consumption of electrons, but a small amount of auto-oxidation
of the ferrous to the ferric state occurs, so *in vitro* assays contain an excess of
ascorbate as a reductant.[26,28–30] Furthermore, methylene blue is added to
mediate heme reduction by ascorbate. Catalase is also present to remove any
superoxide or peroxide by-products that can damage the heme and destroy
catalytic activity. The activity of IDO and TDO is calculated by monitoring the
production of NFK (*N*-formylkynurenine) or related derivative by observing
the formation of a strong absorbance band ($\varepsilon = 3750$ M^{-1} cm^{-1}) at 321 nm.

The rate of catalysis by IDO and TDO is fairly robust, with k_{cat} values for
L-Trp turnover of ~ 1–2 s^{-1} for mammalian proteins (human IDO and TDO,
mouse IDO) and faster rates of ~ 15–20 s^{-1} for most bacterial TDO enzymes
(*Pseudomonas, X. campestris*).[23,31–34] Although both dioxygenases catalyse the

same reaction, they are distinct not only in their ability to utilize different substrates but also in their stereospecificity. IDO catalyses the insertion of dioxygen into both D- and L- tryptophan, whilst almost all TDO enzymes are completely inactive toward D-Trp. Historically this distinction allowed categorization of the enzymes, but two TDO enzymes, from *Bacillus brevis* and rat liver, have been shown to have limited D-Trp-dependent activity.[35] This activity is very low, however, and a large excess of substrate must be added for activity to be observed.

The reasons for the ability of IDO to turnover both D- and L-enantiomers of tryptophan, whilst TDO distinguishes between them, remained unresolved in the literature until the crystal structures were solved. Analysis of the active sites of TDO and hIDO reveals narrower substrate selectivity due to a tighter substrate binding in TDO compared to IDO (Figure 12.4). Many key active-site residues are similar in IDO and TDO (Figure 12.4), but IDO appears to form weaker interactions with L-Trp. Ionic interactions in TDO are important for recognizing the L-Trp ammonium and carboxylate groups, making the enzyme stereospecific for L-Trp. Some of these electrostatic interactions are likely to be conserved in IDO (interactions with Arg_{231} and the heme 7-propionate), but other hydrogen-bonding interactions must be absent. For example, Tyr_{113} of xcTDO is replaced by Phe_{226} in hIDO and additionally the xcTDO Thr_{254} residue, which interacts with the substrate carboxylate, may not have an equivalent in human IDO.[4] The loop region that includes this residue has a highly divergent sequence in IDO compared with TDO, and no analogous interaction may exist. Therefore, IDO appears to form fewer interactions with L-Trp, and indeed this may be the reason why IDO does not distinguish among the indoleamine substrates. This theory also explains differences in the specificity of D- and L-Trp turnover by IDO. IDO D-Trp activity is roughly comparable to that for L-Trp in terms of turnover number, being no more than ~five-fold faster with the L-enantiomer.[31,33] However, the K_m for L-Trp turnover is much (70- to 250-fold) smaller than that for D-Trp, so that although turnover may be as rapid for D-Trp, it does not bind as well to the active site.

The specificity constant (k_{cat}/K_m) gives a measure of the absolute reactivity of an enzyme for a substrate, and for IDO all tryptophan derivatives are poorer substrates, displaying smaller specificity constants than L-Trp. However, for TDO 6F-D/L-Trp has a slightly higher specificity than L-Trp (2.0×10^5 *versus* 1.7×10^5 M^{-1} s^{-1}) and it is probable that catalysis only occurs with the L-isomer of 6F-Trp, indicating that a specificity constant nearer 4.0×10^5 M^{-1} s^{-1} may be a more realistic value. To learn more about the effect of substrate on the reactivity of TDO and IDO, extensive research has been conducted using substituted tryptophan derivatives as mechanistic probes. Turnover of tryptophan derivatives substituted by a halogen or a methyl group in the 4-, 5-, 6- or 7- position (reviewed in Sono *et al.*[36]) display widely varying results between the two enzymes. Turnover of these derivatives by TDO can be explained by considering the electronic effect of the substituent. The electron-donating methyl group in the 6-position decreases k_{cat}, whilst addition of electron-withdrawing fluorine in this position increases k_{cat}, perhaps by decreasing the energy of activation for electrophilic addition of bound dioxygen to the substrate indole ring. Molecular modelling simulations, comparing the activation energy for the conversion of L-Trp, 6Me-D/L-Trp and 6F-D/L-Trp, suggest that the reaction is most favourable for 6F-D/L-Trp, in agreement with the experimental observations.[37] When the activity of IDO with the same substrates is considered however, rates were found that do not conform to these electronic considerations, possibly due to structural effects caused by the presence of bulkier active site amino acid side chains in IDO.

Another interesting feature of the heme dioxygenases is the differing patterns of substrate inhibition displayed. IDO demonstrates substrate inhibition with a K_i of ~ 200 μM, while TDO does not display any substrate inhibition, even at high L-Trp concentrations. There have been several proposals to explain the substrate inhibition of IDO, with early theories noting that, under physiological steady-state conditions, the majority of the enzyme is likely to be in the active, ferrous form, whilst *in vitro* catalysis starts from the inactive ferric form. It was proposed that formation of the [Fe^{III}-L-Trp] adduct, due to high L-Trp concentrations employed *in vitro*, is catalytically unfavourable and the likely cause of the substrate inhibition.[38] However, a more recent study suggested that the binding of L-Trp to the ferrous enzyme is the more likely cause of this inhibition.[39,40] The K_d for O_2 binding to ferrous IDO increases dramatically when L-Trp is bound to the enzyme, whereas when O_2 binds first, L-Trp binds with a weaker, but similar, affinity to that of the ferrous, O_2-unbound enzyme. This could readily cause the substrate inhibition observed; however, it remains curious that other IDO substrates do not display this effect in their turnover.

12.4.2 Inhibition of TDO and IDO

The implication of IDO in several types of cancer (including colorectal, hepatocellular and ovarian carcinomas) has led to a search for potential inhibitors of the enzyme.[41–43] Until now only 1-Me-Trp, a competitive inhibitor for IDO ($K_i = 3$ μM), has been used in clinical trials, without any significant result.[44,45] Findings show that the function of 1-Me-Trp *in vivo* is not fully understood.[46,47] While 1-Me-Trp exists as two stereoisomers (D-, L-), it is unclear what the effect of each isomer is in the body.[45,47] It has been found that of D- and L-Trp, IDO binds and oxidizes the L-enantiomer with greater efficiency.[48] In fact, the inhibitory action of 1-Me-D-Trp went unexplained until the discovery of an isoform of human IDO, designated IDO-2.[49,50] It was subsequently proposed that 1-Me-D-Trp inhibits IDO-2 but it is ineffective against recombinant IDO.[49,51]

1-Me-Trp is ineffective against TDO activity and this can be explained by comparison of the active sites of the enzymes (Figure 12.4). In xcTDO the N^1 atom of tryptophan is directly hydrogen-bonded to the His$_{55}$ side chain and methylation at this position would clearly cause a steric clash with this residue. The presence of Ser$_{167}$ in IDO in the equivalent position to His$_{55}$ of xcTDO creates a small pocket that can accommodate the methyl group, thus allowing binding.

The most potent TDO inhibitors published so far belong to the family of 3-(2-pyridylethenyl)-indoles, with inhibition constants in the 30–100 nM range.[52] However, despite their inhibitory action, these molecules do not have any clinical interest as yet. In contrast with TDO, the list of IDO inhibitors is longer and has greater chemical diversity. One such inhibitor is norharman, which inhibits IDO with an inhibition constant in the micromolar range.[53,54] Methylthiohydantoin tryptophan, a non-toxic molecule, is another IDO inhibitor with similar efficacy ($K_i = 11.4$ μM).[55] Small molecules based on the natural product brassinin have also been reported as IDO inhibitors more potent than 1-Me-Trp.[56] Inhibitors such as exiguamine A and annulin B have been found to have inhibition constants in the nanomolar range, and are some of the most potent IDO inhibitors yet described.[57–59]

12.5 Catalytic Cycle

In this section we seek to describe the various steps and species involved in the catalytic cycle of TDO and IDO. This encompasses the events leading to the formation of the catalytically-relevant ternary complex, the role of electro-chemistry and the binding of O_2 and substrate. We will also detail what is known about the nature of the various species formed transiently *en route* to the ternary complex.

12.5.1 Formation of the Active Ternary Complex

As is common for heme proteins, such as the cytochromes P450 and heme oxygenase, turnover of L-Trp by TDO or IDO occurs with the enzyme in the ferrous state. In the enzyme active site, O_2 and L-Trp must both bind to the active site, forming a ferrous ternary complex ($[Fe^{II}$-L-Trp-$O_2]$) before any reaction can take place. The intermediate species involved in the generation of this ternary complex are depicted in Scheme 12.1. Once the ternary complex is formed, catalysis proceeds *via* an as-yet undefined mechanism (Section 12.5.4), ultimately regenerating the ferrous enzyme and producing the reaction product NFK.

The L-Trp binding affinities of selected IDO and TDO proteins are shown in Table 12.1, and it can be seen that the general trend is for L-Trp binding to ferrous enzyme with a higher affinity than to ferric enzyme. Once the enzyme is reduced, theoretically the enzyme will remain in the ferrous state as no electrons are required to dioxygenate tryptophan; however, IDO (and to a lesser extent TDO) undergoes unproductive auto-oxidation during catalytic turnover, generating ferric enzyme and superoxide.[5,28,32,60] Therefore, IDO and TDO need to be kept reduced *in vitro* and *in vivo* and under physiological conditions it has been proposed that cytochrome b_5 acts as the electron donor.[26]

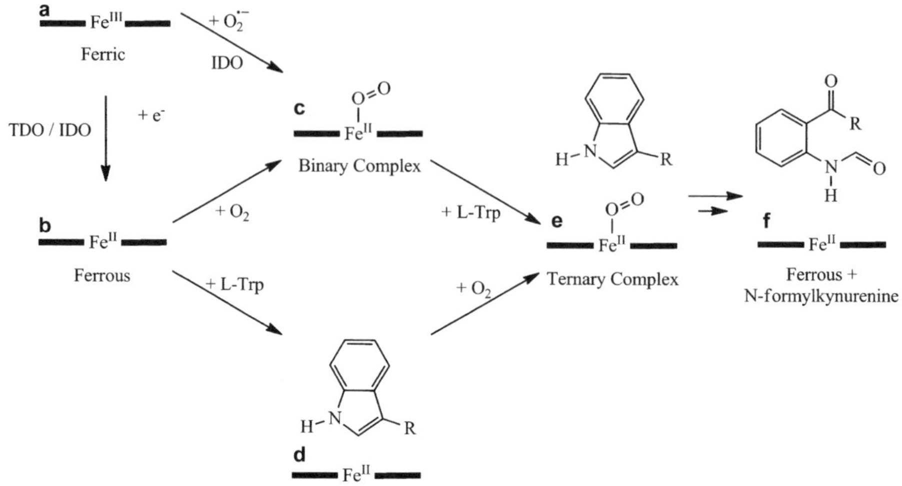

Scheme 12.1 Catalytic cycle of IDO and TDO. Proposed routes for the formation of the active ferrous ternary complex of IDO and TDO: (a) ferric enzyme, (b) ferrous enzyme, (c) [Fe^{II}-O_2] binary complex, (d) [Fe^{II}-L-Trp] binary complex, (e) [Fe^{II}-O_2-L-Trp] ternary complex and (f) ferrous enzyme with the product *N*-formylkynurenine (NFK).

Table 12.1 Kinetic parameters for the heme dioxygenases.[a]

Enzyme	Substrate	k_{cat} (s^{-1})	K_m (μM)	Specific turnover k_{cat}/K_m ($M^{-1}s^{-1}$)	K_d Fe$_{III}$ (μM)	K_d Fe$_{II}$ (μM)	E_{mid} (mV versus SHE)	E_{mid} + L-Trp (mV versus SHE)
H. sapiens IDO[31]	L-Trp	1.4	7.1	730 000	285	<7.0	−63	+16
	D-Trp	3.9	1570	2 500	−	−		
H. sapiens TDO[32]	L-Trp	1.4	220	6 300	∼350	∼200	−92	−76
X. campestris TDO[24]	L-Trp	19	110	170 000	3800	4.1	+8	+144
	6F-D/L-Trp	37	186	200 000	1510	<1.0		

[a]Potentials for *H. sapiens* IDO and *H. sapiens* TDO obtained at pH 7.0, those for *X. campestris* TDO obtained at pH 7.5.

Upon reduction, ternary complex formation proceeds *via* L-Trp and O_2 binding (Scheme 12.1). At least in the case of the *Xanthomonas* enzyme, TDO proceeds *via* L-Trp binding followed by O_2 coordination [Scheme 12.1, (B→D→E)], due to the instability of the ferrous-oxy complex.[5,29,31,38] Indeed it has been suggested that xcTDO has evolved such that the ferrous enzyme favours L-Trp binding ($K_d = 4$ μM) over O_2 binding ($K_d = 119$ μM), thus avoiding auto-oxidation.[5]

In the case of IDO there is some controversy in the literature as to how ternary complex formation proceeds. Does oxygen bind to the enzyme before the substrate or *vice versa*? Early studies showed that when L-Trp is bound to either IDO or TDO, there is an apparent activation towards the binding of diatomics such as CO or NO to the ferrous enzyme and that the K_d for the binding of L-Trp to the ferrous enzyme is smaller than (or similar to) that for binding of dioxygen.[39,61,62] However, a recent study investigating human IDO showed that the K_d for O_2 binding to ferrous IDO increases dramatically when L-Trp is bound to the enzyme, in contrast to CO or NO binding data.[39] When O_2 binds first, L-Trp binds with a slightly weaker affinity than it does to the ferrous O_2-free enzyme, suggesting that O_2 binding may be the initial step.

IDO is unusual as it can also utilize superoxide as an electron donor and concomitant oxygen source to effect catalysis. The ability of IDO to bind superoxide is similar to that of ascorbate peroxidase and catalase, which bind superoxide strongly, but clearly different from hemoglobin, myoglobin and cytochrome P450, which display very low reactivity with superoxide.[63–66] In this way both ferric and ferrous IDO will react with superoxide ($O_2^{\bullet-}$) to produce a binary ferrous oxy complex [Fe^{II}-O_2], which can then bind L-Trp and effect substrate turnover (Scheme 12.1, A → C).[28,29] However, it is not clear if the binary [Fe^{III}-L-Trp] adduct can react with superoxide.[28]

To assess whether superoxide can be utilized by IDO *in vivo*, Sono *et al.* investigated the roles of superoxide in the reductive activation of ferric IDO.[29,67] Turnover assays show that IDO activity is predominantly activated *via* an electron mediator, *without* requiring superoxide. This is consistent with cellular conditions, where utilization of superoxide is minimal, due to its highly unspecific reactivity and extremely low picomolar concentration.[68] Overall these studies provide compelling evidence that superoxide is unlikely to play a significant biological role in IDO activity under normal physiological conditions. However, the use of IDO in cells under extraordinary stress (such as during ischemia), where superoxide conditions are much higher, has not been thoroughly investigated.[69,70]

TDO is not thought to utilize superoxide as an electron donor and oxygen source, and the addition of superoxide to ferric TDO produces no detectable ferrous-oxy or ferric-superoxide complex. This complex is unstable and attempts to isolate either species *via* the addition of oxygen to ferrous TDO generate ferric enzyme and superoxide, with no stable intermediate being formed.[5,32] In this way TDO will generate superoxide rather than consume it but any biological relevance of this phenomenon is unknown.

12.5.2 Electrochemical Control of Substrate Reactivity

The midpoint reduction potentials of members of this family of enzymes show clear trends (Table 12.1). Upon substrate binding there is an increase in reduction potential for both IDO and xcTDO. The observed increase is larger for xcTDO than IDO, and the shift in reduction potential correlates with the increase in L-Trp binding affinity of the enzyme upon reduction. For example, for xcTDO the shift in the electrochemical midpoint potential is +140 mV upon L-Trp binding, and this correlates with a 1000-fold decrease in K_d for L-Trp from 3800 to 4 μM upon reduction.[5] This controlled specificity of binding is most likely caused by the loss of water from the distal site upon reduction of the heme, leaving the active site available for L-Trp and O_2 binding. For IDO turnover in the absence of superoxide, reduction of the heme also promotes substrate binding, and again this is stabilized by a large shift (+78 mV) in the midpoint potential upon L-Trp binding.[31,71] However, if superoxide is present it may also bind, reducing the heme and activating the enzyme. The large positive change in the reduction potential for both enzymes suggests that there is a significant stabilization of the ferrous form when substrate is bound.

Another consideration for the electrochemical properties of IDO and TDO is that the observed increase in the midpoint potential upon L-Trp binding could play a physiological role in keeping the proteins reduced, and therefore active, when L-Trp is present. This would aid coupled turnover of L-Trp, decreasing the requirement for additional electrons to reduce any enzyme that became oxidized by uncoupled binding of oxygen to ferrous IDO and TDO.

Some studies into IDO and TDO turnover have shown very weak catalytic activity by ferric IDO and TDO in the absence of electron donors or superoxide.[32] This activity could (and often has) been assigned to activity by the ferric enzyme, but this seems unlikely considering the electronic requirements of the heme to effect turnover. Instead it appears reasonable to assign this activity to small amounts of the ferrous enzyme being formed transiently in solution due to the relative ease of reduction of IDO and TDO, as evidenced by the comparatively high reduction potential of these enzymes, especially as these studies were conducted in large excesses of L-Trp. Excess substrate would lead to full L-Trp occupancy in the active sites of IDO and TDO, increasing the electrochemical midpoint potentials and favouring transient reduction.

12.5.3 Heme Coordination Environment

Several spectroscopic techniques have been used to study IDO and TDO, including (but not limited to) UV–visible spectrophotometry and electron paramagnetic resonance (EPR), resonance Raman, magnetic circular dichroism (MCD) and nuclear magnetic resonance (NMR) spectroscopies. Early work on the spectroscopy of IDO and TDO focussed on the determination of the heme coordination environment, and recent successes in determining the structure of these enzymes have allowed confirmation and re-analysis of these studies in the context of the actual heme environment. All the intermediates in the formation of the ternary complex (Scheme 12.1), with the exception of the unstable TDO ferrous-oxy complex, have been characterized by multiple spectroscopic techniques. As such, here we will discuss spectroscopic data relating to the heme coordination environment of the species leading up to the ternary complex, and the conclusions drawn as to the facilitation of the reaction by these species.

12.5.3.1 [Fe^{III}]

The crystal structures of both TDO and IDO show a six-coordinate heme iron in the ferric state, with a histidine residue ligating the heme iron in the proximal position.[4,5] A water molecule is bound covalently in the distal site in both crystal structures, with IDO binding water *via* the heme iron alone, whilst ferric TDO coordinates water *via* the heme iron and a hydrogen bond to His$_{55}$.[4,5] EPR experiments on human, rat liver, *C. metallidurans* and *Pseudomonas* ferric TDOs, and combined MCD/UV–visible studies on the rat liver enzyme have shown the presence of two distinct species in solution, a high-spin form and a low-spin form.[32,71–75] At low pH only the high-spin state is reported and on increasing the pH a transition to a mixed-spin state occurs, containing a mixture of low- and high-spin forms. The spin state of ferric TDO at high pH is temperature dependent, with the proportion of high-spin species increasing with elevated temperature. This temperature-dependent spin state change is comparable to that observed for ferric myoglobin at neutral pH and is consistent with water as the sixth ligand to the heme iron.[76,77] The coordinating ligand in the high-spin form of TDO is a water molecule, whilst that of the low-spin species is postulated to be a hydroxide ion. This is consistent with the crystal structure of ferric TDO, which shows a histidine residue in the active site positioned appropriately to be able to deprotonate a bound water molecule at elevated pH.

Spectra of ferric IDO are noticeably different from those of ferric TDO. EPR, MCD and UV–visible spectra of ferric human IDO show the presence of two distinct species in solution, high- and low-spin heme, in approximately equal population at pH 6.0.[71,78] The two species are in equilibrium, and on increasing the pH there is an increase in the low-spin component. The high-spin species is confirmed by MCD and resonance Raman spectroscopy as arising from water ligation in the distal position.[78] However, in contrast to TDO the identity of the sixth ligand in the low-spin species was proposed to be a nitrogenous ligand. Before the determination of the crystal structure it was postulated that a nitrogenous base, such as a histidine side chain, could be present in the active site of IDO within coordinating distance of the heme. This was based largely on the idea that stabilization of the bound oxygen substrate by hydrogen bonding might be necessary, and that participation of an active site histidine, as observed in the globins and the peroxidases, might be involved.[65,79] However, when the crystal structure of hIDO was solved it revealed a hydrophobic active site that contained no active site histidine (Figure 12.4).

12.5.3.2 [Fe^{III}-L-Trp]

As mentioned in Section 12.5.1, ferric TDO and IDO are not able to bind
L-Trp *in vivo* at physiological concentrations ($\sim$50 µM) due to their low
tryptophan binding affinities in this oxidation state (Table 12.1). However,
spectroscopic studies have been performed on the ferric enzymes in the
presence of a large excess of tryptophan. EPR data on rat liver and
Pseudomonas TDOs show that addition of L-Trp produces two distinct species in solution at low pH, these being high- and low-spin heme.[72,73] This is in
contrast to the L-Trp-free data because the pH of the spin state transition is
significantly modified, decreasing by about 1.6 pH units. The spectral changes
indicate that L-Trp binding to the ferric protein causes increased deprotonation of the heme-bound water. This substrate-induced shift could be caused
by the bound L-Trp deprotonating the bound water molecule, as in IDO,
rather than the active site histidine. The non-bonded distance between the
indole nitrogen and the bound water molecule is 3.3 Å, which is consistent
with a role in deprotonation. In addition, a more recent EPR and Mössbauer
study on oxidized substrate-bound and substrate-free *C. metallidurans* TDO
indicates the presence of two dominant inequivalent heme species in multiple
states of the enzyme.[75] This is consistent with a "dimer of dimers" enzyme
structure (discussed in Section 12.3) extended to the electronic properties of
the hemes. The presence of the L-Trp affects the hemes; however, it is proposed that substrate binding is not the cause of the electronic difference
between the hemes.

MCD and EPR spectra of ferric IDO in the presence of L-Trp identify three
distinct species in solution, one of which is high spin and the other two are low
spin.[71,78] The spectra are pH dependent due to an ionizable interaction affecting
the formation of species in the active site, analogous to results observed for
TDO. Parallel to substrate-free IDO, the high-spin component is thought to be
water-bound, while one of the low-spin species appears to arise from a nitrogenous interaction with the heme iron. However, the sixth heme ligand for the
additional low-spin species is hydroxide and this additional component is the
predominant species in solution. These species can be explained by modelling L-
Trp binding to the active site of ferric IDO, in a similar orientation to that
observed in the xcTDO crystal structure. The bound indole nitrogen atom of
L-Trp can provide a hydrogen-bond acceptor to stabilize deprotonation of the
bound water molecule to form hydroxide. This deprotonation makes the
spectrum pH-dependent due to the ionizable interaction between L-Trp and
the bound water molecule affecting the formation of the hydroxide species in
the active site. In contrast, ferric substrate-free human IDO does not form a
hydroxide-bound species because IDO lacks a hydrogen-bond acceptor to
facilitate deprotonation of the bound water molecule.

These results mirror those obtained with TDO; addition of a hydrogen bond
donor to the active site leads to the production of hydroxide-bound heme
(observed for both ferric and ferric L-Trp bound TDO), suggesting that
hydrogen bonding plays a pivotal role in the mechanism of these enzymes.

12.5.3.3 $[Fe^{II}]$ and $[Fe^{II}\text{-}L\text{-}Trp]$

Ferrous IDO and TDO are five-coordinate, as expected for a high-spin ferrous heme in which the Fe moves out of the porphyrin plane toward the proximal His to give square pyramidal geometry. Binding of L-Trp to ferrous IDO and TDO leads to a small shift in the Soret maxima, giving a clear indication of substrate binding.[5,31] Both the IDO and TDO ferrous L-Trp complexes are extremely stable in an anaerobic environment, showing little change over a period of days. The crystal structure of ferrous xcTDO with L-Trp bound reveals an active site environment where water is excluded and the heme is five-coordinate. EPR and Mössbauer spectroscopy of reduced TDO from *C. metallidurans*, with and without substrate, indicate the presence of two dominant inequivalent heme species in multiple sub-states.[75] These data are similar to that obtained for the ferric enzyme (discussed above) but the spectroscopy also indicates that the electronic properties of the hemes change significantly upon L-Trp addition, which is consistent with a change in the protonation state of the proximal histidine to the hemes.[75]

12.5.3.4 $[Fe^{II}\text{-}O_2]$

The formation of an initial oxy complex is clearly crucial to the reaction mechanism, but little is known about the factors affecting the stability of this ferrous-oxy species, and why the ferrous-oxy TDO complex is unstable (at least in the human and *Xanthomonas* enzymes). Unfortunately, the $v(O\text{--}O)$ mode of heme bound dioxygen in histidine-coordinated hemoproteins is, in general, not Raman active (as is the case for IDO), and as a result conclusions on the mode of O_2 binding have been drawn primarily from structural models of ferrous enzymes or from spectroscopic work on the related CO-bound enzyme.[80]

A recent EPR/ENDOR study has characterized the binary $[Fe^{II}\text{-}O_2]$ complexes of human IDO and *Shewanella oneidensis* IDO.[81] Multiple conformations of the $[Fe^{II}\text{-}O_2]$ species of IDO are observed to exist in solution, providing evidence that the ferrous-oxy complexes of these IDOs exist in multiple conformational sub-states. This finding is consistent with a flexible active site structure in the absence of tryptophan. In most conformers the complex is solvent accessible, and overall the results show that the majority of active sites contain a water molecule that is within hydrogen-bonding distance of the distal oxygen of the bound dioxygen.

These data, in tandem with mutagenesis studies on human IDO, suggest that the difference in the stability of the ferrous-oxy complexes of IDO and TDO may be attributed to hydrogen bonding differences in the active site although the structural interactions that define this difference in reactivity remain to be established.[31]

12.5.3.5 $[Fe^{II}\text{-}L\text{-}Trp\text{-}O_2]$ Ternary Complex

The ternary $[Fe^{II}\text{-}O_2\text{-}Trp]$ complexes of both IDO and TDO in the presence of L-Trp and a substrate analogue (1-Me-Trp) have been studied by cryoreduction

EPR/ENDOR and resonance Raman techniques, and a model of the ternary complex of xcTDO, constructed using the ferrous L-Trp bound crystal structure, has been proposed by Forouhar *et al.*[5,80,82]

Cryoreduction of ternary [Fe^{II}-O$_2$-Trp] IDO or TDO complexes generates a ferric peroxy heme species that is not solvent accessible, but a clear H-bonding interaction between substrate and the bound dioxygen is observed. Parallel studies with 1-Me-Trp eliminate the possibility that the indole NH group provides this H-bond donor to the bound O$_2$, and it was proposed that the ammonium group of the substrate fulfils this role. These data show that substrate binding, primarily through this H-bond, causes the bound dioxygen to adopt a new conformation, which presumably is optimized for insertion of O$_2$ into the C^2–C^3 double bond of the substrate. This substrate interaction further helps control the reactivity of the heme-bound dioxygen by "shielding" it from water.

Unusually, recent resonance Raman studies found that the v(O–O) stretching mode of dioxygen was Raman active in the ternary [Fe^{II}-L-Trp-O$_2$] complex of human IDO.[39,82] This is not usually the case for oxygen-bound ferrous heme proteins, and the observation of this signal was attributed to a hydrogen bonding interaction in the active site ternary complex. As above, this hydrogen bonding interaction was attributed to the heme-bound O$_2$ interacting with L-Trp and the surrounding environment in the ternary complex. Interestingly, recent work on the structurally homologous heme- and O$_2$-dependent enzyme PrnB has yielded the crystal structure of a ternary complex ([Fe^{III}-L-Trp-CN$^-$] – with cyanide mimicking O$_2$).[83] However, in this case the enzyme catalyses cleavage of the indole C^2–N^1 bond rather than O$_2$ insertion, and the substrate is bound in a different orientation than that observed in xcTDO, which is consistent with their different catalytic functions and mechanisms.

12.5.4 Mechanism of Oxygen Insertion

The catalytic mechanism employed by this family of enzymes is still uncertain, with little characterization of reactive intermediates reported. This is in clear contrast with the mechanisms of action for the heme-containing monoxygenases (*e.g.* cytochrome P450[64] and heme oxygenases) and non-heme dioxygenases (*e.g.* catechol dioxygenases and naphthalene dioxygenase).[36,84,85] Slow progress in elucidating the mechanism for TDO and IDO can be attributed to the difficulty in observing any reaction intermediates in the transition from the ternary complex to product, and the inability to obtain substrate analogues that trap the reaction at a particular intermediate. Most data on the reaction are from indirect observations, utilizing substrate analogues and several modelling simulations. For example, the use of tryptophan analogues where the indole nitrogen is replaced by oxygen or sulfur leads to inactivity, and the use of slow substrates still does not allow detection of reaction intermediates.[25,36,44,86] As such, the mechanism employed is still debatable; however, progress has been made in recent years in ruling out certain pathways.

Early studies into the mechanism of IDO and TDO during the 1970s and 1980s indicated that catalysis may proceed *via* base-catalysed deprotonation of the indole nitrogen atom of tryptophan, based on the observation that 1-Me-L-Trp is an inhibitor of dioxygenase activity.[36,87] Two rearrangement pathways initiated by base-catalysed attack were proposed, namely, the Criegee and dioxetane mechanisms, which became the most widely accepted possible routes for product formation. These two mechanisms are both ionic and utilize the nucleophilicity of the indole ring to attack an electrophilic heme-bound dioxygen species in promoting the reaction.[36]

At this stage, progress into elucidating the mechanism of TDO and IDO halted, mainly due to lack of biochemical information into intermediates or the active site environment of the heme dioxygenases. The recent solution of the structures of IDO and TDO rejuvenated investigations into these enzymes, and it has become apparent that these initial mechanistic proposals are inadequate. The crystal structures of IDO and TDO in tandem with mutagenesis studies have revealed that no active site base is necessary for catalysis by either dioxygenase.[24,31] IDO does not contain a polar active site residue close enough for proton abstraction at the indole nitrogen (Figure 12.4, discussed in Section 12.3.2), while although TDO enzymes are thought to contain a histidine (*e.g.* His$_{55}$ in xcTDO) in the correct position to facilitate deprotonation, catalytic activity is retained upon substitution of this histidine with alanine.[24] These data indicate that base-catalysed deprotonation of tryptophan is not essential for catalytic activity, in contrast to previous proposals.

These discoveries have generated significant speculation upon the route of catalysis, enlivening debate and reviving biochemical studies into the reaction mechanism. Evidence supporting particular mechanisms and apparently excluding others has recently emerged. Studies by Chauhan *et al.* have showed that 1-Me-L-Trp, (previously described as an IDO inhibitor) is in fact a slow substrate for IDO.[44] 1-Me-L-Trp cannot undergo base-catalysed deprotonation of the nitrogen atom, confirming that an active site base is not necessary for catalysis.

In addition, EPR/ENDOR studies on the ternary complexes of IDO and TDO have shown that the active sites of both enzymes are shielded from solvent, and that the ferrous-oxy species is a hydrogen-bond acceptor. However, this H-bond is not donated by the substrate indole moiety as proposed previously for IDO, and taken together these data seem to rule out mechanisms involving a direct hydrogen bonding interaction between the bound O_2 and the substrate and, furthermore, indicate that bound dioxygen does not become easily protonated in the ternary complex, which would be consistent with a reaction mechanism that does not require external protons.[81,88]

The results of resonance Raman spectroscopy experiments have now been published independently by two different groups, and these provide suggestions as to the nature of the species involved in catalysis.[80,82] These data are consistent with transient generation of a ferryl intermediate during catalytic turnover of human IDO, indicating that dioxygen insertion into tryptophan might occur in sequential steps. Both groups subsequently propose that the reaction proceeds *via* the formation of a ferryl-oxo (Compound II) species with concomitant production of a tryptophan epoxide intermediate, followed by subsequent addition of the ferryl oxygen to generate NFK (Scheme 12.2).

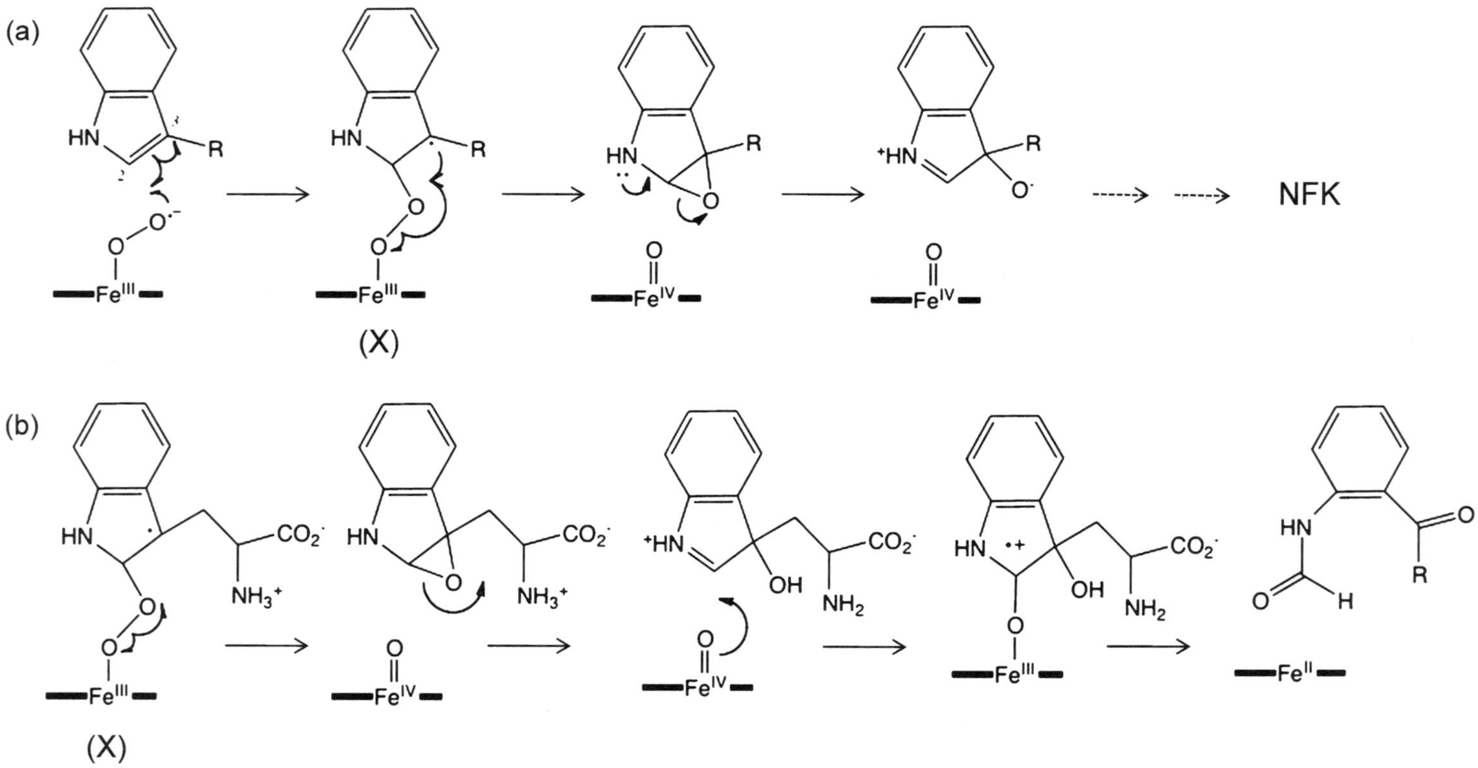

Scheme 12.2 (a) Radical addition (rather than base-catalysed abstraction) leading to species X is followed by formation of a ferryl heme species (Compound II) and a proposed epoxide species.[80] Formation of an epoxide has also been suggested by computational work (in fact, it was considered in an earlier study, but initially considered energetically unfavorable in the gas phase).[37,90] Electrophilic addition could also involve epoxide formation through a similar (two-electron) mechanism. Possible ring opening of the epoxide is also indicated. (b) Proposed mechanism for subsequent formation of NFK, starting from the same species X as in panel A, and also involving initial ring opening.[90]

Such a mechanism could occur initially *via* either ionic or radical attack at the C^2 position of tryptophan to generate Compound II and an epoxide, followed by ferryl-oxo attack and facile C^2–C^3 bond cleavage to form NFK. Such a mechanism appears plausible, and is currently the most accepted model for catalysis in IDO and TDO. As such, this marks a significant change in thinking, placing the mechanism in-line with that of the heme monoxygenases, which also utilize a ferryl-oxo intermediate during catalysis.

That said, notably, at the present time a ferryl intermediate has only been observed transiently for IDO, and parallel studies on human TDO did not detect a ferryl intermediate. As such, evidence for this mechanistic scheme remains scarce. However, molecular modelling studies utilizing several advanced computational techniques, and based on the active-site structures of human IDO or xcTDO with L-Trp bound, have provided separate information that supports the ideas emerging from the experimental work. Most of the data do not support base-catalysed abstraction, and either direct electrophilic or radical addition without deprotonation have been proposed.[37,89,90] There is also independent computational evidence for ferryl heme formation.[38,90] However, further investigation of the species involved in catalysis is clearly required to characterize the oxygenating species and confirm the exact nature of the mechanism of action.

12.6 Summary and Conclusions

What makes a heme dioxygenase? And, how do the heme dioxygenases utilize heme to control the chemistry of tryptophan?

Recent evidence detecting a transient ferryl intermediate in IDO turnover and studies excluding several other mechanistic routes to NFK formation have led to the current thinking that the mechanism of the heme dioxygenases may employ two sequential monooxygenation steps, with no involvement from an active site base. The proposed sequential oxygen insertion mechanism of the heme dioxygenases resembles that of several heme monoxygenases, and a few non-heme dioxygenases (such as homoprotocatechuate 2,3-dioxygenase), although there are significant differences between these enzymes. In heme monoxygenases and non-heme oxygenases precise control of the reaction intermediates is necessary for the reaction to be "channelled" down the right path. Numerous factors regulate the behaviour of these dioxygenase enzymes; however, the main regulators of catalysis in the active-site appear to be precise control of proton and electron availability, the oxidizing species utilized and its spin state, and stereochemical control of the substrate to effect an optimal configuration for reactivity. If indeed there is a preference in the heme dioxygenases for two sequential monoxygenation steps rather than a concerted mechanism, it is presumably a result of these factors combining to effect addition of dioxygen to a relatively inert substrate in an efficient manner.

Clearly, significant progress has been made into identifying the mechanism of the heme dioxygenases; however, further investigation of the species involved in catalysis is clearly required to confirm the exact nature of the mechanism of action.

References

1. T. D. H. Bugg and S. Ramaswamy, *Curr. Opin. Chem. Biol.*, 2008, **12**, 134.
2. E. G. Kovaleva, M. B. Neibergall, S. Chakrabarty and J. D. Lipscomb, *Acc. Chem. Res.*, 2007, **40**, 475.
3. E. G. Kovaleva and J. D. Lipscomb, *Science*, 2007, **316**, 453.
4. H. Sugimoto, S.-I. Oda, T. Otsuki, T. Hino, T. Yoshida and Y. Shiro, *Proc. Natl.Acad. Sci. USA*, 2006, **103**, 2611.
5. F. Forouhar, J. L. R. Anderson, C. G. Mowat, S. M. Vorobiev, A. Hussain, M. Abashidze, C. Bruckmann, S. J. Thackray, J. Seetharaman, T. Tucker, R. Xiao, L. C. Ma, L. Zhao, T. B. Acton, G. T. Montelione, S. K. Chapman and L. Tong, *Proc. Natl. Acad. Sci. USA*, 2007, **104**, 473.
6. W. De Laurentis, L. Khim, J. L. R. Anderson, A. Adam, R. S. Phillips, S. K. Chapman, K.-H. Van Pee and J. H. Naismith, *Biochemistry*, 2007, **46**, 12393.
7. S. Yamamoto and O. Hayaishi, *J. Biol. Chem.*, 1967, **242**, 5260.
8. Y. Kotake and I. Masayama, *Z. Physiol. Chem.*, 1936, **243**, 237.
9. G. Allegri, E. Ragazzi, A. Bertazzo, C. V. L. Costa and R. Rocchi, *Adv. Exp. Med. Biol.*, 2003, **527**, 481.
10. H. Ohashi, K. Saito, H. Fujii, H. Wada, N. Furuta, M. Takemura, S. Maeda and M. Seishima, *Arch. Biochem. Biophys.*, 2004, **428**, 154.
11. K. Saito, S. Fujigaki, M. P. Heyes, K. Shibata, M. Takemura, H. Fujii, H. Wada, A. Noma and M. Seishima, *Am. J. Physiol.*, 2000, **279**, F565.
12. A. Tankiewicz, D. Pawlak and W. Buczko, *Postepy Hig. Med. Dosw.*, 2001, **55**, 715.
13. J. A. Aquilina, J. A. Carver and R. J. W. Truscott, *Exp. Eye Res.*, 1997, **64**, 727.
14. U. Grohmann, F. Fallarino and P. Puccetti, *Trends Immunol.*, 2003, **24**, 242.
15. D. H. Munn and A. L. Mellor, *J. Clin. Invest.*, 2007, **117**, 1147.
16. A. L. Mellor, D. Munn, P. Chandler, D. Keskin, T. Johnson, B. Marshall, K. Jhaver and B. Baban, *Adv. Exp. Med. Biol.*, 2003, **527**, 27.
17. C. R. MacKenzie, K. Heseler, A. Mueller and W. Daeubener, *Curr. Drug Metab.*, 2007, **8**, 237.
18. G. Neurauter, B. Wirleitner, K. Schroecksnadel, B. Frick and D. Fuchs, *Aktuelle Ernaehrungsmedizin.*, 2004, **29**, 171.
19. M. Platten, P. P. Ho, S. Youssef, P. Fontoura, H. Garren, E. M. Hur, R. Gupta, L. Y. Lee, B. A. Kidd, W. H. Robinson, R. A. Sobel, M. L. Selley and L. Steinman, *Science*, 2005, **310**, 850.

20. A. L. Mellor, P. Chandler, K. Lee Geon, T. Johnson, D. B. Keskin, J. Lee and D. H. Munn, *J. Reprod. Immunol.*, 2002, **57**, 143.
21. S. Suzuki, S. Tone, O. Takikawa, T. Kubo, I. Kohno and Y. Minatogawa, *Biochem. J.*, 2001, **355**, 425.
22. A. Tankiewicz, D. Pawlak, J. Topczewska-Bruns and W. Buczko, *Adv. Exp. Med. Biol.*, 2003, **527**, 409.
23. Y. Zhang, S. A. Kang, T. Mukherjee, S. Bale, B. R. Crane, T. P. Begley and S. E. Ealick, *Biochemistry*, 2007, **46**, 145.
24. S. J. Thackray, C. Bruckmann, J. L. R. Anderson, L. P. Campbell, R. Xiao, L. Zhao, C. G. Mowat, F. Forouhar, L. Tong and S. K. Chapman, *Biochemistry*, 2008, **47**, 10677.
25. J. M. Leeds, P. J. Brown, G. M. McGeehan, F. K. Brown and J. S. Wiseman, *J. Biol. Chem.*, 1993, **268**, 17781.
26. G. J. Maghzal, S. R. Thomas, N. H. Hunt and R. J. Stocker, *J. Biol. Chem.*, 2008, **283**, 12014.
27. E. Vottero, D. A. Mitchell, M. J. Page, R. T. A. McGillivray, I. J. Sadowski, M. Roberge and A. G. Mauk, *FEBS Lett.*, 2006, **580**, 2265.
28. T. Taniguchi, M. Sono, F. Hirata, O. Hayaishi, M. Tamura, K. Hayashi, T. Iizuka and Y. Ishimura, *J. Biol. Chem.*, 1979, **254**, 3288.
29. M. Sono, *J. Biol. Chem.*, 1989, **264**, 1616.
30. J. T. Pearson, S. Siu, D. P. Meininger, L. C. Wienkers and D. A. Rock, *Biochemistry*, 2010, **49**, 2647.
31. N. Chauhan, J. Basran, I. Efimov, D. A. Svistunenko, H. E. Seward, P. C. E. Moody and E. L. Raven, *Biochemistry*, 2008, **47**, 4761.
32. J. Basran, S. A. Rafice, N. Chauhan, I. Efimov, M. R. Cheesman, L. Ghamsari and E. L. Raven, *Biochemistry*, 2008, **47**, 4752.
33. C. J. D. Austin, F. Astelbauer, P. Kosim-Satyaputra, H. J. Ball, R. D. Willows, J. F. Jamie and N. H. Hunt, *Amino Acids*, 2009, **36**, 99.
34. Y. Ishimura, M. Nozaki, O. Hayaishi, T. Nakamura, M. Tamura and I. Yamazaki, *J. Biol. Chem.*, 1970, **245**, 3593.
35. Y. Watanabe, M. Fujiwara, R. Yoshida and O. Hayaishi, *Biochem. J.*, 1980, **189**, 393.
36. M. Sono, M. P. Roach, E. D. Coulter and J. H. Dawson, *Chem. Rev.*, 1996, **96**, 2841.
37. L. W. Chung, X. Li, H. Sugimoto, Y. Shiro and K. Morokuma, *J. Am. Chem. Soc.*, 2008, **130**, 12299.
38. M. Sono, T. Taniguchi, Y. Watanabe and O. Hayaishi, *J. Biol. Chem.*, 1980, **255**, 1339.
39. C. Lu, Y. Lin and S.-R. Yeh, *Biochemistry*, 2010, **49**, 5028.
40. C. Lu, Y. Lin and S.-R. Yeh, *J. Am. Chem. Soc.*, 2009, **131**, 12866.
41. G. Mazzolini, O. Murillo, C. Atorrasagasti, J. Dubrot, I. Tirapu, M. Rizzo, A. Arina, C. Alfaro, A. Azpilicueta and C. Berasain, *World J. Gastroenterol.*, 2007, **13**, 5822.
42. K. Pan, H. Wang, M.-s. Chen, H.-k. Zhang, D.-s. Weng, J. Zhou, W. Huang, J.-j. Li, H.-f. Song and J.-c. Xia, *J. Cancer Res. Clin. Oncol.*, 2008, **134**, 1247.

43. M. Takao, A. Okamoto, T. Nikaido, M. Urashima, S. Takakura, M. Saito, M. Saito, S. Okamoto, O. Takikawa, H. Sasaki, M. Yasuda, K. Ochiai and T. Tanaka, *Oncol. Rep.*, 2007, **17**, 1333.

44. N. Chauhan, S. J. Thackray, S. A. Rafice, G. Eaton, M. Lee, I. Efimov, J. Basran, P. R. Jenkins, C. G. Mowat, S. K. Chapman and E. L. Raven, *J. Am. Chem. Soc.*, 2009, **131**, 4186.

45. F. Qian, J. Villella, P. K. Wallace, P. Mhawech-Fauceglia, J. D. Tario, Jr., C. Andrews, J. Matsuzaki, D. Valmori, M. Ayyoub, P. J. Frederick, A. Beck, J. Liao, R. Cheney, K. Moysich, S. Lele, P. Shrikant, L. J. Old and K. Odunsi, *Cancer Res.*, 2009, **69**, 5498.

46. B. J. Van den Eynde, I. Theate, C. Uyttenhove, D. Colau, L. Pilotte and V. Stroobant, *International Congress Series* (Interdisciplinary Conference on Tryptophan and Related Substances: Chemistry, Biology, and Medicine, 2006) 2007, **1304**, 274.

47. D.-Y. Hou, A. J. Muller, M. D. Sharma, J. DuHadaway, T. Banerjee, M. Johnson, A. L. Mellor, G. C. Prendergast and D. H. Munn, *Cancer Res.*, 2007, **67**, 792.

48. A. J. Muller, R. Metz and G. C. Prendergast, *International Congress Series* (Interdisciplinary Conference on Tryptophan and Related Substances: Chemistry, Biology, and Medicine, 2006) 2007, **1304**, 250.

49. H. J. Ball, H. J. Yuasa, C. J. D. Austin, S. Weiser and N. H. Hunt, *Int. J. Biochem. Cell Biol.*, 2009, **41**, 467.

50. H. J. Yuasa, H. J. Ball, Y. F. Ho, C. J. D. Austin, C. M. Whittington, K. Belov, G. J. Maghzal, L. S. Jermiin and N. H. Hunt, *Comp. Biochem. Physiol., Part B: Biochem. Mol. Biol.*, 2009, **153**, 137.

51. R. Metz, J. B. DuHadaway, U. Kamasani, L. Laury-Kleintop, A. J. Muller and G. C. Prendergast, *Cancer Res.*, 2007, **67**, 7082.

52. D. J. Madge, R. Hazelwood, R. Iyer, H. T. Jones and M. Salter, *Bioorg. Med. Chem. Lett.*, 1996, **6**, 857.

53. M. Sono and S. G. Cady, *Biochemistry*, 1989, **28**, 5392.

54. N. Eguchi, Y. Watanabe, K. Kawanishi, Y. Hashimoto and O. Hayaishi, *Arch. Biochem. Biophys.*, 1984, **232**, 602.

55. T. Okamoto, S. Tone, H. Kanoichi, F. Ohyama and Y. Minatogawa, *International Congress Series* (Interdisciplinary Conference on Tryptophan and Related Substances: Chemistry, Biology, and Medicine, 2006) 2007, **1304**, 352.

56. B. Van Den Eynde, L. Pilotte and E. De Plaen, Application: *WO Patent* No. 2009-US22502010008427 (20090410).

57. H. C. Brastianos, E. Vottero, B. O. Patrick, R. Van Soest, T. Matainaho, A. G. Mauk and R. J. Andersen, *J. Am. Chem. Soc.*, 2006, **128**, 16046.

58. M. Volgraf, J.-P. Lumb, H. C. Brastianos, G. Carr, M. K. W. Chung, M. Muenzel, A. G. Mauk, R. J. Andersen and D. Trauner, *Nat. Chem. Biol.*, 2008, **4**, 535.

59. A. Pereira, E. Vottero, M. Roberge, A. G. Mauk and R. J. Andersen, *J. Nat. Prod.*, 2006, **69**, 1496.

60. K. Kobayashi, K. Hayashi and M. Sono, *J. Biol. Chem.*, 1989, **264**, 15280.

61. M. Sono, *Biochemistry*, 1990, **29**, 1451.
62. M. Sono, *Biochemistry*, 1986, **25**, 6089.
63. P. Vernet, R. J. Aitken and J. R. Drevet, *Mol. Cell. Endocrinol.*, 2004, **216**, 31.
64. T. M. Makris, R. Davydov, I. G. Denisov, B. M. Hoffman and S. G. Sligar, *Drug Metab. Rev.*, 2002, **34**, 691.
65. K. Shikama, *Prog. Biophys. Mol. Biol.*, 2006, **91**, 83.
66. J. M. Rifkind, S. Ramasamy, P. T. Manoharan, E. Nagababu and J. G. Mohanty, *Antioxid. Redox Signal.*, 2004, **6**, 657.
67. F. O. Brady, H. J. Forman and P. Feigelson, *J. Biol. Chem.*, 1971, **246**, 7119.
68. B. Chance, H. Sies and A. Boveris, *Physiol. Rev.*, 1979, **59**, 527.
69. M. Hoshi, K. Saito, Y. Murakami, A. Taguchi, H. Fujigaki, R. Tanaka, M. Takemura, H. Ito, A. Hara and M. Seishima, *Neurosci. Res.*, 2009, **63**, 194.
70. H. Fujigaki and K. Saito, *International Congress Series* (Interdisciplinary Conference on Tryptophan and Related Substances: Chemistry, Biology, and Medicine, 2006), 2007, 1304, 314.
71. N. D. Papadopoulou, M. Mewies, K. J. McLean, H. E. Seward, D. A. Svistunenko, A. W. Munro and E. L. Raven, *Biochemistry*, 2005, **44**, 14318.
72. R. Makino, K. Sakaguchi, T. Iizuka and Y. Ishimura, *J. Biol. Chem.*, 1980, **255**, 11883.
73. F. O. Brady, P. Feigelson and K. V. Rajagopalan, *Arch. Biochem. Biophys.*, 1973, **157**, 63.
74. P. Feigelson, F. O. Brady and J. A. McCray, *J. Biol. Chem.*, 1973, **248**, 5267.
75. R. Gupta, R. Fu, A. Liu and M. P. Hendrich, *J. Am. Chem. Soc.*, 2010, **132**, 1098.
76. R. Davydov and B. M. Hoffman, *J. Biol. Inorg. Chem.*, 2008, **13**, 357.
77. R. Davydov, R. L. Osborne, S. H. Kim, J. H. Dawson and B. M. Hoffman, *Biochemistry*, 2008, **47**, 5147.
78. M. Sono and J. H. Dawson, *Biochim. Biophys. Acta, Protein Struct. Mol. Enzymol.*, 1984, **789**, 170.
79. C. E. Castro, *Pharmacol. Ther.*, 1980, **10**, 171.
80. A. Lewis-Ballester, D. Batabyal, T. Egawa, C. Lu, Y. Lin, M. A. Marti, L. Capece, D. A. Estrin and S.-R. Yeh, *Proc. Natl. Acad. Sci. USA*, 2009, **106**, 17371.
81. R. M. Davydov, N. Chauhan, S. J. Thackray, J. L. R. Anderson, N. D. Papadopolou, C. G. Mowat, S. K. Chapman, E. Lloyd Raven and B. M. Hoffman, *J. Am. Chem. Soc.*, 2009, **132**, 5494.
82. S. Yanagisawa, K. Yotsuya, Y. Hashiwaki, M. Horitani, H. Sugimoto, Y. Shiro, E. H. Appelman and T. Ogura, *Chem. Lett.*, 2010, **39**, 36.
83. X. Zhu, K.-H. van Pée and J. H. Naismith, *J. Biol. Chem.*, 2010, **285**, 21126.
84. T. D. H. Bugg, and G. Lin, *Chem. Commun.*, 2001, 941.

85. M. Palaniandavar and R. Mayilmurugan, *C. R. Chim.*, 2007, **10**, 366.
86. S. G. Cady and M. Sono, *Arch. Biochem. Biophys.*, 1991, **291**, 326.
87. M. Sono and O. Hayaishi, *Biochem. Rev.*, 1980, **50**, 173.
88. A. C. Terentis, S. R. Thomas, O. Takikawa, T. K. Littlejohn, R. J. W. Truscott, R. S. Armstrong, S.-R. Yeh and R. Stocker, *J. Biol. Chem.*, 2002, **277**, 15788.
89. L. Capece, A. Lewis-Ballester, D. Batabyal, N. Di Russo, S.-R. Yeh, D. A. Estrin and M. A. Marti, *J. Biol. Inorg. Chem.*, 2010, **15**, 811.
90. L. W. Chung, X. Li, H. Sugimoto, Y. Shiro and K. Morokuma, *J. Am. Chem. Soc.*, 2010, **132**, 11993.

Subject Index

Page numbers in *italics* refer to entries mainly, or wholly, in figures or tables.

alkanes (*continued*)
  hydroxylation, CYP 450$_{cam}$
    339–57
  oxidation 155–75, 198–201,
    381
  stereospecific oxidation 162, *163,
    164*
  see also ethane oxidation; methane
AlkB repair enzymes 10, *11*
alkenes
  cis-dihydroxylation 183–7,
    189–98
  epoxidation 150, 151, 156, 176–82,
    189–98, 386
    by compound I 312–15
  oxidations *164*
alkoxyl radical intermediates 101–3
  reactions 110–12
alkylarene oxidation *170*
Alzheimer's disease 22
amines
  in alkene epoxidation *181*
  see also indoleamine dioxygenase
    (IDO); tpa complexes
δ-(L-α-amino-adipoyl)-L-cysteinyl-D-
  valine (ACV) 27, 28, *29*
1-aminobenzotriazole *274*
1-aminocyclopropane-1-carboxylic
  acid (ACC) 30, 31
1-aminocyclopropane-1-carboxylic
  acid oxidase (ACCO) 30–2,
    90, 92
  active site *31*
  ethylene formation 108
  oxygen binding 98
tert-amyl alcohol 171
anastrozole *274*
animal CYP450s 283
annulin B 411
antibiotics
  biosynthesis 91
  coupled epimerization and
    desaturation 108–10
  αKDD in biosynthesis of 3
  ring closure reactions 110
  see also individual antibiotics

apocarotenoid oxygenase (ACO) 93,
  97, 103
  alkoxyl radical intermediates
    111
  oxygen activation *149*
Arabidopsis thaliana 263
arachidonic acid 55, 258, 265, 366
ARD dioxygenase 50, *53*
arenes
  alkylarene oxidation *170*
  cis-dihydroxylation *185*
  dihydroxylation 59
    extradiol/intradiol
      dioxygenases *35*
    Rieske dioxygenases 151, *153*
aromatic hydroxylation 317–19
  with oxoferryl species 106
  polycyclic aromatic
    hydrocarbons 344–5
  QM/MM studies 387–90
aromatic oxygenation *89,* 90
(+)-artemisin *172,* 173
ascorbate 30, 408
ascorbate peroxidase 296–7, 298
Asp$_{297}$ *382, 383*
Aspergillus 270
axial ligands 230, 291
  compound I 293, *294,* 295–301,
    308, *309*
azole inhibitors 259, 273, *274,* 370

B-term 213–14, 215
B3LYP method 93, *132, 136*
  B3LYP/CHARMM 319
  B3LYP/CHARMM22 374, 375,
    376, 377
  B3LYP/LACV3P *302, 305*
  B3LYP/LACVP *320*
  benzene hydroxylation *390*
  camphor hydroxylation 381, 382
  in QM/MM studies 372
Bacillus megaterium 258, 391
bacterial CYP450s 283
Badger's rule 78
bands I, II and III, MCD spectra
  231, 232–3, 234, *235*